CANDID SCIENCE V

Conversations with Famous Scientists

Other books by István Hargittai

Candid Science IV: Conversations with Famous Physicists, Imperial College Press, London, 2004 (with M. Hargittai).

Our Lives: Encounters of a Scientist, Akadémiai Kiadó, Budapest, 2004.

Candid Science III: More Conversations with Famous Chemists, Imperial College Press, London, 2003.

The Road to Stockholm: Nobel Prizes, Science, and Scientists, Oxford University Press, Oxford, 2002 (paperback edition 2003).

Candid Science II: Conversations with Famous Biochemical Scientists, Imperial College Press, London, 2002.

Candid Science: Conversations with Famous Chemists, Imperial College Press, London, 2000.

In Our Own Image: Personal Symmetry in Discovery, Kluwer/Plenum, New York, 2000 (with M. Hargittai).

Upptäck Symmetri! (Discover Symmetry!, in Swedish), Natur och Kultur, Stockholm, 1998 (with M. Hargittai).

Symmetry through the Eyes of a Chemist, Second edition, Plenum, New York, 1995.

Symmetry: A Unifying Concept, Shelter Publications, Bolinas, CA, 1994 (with M. Hargittai).

The VSEPR Model of Molecular Geometry, Allyn & Bacon, Boston, 1991 (with R.J. Gillespie).

The Structure of Volatile Sulphur Compounds, Reidel, Dordrecht, 1985.

The Molecular Geometries of Coordination Compounds in the Vapour Phase, Elsevier, Amsterdam, 1977 (with M. Hargittai).

Edited books

Strength from Weakness: Structural Consequences of Weak Interactions in Molecules, Supermolecules, and Crystals, Kluwer, Dordrecht, 2002 (with A. Domenicano).

Symmetry 2000, Vols. I–II, Portland Press, London, 2002 (with T.C. Laurent).

Advances in Molecular Structure Research, Vols. 1–6. JAI Press, Greenwich, CT, 1995–2000 (with M. Hargittai).

Combustion Efficiency and Air Quality, Plenum, New York, 1995 (with T. Vidóczy).

Spiral Symmetry, World Scientific, Singapore, 1992 (with C.A. Pickover).

Fivefold Symmetry, World Scientific, Singapore, 1992.

Accurate Molecular Structures, Oxford University Press, Oxford, 1992 (with A. Domenicano).

Quasicrystals, Networks, and Molecules of Fivefold Symmetry, VCH, New York, 1990.

Symmetry 2: Unifying Human Understanding, Pergamon Press, Oxford, 1989.

Stereochemical Applications of Gas-Phase Electron Diffraction, Vols. A–B, VCH Publishers, New York, 1988.

Crystal Symmetries, Shubnikov Centennial Papers, Pergamon Press, Oxford, 1988 (with B.K. Vainshtein).

Symmetry: Unifying Human Understanding, Pergamon Press, Oxford, 1986.

Diffraction Studies on Non-Crystalline Substances, Elsevier, Amsterdam, 1981 (with W.J. Orville-Thomas).

CANDID SCIENCE V

Conversations with Famous Scientists

Balazs Hargittai

István Hargittai

Imperial College Press

ICP

Published by

Imperial College Press
57 Shelton Street
Covent Garden
London WC2H 9HE

Distributed by

World Scientific Publishing Co. Pte. Ltd.
5 Toh Tuck Link, Singapore 596224
USA office: 27 Warren Street, Suite 401-402, Hackensack, NJ 07601
UK office: 57 Shelton Street, Covent Garden, London WC2H 9HE

Balazs Hargittai
Saint Francis University
117 Evergreen Drive
Loretto, Pennsylvania 15940, USA

István Hargittai
Budapest University of Technology and Economics
Eötvös University and Hungarian Academy of Sciences
H-1521 Budapest, Pf. 91, Hungary

Library of Congress Cataloging-in-Publication Data
Hargittai, Balazs.
    Candid science V : conversations with famous scientists / Balazs Hargittai, István Hargittai.
       p. cm.
    Includes index.
    ISBN-13 978-1-86094-505-2 -- ISBN-10 1-86094-505-8 (alk. paper)
    ISBN-13 978-1-86094-506-9 (pbk) -- ISBN-10 1-86094-506-6 (pbk. : alk. paper)
      1. Scientists--Interviews. 2. Scientists--Biography. 3. Scientists--History--20th century.
    4. Mathematicians--Interviews. 5. Mathematicians--Biography. 6.
    Mathematicians--History--20th century. I. Title: Candid science five. II. Title: Candid
    science 5. III. Hargittai, István, IV. Title.

    Q141.H264 2005
    509'.2'2--dc22
    [B]

                                                                            2004062538

British Library Cataloguing-in-Publication Data
A catalogue record for this book is available from the British Library.

Printed in Singapore

FOREWORD

I feel greatly honored to be given the task to write this Foreword. It has been a great pleasure to go through the fascinating interviews of yet another nearly two scores of scientists by the Hargittais. What strucks me this time as much as before is the enormous individual variation of the characters exposed, making every story unique. Whether such variation is peculiar to scientists, or to human beings in general, or perhaps even to other species, is beyond me. However, one can easily identify an important element that all these individuals have in common: curiosity. Again, one can ask if this is something peculiar to scientists. In this case I feel inclined to answer yes. Admittedly, the exploratory drive rests on a fundamental instinct of profound survival value that all human beings and a great number of other species have in common. In childhood the response to novelty is a lot more dramatic than later in life. As we grow older our curiosity loses some of its intensity, which is perhaps a sign of maturity. Maybe a common feature of scientists is a slow maturation process, at least in this regard.

An interview, like an autobiography, is of course not an impartial statement. As time goes by we tend to remodel our reminiscences, perhaps to make them more palatable for our self-esteem. People involved in one and the same event will thus often describe it and their role in it differently. For the historian it must therefore be of utmost value to have access to as many personal accounts as possible of a scientific discovery. In this regard autobiographies and interviews are complementary. An advantage of the interview is that the interviewer can bring aspects into focus that the interviewee might otherwise tend to pass by. In any event it will remain for the historian to scrutinize all relevant documents in order to come as close to the objective "truth" as possible.

It is remarkable how the Nobel Prize has been able to keep its top position over the years. One may wonder why. At the outset the announcement of the Prize must have been astounding, considering its size and its scope. Subsequently, the Nobel Foundation and the institutions involved in the evaluation process have apparently done a sufficiently good job to keep up the reputation. The existence of such a Superprize is, however, not unproblematic. To some extent this has to do with Alfred Nobel's Testament. First of all, it brings into focus a distinct, prizeworthy discovery. In fact, the discovery should preferably have been made during the year preceding the award, even though earlier discoveries could be taken into account provided their importance were not immediately obvious. These stipulations should perhaps be viewed against Nobel's own astounding discoveries of dynamite and the like, even though it would be unfair to blame the richly gifted Alfred Nobel for simple-mindedness. Nevertheless, the emphasis of a distinct discovery has probably left out a number of outstanding pioneers who have opened up new important fields without necessarily contributing with any specific discovery. It may have been difficult for the various Nobel Committees to deal with this problem in some cases. In any event it is obvious that the prizeworthy candidates outnumber the laureates and that the actual outcome will often depend on a number of more or less relevant circumstances. It is regrettable that this fact is not always considered enough and that consequently some prizeworthy candidates feel unnecessarily disappointed.

Once again, the Hargittais are to be congratulated on yet another masterful Candid Science volume. It will certainly be enjoyed by a great number of enthusiastic readers.

Göteborg, January 2005   Arvid Carlsson

PREFACE

In this fifth volume of the *Candid Science* series, there are some departures from the previous volumes in the approach of the compilation of the material. The present volume as a whole is not classified as physics, chemistry or bio-medical sciences, but keeping with cross- and inter-disciplinarity, it contains entries from all these disciplines and, in addition, from mathematics. There is only a loosely-followed sequence in the volume, going from mathematics to physics to chemistry and to the biomedical sciences.

Another peculiarity of this volume is that it includes nine interviews from another project, the Larson Tapes. The story behind it is as follows. In 1998, I recorded a conversation with Clarence Larson (contained in this volume) and learned about the project he and his wife Jane had been doing, called "Pioneers of Science and Technology". Clarence and Jane (video) recorded conversations with famous scientists and technologists. By the time we met, they had collected over sixty recordings. They gave us a copy of a few conversations and my wife (Magdolna Hargittai, Magdi in short) and I published some edited transcripts from them in addition to my Larson interview in the magazine *The Chemical Intelligencer*.

In 1999, Clarence died and Jane donated all the original tapes in their collection to us, encouraging us to use them to the benefit of a wider readership. She stated in her letter of April 14, 1999:

> In recognition of your activities in recording interviews with outstanding scientists in *The Chemical Intelligencer* and elsewhere, including the interview with my late husband Clarence E. Larson incorporating excerpts from Clarence's interview with Luis

Alvarez and the article based on Clarence's Wigner interview, I am giving you all of Clarence's interview tapes. I am doing so with the understanding that you and Magdi will try to bring out articles using this interview material, possibly even producing a book based on this material. I am very much in support of your doing this in the interest of disseminating the knowledge and information Clarence had accumulated on these tapes. You have my permission and my blessing for your activities related to these tapes.

We started a series from the Larson Tapes in the magazine, but the magazine folded soon and the tapes laid idle for some years. Recently the idea came up to use some entries of the Larson Tapes in the *Candid Science* series and hence nine of the Larson interviews are included in the present volume. They are dispersed among the Hargittai interviews as found reasonable in the sequence of the volume, but clearly identified as Larson Tapes.

The original Larson interviews are of a different character from our interviews; they greatly differ also from each other, and we did not try to make a uniform presentation out of them. In some cases we merely produced an abbreviated narrative, in other cases some annotated excerpts and in yet other cases we tried to reproduce the whole interview in near completeness. We are grateful to Charles Townes, who — in addition to his own Larson interview — agreed to review the material of the Schawlow Larson interview as well. In the case of Dr. Townes, we also recorded our own interview, and the two interviews — twenty years apart — nicely augment each other. For the Fowler interview, we received help from his two former associates and especially from Charles Barnes of the Kellogg Laboratory of the California Institute of Technology. The Pauling transcripts were reviewed by Zelek Herman, Linus Pauling's long-time associate. There has already been a Pauling entry in the *Candid Science* series, viz., in the very first volume; however, that was a very brief interview, one of the last, if not the very last, Linus Pauling granted before his death. We are happy to have a longer exposure of this great scientist in this volume. To augment the Dulbecco interview, we asked Paul Berg to share his experience with him. In his own interview (see, *Candid Science II*, pp. 154–181), Dr. Berg mentioned Renato Dulbecco's impact on his research career and this was a good occasion to ask Dr. Berg to tell us more about it.

For our original contact with the Larsons, we have to thank Arnold Kramish with whom I had come into contact when I was editing *The Chemical Intelligencer*. Arnold Kramish served in the Manhattan Project and, later, with the U.S. Atomic Energy Commission. He served as a consultant to the U.S. government and industry. He is the author of many books and articles on nuclear history. In his letter of May 27, 1997, he suggested to me and my wife to get in contact with the Larsons, and I am quoting from his letter:

> On September 2, 1944, I suffered a near-fatal accident at an installation of the Manhattan Project. Colonel Stafford Warren, chief physician of the Manhattan Project, intervened on my care, and to him I credit much of the fact that I am here today. By chance, his daughter, Jane became my valued and efficient secretary after the war at RAND.
>
> Enter Clarence Larson, who, during the war, was an associate of E. O. Lawrence at Berkeley in developing the electromagnetic method of isotope separation. He then became head of that project at Oak Ridge. After the war, he was placed in charge of all Manhattan Project — then Atomic Energy Commission (AEC) — isotope separation. Eventually, he became an AEC Comissioner. He married my secretary, Jane, the daughter of Colonel Warren.
>
> Through the years, Jane has become an accomplished and internationally-known ceramicist, specializing in murals for scientific buildings. For example, the mural at the AAAS [American Association for the Advancement of Science] building is hers. She has just been awarded the commission to do the mural for the chemistry building at the University of Maryland.

We are happy that with the nine Larson interviews communicated in this volume, we are able to pay tribute to Clarence Larson's memory and express our appreciation to Jane Larson's generosity.

Readers will notice that a new Hargittai name appears on the cover of the present volume. Balazs is our son and although his research interest, peptide chemistry, is far from our structural chemistry, he has shared our general interest in science history and scientists.

The technique of the Hargittai interviews has been described repeatedly in the Prefaces to previous volumes and it has not changed. We record an informal conversation and later submit the slightly edited transcripts to the interviewee for checking, changing, and augmenting. Very few interviewees do not respond to our sending them the transcripts; most do and improve the presentation without changing the flavor of live conversation. In some cases there were repeated cycles of improvement. My interviewing approach is different from that of a journalist. Not only am I not trying to press for answers in cases where I sense reluctance on the interviewee's part, I do not mind the interviewee's changing of what had been said, in the subsequent exchange. I want the interviewee to feel relaxed during the interview and comfortable with the final product. I know I lose some following this approach, but I believe we all gain too because the interviewee senses this interviewee-friendliness and collegiality, and in most cases opens up more to a sympathetic colleague than to an aggressive journalist. The interview is a joint product in the final account; it reflects on both of us; and I believe that we both try to do our best in bringing forth an insight into both science and the individual scientist.

With this, I am not saying that on occasions I would not have liked to get more information than what had been offered. A case in point was the Damadian controversy in the Lauterbur interview. I knew very little about the story, I knew primarily about the unprecedented newspaper ads protesting the Nobel decision in the Physiology or Medicine Prize for the year 2003, the fact that Dr. Damadian was left out from it. Some colleagues had warned me not even to try asking Paul Lauterbur about it. However, I did, and my questions were not brushed away, rather, I understood that this was a topic that Paul had been exposed to saturation and the long story behind it had caused him and his family pain and much unpleasantness. Nonetheless, Paul gave me meaningful responses and he pointed to a book that helped me assess of what had happened. When I sent him the transcripts, he barely changed anything, and found my treatment of the issue sensitive.

The book Paul Lauterbur gave me is Donald P. Hollis's *Abusing Cancer Science: The Truth about NMR and Cancer* (The Strawberry Fields Press, Chehalis, Washington, 1987). From this book I would like to quote a few passages that are appreciative with respect to Damadian's contributions. I am doing this because in my reading, the newspaper ads and the hoopla around them have created a notion that might mask the value of Dr. Damadian's real contributions. In addition, it is a sad consequence of the Nobel Prize that those who are left out in most cases are not even accorded an "honorable mention" in Nobel dealings. The Nobel Prize — even if it may not be the intention of the Nobel Prize committees — often contributes to rewriting science history. From this point of view, it is worthwhile to quote a few brief excerpts from Hollis's book what seems to me an objective evaluation of Damadian's contribution (pp. 174–175):

Damadian had clearly made a provocative contribution to medical NMR when he showed the differences in relaxation times between the normal and malignant rat tumors and suggested that they might be used to diagnose cancer in humans. That provided a reason for further study of cell NMR and a goal for the early imagers. No one that I know of has ever denied Damadian credit for that contribution. Most people, sensitive to Damadian's desire for credit, emphasize that this idea, at least with respect to cancer, is Damadian's and his alone.

...

Damadian's cancer work gave Lauterbur a reason to invent imaging.

...

Lauterbur had the idea of making pictures of the human body by NMR and he invented and quickly demonstrated a practical way to do it. It is not difficult to imagine the feelings and disappointment of a person like Damadian when he realized that his observations had played a role in calling Lauterbur's attention to the fact that tissues of the body had different NMR properties and that had he possessed Lauterbur's education, background and more general view of science, he, himself, might have been the one to invent NMR imaging.

I would have been interested in meeting Dr. Damadian and including a conversation with him in this volume; I wrote to him inviting him for an interview. I was willing to visit him; alas, I received no response.

There was a controversy of a somewhat different kind around the 2000 Nobel Prize in Physiology or Medicine and I am pleased that all three laureates and also Dr. Hornykiewicz agreed to record a conversation with me. In this case, the person who was left out, Dr. Hornykiewicz, did not wage a protest to be sure, but many on his behalf did. Whereas in the case of the 2003 Nobel Prize in Physiology or Medicine where there was an unused third slot, in the case of the 2000 prize, all three slots were filled; nobody questioned the prize-worthiness of any of the awardees; however, the prize could have been formulated in other ways as well beside the way it was formulated and with other compositions of the laureates.

There was no controversy in the 2000 Chemistry Nobel Prize, yet I would have loved to interview Hideki Shirakawa in addition to Alan MacDiarmid and Alan Heeger; alas, it did not happen. The most intriguing question to

Shirakawa would have been about his Korean colleague who for the first time – quite by accident – produced the conducting polyacetylene polymer. There have been various versions of this accidental discovery, which was the foundation from which the work emerged that was ultimately awarded the Nobel Prize to Heeger, MacDiarmid, and Shirakawa. In his e-mail of April 30, 2002, Shirakawa gave me the name of the Korean scientist, Dr. Hyung Chick Pyon and clarified what happened: "It is rumored that the silvery form of polyacetylene film was discovered as a consequence of a linguistic mis-understanding between the visiting Korean scientist and me. But this is totally incorrect. He was a fluent speaker of Japanese because he was educated in Japanese while Korea was occupied by Japan for 35 years before World War II."

As I have written, this interviews project is merely a Hargittai hobby, being a side product of our main activities. As travel takes us to places, mostly lecture invitations and, to a smaller extent, family vacations, we try to use them to cultivate our side interests. In the [European] Summer of 1999 I was invited to a visiting professorship at the University of Auckland in New Zealand. Magdi and I organized the trip in such a way as to include a stopover in Bangkok and to interview Princess Chulabhorn, who is also a research chemist. Herbert Brown [*Candid Science I*, pp. 250–269] made the connection between us. Hence a most unusual encounter took place.

In addition to the acknowledgments already expressed above, I would like to thank the following for invitations and hospitality extended to us that made some of the interviews possible: Richard Henderson and the MRC Laboratory of Molecular Biology in Cambridge in 2000, James Watson at the Cold Spring Harbor Laboratory in 2002, Ingemar Ernberg at the Karolinska Institute in 2003, Alex Varshavsky at the California Institute of Technology in 2004, and Gunther Stent at Berkeley, California, in 2004.

Balazs would like to mention his postdoctoral stay at the University of Arizona in Tucson that brought us to the meeting with Donald Huffman (and Wolfgang Krätschmer) and express his appreciation to Miss Amy K. Croskey, a student of Saint Francis University for her assistance in transcribing some of the Larson interviews. We are both grateful to Magdi for her letting us include a couple of her interviews in this volume and for her untiring support and help in bringing this project to completion. I am grateful to the Budapest University of Technology and Economics and the Hungarian Academy of Sciences as well as to the Hungarian National Scientific Research Funds for their support of our research activities in structural chemistry.

Balazs is grateful to Saint Francis University in Loretto, Pennsylvania, for support and encouragement.

We note with appreciation that some of the entries — often in a somewhat different form — had appeared — as always duly noted — in *The Mathematical Intelligencer*, *The Chemical Intelligencer*, *Chemistry International*, and *Chemical Heritage* (in the latter with Alfred Bader's generous support).

Budapest   István Hargittai
Loretto, Pennsylvania   Balazs Hargittai

CONTENTS

H. S. M. (Donald) Coxeter, 2000 (photograph by I. Hargittai).

1

H. S. M. (Donald) Coxeter

Harold Scott MacDonald Coxeter (1907, London – 2003, Toronto) was Professor Emeritus at the Department of Mathematics of the University of Toronto when my wife and I visited him on August 1, 1995 and I recorded the following conversation with him.* We met on other occasions as well, mostly in symmetry meetings in Northampton, Massachusetts, Stockholm, and Budapest, but this was the only occasion when we did such a recording.

It would be difficult to give a better characterization of Professor Coxeter's activities than what Buckminster Fuller wrote about him as he dedicated his opus magnum, *Synergetics*, to H. S. M. Coxeter: "By virtue of his extraordinary life's work in mathematics, Dr. Coxeter is *the* geometer of our bestirring twentieth century, the spontaneously acclaimed terrestrial curator of the historical inventory of the science of pattern analysis."

You have three first names. Which is the one you like most?

I prefer to be known as Donald. The original intention of my parents was to call me MacDonald Scott Coxeter but some stupid godparent said that I should be named after my father and they added Harold at the beginning. That made Harold MacDonald Scott. The initials then would

---

*In part, this interview has appeared in *The Mathematical Intelligencer* **1996**, *18*(4), 35–41.

Donald Coxeter with his daughter in the Hargittais' home in Budapest, 2000 (photograph by I. Hargittai).

look like a ship, H. M. S., Her Majesty's Ship. This is why they switched the two names, and it became Harold Scott MacDonald. What I have done lately is to use H. S. MacDonald Coxeter.

You have a son and a daughter. Did they follow your footsteps?

Not at all. My son got interested in the church and took a degree, Master of Theology. As a minister he did not fully enjoy anything except the parish visiting, looking after unfortunate people. Eventually he gave that up and got a second degree as Master of Social Work. He did something about rehabilitation of drug addicts, then got interested in geriatric hospitals and getting supplies for them and he is still in that position now in the state of New Jersey. My daughter married an accountant. She is a Registered Nurse and she lives in a small place between Toronto and Hamilton. We can visit her more easily than our son who is 800 km away.

Grandchildren?

I have five grandchildren and four great-grandchildren.

You wrote somewhere that your hobby was music and travel. When you listen to music, do you relate it in any way to geometry?

Not directly, but the artistic feeling that one has is very much the same in both cases. Before I took up mathematics, I was very interested in music, to the extent that I tried to compose. Between the ages of 7 and 14 I did a lot of musical composition, under the guidance of Tony Galloway, an old friend of my family who was a very expert violinist and a sadly neglected composer. He taught me a lot about the theory. I wrote a lot of piano pieces, and songs that my father used to sing. I was even so ambitious as to write a string quartet. However, very few of them are worth preserving. Two samples can be seen after the biographical sketch at the beginning of my new book *Kaleidoscopes*. It was edited by F. A. Sherk, one of my former students. He collected 26 of my papers that had to do with symmetry.

Who turned your attention to geometry?

It was pretty much by myself. I was always interested in the idea of symmetry. When I was 14, I was in a boarding school in England, and happened to have some trivial illness. In the school sanatorium I was put in a bed next to a boy called John Flinders Petrie and he became a firm friend. (He was the only son of Sir Flinders Petrie, the great Egyptologist.) He and I looked at a geometry textbook with an Appendix on the five Platonic solids. We thought how interesting they were and wondered why there were only five, and we tried to extend them.

   He said, if you can put three squares around a corner to make a cube, what about putting four squares around a corner? Of course, they'd fall flat, giving a pattern of squares filling the plane. He, being inventive with words, called it a "tessarohedron". He called the similar arrangement of triangles a "trigonohedron". Later on he said, what about the limitation of putting four squares around the corner and why not more than four? Maybe you can put six squares around the corner if you don't mind going up and down in a zigzag formation. Thus he discovered a skew polyhedron with "holes", a kind of infinite regular sponge. He also noticed that the squares in this formation belong to the cubic lattice. He saw that it can be reciprocated so that instead of six squares at each vertex you have four hexagons. He noticed that this could be obtained from the uniform honeycomb of truncated octahedra fitting together to fill space. The hexagons of the truncated octahedra come together, four at each vertex, and continue to form a sponge filling all space; so this was a second skew polyhedron. Then I said if you can have six squares and you can have four hexagons, why not even more: why not have six hexagons at the vertex as in the space filling of tetrahedra and truncated tetrahedra? Then we extended the Schläfli symbol by which the cube is called {4, 3} and we called these new polyhedra {4, 6|4} and {6, 4|4}, and {6, 6|3}, the number after the stroke indicating the nature of the holes one sees in the sponge.

Before we left school, we went on to consider what'd happen in four or more dimensions and other things which later we learned had been discovered before, by L. Schläfli in Switzerland.

Did your friend also continue in geometry?

He did, and became quite clever at it. Unfortunately, because his father belonged to University College London, and my teacher wanted me to go to Cambridge, we went to different universities. He did quite well at University College and then the war came, World War II; he enlisted as an officer and was taken prisoner by the Germans. He organized a choir there. After the war ended and he was released, he went to a well-known school in southwest England, Dartington Hall, and he had a rather trivial job there. He never seemed to fulfill his early promise. He just became a tutor who looked after children who were not doing well in school.

But he still corresponded with me, and it was he who noticed that when you take a regular polyhedron and look at the edges, you see that there is a zigzag of edges that go round and close up; for instance, if you take those edges of a cube that do not involve one pair of opposite vertices, they form a skew hexagon. We call this the "Petrie polygon" and it is now a well-known property of a regular polyhedron to have a Petrie polygon: a skew polygon in which every two consecutive edges, but no three, belong to a face.

Is he retired now?

No, he died. A very sad story. He married a very lovely lady and had a daughter and all went well. Then somehow his wife got a heart attack and died. He was so distraught, missed her so terribly that he didn't know where he was going, and he walked into a motorway in England where the cars were going at a huge speed and he just didn't know what was happening and one of them killed him, just two weeks after his wife died. This was about 24 years ago.

Buckminster Fuller called you "the geometer of the twentieth century". How did you get to know each other?

This is a terribly exaggerated statement, but he was given to that sort of writing and speaking. He was a dear old man, and I was quite fond of him but he had overblown his stars as a mathematician. He was really a very good architect and a very good engineer. His geodesic domes are really a wonderful thing. But when he got into mathematics he was a little bit amateurish.

Did he claim that he was a mathematician?

I think so, yes. He liked to invent different names for things. For instance, the cuboctahedron he called "vector equilibrium" or something like that.

How much interaction did you have with him?

Very little. Once Hendrina and I visited him in his home in Southern Illinois. I have a friend who is a Professor of Philosophy in Carbondale, and while we were there, we visited Bucky's polyhedral house. As people passed by, they were very curious, and he finally had to build a high fence around the house so that people couldn't see it and he could have some peace.

How far back can we detect the regular polyhedra in human history?

Of course, Plato wrote about them and this why they are called Platonic solids. Obviously the Pythagoreans knew them before that. Sometimes the archeologists find dodecahedral dice. That sort of thing is what I mean when I say that we don't know how far back they go.

In some of your writings you distinguish between crystallographic solids and others such as the icosahedron and dodecahedron. Nowadays, however, this distinction is quite blurred.

That's true. Just look at the writings of Professor Marjorie Senechal. I'm just reading her lovely book *Quasicrystals* which refers to some recent papers of mine.

So even geometry is changing and evolving.

As in all branches of mathematics, there is a tremendous increase in productivity. Research goes on and much of it I have no inkling of. If you only look at the development of *Mathematical Reviews*, when they first started about 1940, it was quite a thin volume, and each month they got more and more and eventually there were hundreds of thousands of

Magdolna Hargittai and Donald Coxeter in his office at the University of Toronto, 1995 (photograph by I. Hargittai).

papers being written and so the later volumes are ever so much thicker than the original ones.

Geometry is very important in chemistry. We have simple but very helpful models of molecular geometry, but teaching them in a freshman chemistry course in the U.S. is rather hindered by the students' lack of knowledge of basic geometry.

It's even worse in England, where in school they teach almost no geometry.

Your books are full of quotations. How do you collect them?

Just by noticing. I must have read a lot, and I just remember them.

Do you return to books that you'd read before or you just keep moving on to other books?

I just move on to other books. When I was young I was very interested in stories by H. G. Wells and when I was a student I was very interested in the plays of G. Bernard Shaw.

You have had some connections with M. C. Escher.

First, at one of the International Congresses of Mathematicians which took place in Amsterdam, there was an exhibition by M. C. Escher. My wife, being Dutch, naturally talked to him when he was exhibiting his art to the mathematicians. So she got to know him and that was very helpful; we kept up correspondence. Later I wrote an article for the Royal Society of Canada: my Presidential address for Section III, on symmetry. It included a Poincaré-style model of the tessellation of (30°, 45°, 90°) triangles filling the hyperbolic plane so as to form a black and white pattern. Escher saw this and thought it was just what he wanted. In some of his work he had got tired of filling the plane with *congruent* figures, fitting together, and he thought how nice it would be if they were not congruent but just *similar* and changed size while keeping their shape. Escher liked these things because they fulfilled his wish to make a pattern in which he had fishes, for instance, of a good size near the center but getting smaller and smaller as he went towards the circumference. He made *Circle Limit I*, and then *Circle Limits II, III*, and *IV*. *Circle Limit III* was particularly interesting because it had four colors besides black and white. It was closely related to the hyperbolic reflection group that I'd described.

Did you inspire him to this work?

That's right. He was very pleased with this idea. After he had seen that paper of mine he did *Circle Limits III* and *IV*. He had done *Circle Limits I* and *II* before.

Did he construct his drawings with precision?

Extraordinarily well, yes. There was a very interesting apparent exception because in *Circle Limit III*, if you look at the rows of fishes following one another, they have white stripes along their backs so that the circle is filled with a pattern of white arcs that cross one another. It is remarkable that the spaces between the white arcs appear to form a tessellation of hexagons and squares. Yet the white arcs cross one another, three going through each vertex; therefore they cross at angles of 60 degrees. In particular, you seem to have triangles all of whose angles are 60 degrees, and that, of course, is wrong because such a triangle would be Euclidean and not hyperbolic. Bruno Ernst, in his book about Escher, *The Magic Mirror*, page 109, was similarly disturbed, saying, "In addition to arcs placed at right angles to the circumference (as they ought to be), there are also some arcs that are not so placed." I was interested in this and looked at it for a long time and at last I realized what had happened. By careful measurement, I saw that *all* those white arcs meet the circumference at an angle which is very close to 80 degrees instead of 90 degrees. In fact, each of the white arcs does not represent a straight line in the hyperbolic plane but one branch of an *equidistant curve*. When you put it that way, everything falls into place, and you see that Escher did those drawings with extraordinary accuracy: when I worked it out trigonometrically I found that the angle of 80 degrees is actually arc cos $[(2^{1/4} - 2^{-1/4})/2] \approx 79° 58'$.

Was he aware of this?

Absolutely unaware. In his own words: "… all these strings of fish shoot up like rockets from the infinite distance at right angles from the boundary and fall back again whence they came."

Was it intuition?

True intuition. He came to hear me give a lecture once, and I tried to make it as simple as possible; he said he didn't understand a single word.

Mathematicians and crystallographers recognized Escher before anybody else. What was his main appeal?

It was the appeal of symmetry.

You give a definition of symmetry in one of your books and that definition, very geometrical, is based on congruency. How far, do you think, such a rigorous definition can be relaxed?

With Escher we've relaxed it to considering shapes that are similar instead of congruent. Groups of similarities are more general than groups of isometries. More precisely, groups of isometries occur as normal subgroups in groups of similarities. Part of the fascination for me was to look at *presentations* of groups. The groups have generators which satisfy certain relations. There is actually something they call a "Coxeter group" which means you have a certain number of generators of period two and you specify the periods of their products in pairs. Such a presentation they now call a Coxeter group. It's a very simple idea but apparently nobody had put it like that as defining a particular family of groups. Then it turned out that some of the Coxeter groups have a relationship with Lie groups which I don't understand at all. I am very pleased though to see that these ideas have an application.

You mentioned before your wife's role in the contact with Escher. What does she do?

She is very artistic and appreciates music very much. She's been a wonderful wife to me, looking after me very carefully, and bringing up our children.

Did you know D'Arcy Thompson?

He visited us about 1940. He had a tour of Canada and actually stayed at our house. He was a wonderful man. His book *On Growth and Form* was very influential and he brought out a huge second edition when he was 70 years old. He was extraordinary in combining interest in so many different things: in geometry, biology, and classical literature, languages, everything. Very remarkable.

How about Kepler?

I've been an admirer of Kepler ever since I read that it was he who invented names for all the Archimedean solids, such as the cuboctahedron. Although the names of the Platonic solids are ancient, these less regular figures were only named later.

One of the Archimedean solids, the truncated icosahedron, has now become very conspicuous as buckminsterfullerene, the name of the C_{60} molecule. Unfortunately the chemists who discovered it were not familiar with Kepler's work.

They thought this shape was discovered by Buckminster Fuller.

The story of the discovery shows how useful geometry is, even for chemists.

It also illustrates the fact that people who don't know any mathematics, if they happen to play with hexagons and pentagons, inevitably make that figure. This fact was demonstrated very well by a present that I once received from Mrs. Alice Boole Stott: a lampshade made of 12 glass pentagons and 19 glass hexagons, joined together by strips of lead, as in a stained glass window. I may as well tell you a little more about her.

About 150 years ago an Englishman, George Boole, started what is known today as Boolean Algebra. He wrote a famous book on finite differences. He had five daughters and they were all distinguished in various ways. The youngest daughter, Ethel, married a Pole called Wojnicz so she is known as Ethel Lillian Voynich. She wrote novels and one of these novels was called *The Gadfly*. That novel somehow appealed very much to the Russians at the time of the Soviet Union, and they made a movie of this book. The music for it was composed by Shostakovich. Sometimes one hears excerpts from this music; it's quite fascinating.

Another one of Boole's daughters, the middle one, was called Alice; she married an actuary, Walter Stott, so she became Mrs. Stott. I got to know her very well, as it happened, through her nephew, Geoffrey Taylor, who was a mathematician and a Fellow of the Royal Society of London. He was in Cambridge when I was a student there and he introduced me to his aunt, Mrs. Stott, because he realized that she was interested in Archimedean solids as I was. She visited me and my mother, and I visited her very often in London. She was quite elderly and I was a student, so I called her "Aunt Alice". She got to know Dutch mathematicians because her husband happened to notice some articles by a Dutchman called Pieter Hendrik Schoute. Schoute was an expert concerning regular and semiregular polytopes in any number of dimensions, following in the footsteps of Schläfli. She was helpful to him and he was helpful to her. Between them they made a complete classification of uniform polytopes in four dimensions. He invited her to Holland and she was given an honorary degree by the University of Groningen. She didn't have a formal education. She was self-taught until she was taught by Schoute. Quite amazing. She had such a feeling for four-dimensional geometry. It was almost as if she could work in that world and see what was happening. She was always very excited when I had things to tell her about what was happening, and she helped me in what I did. Through her I was introduced to some of the Dutch mathematicians.

Did you keep up your interest in Archimedean solids?

Yes. In 1950 I was one of the three authors of a paper on uniform polyhedra which dealt with a generalization of Archimedean solids, the idea being that you have regular faces of two or more kinds and the same arrangement at every vertex. This is characteristic of the prisms and antiprisms as well as the Archimedean solids. If you allow the faces to cross one another, as Kepler did, then you get many more: 53 of these non-convex uniform polyhedra. I wrote a joint paper on these things with Jeffrey Miller (who died long ago) and Michael Longuet-Higgins. There are two brothers Longuet-Higgins: Hugh-Christopher is a psychologist and Michael is an oceanographer. It is a unique case. The two brothers are not only Fellows of the Royal Society of London but for five years they were Royal Society *Professors*, both of them at the same time!

Alan Mackay and Donald Coxeter in Stockholm, 2000 (photograph by I. Hargittai).

So we wrote this paper in 1953, enumerating the uniform polyhedra, allowing them to be non-convex. S. P. Sopov in 1968, and J. Skilling in 1975, using electronic computers verified that our list is, in fact, complete.

When did you leave England?

In 1936, when Hendrina and I moved to Canada. Before that I was a fellow of Trinity College. At one point the Princeton topologist, Solomon Lefschetz came to visit Cambridge and talked to Professor M. H. A. Newman who knew me. He happened to mention to Lefschetz that I showed promise in geometry. Lefschetz said that he would arrange for me to get a Rockefeller Foundation Fellowship to spend a year in Princeton. So I went there and was influenced a lot by his colleague Oswald Veblen, who had written a wonderful book on *Projective Geometry*. While at Princeton I thought about kaleidoscopes, groups generated by reflections, and what sort of fundamental region such a group would have.

During a second fellowship in Princeton I was invited to Toronto by Gilbert Robinson, who had earlier been with me in Cambridge. He was a Canadian and had a job at Toronto. So I gave a lecture and Samuel Beatty, Chairman of the Mathematics Department, must have liked my talk. For quite unexpectedly in 1936, back in England, I received a telegram from him, asking if I'd like to come to Toronto as an Assistant Professor. That was quite startling because usually one starts as a Lecturer and not as an Assistant Professor, so it was very flattering to be asked. I consulted Professor G. H. Hardy and my father; they both said that this was an offer one shouldn't turn down: you never know what's going to happen.

In 1936, people already thought that war was possibly coming; so they said, take your newlywed wife and go to Canada, which we did.

I met my wife in 1935 in an English village called Much Hadham, where she was visiting from Holland a certain Mrs. Lewis. My mother introduced me to her neighbor, Mrs. Lewis. A beautiful young Dutch lady was there: Hendrina Brouwer. We liked each other, and I invited her to come to Cambridge to see my rooms. Later she got a job in Cambridge, and we became engaged, and finally married in 1936.

We thought that we would be going back to England in a few years, but then the War came and we remained in Canada.

What was your father's profession?

He was a businessman, and at heart an artist. He belonged to the firm of Coxeter and Son, founded by his father and grandfather. They were manufacturers of surgical instruments and compressed gases, especially anesthetics. Nitrous oxide, N_2O, was their specialty. My father and his partner, Leslie Hall invented a machine that had a controlled mixture of nitrous oxide and oxygen to give to a person undergoing an operation. The anesthetist would watch the patient and gave him more oxygen if he seemed to be failing and more nitrous oxide if he seemed to be coming awake. That has been used ever since by some hospitals. I wish it were used more; it is a wonderfully safe anesthetic.

Did you ever meet J. D. Bernal?

I visited his laboratory in London. He worked with little balls of plastic clay; rolled them up, dusted them, put them together in large numbers and squeezed them to see what shapes they formed. I visited him because of my interest in sphere packing. He was a very fascinating person.

Another man in the same direction was Frederick Soddy. He was the man who invented the name "isotope". I knew him because of his interest in the Descartes circle theorem.

Soddy was a chemist.

Yes, but he was also interested in geometry, just like you. I met Soddy around 1933. I visited him in his house on the south coast of England, and had a wonderful walk with him along the beach. He wrote an article for *Nature* about the problem of putting circles in contact with one another. The particular problem that started it was about four circles in an ordinary plane all having contact with one another. It's very easy to make three circles have contact; the fourth one will go in between in the middle or outside. So you have four circles in mutual contact.

Soddy noticed that if you don't consider the radii themselves but their reciprocals, the curvatures of the circles, then the four curvatures satisfy a nice quadratic relationship: the sum of the squares of the curvatures is half the square of their sum. He didn't know that this was already discovered hundreds of years before by Descartes. Soddy wrote the theorem and the proof in the form of a poem and sent it to the magazine, *Nature*, where it was actually published. Somehow I got to know about this, and became fascinated by it, and generalized it in an article called "Loxodromic sequences of tangent spheres".

How about your pupils?

I've had 17 graduate students who went on to get their Ph.D.s and most of them have done quite well. Thirteen of them are professors.

You have been retired for some time now, but stayed very active.

I have been retired for 23 years but the University is kind enough to let me have this little office and so I go on.

Do you need any support for your work?

No, just this office. Of course, I have a pension which is an annuity. Then sometimes I get a hundred dollars for a lecture and recently I was awarded a prize for research by the Fields Institute in Toronto and the Centre de Recherches Mathématiques in Montreal.

You don't use a computer.

No, I never used a computer. I'm too busy writing with pencil and paper. Fortunately, they have a very good secretary here who does word-processing. Then she says I mustn't mind that I am fourth in line and she may have the paper typed by next Monday, but that's all right.

What's your next paper about?

At the moment I'm writing a paper on the trigonometry of hyperbolic tessellations. Escher may have known the solution intuitively, by trial and error, and I suppose he might have been interested in seeing precisely how to find the centers and radii of all his circular arcs.

John H. Conway, 1999 (photograph by I. Hargittai).

2

JOHN H. CONWAY

John Horton Conway (b. 1937 in Liverpool, England) is John von
Neumann Professor of Applied and Computational Mathematics at
Princeton University. He received his B.A. and Ph.D. from the University
of Cambridge, England, in 1959 and 1962. He was Lecturer in Pure
Mathematics, then Reader, and finally, Professor at the University of
Cambridge before he joined Princeton University in 1987. He was elected
Fellow of the Royal Society (London) in 1981, received the Pólya Prize
of the London Mathematical Society in 1987, and the Frederic Esser
Nemmers Prize in Mathematics in 1998. We recorded our conversation
on August 5, 1999, at the University of Auckland, New Zealand where
both of us were Visiting Professors for a brief period of time (John in
mathematics and I [IH] in chemistry).*

What does it mean to you to be von Neumann Professor at Princeton?

Von Neumann himself was a professor at Princeton at one time. He did
a tremendous amount of different things in mathematics, many of them
revolutionary. The most famous one is the idea of the computer. He not
only theorized about it, he was also involved in the building and use of
one. Earlier in his career, when he became established, he designed a system
of axioms of set theory. He had this idea of continuous geometry in which
the dimension function took continuous values. With Morgenstern, he wrote
The Theory of Games and Economic Behavior. Many of the things von

---

*In part, this interview has appeared in *The Mathematical Intelligencer* **2001**, *23*(2), 7–14.

Neumann was interested in, I'd been interested in — such as set theory, finite numbers, games, abstract computation, and this helped me to accept the job. It amused me that von Neumann's interests and mine were so closely related, with the exception of making bombs. Also, I have never moved into his system of continuous geometry.

What is your main interest?

I've had so many. As you know, I've been interested in symmetry for a long time, and that comes out as group theory. I spent a good twenty years of my mathematical life working very intensively with groups, but I'm not really a group-theorist. All the time I was attending these group-theoretical conferences I felt myself a little bit of a fraud because all the participants were concerned with the really big problem of understanding all the simple groups, the building blocks of group theory. They also had a lot of technical knowledge that I didn't have. My interest is only in studying and appreciating all the beautiful patterns, whenever you have a group, and I was interested in studying the associated symmetrical objects.

I had a long odyssey. When I was a graduate student I was interested in number theory and my advisor was a famous number-theorist Harold Davenport. Then, while I was still officially his student I became interested in set theory, and that's what I wrote my thesis on. After that, suddenly, these large groups began to be discovered, and I jumped into that field and made my professional name in it. That interest lasted for many years.

When I moved from Cambridge to Princeton, I didn't have anybody group-theoretical to talk to, and I became much more of a geometer, and that's what I consider myself now. In this, of course, I interact with others studying symmetry, but it doesn't have to be symmetry. The net effect of this long journey has been that I've been in a good position to notice certain things. For instance, I've always been interested in games and regarded it as a mathematical hobby, but then the theory of games led to my discovery of surreal numbers. I wish I'd invented the name but I didn't. There's a bizarre aspect to the surreal numbers. You take a definition *a priori* and it looks as though it's sort of tame, giving you ordinary real numbers, one and a half, root two, pi, and so on. But the same definition gives you infinite numbers and infinite decimal numbers. I stumbled on these things as a consequence of studying game theory. The fact that I had already studied infinite numbers, as part of my mathematical development, meant that I was able to recognize that what I had come upon was a far-reaching generalization of various notions of numbers: Cantor's infinite numbers, the classical real numbers, and everything else. So because I've done so many subjects, I was able to grasp this, and I wrote a book called *On Numbers and Games*. It founded the theory of surreal numbers and that includes the ordinary real numbers; this method of thinking of them as games turned out to give a simpler, more logical theory than anybody had found before, even for the real numbers. That sort of thing has happened to me a number of times. For instance, one of the big discoveries in group theory was recognition that the monster group, which is an absolutely enormous beautiful group, was connected with various things coming from classical nineteenth century number theory. As somebody who had done both, I was able to see those connections.

Martin Gardner once told me that, while he was editing the mathematical column of Scientific American, *whenever he stumbled on a new problem and asked you about it, it turned out that you had already dealt with the problem, mostly had solved it, and yet hadn't bothered to publish the solution.*

It's a big job writing something for publication, and I'm lazy. I'm not ambitious anymore. When I was a young man I was ambitious to be recognized as a great mathematician. I haven't lived up to that ambition because the kind of mathematics I'm doing is not the kind that had my ambition. In some sense I've lowered my sights; I've pulled in my horns. But I'm enjoying myself. I've got a good job, although I don't fit in the Princeton set. I'm not the typical Princeton mathematician, yet I'm recognized as such. I'm at the top of the mathematical tree, not the top person but near the top. I don't feel any compulsion to justify myself anymore. What I think is this: "Princeton bought me, and whether it was a good buy or not is no longer my concern."

In my late twenties I was quite worried that I didn't seem to have justified myself. I had a job at Cambridge, and I got my job very easily. Then a few years later there came a sort of crunch and nobody could get a job. There were very good people who were my near contemporaries, who came just a year or so later than I and who had done better work than I had, and they would not be getting anything. That made me feel guilty, and the guilt was exacerbated by the fact that I didn't seem to have done any mathematics worth noticing after I got the job. That made me feel depressed.

Then something very nice happened. I worked out this simple new group called the Conway group, which at the time was a really exciting contribution to knowledge. As soon as it was done I started traveling all over the world. I crossed the Atlantic, gave a twenty-minute talk, and flew back. That was around 1970. The upshot was that suddenly I started producing things. The next year I produced the surreal numbers, and then something else. Not only did I become successful, but I also deserved the success.

I remember thinking one day, asking myself, "What's happened? Why is it that I suddenly produced three or four really good things and nothing in the previous ten years?" I suddenly realized that the lack of guilt feelings was a good thing about it. Once I had justified myself that I deserved the job, I found the freedom to think about whatever I was interested in and not worry about how the rest of the world evaluated this.

What lifted you out of your depression in the first place?

Just this tremendous ego trip of discovering this new thing, which put me into the forefront. From then on it took a little while to convince myself that I was not going to worry and that I was going to study what seems interesting to me without worrying what the rest of the world thinks about it.

It's been rather hard to live up to it at times. For instance, when I moved from Cambridge to Princeton I started giving some graduate lectures about what I'd been doing the last few years. There, in the audience, were very famous mathematicians at Princeton who were all coming along to hear me. My style of lecturing in Cambridge was always elementary. Also, Cambridge is an informal place with a tradition of tolerating eccentrics. You're almost expected to be a little bit odd. In Princeton, however, I felt inhibited by the presence of these *big* people. I started to lecture more as a formal mathematician, as everyone else does, and then I realized that it was a disaster because it wasn't me. It took some effort to get me back to my own style. By the way, those famous mathematicians are no longer in my audience; the audience consists of graduate students or undergraduate students, depending on who I am lecturing to. When I am outside that climate, giving a lecture to a big international meeting, then the audience is always mixed, and that's a wonderful thing because then I can lecture at whatever level I want to.

I meant to ask you about your lecturing style. I remember when we were both giving lectures on symmetry at the Smithsonian Institution and you were jumping on top of a table and then hiding beneath it.

There's a certain amount of almost cynicism in this. Every now and then a joke appears to me spontaneously while I'm lecturing, and I incorporate it. If it's good then it stays in that lecture forever. If I give a lecture 20 or 30 times, the jokes just accumulate. The net effect is that the lecture gets better. I remember a terrible time when I was lecturing in Montreal and they asked me to let them videotape it. After the lecture it turned out that the man with the video camera didn't arrive, but he arrived after the lecture was over and they asked me to give the lecture again. I asked them to drag up an audience that was disjoined from the previous one. So I gave the lecture again. However, the audience was not disjoined because some of the same people still attended. This inhibited me tremendously because a joke that looks as though it occurs to you on the spur of the moment, you can't tell a second time.

Did you come across Paul Erdös?

He was a bit strange. I met him when I was an undergraduate. He used to pose problems, and I got involved in some of them. He did a lot of traveling and I did a lot of traveling myself, though nowhere near Erdös, but I tended to meet him sometimes. I would meet him in Montreal and a few days later in Vancouver or in Seattle. I walked into the cafeteria at Bell Telephone one day and sat next to Erdös. My Erdös number is 1.

Donald Coxeter?

Coxeter has been one of my heroes. When I was still at high school in England, grammar school, I wrote to Coxeter. He was the Editor of Rouse Ball's *Mathematical Recreations.* I was absolutely delighted by that book. That was 1953-ish, and I have known him ever since.

Buckminster Fuller stated that Coxeter is the geometer of the twentieth century.

This must be one of the very few things I would agree with Bucky about. Coxeter is my hero. I remember a story at one of the conferences in Coxeter's honor and people were telling how this wonderful man had turned them into mathematicians. I thought I must say something different. So when

I got up, I said, "Lots of people have come here to thank Coxeter; I've come here to forgive him." I told them that Coxeter once very nearly succeeded in murdering me. His murder weapon was something that even Agatha Christie would never have thought of: a mathematical problem. Then I told the story, which is actually true.

Coxeter came to Cambridge and gave a lecture. Then he had this problem for which he gave proofs for selected examples, and he asked for a unified proof. I left the lecture room thinking. As I was walking through Cambridge, suddenly the idea hit me, but it hit me while I was in the middle of the road. When the idea hit me I stopped and a large truck ran into me and bruised me considerably and the man considerably swore at me. So I pretended that Coxeter had calculated the difficulty of this problem so precisely that he knew that I would get the solution just in the middle of the road. In fact I limped back after the accident to the meeting. Coxeter was still there, and I said, "You nearly killed me." Then I told him the solution. It eventually became a joint paper. Ever since, I've called that theorem "the murder weapon". One consequence of it is that in a group if $a^2 = b^3 = c^5 = (abc)^{-1}$, then $c^{610} = 1$.

Wall painting in John Conway's office at Princeton University (photograph by Magdolna Hargittai).

Other heroes?

Archimedes. Two thousand years ago, he had very clear ideas about difficult, subtle problems, the nature of the real numbers. In my office I have painted on the wall all my friends. There was a young man who painted a café in Princeton, and I got him to come and paint pictures on my wall. Archimedes is there and Leonard Euler is there. Johannes Kepler is also one of my heroes. He was the greatest mathematician of his age and a very interesting guy, too. There are some people about whom I have ambivalent feelings, Isaac Newton and Karl Friedrich Gauss, for instance. They were really great mathematicians and great physicists too, but they don't seem to be such nice people and that rather distances me from them. I would like to have the opportunity to have a 20-minute chat with Archimedes or Kepler, and I'm not sure about Gauss though he might be able to tell me more. I wouldn't enjoy the interview with him so much.

Of the living heroes, I don't think there's anybody to match up Coxeter as an intellectual hero for me. The work he does is elegant and he writes beautifully. There is a paper by Coxeter, Miller, and Longuet-Higgins, and I know Coxeter wrote it, and I admire how beautifully it was written. If you look at any of Coxeter's papers you will find this beautiful craftsmanship in the design of his papers. That means his papers can just be read smoothly. The really important thing about Coxeter is that he kept the flame of geometry alive. There was a terrible reaction against geometry in the universities 30 or 40 years ago, which has had tremendously bad effects. So geometry was not a popular subject, and Coxeter all the time did his beautiful geometry. And he is a lovely man. I remember him at meetings; there's often this embarrassing time at the end of a lecture when the chairman asks for questions and comments, and there may be none. Coxeter always had something to say, complimenting the speaker. He's a true gentleman.

What's your principal problem with Buckminster Fuller?

His way of saying things is so obscure. To me, geometry is nothing if you don't also have precise proofs and clear enunciation and logical thoughts. There isn't any logical thought in Fuller, only a sort of simulacrum of logical thought. You don't know what the rules are in manipulating the words in the way Bucky does.

Don't you think he deserves credit for having enhanced interest in geometry and what is called today "design science"?

He's certainly had a positive effect in that sense. On the other hand, he says somewhere that you can't pack spheres with higher density than you get in face-centered cubic packing. I don't think he thought he had a proof, but he has some sort of plausible argument for why this is true. But countless people say that Buckminster Fuller had proved this, years ago. I experienced this when I was involved in a dispute over densest packing. And you look back at these words and find that he just sort of asserts it. Then they say, "Bucky wouldn't assert something unless he could prove it." They say this because, to them, Bucky is a god who could do no wrong.

What is the situation today with the packing problem?

The situation is that in 1990 someone produced what he called the proof, which never was a proof and which was heavily attacked. He had some good ideas. He has now actually withdrawn his claim to have a proof but he still thinks he can patch it up.

One year ago now, Tom Hales announced that he'd finished his work on this. He has a 200-page paper supplemented by computer logs of hours of interrogation between him and the machine. His student, Samuel Ferguson, is also involved. My view is, yes, this is probably a proof. On the other hand, since it involves so much interaction with the machine, it will be very difficult to referee it.

So how can you assess it?

I can best do that in a rather invidious way, by comparing it with the previous claim. There the criticism was that he sort of tended to wave his hands, he had some inequality he had to prove, and in one notorious case he evaluated the inequality at one point and then he maintained that it was true everywhere. Hales's way of proving inequalities are so much tighter, it's amazing. He cuts the integral into lots of little pieces, and in each place he replaces the functions that he is dealing with by a linear approximation based on the derivatives; he then reduces the problem to a linear programming problem and uses the computer to show the inequality. In the arithmetic of the computer he uses what is called "interval arithmetic", which means that you at any time say, "This real number is definitely greater than this and less than this." You don't just round it to the nearest ten places of decimals. You have explicit upper and lower bounds, and so on. Every inequality that Hales wants to prove — and the thing boils down to proving a large number of inequalities — is getting "interrogated" with the machine and examined by Hales. He shows that everything is one of the 2000 cases. The inequalities are proved not just by getting some rough idea how the functions are arrayed but getting very precise ideas that the function is being between this number and this number, and so on. The whole thing is a lot tighter and Hales has taken considerable pains to provide an audit trail. Anybody who disbelieves any assertion, can follow it through the tree and find that this was actually shown on this and this day by the following computation, which you can do again. Obviously, it would take a tremendous amount of work to check this. On the other hand, the feeling of reliability it gives to you is enormous.

Couldn't there be a simpler way of proving this?

My attitude to this is, "I don't want to get involved, even reading it." My other feeling is that this isn't part of the permanent furniture of mathematics, this type of proof. My feeling is that eventually some simpler proof will be produced. I have this idealistic viewpoint: I am prepared to wait. The waiting may mean that I die before I see the simpler proof but still I'm not interested in anything that isn't going to be permanent. This is an aristocratic viewpoint.

May we move now to fivefold symmetry, Penrose, quasicrystals? You designed the cover illustration for Scientific American *when Martin Gardner wrote about the Penrose tiling, which then became an influential paper.*

It's funny that you have quoted Martin Gardner's saying that I had always been there before. I attend the Art and Mathematics Conferences in Albany, organized by Nat Friedman and the participant list has the information about the participants' fields of interest. Somebody said to me once that he admired what I'd written somewhere and I asked him what was it about and he said, "Everything." Then it turned out that Nat hadn't anticipated my responding to his question about my field of interest and he filled it out for me, including everything. This idea of being interested in everything is something I almost consciously try to be.

But what you are asking me about is a very simple geometrical problem: can you have some tiles tile the plane only aperiodically? I had already been interested in that problem and when Penrose came up with his solution, I became tremendously excited and started making the damned things and drawing them. I was staying with Martin Gardner one time, and I drew out rather carefully a small page full of the tiles. Gardner had his own old-fashioned copying machine and we ran off a number of copies of this and pieced them together to produce a larger mosaic and later when I was back in Cambridge, we photocopied these smaller and then made still larger ones, and so on. Martin took the initial version I had made at his house in to the *Scientific American* office, where the graphics people redid it properly, and it became the cover.

I've always felt rather sad about our dining room table. We had a rather nice dining room table and we couldn't use it for about six months, and my wife was furious with me because it was covered with thousands of Penrose pieces, making a really beautiful pattern and I never wanted to disturb it. I remember having discussions about the possibility that chemicals might crystallize in that sort of manner, and I wish I had come out with that speculation in print because seven years later people found such crystals.

Alan Mackay did come out with such a suggestion in print prior to the experimental discovery in 1982.

Martin Gardner's *Scientific American* article appeared in 1974, and we conjectured at that time about the possibility of crystallization, and I wish we had come out with it in print. I remember that I wondered to myself how many different substances have been studied with respect to crystallization, and my guess was less than ten to the seventh power. Then I thought what was the probability that something will crystallize in this manner and one in ten to the seventh power seemed a reasonable guess; therefore such crystallization should happen.

Did you ever discuss this with Roger Penrose?

No, I didn't. When I was a student in Cambridge we got together; he and I were both interested in puzzles. Then he went off to Oxford, and in the early seventies I didn't see him often. Soon, I didn't see him at all. I knew about his "pieces" from Gardner, and I re-proved some of the things that he had proved, but I didn't know about the Penrose pattern. Gardner's *Scientific American* article was largely based on what I'd done in Cambridge.

I didn't meet Roger again until a few years after the quasicrystals had been discovered. I still think the situation is still rather funny: we still don't know that the actual physical stuff is really behaving like the Penrose pieces. To my mind this is annoying. It enables some people to deny this possibility. Linus Pauling was a big holdout, but in his case he just didn't understand what the new configurations were. Certainly, it's ridiculous to deny the possibility because these things exist geometrically — why shouldn't they exist physically? Those were interesting times for me.

Concerning the broadening interest in symmetry, as a mathematician, don't you feel sometimes that it's an infringement on your territory that physicists, let alone chemists and biologists speak about symmetry?

No. I don't have any territory. If I'm claiming for my territory the entire world, I can't very well complain if people tread on some of it. What I do feel in this respect is this: the physicists and chemists have this tremendous investment in all sorts of things. Take, for example the names for these groups. The crystallographic point groups were enumerated ages ago, the space groups were enumerated in the 1890s, and they've got into the *International Tables* so people all over the world use the existing notations. There is no prospect of changing it to a rational system. If I propose a new system of naming, this means that I have to just throw away that community because I can't get to them. I perfectly well understand the reasons and wouldn't even want to argue about them, they're just too invested in the system as it is.

And it works.

And it works, yes. But the point is, as a mathematician, my aims are different. I want to understand the thing. Let me give you an example. There are these little shells in the electronic structure of the atom, the *s, p, d, f* shells, where *s, p, d, f* are the initial letters of various words, which indicate various properties of the spectra. But if you'd start it rationally you'd never use this sequence of letters. I would start calling them 1, 2, 3, or *a, b, c*. I don't want to be constrained by having to agree by some pre-existing usage, even if I understand historically how this usage came about.

Let's take the particular case of symmetry. The most recent thing I've done is a joint work with several colleagues. We have completely re-enumerated the 219 space groups *ab initio*, and it takes only ten pages. We were held up in doing this by the feeling that we had to provide a dictionary to the international notation. Understanding the international notation for us was much more difficult than understanding the groups. It actually held up the completion of our paper for ten years.

My aim is to understand something for me. I'm less interested in publication. We're going to publish this paper, of course, but I want to understand it myself. In doing that, I can throw away the international convention. It's a pity. Here I see this chemist or physicist and I can see he is talking about the same things but I see him as limited by having to accept the baggage; he doesn't annoy me, rather, I pity him.

Physics and chemistry are full of historical notations.

And so is mathematics but we're less reluctant to give up old notations in mathematics, since the whole aim of mathematics is to get some kind of understanding of what's going on.

You have introduced the term gyration when speaking about rotation.

That's a good example because gyration isn't just a rotation. Gyration is a rotation about a point that is not on the mirror line. It's really rather important. What is made clear by the new way of thinking about things is that you should distinguish between rotations. Rotation means rotation, any rotation but gyration is a rotation when the axis of rotation doesn't go through a mirror line. We're talking about a plane pattern or a pattern on a surface.

A number of crystallographers have learned about the new notation but it's an uphill struggle. My feeling is, in two hundred years they'll be thinking the correct way. I'm not saying that my notation will be exactly what it is but eventually the baggage will be thrown away.

This way of thinking about the groups is really Bill Thurston's idea. What actually happened was rather funny. We were discussing the 17 groups and I said, "Let me show you my way of thinking about it," and he said, "No, let me show you my way of thinking about it." We agreed upon giving him ten minutes. When he explained his idea to me in ten minutes, I didn't bother to show him mine, and I have got quite a big ego. As soon as I saw his way of thinking about things, I realized it was the correct way. Then I said, "We need a notation that conveys this way of thinking about things." I set off for about two weeks to think what the notation should be because to my mind notational matters are tremendously important. I finally designed the new system, which is very simple and which conveys Bill Thurston's philosophy.

I haven't written it up very well, I've only written one brief paper about it, but that situation is going to be changed soon. It's already on its way to becoming the standard notation for mathematicians. There's also a good chance that I can reach the so-called arty community, that part of the art community that's interested in mathematics. It'll take a long time to get through to the crystallographers, the genuine chemists and physicists who have to use a little bit of this stuff, and I don't see much point in trying but we'll publish some papers. I have a young colleague at Princeton, Daniel Huson, who is the person who most helped to complete the re-enumeration of space groups. He is a young man and he needs published papers to advance his career, to say what he has been doing for the last year or so, so he's very keen to get these things published. The three-dimensional thing depends on the two-dimensional thing. We wrote the three-dimensional paper knowing that we'd have to write — paying a hostage to fortune — the two-dimensional paper. In the last few weeks before I left Princeton, we wrote the two-dimensional paper. We have a plan to write a much longer paper, lavishly illustrated with arty pictures, and address it to a much wider community. I also have a plan to write a book on these things. That's a real trouble for me. So many of the things I do are elementary that publishing papers is not the right way to do it. I want to reach a wider audience, I want to re-found some subjects. That demands writing a book, but writing a book is such a big hassle. There are about five books I ought to write some time.

Does it bother you that physicists talk more about broken symmetries than symmetries?

It does worry me about the Universe. If we depend on the breaking of symmetry, it's not as nice as it would be if symmetries were there. It does seem to be what the Universe does. I don't fault the physicists for talking about what's true. I understand that, and it happens also on a very elementary level. In Aristotelian physics there was a concept of "down", that was invariant. The direction down is different from the direction up. If you're prepared to jump, in other words, if you go to high enough energies, up becomes rather more similar to sideways. This is a very simple instance of symmetry breaking. If you really want to travel as easily upwards, rather than horizontally, you need a tremendous amount of energy and to build yourself a rocket.

This is a paradigm; this is something all over the place. If you want to see the symmetry between space and time you have to travel at speeds close to the speed of light. So I recognize that the symmetry breaking has actually happened. I deliberately chose some examples that are easier and prior to the examples worked out by the physicists.

Would you care to tell us something about your background?

I was born in Liverpool in not a terribly well-off district. My father was a laboratory assistant who also did some minor teaching at the school where two of the Beatles went too. I was interested in mathematics from a very young age. My mother always used to say that she found me reciting the powers of two when I was four. I tended to be top or nearly top in most subjects until I became an adolescent, when I went down and got interested in other things. But somehow mathematics was always there. The interest in other subjects was also always there, but I don't call myself a "somethingelsist". When they couldn't teach me anything new at school, I decided to become a lightning calculator. That's a little hobby that I'm getting back to now. Tell me the date when you were born.

August 11, 1941.

OK. That was a Monday. Now, give me a three-digit number.

999.

That's three times, three times, three times thirty seven.

How did you develop this ability?

I practiced it during the six months when I was still in Liverpool after I'd been accepted to go to Cambridge as a student on a scholarship. Then I went to Cambridge. I found it very hard because most of the students were from rather posh homes, well off, had been to public [i.e., private] schools and I was a poor boy. However, I sort of gradually adapted to the life. One thing that did happen was that there were other people who were interested in mathematics there, whatever their backgrounds. Then I got married at quite an early age and had four daughters by my first wife. My personal life has been decidedly unhappy. I had two boys by my second wife. Over the break-up of my second marriage I went suicidal and I attempted suicide and I was in hospital for a week after it. This was about five years ago. It's taken me a long time to recover from that. Now I'm getting better and hoping to remarry again when everything can be sorted out.

Did you become an insider in Cambridge society?

My old college, Caius, in Cambridge made me an honorary fellow last year. That was very nice. But still I know that I won't use this fact very much because I still feel faintly uneasy; I don't feel that I belong in this particular social grouping.

Is there a sizzling intellectual social life in Princeton?

I've attended a few dinner parties in Princeton and a few party parties where things happen, and there is plenty of intellectual discussion going on there. I've always lived in some intellectual center like this since I grew up. It's nice when the newspapers are saying something about some new discovery in astronomy, to be able to ask my neighbor who is a famous astronomer about it. Something I didn't know until very recently is that I'm rather well known in Princeton. I was trying to get Princeton to buy these famous manuscripts of Archimedes that were up for sale at Sotheby's. This involved going around various departments soliciting opinions. I saw a number of people in the Classics Department, Hellenic studies, and others, and I found that a large number of these people knew me or knew of me. I would've expected this of the mathematicians or the physicists, maybe a few chemists, but to find that the classicists know who I am was a surprise. So I am part of the society there although I still don't feel like a Princetonian.

John Conway with István Hargittai in Princeton (photograph by Magdolna Hargittai).

When I became professor in Cambridge, and it means a lot more than it does in the States, I was rather hoping that somebody would approach me and say something like, "Excuse me, Professor," and people would look around to see who this god-like figure was. It never happened. The students went on just calling me "you" or "John", and that was that. But when I went to Princeton, people started calling me Professor and the students did, and then I found it rather annoying because it distanced them. One of the secretaries got it absolutely beautifully right. If I came in by myself in the morning, she said, "Hi, John." If I came in with someone else, she said, "Good Morning, Professor Conway."

I've also changed my appearance a bit. My hair used to be longer and my beard used to be longer. After I got my haircut I went into the local ice cream shop next door and the girl said, "Oh, you look a lot younger." The secretaries in the department said the same thing.

So you care what other people say.

I always thought that I didn't care about appearance, and I didn't care until recently, but I am getting a bit worried about getting old.

You have said that you no longer had ambition. Aren't you looking forward to something?

I don't think I am. What's there at the end, death, and I don't like that very much. I am thinking about how much time is there to go. I don't want to grow old. I don't feel old in my mind. On the other hand, I see myself behaving in various ways I wouldn't have behaved when I was twenty. Growing old is a bit upsetting. This is one of the reasons I'm taking up this lightning calculation again. I envisage myself in twenty years time hobbling in with a stick, sitting down painfully. In an academic environment you're always surrounded by young, very bright people, and I envisage one of them looking over at this old fool, saying, "Oh, yes, he did some interesting stuff once." But now he mentions the date he was born and I instantly say it was a Friday and I do this even though my physical frame is so fragile and he thinks, "There must be something in there still working."

Are you vain?

Very. I would like to think that I don't care what other people think, but it's not true, as the haircutting episode showed. I do care what people think. But I don't care very much. The conventions about the ways you act or dress don't impress me at all. I just prefer to be comfortable. I'm prepared to go to some length to defend this. I sometimes deliberately think: "What would Conway do here?" and then do it. By behaving in some unexpected way you give yourself the right to behave in an unexpected way. That's very, very nice. Here's a little thing I remember. I attempted suicide and, in fact, I usually think to myself I committed suicide but didn't quite succeed; I woke up in a hospital and then I was very glad. But then came the problem of coming back to life. I was rather worried, I didn't want people whispering behind my back. I thought, "What would Conway do here?" Conway would make it perfectly obvious that he knew. So I borrowed from Neil Sloan a T-shirt he had, which said "SUICIDE" in very large letters and then "Rock" underneath it. It indicated that he had climbed this rock "Suicide". He is a very keen rock climber, and "Suicide" is the second most difficult rock to climb in the United States. So I re-emerged in society after my suicide attempt and went around for three days in this shirt. I just met the problem head-on, splash. That was a case when I actually thought about it because it would've been painful for me to ignore this problem. I also remember when starting lecture courses at various times, I've felt, "What can we do to just shake these students?" So I'd just burst into the room with a big scream or a jump. It's the same sort of thing.

You obviously do care.

I really do love my subject, and that includes the teaching of it. It's not just developing the subject, but the teaching is as important. I often say that I consider myself a teacher more than a mathematician. I spend a lot of time thinking how to teach, I really do. Are we done?

Is there a message?

There is something. It's how I feel about mathematical discovery. You're wandering up and down, it's like wandering in a strange town with beautiful things. You turn around this corner and you don't know whether to go left or right. You do something or other and then, suddenly, you happen to go the right way, and now you are on the palace steps. You see a beautiful building ahead of you, and you didn't know that the palace was even there. There's a certain wonderful pleasure you get on discovering a mathematical structure. It happened to me tremendously when I discovered the surreal numbers. I had no idea that I was going to go in there at all. I had no idea of what I was doing. I thought I was studying games, and suddenly I found this tremendous infinite world of numbers. It had a beautiful simple structure, and I was just lost in admiration of it, and in a kind of secondary admiration of myself for having found it. I was so pleased that I'd found it. For about six weeks I just wandered about in a permanent daydream. What happens after that is that I'm vainly trying to re-create that in the people I'm trying to talk to about it, trying to show what this wonderful thing is like and how amazing it is — that you can reach it by studying something else. I'm perennially fascinated by mathematics, by how we can comprehend this amazing world that appears to be there, this mathematical world. How it comes about is not really physical anyway, it's not like these concrete buildings or the trees. No mathematician believes that the mathematical world is invented. We all believe, it's discovered. That implies a certain Platonism, implies a feeling that there is an ideal world. I don't really believe that. I don't understand anything. It's a perennial problem to understand what it can be, this mathematical world we're studying. We're studying it for years and years and years, and I have no idea. But it's an amazing fact that I can sit here without any expensive equipment and find a world. It's rich, it's got unexpected properties, you don't know what you're going to find, you might just turn the corner, you might find yourself on the steps of a palace, and you might not.

I can't comprehend how this can be. I don't know what it means. I don't know whether there is such an abstract world, and I tend not to believe there is and to believe that we are fooling ourselves.

We used to think that the earth is flat and it was inconceivable that it could be round. It was only some very painful facts that eventually forced us to believe that the earth is roughly spherical. What's happened continually in the physical sciences is that the truth was not one of the possibilities that was considered and then rejected, not even that. It was one of the possibilities that couldn't even be considered because it was so obviously impossible.

In mathematics our development has come a little bit later, but the same sort of thing happened with Gödel's theorem and so on. What we thought was the truth was just a kind of approximation to the truth. Newtonian dynamics is an approximation to relativistic dynamics, and it's not literally true if you go to high speeds and high energies; if you go to small distances it doesn't work quite either, according to the quantum theory. In mathematics we have these beliefs that there are infinitely many integers and so on.

Any belief like that about the nature of things that are arbitrarily far away has turned out to be false in physics. So I think it is false in mathematics too. I think that eventually we'll find something wrong with the integers and then the classical integers will be just an approximation. That's a big puzzle for me. I don't quite believe in this artificial mathematical world. There appears to be a wonderful consistency about it, which means that I can think of something in some way and someone else can think about it in a different way and we both come to the same conclusion. If we don't, there must be a mistake — at least it has been so, so far. But I don't see why there should be this consistency in a world that I don't really believe exists. So to me it's a sort of fairy tale and fairy tales don't have to be consistent because they are human creations ultimately. But this mathematical world is consistent and I wonder, "What the hell is it?" without implying the works of anything supernatural. I'm non-religious.

Roger Penrose, 2000 (photograph by I. Hargittai).

3

ROGER PENROSE

Roger Penrose (b. 1931 in Colchester, Essex, England) is the Rouse Ball Professor of Mathematics, Emeritus, at the University of Oxford. He received a B.Sc. degree from University College London and a Ph.D. in algebraic geometry from Cambridge University. He has been a Fellow of the Royal Society (London) and a Foreign Associate of the National Academy of Sciences of the U.S.A. His awards include the Wolf Prize (Israel), the Dannie Heinemann Prize, the Royal Medal of the Royal Society, the Dirac Medal, and the Albert Einstein Prize. He is also a critically-acclaimed science writer. In 1994, he was knighted for services to science. We recorded a conversation in his office at Oxford University in March 2000.

The Penrose tiling has made a tremendous impact in science within years of its appearance. It originated almost from doodling. On the other hand, you have written that sometimes it takes centuries before some new ideas related to symmetry find their applications.

I wouldn't like to draw any clear line between doodling and what might be called my serious professional activity. I don't think there is any distinction really. I've often been playing around with tile shapes, tiling problems, just for fun. I was interested, for example, in the shape Escher used in one of his last pictures called "Ghosts". It's based on a tiling arrangement, which I showed him. I visited him on occasion and left with him some pieces of puzzle shapes, based on equilateral triangles fitted together. It's one shape and the problem was to cover the entire plane with that one shape. It has 12 different orientations before it repeats itself. It's a little bit difficult to find the arrangement. I had this more as a puzzle. Somewhat later he wrote to me asking me what the principle was. Then, somewhat later I responded to him. Then he designed his little ghosts based on that. That just indicates that I had been interested in tiling before producing these non-periodic ones, which came later. One of the things, I think, I had had in the back of my mind — although all these things are interconnected — was the Universe itself, if you like, something with extraordinary complication. Yet one likes to believe that the laws underlying it all are something very simple. I was trying to think of some example of this kind of thing, where you had something simple but yet which produced great complication on a big scale.

The rigorous rules of crystallography didn't seem to intimidate you.

At this time that wasn't the question. I was thinking of hierarchical systems. I produced a number of hierarchical tilings. Somewhat later I was responding

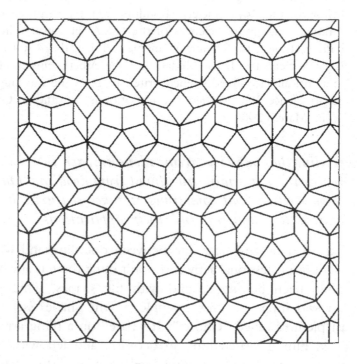

Penrose pattern.

to David Singmaster, who had written to me asking me for a seminar, and there was a logo in the corner of his letter. The logo had a pentagon in the middle, surrounded by five others within a larger pentagon. I thought of what happens if you iterate this? I made a pattern iterating it, but you had to fill the gaps in a certain systematic way. The only interesting thing is how you fill the gaps up. Thus I produced this pattern, which I designed partly to show to somebody, who'd been in hospital, just as an amusement. A little later I realized that you could actually force that pattern by making it a jigsaw. There are pentagons, little rhombuses, five-sided, what I call jester's caps, which are half of them. The problem was to find a way forcing that pattern by some local matching rules. Having three versions of the pentagons and one of each of the others you could force it, so it was a six-piece tiling, which was non-periodic and which happened to have this fivefold quasisymmetry. But I wasn't thinking particularly of trying to refute crystallography. It was just like an amusement.

When was this happening?

It was about 1972.

What was next?

I had this tiling pattern and I'd actually published an article about it in the *Bulletin of the Institute of Mathematics and Its Applications*. This was based on a lecture at a meeting focusing on aesthetics.

In my first question I was referring to this paper. Did you think about possible applications?

Yes, because when I was giving lectures about these tilings people asked me about it. They asked me whether a generalization was possible to crystals and whether fivefold symmetry and icosahedral symmetry might occur. My response was, yes, that's certainly possible, but the problem that I had seen in this was that the assemblies are non-local. You can never be sure that you don't make bad mistakes, which would forbid you from continuing. Once you start making mistakes it tends to get out of hand. If you wanted such a quasicrystal, as they subsequently came to be known, to be perfect, there is no local assembly rule. I thought this as being an obstruction to this being actually found in nature. Of course, they hadn't been found in nature so I didn't think of it as something that you would likely discover.

What came after the Bulletin *paper?*

The kites and darts and rhombuses, a six-piece non-periodic set, an aperiodic set as we now say. This happened when Simon Kochin was visiting from Princeton. We were discussing Raphael Robinson and how he'd done various things. We would like to get the minimum solutions to problems and he had a set of six tiles, which was based on squares, which is an aperiodic six-piece set. I said, I could do better than that. I knew that with my six-piece set I could get it to five, simply by gluing two pieces together. It was obvious. Then I went thinking about it and I got it to four and then to two, very quickly. It only took me an evening to realize that I could get it down to two. My initial response was that I was somewhat disappointed because it was too easy.

Were you aware of Kepler's and Dürer's attempts?

I was not aware of Kepler's but I had seen Dürer's picture at some stage. It shows how you can't get it done so it's not very encouraging. But I had seen the Kepler picture in one of my father's books. The effect on me of that picture was not that it directly influenced what I did, but psychologically I was better disposed towards pentagons. Most people would've thought pentagons were useless for tiling. My impression had been, yes, you can get some way with pentagons, it is interesting what you can do with them. I hadn't remembered the Kepler picture when I produced my own tiling, so it was just a feeling, but it had an influence behind the scenes of predisposing me towards this.

What I hadn't realized until much later was that the largest pattern in Kepler's famous picture can be directly superimposed on my original tiling; it's very, very close, much closer than the impression you'll get when you read the Grünbaum and Shephard book. They show you this picture and they have a scheme for making it tile periodically because it's only a finite portion. You don't know what Kepler had in mind, but I suspect Kepler had it actually much closer to what I was doing in my mind, something non-periodic. I believe there are some writings that he had on this, which explain more about it, but I've never seen them, so I'm not sure in detail what he had in mind. I don't think he was trying to do something periodic like the picture in the Grünbaum–Shephard book. I suspect he was doing something much closer to what I was doing, but it was not completely the case. What he had was regular decagons, which were sometimes overlapped partially and sometimes were separate. To get my pattern, you'd fill those in a systematic way, which he hadn't actually done in his drawing. If he had, it would've been my first aperiodic tiling, very, very close to that.

Was Buckminster Fuller an influence?

Not so close. Kepler was a much bigger influence. I'd certainly known, I'd seen his things.

How about J. Desmond Bernal's influence?

Yes, that's very interesting. First of all, I'd met Bernal a long time before. Curiously, he came to see me once when I was a research student in Cambridge. He came to see me completely out of the blue, just because he was looking for people who might have ideas, to do with these pentagons, and so on. I don't know why he thought of me, I was not known as somebody with an interest in polyhedra. I don't think anything particular came out of our discussion, but I was very struck by the fact that he would be prepared to come and see a young graduate student who knew nothing. This was in the mid-fifties.

How important was your father's influence?

It was very important, no question about that. He was interested in human genetics and the inheritance of mental illness. His first name was Lionel, he was at University College London, and he was very well known. The thing that influenced me particularly was his general attitude to science and mathematics and fun puzzles. Again, with him there was no clear line between what he did for his serious work and what he did for pure enjoyment. He used to be a chess problemist. There was a lot of chess-playing in the family, my younger brother particularly, he was British chess champion ten times, but I was not particularly interested in chess. The mathematics in the puzzles was very important to me and, also, we used to go for long walks and look at plants to see how they grew and look at the Fibonacci numbers, which brought these things to life for me.

Your father was Galton Professor of Human Genetics. This is a controversial name. How did he feel about it? Was his interest and his work related to Galton?

I suppose, to some degree, because Galton was an influential early figure in measuring biometry, measuring biological features. There's a story behind the name of his Chair. When my father first took it at University College, it was called the Galton Chair of Eugenics. This has connotations, which my father was very much against. One of the first things he did was to have the name changed so it became the Galton Chair of Human Genetics. Although he hadn't agreed with everything Galton stood for, he nevertheless thought that Galton was an important figure, so he was happy with the Galton part, but he was very unhappy with the eugenics part. He was very insistent that this should be changed and the name of the journal too. The journal was called *Eugenics* and he changed it to *Human Genetics*.

I would like to return now to the story of the Penrose tilings. We've been through Kepler and Dürer but haven't reached Martin Gardner yet.

I don't quite remember the order of these things. Either Martin Gardner wrote to me first or I wrote to him. I think he must have asked me about the tiles that I'd given to Escher, which had the twelvefold cell. He set this up as a puzzle in his column in *Scientific American* and called it the loaded wheelbarrow. In my correspondence with him I said that I had some other things that might interest him, but I didn't tell him in detail. I may've said that it was something non-periodic. In the meantime he had heard about it from Richard Guy who had heard about it from John Conway. I'd shown these shapes to John Conway at the British Mathematical Colloquium meeting and he had then picked up on it, and the information came partly through that route to Martin Gardner. He got very interested and produced his article in *Scientific American* largely based on his discussion with John Conway.

This is how the Penrose pattern became well known. Even the cover illustration of that Scientific American *issue showed a Penrose pattern. People very often refer to Gardner's article rather than to your original paper in the* Bulletin.

I've often had problems with publishing things that I'd thought of and this was a good example. The *Bulletin* paper didn't have kites and darts. It was simply the original pentagon pattern.

There's another curious historical thing, which was the influence of Robert Ammann. He must have seen Martin Gardner's article, but he'd rediscovered

Robert Ammann
(courtesy of Esther Ammann).

the rhombus shapes with the matching rules. I was very impressed because these shapes had not been published at that stage, he only knew of their existence. It's like Galileo not knowing how the telescope was constructed, just knowing it was there, rediscovered how it was constructed.

Did you meet him?

I met him much later. He had a rather tragic existence. He was not mathematically trained. He worked in a post office for a while and then was unemployed for a long time. He didn't like to talk to people. The only time I met him was in Bielefeld in Germany where he was invited to come. Then he rather lost interest in tiling and had some other strange ideas how the dinosaurs were wiped out. Somewhat later he died. It's very sad because he produced a number of extraordinary ideas, not just rediscovering my ideas. He had ideas about three-dimensional tilings and he discovered the eightfold quasisymmetry, the Ammann bars as patterns, and so on.

You were also in contact with Alan Mackay about your tilings.

He'd taken up the idea that this might seriously be important in crystallography, which I was flattered by, somebody taking this attention, but I was not terribly optimistic at that stage that these puzzles could be important in crystallography. The only reason that I hadn't thought so was because people hadn't actually seen them in caves or anywhere else naturally, where people find crystals normally. I'd known about Alan Mackay's interest, but I wasn't quite sure what to make of it — it seemed to me a little too fanciful. He was also interested in hyperbolic tiling, which is fine as

Roger Penrose and Alan Mackay (courtesy of Alan Mackay, London).

mathematics, but whether they were relevant to actual crystals, I was never quite sure about.

The next thing, from my own point of view, was when I met Paul Steinhardt at a conference in Jerusalem on cosmology and relativity and we were each giving lectures on different topics, nothing to do with this. Then he showed me his pictures. He was aware of Shechtman's work.

We've talked about Kepler's influence on me. It's interesting that later I talked with Shechtman and asked him if he knew about my tiling patterns. He said he'd seen them but he was not thinking about them when he made his suggestions about the quasicrystals. There's something similar when one is psychologically predisposed a little more about thinking in a certain direction even though one may not realize at the time that this is an influence.

I didn't think Shechtman was considering tilings in the wake of his experimental observation.

He wasn't and he told me so. What he had said was that he knew about them.

Steinhardt seems to have been underplaying the importance of Shechtman's experimental observation in the quasicrystal story and ignoring Mackay's predictions.

People always shift the balance a little. Mackay had these things, he already knew, he had produced simulated diffraction patterns from my tilings. Steinhardt had looked at it more theoretically. The difference is somewhat subtle.

Mackay was not afraid to come out with suggestions that were rather risky at the time.

That's true.

Steinhardt kept his findings in his drawer until the experiment happened.

One has to give Mackay a lot of credit. That's important.

Do you consider the quasicrystal discovery a real discovery or is it so novel only because we were so narrowly focused on the rules of classical crystallography?

I think there is something surprising about the patterns. It is remarkable how close they are to being periodic. You might've taken the view that we've been too narrow about our crystallographic notions, but it's very surprising how close my patterns are to being crystalline. To suggest that we could've just broadened our views of crystallography wouldn't reflect the remarkableness of these patterns. I'm not giving myself any particular credit because it's there in nature, meaning mathematical nature. These patterns are there in mathematics and this is a striking fact. I don't think one could've anticipated this just from saying that we've been a little too narrow in our definitions.

I'm aware of your other activities and I'd like to ask you about them, but first this: would it bother you if you'd go down in science history as the one who created the Penrose pattern?

I don't think anything particularly bothers me one way or another here. It's certainly not the most difficult or sophisticated thing I've ever done. On the other hand, it's the most easily accessible and that's an important thing. In some ways, I'd find it rather flattering because things that you can explain to people, to get them interested in mathematics, is very important. So I have no problem with that, really.

May I ask you about the lawsuit in connection with the allegedly unauthorized use of the Penrose pattern on toilet paper?

I can't say very much about that because there was an out-of-court settlement and, as part of the conditions, I'm not allowed to say anything more than the fact that there was an out-of-court settlement. I can explain it more in general terms. The question was whether a certain pattern appears, and it still appears because now it appears with agreement, on certain high-quality toilet paper. The issue is not whether that pattern has been produced by some mathematical formula, for example, which is public property. Mathematics is public property, but the question in this instance was whether a certain pattern had been directly copied. There is an issue of copyright involved in copying specific patterns. That is the issue, which is relevant here, not whether there is a piece of mathematics, which is judged as being private or public property. This is an important point.

Were you upset?

I was not upset at all and I would like to stress this too. This issue had to do with a certain company, which is called Pentaplex, which makes things based on my designs. Their livelihood depends on producing things, which have certain patterns. It's unfair on them if other people encroach on this without agreement.

Would you care to define yourself?

I don't know if I can define myself if I am a mathematician or a physicist; I'm certain I'm not a businessman. What interests me most is what's true, what is there out there and to try to understand. When I talked about my father not having divisions between what he did for fun and his serious work, that applies to me too. I don't have a division between mathematics and physics or between what I do for enjoyment. I'm better suited to academic life here than in many other countries in the sense that mathematics here includes what is called theoretical physics in other countries. If I'm in a company of physicists, they think I'm a mathematician; if I'm in a company of mathematicians, they think I'm a physicist. I feel as if I don't completely belong to any group.

What was the origin of your interest in science?

My father was one but by no means the only one. I must mention Dennis Sciama who was very important. He died recently. When I first went to Cambridge as a research student in pure mathematics, I was working in algebraic geometry. Sciama realized that I had interest in physics and he knew everything that was going on in physics at the time and he was excited about it and was a very original person. We had long discussions, used to go to Stratford to watch Shakespeare plays, and all the way in the car we would talk about physics. He developed partly my excitement about physics and partly my knowledge of physics.

In addition to that, there were three important factors. When I was a graduate student in Cambridge, I went to any courses, which were not directly connected with my work. One of these was a course by Dirac on quantum mechanics, which was absolutely stunning. In a completely different way, the lectures on general relativity of Bondy were also wonderful. My interest in general relativity and quantum mechanics in detail came a lot from those courses. Then there was a course by a mathematician called Steen on mathematical logic; it was not of the same caliber as the other two, but for me it was very influential. We learned about Turing machines and Gödel's theorem. Schrödinger was also an important influence. I read everything he had written on the semi-popular level.

You are also a science writer, aren't you?

I certainly write articles and, occasionally, books. They're semi-popular.

They sell a lot of copies. Do you think all of them are read?

That's a good question. I worry about that sometimes. They are rather uneven. Sometimes they are quite technical, sometimes quite general. I also have another book about the interrelation of physics and mathematics and it's somewhat philosophical too, but no less technical than my other so-called popular books. It's called *The Road to Reality*.

What do you consider to be your most important contribution?

Twistor theory.

Would you, please, give us a popular introduction about it? Where does the name come from?

The subject has become very mathematical, but my interest was physics. I was interested in trying to find some way of bringing together space-time structure and the basic rules of quantum mechanics. An important factor is complex numbers or holomorphic structure, complex geometry, which has always seemed to me the underlying ingredient of quantum mechanics. People usually think of both quantum mechanics and complex numbers as being very abstract. To me they are very real geometrical entities, which relate to space-time structure. In order to see how the physics of the large, basically Einstein's general relativity, and the physics of the small, basically quantum mechanics, can come together, I was trying to see whether these complex structures could underlie the structure space-time.

For certain reasons I was interested in massless particles as more fundamental in space-time than massed particles, in particular a photon, although I don't necessarily mean a real photon. Think of an idealized photon, just of its path in space-time, which we would call a v-ray or a light-ray, which is along a light cone. If you regard the space of those objects as being more fundamental than space-time itself that will be your first step in twistor theory. Then you have to bring in not just light-ray but its angular momentum structure. The photon, for example, has a spin; it has no angular momentum. You want to incorporate that into this geometry. When you do that you find a certain configuration. I'm slightly distorting the history here because it didn't come quite this way around. I didn't realize that it was to do with angular momentum when I first thought of these structures. It was a piece of geometry where I could see the role of complex numbers. It was the complex holomorphic geometry as seeing how that related to space-time structure. The name twistor comes from the little circles in the configuration, which link each other. The configuration is in four-dimensional geometry of a three-sphere, which was known to Clifford about 150 years ago. He realized he had these parallels on a three-dimensional sphere, they all link each other and if you project that configuration into three dimensions, you have this thing, which actually describes the angular momentum structure of a massless particle. He didn't know that and I didn't know that at the time either. These things can be described by complex numbers and the freedom they have is a complex freedom.

The motivations were complicated and come from many different directions all together, partly to do with quantum theory, partly to do with relativity, and partly to do with mathematical elegance. I felt that having space-time as the basic ingredient, these spinning photon-like structures were more fundamental, mainly because they have this analytic complex geometry.

It took many years to see how Maxwell's equations for electromagnetic fields and other basic equations of physics relate in interesting ways to this kind of geometry. It's been a long, complicated development. The most important thing from the point of view of the development of the subject was probably how to start to bring Einstein's general relativity into it. I had a scheme nearly 25 years ago, which describes half of Einstein's general relativity. This is one of the tantalizing things. It's asymmetrical with respect to left and right and graviton. The particle of gravity, if you like, can be split into a left-handed and a right-handed part. This construction that I found nearly 25 years ago describes the left-handed part of the graviton with the full non-linearities of Einstein's general relativity. But the right-handed part was mysterious and I'm still trying to understand how to bring the left and right parts together.

Is there no parity between them?

It's one of those curious things. You know in physics that parity is violated, that is to say, weak interactions, left and right, behave differently. Deep down in nature there is a handedness, there is chirality. But you don't normally think of that as showing itself up in Einstein's general relativity, because Einstein's theory is completely symmetrical, left and right. This is why this is a strange idea that this left/right asymmetry would be important also in how quantum theory should relate to gravity, and this is an essential feature of twistor theory.

Can there be a connection to the origin of life?

Maybe. It has been suggested that chirality may actually be important, but it might be an accident. For the moment, this is a little bit of a long shot.

Getting back to the story of your life. You did postdoc in Princeton and this was due to a NATO decision to catch up with the Soviet Union.

That's how Wheeler puts it. From my own point of view, I was just interested in what Wheeler was doing. It was, again, Dennis Sciama who suggested that I do this, but I found Wheeler's ideas very stimulating.

Which ones?

At that time he had this idea of geometro-dynamics, that somehow all of physics was to be incorporated into geometry, which is appealing to me. Later on I was not so convinced by that particular idea but at the time I found it especially attractive to be with people who were interested in physics of that kind of level. It was valuable to me, because I became acquainted with the relativity community, and a lot of ideas have developed from my acquaintance with them.

You conduct debates with Stephen Hawking.

Not really. I knew him when he was a young research student and the early ideas that he had in singularity theories were certainly influenced by my ideas and we did things together. The papers we wrote together were on developing ideas on singularities, how you use topological arguments to show that gravitational collapse will produce a singular state and how the Universe as a whole produces the Big Bang and how introducing new regularities don't get you out of it. Then he started developing his more quantum mechanical ideas, which is fine, but then they lead you in certain directions, which have to do with how seriously you take quantum mechanics. I think our main disagreement is that I'm much more prepared to see quantum mechanics change whereas he is more resistant to that notion. But it's wrong to say that we have debates. There's this book, which was a series of lectures where we alternately gave lectures and the disagreements between us emerged later on in the book. There is a discussion at the end, which makes references to our debate and we have different views on things.

How do you feel when it is compared to the Einstein–Bohr debate?

That's probably my fault. Not that I was trying to say that this is of the same magnitude as the Einstein–Bohr debate. I was trying to say that the positions we took were similar. His position was similar to Bohr's position in the view that in a sense there is no reality at the sub-microscopic level and you just use the formalism of quantum mechanics as formalism; you don't attribute it any reality. This is a positivist position. Einstein took a view that there is a reality at the quantum level. We don't know what it is; it's not exactly described by quantum mechanics, but it's something that we have to discover. My position is more in tune with that. There is reality even though we don't know what it is. The rules of quantum mechanics must change at some stage in order to make that sense.

Roger Penrose in front of the blackboard in his office, 2000 (photograph by I. Hargittai).

You don't need the presence of an observer for something to be there.

It's an objective process, taking place not depending on observers, not dependent on conscious individuals looking at it.

How can this be decided?

There are experiments. I have a proposal. You might think it strange for a mathematician. I have a proposal for an experiment that I hope will be performed in space. I'm trying to get NASA, GPL, and the European Space Association to try to get it done as a real experiment. I'm doing this with a colleague. It could be done as a part of another project that would take about 10 years. It would test whether the rules of quantum mechanics persist at all levels or whether there is some level at which something else happens. You can give a clear estimate for this level on the basis of a conflict between the rules of quantum mechanics and the rules of Einstein's theory of general relativity.

We propose in this experiment to have a little crystal, which is a little bigger than a speck of dust and put it into a superposition of being in two slightly different places at the same time. You hit it with an X-ray photon, which is being split into two beams so the photon shares these two roots. When it hits the crystal one of them is deflected, slightly displaced by the photon, only about a nuclear diameter, not very much, and then you have to keep both parts of the photon coherent for about a tenth of a second. You do this by reflecting them from one satellite to another where these satellites are about the diameter of the Earth distant from each other. It takes a photon about a tenth of a second to go the diameter of the Earth and back again. If you can keep your crystal for a tenth of a second, according to me, it will become one position or the other position spontaneously, with no observer, by itself, in about a tenth of a second. If that happens, you'll be able to see when the photon comes back, you'll see if the coherence has been lost. It's a very difficult experiment but there are clear predictions on the basis of what I claim, which could be refuted by such an experiment or supported. You could pick out the signal from other things, which might mess it up so it's a clear signal that one should be able to pick out.

How far are you in your negotiations?

It's hard for me to say. Interest has certainly been expressed. My colleague, Anders Hansson, has been doing all the negotiating.

You have had a tremendous career and you and your wife are expecting a new baby soon. What are you looking forward to?

Certainly my child, which is due to be born in the middle of May.

You have older children, too.

Yes, three, from a different marriage. One of them is a mathematician, but they all have problems with permanent employment. This is a difficulty of these days.

Is it hard to be a Penrose for them?

You're asking difficult questions.

How long have you been married this time?

For 11 years, but this will be our first child.

What does your wife do?

She started as a mathematician and has a degree from King's College London. Then she worked here a bit as a research student. She's now doing research in education.

Charles Coulson was your predecessor in this Chair. Would you care to say something about him?

I only met him very briefly. His interest was quite different from mine. He was a theoretical chemist.

Who are your heroes?

Archimedes, Galileo, Rieman, Newton. When I was young, perhaps Galileo influenced me most. There was something about his being against the prevailing thought.

Are you a maverick?

Some people think so. They get upset by my work related to consciousness.

What is your view?

There is a view that either you believe that we are just computers or you are mystical or religious or you have some view, which is regarded as unscientific.

Is there a contradiction between your being a realist who says that the world is there regardless whether an observer is present or not and your being an anti-reductionist who says there must be something else in addition to physics and chemistry?

It's part of the same philosophy because I'm saying that conscious phenomena are real things. There is a real phenomenon there; it's part of the real world.

Can you explain it?

Some day, yes. People haven't got the explanation yet. I'm only interested in what's true.

Have you met this view when some say that we may not understand quantum mechanics because it's part of divine reality?

Yes, I have. Of course, there is fuzziness in boundaries here. When I say I'm not religious, it means that I don't believe any religious doctrine that I have seen. It doesn't mean that I don't think there's something more than what is described by a purely reductionist view of the world. We have to discover what's going on. The word reductionist doesn't have a clear meaning. Sometimes it just means scientific and I have nothing against that. But if it means that you can explain large things in terms of the behavior of small things, I don't necessarily believe that. There's a lot we don't understand about the world.

What's the next step in learning more about it?

Quantum mechanics is very important and we have to discover what's really going on there. We don't have the right view yet. We need a new physics, which we don't have. Consciousness is not independent of this question; it's just the wrong way around. There are people who say that quantum mechanics needs to be completed by having consciousness, but that I don't believe. What I believe is that consciousness is a physical phenomenon, it takes place in the world, and we don't understand the basis of it yet. There's this missing borderline between the small quantum physics and the large-scale classical physics. This missing ingredient is much more important than almost anyone would claim. You talk with physicists and they think it's a minor problem that we just need to understand quantum mechanics better. What I think we need is another theory. When we have that theory, it will only be a little step towards knowing what consciousness is. So you see what I mean by saying that I'm going the other way around. I'm not saying that consciousness is needed for quantum mechanics, what I'm saying is that quantum mechanics needs improvement to understand what consciousness is. And it's not even the whole story. You'll need to know more things too. There is a lot more we don't know in the way the world operates than most physicists would claim.

You asked me about being a maverick. When I think of how quantum mechanics could be relevant in brain processes, that is the question I picked up on the suggestion by Stewart Hamaroff. There are microtubules, little tubes in neurons and in almost all cells in the body, but in neurons they have a particular role. If my ideas are going to make any sense you need something, which is beyond the level of neurons for large-scale activity in quantum mechanics. This is a very controversial idea. People say that if you're going to have large-scale quantum coherence at the temperature of the human body, this is ridiculous, they would say. You can only get these large-scale quantum activities at very low temperatures, like a superconductor. I think this is a very narrow view because there are lots of structures we don't know. Even high-temperature superconductors are at much lower than body temperature, and we don't know why they work. It's premature to make too strong claims of what can be going on there. But something must be going on, which is not explicable in ordinary classical terms.

What's going on in the brain? People say, it's just some kind of computational process. That's why I'm a maverick because I don't think this is an explanation.

Alan L. Mackay, 1982 (photograph by I. Hargittai).

4

ALAN L. MACKAY

Alan L. Mackay (b. 1926 in Wolverhampton, England) is Professor Emeritus at the Department of Crystallography, Birkbeck College, London University and he is a Fellow of the Royal Society (London). He was a student of Trinity College in Cambridge 1944–1947 and received his B.Sc. and Ph.D. degrees in physics and his D.Sc. degree in crystallography and studies of science from London University. He has been associated with the Department of Crystallography of Birkbeck College, London University, since 1951 where he became Professor Emeritus in 1991. Among his scientific achievements is that he predicted the existence of what are called today quasicrystals some time before they were experimentally discovered. He has expanded the realm of crystallography, has broken out of the rigid rules of classical crystallography, did pioneering work in icosahedral packing, and discovered what is known today as the Mackay icosahedron. We recorded several conversations in London in October, 1994, and what follows is a compilation from these conversations.*

The first question is about family influence in your life and career.

My parents were both doctors and they were both born in Glasgow. In 1923 my mother came south to Wolverhampton, to work for her uncle, who was in general medical practice there. My father came down a year later to join her. With difficulty he eventually bought out the owner of

---

*In part, this has appeared in *The Chemical Intelligencer* 1997, 3(4), 25–49.

the practice and in 1938 sold it again to set up as a consultant. After the war he became a consultant working for the NHS until he retired about 1962. My mother had a particularly strong social conscience and had been involved in all kinds of projects. In particular she was a magistrate and eventually became Chairman of the Juvenile Court in Wolverhampton.

My father had been an infantry officer of the Argyll and Sutherland Highlanders in the First War and had served with distinction on the Western Front from September 1916 until the Armistice of 11 November 1918. During the war the 15th Scottish Division, with a nominal strength of about 15,000, suffered 45,000 casualties. The war was the background to my childhood. How could such a thing have happened? When we walked in the country my father would pass on various maxims, such as: "Never go home the way you went out. The enemy may be waiting for you." "Cover from view is more important than cover from fire." He had then gone back to Glasgow University, changing over from chemistry to medicine. My mother was also studying medicine and in 1920 my father saw her one day in the dissecting room and said: "That's the girl for me!"

I had also been thinking about influences and my question to myself was how did I acquire a skeptical attitude? I remember one incident, which must have been when I was about five or six, when I demonstrated to a girl along the road of the same age that her parents had been lying to her over the nature of Father Christmas. I was very surprised to find how annoyed people were. It was like Gandhi or H. G. Wells' experiments with truth. I discovered that you should not believe everything that grown-ups tell you nor say what you actually think.

I also remember asking my mother if the minister at the local non-conformist church really believed all he said, most of which was incredible to me. Mother did not answer and she continued to attend the local church throughout her life, I think more for social solidarity than for belief. In the trenches my father had often argued about religion against George MacLeod, who later became the Moderator of the Church of Scotland. The tradition of my ancestors was to listen to what authority said and keep their doubts to themselves.

Both my parents were extremely busy with work. There was always medical talk at lunch time. For example, enormous letters used to arrive with chest X-rays and these would be held up, and Father would explain that this was tuberculosis, or a broken bone or something such.

Did you resent having so much medical talk at home?

No. It was interesting. Nothing is more interesting than professional shop talk. It was always understood that there would be talk about professional matters and that it was not to be retailed outside. We often knew who committed suicide or was in trouble or had made a mistake. The surgery and dispensary were in the house and Mother often made up the medicines. I was allowed to put the sealing wax on to the bottles of medicine which kept the magic inside. I hated the telephone on the accurate answering of which Father's professional income depended. There were also sometimes newspaper people after inside information. The standing order was that if someone came bleeding on the doorstep for emergency help, first take up the hall carpet.

Brothers and sisters?

I am the oldest. I have one brother and two sisters. The youngest is my brother Murray, 10 years younger than me and now Professor of Transport Safety in Birmingham. The elder of my sisters, Sheila, is an architect in California. My younger sister, Mary, lives in Tasmania, in Australia. She is the Australian expert on rape and sexual assault. She is a doctor and her husband is a surgeon. They emigrated about 30 years ago when even public discussion of contraception was forbidden in Tasmania. She began contraceptive clinics and gradually she got on to the rape business and has became well known as an expert.

Emigration has always been a factor in our family history and on average one third of each generation always emigrated. There was just no future in Scotland. Father handed on the traditional hates following on the land clearances in the Highlands of about 1820 when the landowners replaced the people with sheep. People were spread over the world. My father wrote a small book on family and clan history with the general conclusion that they had almost always been badly served by their leaders.

My (maternal) grandmother's house in Glasgow was just a few hundred yards from a shipyard where gigantic ships were built so everybody was outward looking in that way. The Queen Mary was built in John Brown's Yard, just down the road and when there was work the riveting could be heard all over the city. My mother went up for the launching in 1934 and, after they retired, my parents, as a life's ambition, crossed the Atlantic in the Queen Mary.

So there was a professional atmosphere in your family.

Yes, but rather tacit, not explicit. It was understood that we would get educated. I didn't choose my school, I just went there. It was chosen for me by my parents. They pushed, but not too much.

I went to the Wolverhampton Grammar School from 1935 to 1940. The Second World War began for us on 3 September 1939. We listened to Chamberlain's broadcast while we were on holiday in Wales. My father joined up again and went to the Middle East in 1941 as second in command of a military hospital. He left my mother with four small children and a pistol with 20 rounds of ammunition in case the Germans came. At 13 I was a messenger in the Auxiliary Fire Service but never went to a fire since from 1940 I went away to boarding school and missed the big air raid on Coventry which my unit attended. There were only minor raids on Wolverhampton. The Battle of Britain also took place during my first term at boarding school which was near Northampton and the defeated army came through the town on their way to dispersal camps after being taken off from Dunkerque.

There were several American aerodromes near the school and we could hear the Fortresses warming up at four in the morning and then they came straggling back, with pieces hanging off them, about 12 hours later. The town was full of airmen, some of whom expected to die the next day. A bugler of the RASC sounded the last post every night from an army depot nearby. I was at school until 1944 when I went to Cambridge with a scholarship. My secondary school time was thus entirely during the war, and conditions therefore were quite different. Everything was serious. The Germans might come and the school cadet corps would have to fight them. Boys who were a year or two ahead of me began to get killed in the services.

Did you get a poorer education because of the war?

No, not at all. We got as good an education as you could get before the war. Although ideologically slanted, the school teaching was often excellent and the teachers had first class degrees in science and mathematics. During and after the Depression first-class graduates were glad to get a job in teaching, and this produced the immediate post-war generations of scientists. Today the educational level of teachers has gone down a lot, although there may be many more teachers today.

Did your education cost a lot of money to your parents?

The boarding school cost a lot in contemporary terms, I do not know quite how the family afforded it. But the best available was thought necessary. There was not really much talk about money at home. Mother talked about the difficulties in the early days when I was very small. During the Depression people couldn't pay the doctor and there were bad debts to be written off. My parents were not obsessed with money but managed it well. We were not short of anything but waste was wicked. I still do not like to throw away potentially useful bits of metal.

Was there anybody in the family who would have liked to go to university but couldn't?

No. Education was fairly available. I got scholarships of various kinds. One characteristic was that already at 12 and a half, going to an upper class school as the first in the family, I was conscious that I was an internal emigrant. I didn't believe in their ideology and I felt sorry for the children who had been there all the time in the preparatory school system and didn't know any better than cricket and football.

How did you become an internal emigrant? Was it from home?

Just thinking about it. The boys I met at boarding school had already been brainwashed. I do not have any friends now from school times although at the time I got on quite well. If you live in a room with four other people you have to decide what is public and what is private.

Did you try to explain this to them?

No. Also, during the war nobody would ask you whether you were happy or not. It was just not a relevant question. We were very privileged compared to most.

Have you remained an internal emigrant to this day?

Yes. I can identify or not with the particular group around me.

How about Sheila, your wife?

No, I don't think so.

Let's return to schooling. What did you study in Cambridge?

Sheila and Alan Mackay in the Mackays' home, London, 1998 (photograph by I. Hargittai).

Natural science, physics, chemistry. Then also electronics, mineralogy, and mathematics as half-subjects. Because of the good school education, the first and second years were too easy and the third year was too hard. This was Part II Physics: classical physics, thermodynamics, quantum mechanics.

Any famous professors?

Quite a lot. Sir Lawrence Bragg, for example. I remember particularly the first year physics classes by Alex Wood whose lectures were full of demonstrations. R. W. G. Norrish, Fred Dainton, H. J. Emeleus, S. G. Mann, and others lectured in chemistry. Real demonstrations of real chemistry and real physics were a very important part of the course. You did not have to take the theory on trust.

How many students were in the class?

A hundred or so. The lecture rooms were almost full.

What was the rate of attendance?

Everybody attended everything. You hear that in the Arts Departments and the English Department, they go to lectures only if they feel like it. In the sciences everyone went to everything.

Were you well off with money?

I stayed in lodgings and for 35 shillings, that is, for 25% of my earnings I got breakfast, supper, lodging, and full board at the weekend. There was also a man there in the same house who worked in a bookshop. He used to bring half a dozen books from the shop on Saturday and take them back to the shop on Monday morning. That went on for two years when I read three books a week — all non-fiction, but all kinds of books.

Did you discuss the books?

Not particularly.

What happened then?

Gradually I moved to Birkbeck and remained there for the rest of my time. First I was part-time, then became a research assistant and got a salary. Sheila and I got married in 1951, when I got a job as a research assistant. It was in the crystallography laboratory which was a section of the Department of Physics then. J. D. Bernal was Professor of Physics and Head of Crystallography. I knew about him because I'd won a prize in Cambridge and I could choose a book for it and I chose his *Social Function of Science*, which was, and still is, immensely important.

Did his leftist political views have anything to do with your going to Birkbeck?

Yes, I think so. The place was very socially conscious, but then so was the whole climate of the times. But it was also that Birkbeck was almost the only place where you could do part-time work.

Were there discussion of politics at Birkbeck?

Yes, all the time. There were quite a lot of left-wing people working in the Department. Also, the 1945 election had been a big turnover; the Labour Party swept the Conservatives away. It was believed then that there need never be another Conservative government. There were tremendous movements of social reconstruction, particularly based on the establishment of the health service and social security system, laid out in the Beveridge Report. In 1948 Yugoslavia was expelled from the Cominform and the Cold War became very evident.

Incidentally, Bernal's *Social Function of Science* has a very important chapter on scientific information, essentially foreshadowing the Institute for Scientific Information, that Eugene Garfield founded in Philadelphia. Bernal took part in the Royal Society conference on scientific information in 1948, and he put forward various revolutionary ideas, like regarding the scientific paper as the unit of information rather than the journal. So apart from actual science, there was continuing discussion all around the department at tea-time not only on Labour Party politics but on the actual planning of the constructive use of science and technology.

C. P. Snow was also very active and assumed some position under Wilson's Labour Government later [in 1964]. He was a writer but concerned with his own ego a great deal. A good deal in his books is autobiographical in various ways. He used his colleagues as material for his novels, which is all right, of course. A list of which characters are based on whom has been published. During the war, Snow was in charge of the scientific and technical register. People who had scientific qualifications were allocated to more or less appropriate jobs through this register. This was in distinction from the First War when this did not happen. Smart people from the University who had done classics, for example, were given the choice of doing Japanese or radar. There were military language classes going on in Cambridge. They had a short course in Russian, for example, which took inside a year and they did nothing but Russian, day and night. They produced Russian interpreters. When they had 5% mental casualties, that meant that the intensity was right. For relaxation there were Russian films which other students could attend.

You studied Russian, too.

Yes, but that was after the war in 1946 and it was related with left-wing politics. It was a summer school and I met Sheila there. She had already done a degree in Russian at Glasgow University during the war. Bernal invited me to accompany him to Moscow in 1956 when he gave lectures on the origin of life at Oparin's institute. I briefly met Kapitsa, Landau, Tamm and Fock at the Institute for Physical Problems besides Oparin and Shubnikov. In 1956 Vainshtein and other crystallographers came to a meeting in Madrid. The International Union of Crystallography meeting in Paris in 1954 began the network of personal friendships with crystallographers around the world. Later I worked rather unsuccessfully at the Institute of Crystallography in Moscow for five months in 1962.

What was your first research project?

The structure analysis of a particular calcium phosphate. It was a material used in fluorescent tubes. So it was of interest to Philips. They had various projects involving X-ray crystallography that were related to their practical applications. Then at Birkbeck I got involved with the section on inorganic materials. This section was dealing with cement which is an immensely complicated material and provided topics for lots of people. However, I don't think this area was satisfactory for me and in retrospect I think I ought to have changed to molecular biology. But I didn't. It would have been better though because from the time of Rosalind Franklin and Aaron Klug a very strong group developed there. Then they all went to Cambridge, around 1964. But there was a very active life at Birkbeck. Everybody visited Bernal. Linus Pauling visited him twice. H. S. M. Coxeter also came. I remember particularly having supper with André Lwoff, Coxeter, Klug, Bernal, and Bernal's technician, John Mason.

Why did Coxeter come?

Maybe to discuss icosahedral geometry.

Were you already interested by then in icosahedral geometry?

Even before that. Incidentally, I have just rescued some interesting old papers from the department. I've begun to look through them. There are some scrappy notes by Bernal about icosahedral things, to do with his theory of water. They may be from about 1952. The icosahedron though came up in several contexts. I certainly knew about beta-tungsten, because

J. Desmond Bernal, about 1962
(courtesy of Alan Mackay, London).

of the work at Philips. It has an essentially icosahedral structure. Bernal considered early the icosahedral coordination, because it would not crystallize, as key to the structure of liquid water. I drew the figures for Bernal's 1956 Budapest lecture, later published. By then, of course, Pauling also got a number of icosahedral structures. Then there was the polio virus, consisting of icosahedral particles. It always figured large and people knew about this. I think that really Buckminster Fuller didn't have any part in it. He came round at that time and there was a lot of interesting discussion. He showed us a number of things to do with his tensegrities. Fuller visited Birkbeck at least twice. He probably came to see Bernal but Aaron Klug was the one who spent the most time talking to him.

Caspar and Klug mention explicitly Buckminster Fuller's influence on their virus work in their 1962 paper.

I don't think though that he actually contributed anything to it, but simply got them thinking. Fuller spoke in a very curious way with his own eccentric terminology. Klug took the trouble to learn what Fuller was actually talking about. When you learned what his words meant much of it was actually sense. In particular, he had these tensegrity structures which Fuller had taken over from Kenneth Snelson and developed further himself. These were spherical structures where there were compression members, rods, which were connected to each other by tension in a very ingenious way. The rods were not touching each other. They were held by tension. Obviously the rods transmitted only compression, and the wires transmitted only tension. But you could make the structure. Some of these were icosahedral. What was especially interesting was that they could also move a bit and adjust themselves to take up some optimal arrangement. That is the mutual arrangement of particles on a sphere and that was what Klug was interested in and Klug's interest was stimulated by the way in which these structures could essentially transform themselves. So, in summary, I think that Fuller did not contribute anything specific, but he got people thinking backwards and forwards on the topic. I went to a lecture by him in London, and came away after about two hours, and he was still going on. I also reviewed his book *Synergetics*, and it was basically rubbish, but there were bits of interest embedded in it. I think he was considerably overblown really, but he was a stimulating influence. Fuller said on occasion that the geometry of geodesic domes, this rather complicated structure, was beyond the ability of science to analyze it, which, I think, was nonsense. This was after the

Montreal Expo Dome, and I was pretty sure that the stress people had done their calculations on it.

Returning to your work at Birkbeck, when did you complete your Ph.D.?

In 1951.

How about the D.Sc.?

Much later and it is not something very highly regarded. You get it by presenting a large pile of papers. Applying for a D.Sc. is something normally done only by chemists. It is also something like making a mark on the wall and indicating that you think you have been passed over for promotion to Professor. I failed in an application for a readership about 1967 and Bernal told me then that having papers in other subjects was a positive disadvantage. I thought so much the worse for the university system if it denied the word "university".

Bernal didn't have a doctorate. Why?

It wasn't necessary or fashionable at that time.

Then you got elected to the Royal Society in 1988. What is the mechanism of the election?

You have to be nominated and the nominator collects at least half a dozen signatures. The candidate used to submit a complete collection of papers. It has changed since and now you submit your best 25 papers only. The election takes place each year. There is a general tendency to keep the number of Fellows constant but it is just custom. About 40 people are elected each year. You can stay as a candidate for 7 years (if you are not elected), then you cease to be a candidate for three years, and then you can become a candidate again for another period. The list of candidates is confidential.

Who were your nominators?

You're not supposed to know. But you can see the circles of influence.

Have you already acted as nominator?

Yes.

What are your criteria to nominate someone?

Must be a distinguished person who can contribute to science and to the Society, and has already done so in a substantial way.

Does the Royal Society support scientific research directly?

It does, the activity taking up the bulk of its money is the maintenance of research fellows. These are high flyers who are identified early and supported with salaries and small amounts of running costs. It's aim is to produce unexpected discoveries by funding individuals, rather than projects. The individuals come with projects but they may not stick to them.

Do you think it is meaningful to do fundamental research only in highly developed places?

No, I think fundamental research can be done anywhere because the questions of fundamental research come out of applied science and actual observations. However, a research worker needs to be in touch with the leaders of the appropriate research front.

How can you identify researchers or areas for support?

With the money culture the most effective strategy for the money-giving bodies is simply to fund the successful places which they funded before. It's an extremely conservative strategy. The alternative is to put a certain amount of money into pure speculation. British Petroleum (B.P.) have a system for funding this kind of thing in the industrial area. They have a venture company where people can try out ideas which may be of commercial interest. The key thing is that it has to be done on a big enough scale since, if you have a very low probability of fundamental discovery, then you have got to have a lot of lines going at once in order to make the statistics manageable. B.P. showed one result of this work, which was an essentially geometrical method, based on random packing. You take an oil well and you want to expand the cracks between the strata to promote the flow of oil. The process was to force an essentially colloidal suspension of little spheres into the cracks. Then the chemical trick is to have the medium between the spheres sufficiently fluid that it will run out again, and the crack is therefore held open, and the oil can run out of the geological formations, through the spheres, and into the main tube. This was a big experiment. They had a special converted ship with super high-pressure pumps for pumping the stuff in. It's a big bet, and it turned out to be worth the money.

Speaking about how to award support for fundamental research I have an example. In the British Royal Navy you either had a beard or were clean-shaven. Suppose you started as clean-shaven, then you were classified as clean-shaven. You could go to an officer and beg permission to grow a beard. The officer may say, right, permission granted. But you had to report back after one month. If the beard was satisfactory, a credit to the Navy, then it became established and you were classified as bearded and you must not shave it off without permission. If the beard was unsatisfactory you had to shave it off and go back to being clean-shaven. In other words, people funded by any research program ought to be able to spend a small amount, let's say 5% of their time on random search on whatever they like. On the basis of this they may have an idea. Then they ask for support for this idea to the extent of some strictly finite amount of time and money. Then you look at the idea at the end and ask whether it's good or not. If it's good then you decide whether to support it and if it is not, then you just go back and start again. Every research contract should have written in a certain amount of essentially unallocated resources which are totally at the disposal of the individual.

Crystallography seems to attract women scientists.

There was a comment by Anne Sayre in this connection about the idea of competition in crystallographic laboratories. It did not really get moving until the sixties, and the absence of competition provided a good atmosphere for women to work in crystallography. There was a survey in 1984 of the percentage of women crystallographers in different countries. It showed striking differences. Japan came at the bottom. I think that the fathers of crystallography rather liked women.

Let's change the topic to something lighter. Let's talk about polywater. Bernal was quite involved in the story. He called it a very important discovery.

Bernal didn't have any choice. He was not at first in the position to criticize Deryaguin's claim. What he did was to ask John Finney to check it over, and John did a very careful study, and it wasn't polywater, it was contamination of the tubes. Then the problem died off and this is how things should work. The amusing thing was that the theoreticians ran off in all directions explaining it before it turned out to be a false alarm. The cold fusion story is another extreme example of this.

There are, of course, other examples, where a scientific discovery truly shatters old dogmas. Consider just the case of fivefold symmetry.

Fivefold symmetry does not really violate anything, it just slips through the exact definitions. All the exact definitions are OK, they just were not designed to cover this case.

You gave two lectures on fivefold symmetry in September 1982 in Budapest. You said then that we should be aware of the possibility of such extended structures because if we thought them impossible, they might go by us unnoticed and unrecognized. As you were saying this, the first quasicrystals had already just been observed without your knowing about it. It happened in April 1982 and nobody outside NBS knew about it. You had actually published a simulated electron diffraction pattern which then turned out to be similar to Shechtman's experimental observation. So it was really you who predicted what we call today quasicrystals.

Yes, something like that, anyway. I used to do science abstracts — for ten years I abstracted all the Russian papers on crystallography — and I remember abstracting a paper on the incommensurate arrangements of spins in iron oxides, in hematite. The period of the helical magnetic spin is not the same as the crystallographic period. So incommensurate structures were current before that time.

Even much longer before that I thought of a simple thing about printing wall paper. Suppose your wall paper is simply printed from a roller. But suppose you are printing two motifs from two rollers of different diameter. Then you get a non-repeating pattern. I wasn't able to think of producing an aperiodic two-dimensional pattern in this way. I was only aware of the possibility of one-dimensional incommensurate patterns. I was really interested in hierarchic patterns, and not in aperiodicity as such. It came directly from Bernal's suggestions and the polio virus project.

I produced a hierarchic pattern, a hierarchic packing of pentagons. Then in 1974 I was getting some help in computing from Judith Daniels at the University College Computing Centre and, incidentally, showed her these patterns. She said that Roger Penrose had something like them. So I made an appointment with Roger Penrose and Robert, my son, and I went to see Penrose in Oxford, and he showed us the jigsaw puzzle, with the kits and darts and so on. Basically his concern was with forcing aperiodicity, and my concern was with hierarchic structures. It turned out to be very similar.

Just after this meeting Robert went back to his school at York and plotted a Penrose tiling on his pen-plotter. He was then doing computer science in York. Then there was also correspondence with Penrose, who had been preparing the article for *Scientific American* with Martin Gardner. From Penrose I got word back about this mythical person Robert Ammann. He had a dissection for three-dimensional tiles. Nobody has actually met Ammann and I don't even know where he was. Penrose had some letters from him. Ammann was a sort of dropout from the academic world. He is an American mathematician. He does something else. Martin Gardner has actually met him but he is a mysterious person.

Do you think Shechtman should get the Nobel Prize?

I think there are three kinds of Nobel Prize. The first is that there is some individual, typically like Bernal or Aaron Klug who had made important

Dan Shechtman and Alan Mackay in the Hargittais' home, Budapest, 1995 (photograph by I. Hargittai).

contributions in lots of directions in the same general area but perhaps nothing totally striking, but has personally been important in shaping the development of a whole area. Klug is an outstanding example. He had lots of things to do with molecular organization. Bernal did not get the Nobel Prize, but he could have if he had not been busy with other things, if he had forgotten about the peace movement and politics and had got on with molecular biology. A number of his associates got it, Aaron Klug, Dorothy Hodgkin, and Max Perutz. A great deal of it comes from the Club of Theoretical Biology in Cambridge. Bernal had a marvelous foresight to see how molecular biology would develop. This was about 1936. The people involved were Needham, Haldane, Waddington, Bernal, and others. They didn't have any special resources. It was just talking. On behalf of this group Needham applied to the Rockefeller Foundation for money to start a new Institute for Morphology or Morphogenesis or something like that. It was something more general than molecular biology. It was not funded then, but documents exist and the phrase molecular biology had already appeared in these discussions. They were quite clear about the application of physical methods and about the idea that defining the arrangement of atoms was going to revolutionize biology.

The second type is someone like Perutz who starts off on hemoglobin, and takes 30 years to do it, gradually developing the techniques. He has a clear objective, which is very difficult to achieve, but eventually he gets there, and it's a really significant achievement. I would put Karle and Hauptman's prize into this category, for instance, and Isabella Karle should have been included because it was her work that made the whole thing believed. The Karles and Hauptman realized that the information about the phases resides in the excessive numbers of amplitudes and, if there were enough of them, phases could be extracted.

The third kind is someone who turns over a stone and finds something really important, and recognizes that he has got something really important, maybe like superconductivity or the scanning tunneling microscope or the Mossbauer effect. There isn't any enormous amount of work but someone was in the right place at the right time, and recognized what he's done. I think Shechtman would come in the third category. There is actually some new evidence that Shechtman's discovery may be more important than it had been believed. It has been mostly followed by a tremendous amount of mathematics, an Ivory Tower of mathematics and little more. Now it appears, however, that the very low thermal conductivity of quasicrystals may be useful for something more than the non-stick frying pan but also as important as turbine blades, internal combustion engines, and so on. People are producing effectively quasicrystal surfaces by glazing metal with a laser. So Shechtman's discovery may be eventually related even to a process of great economic importance.

Looking back, do you consider your prediction of quasicrystals and your simulated electron diffraction pattern among the most important things you've done?

Yes, if you like. We have also predicted graphite with negative curvature but there is hardly any sign of it yet.

Lately you have been involved with what you call flexi-crystallography. What is flexi-crystallography?

The idea is that in ordinary, orthodox crystals the units are arranged in planes, sheets stacked up and the three-dimensional structure can be regarded as layers of two-dimensional planar sheets. Our idea is to look at essentially two-dimensional manifolds, sheets of atoms typically graphite-type sheets or silicate-type sheets or lipid-type sheets. We consider the possibility that these sheets may no longer be flat, but have some other curvature. In particular, if the curvature is positive then the sheets wrap up, and make spheres. The more interesting case is when the curvature is negative and you can get infinite sheets, which are crinkly like seaweed.

When [Daedalus] David Jones described the folding of graphite sheets in 1966, is this something to do with flexi-crystallography?

Alan Mackay in London, 2000
(photograph by I. Hargittai).

Yes, I think so. He described graphite spheres, graphite balloons and so on. That's a good case. I read his note at that time although I didn't recognize its significance at that time.

What is the Mackay icosahedron?

Just icosahedral shells. Bernal was extremely keen on hierarchy as a principle of building things and generalizing crystallography. When there was the discussion related to polio and so on, he asked what happens if you take 13 atoms to make an icosahedron, and then take 13 of this group to make a bigger one, and 13 of those to make a still bigger one? This is now known as a fractal. Obviously the gaps get bigger, so how do you fill up the gaps? It turns out that you get something like the Penrose tiling. You can see the structure of these icosahedral shells as a twin of 20 ordinary face-centered cubic packed crystals. They have to be flattened slightly, by about 5% to fit.

You have given a lot of attention to fields that are outside of classical crystallography.

If you go back to Joseph Needham's book *Order and Life*, 1936, and to the Theoretical Biology Club, you see that these people were perfectly clear on the existence of lots of ordered structures other than crystals, namely, liquid crystals and fibers, and others. They were quite clear that the majority of living things had all these other different kinds of order. The present position is that the success of X-ray crystallography in structure determination has overemphasized the role of actual crystals, and people have forgotten all the other things. People use religious language, saying that things that are not crystalline are defective or imperfect and disordered and other pejorative words. Just because they don't fit into the classification, which is not a natural one but one which was set up artificially. From an oriented fiber you can now get almost as much information as from a crystal. The handling of the diffraction data has steadily improved during the past 30 years.

You often quote Democritus as saying that, "There exist only atoms and empty space, all else is opinion."

Of course, almost nothing of the writings of Democritus survived. They were suppressed by his enemies who wanted a theological rather than a rational world. The actual phrase that you mentioned is perhaps extended a little bit from the original meaning. Still, it's a good argumentative phrase. I am just struggling at this moment to deal with this sort of question, namely, with the generation of structures. There is a duality or conflict between regarding structures as arising from local interactions and as arising from global considerations. The principle of least action is another interesting concept, or minimization of energy. How can a system actually minimize its energy unless it's able to explore all the possible states which have different energies. Some of these minimum principles like the minimum surface are equivalent to some local differential condition like curvature being zero. According to Feynman the system does actually know all the possible paths, and chooses the minimum. But there is still this duality between extended and local considerations. The most important question at the moment is to ask how we can describe structures, then to ask how structures describe structures. We realize this explicitly for the genetic code but how can we apply variants of the genetic algorithm to inorganic or indeed engineering structures?

Can any of this be translated into considerations about society?

Obviously, because states and economies and so on mean nothing. They are entirely composed of interactions between individual people. There is no such thing as the State. It's a kind of global concept for referring to some of the emergent properties of large numbers of individuals taken collectively. A lot of what we learn about structures could be applied to society but I doubt whether it would be of any use to try the reverse. This kind of thing was discussed long ago by Denis Diderot in *The Dream of D`Alembert*. We must try to understand what is happening in Yugoslavia and elsewhere. I am sure that the world is in for a bad time with the victory of the global market but I still have hopes that science can help us to understand human society. I took part in the Bonnington group planning before the 1964 Labor government when Wilson was going to implement the "white-hot technological revolution", but nothing much happened. Mrs. Thatcher's regime was against science and technology, but it must come round again. The most hopeful aspect of science is the way in which it is moving into domains previously the preserves of the humanists and turning them into proper science.

Dan Shechtman, 1995 (photograph by I. Hargittai).

5

DAN SHECHTMAN

Dan Shechtman (b. 1941 in Tel Aviv) is Philip Tobias Professor of Materials Engineering at the Department of Materials Engineering of the Technion — the Israel Institute of Technology. He is most famous for his discovery of quasicrystals. He is a member of the Israeli Academy of Sciences, and has been awarded the Israel Prize and the Aminoff Prize (in 2000 by the Royal Swedish Academy of Sciences), among other recognitions. We recorded our conversation during an international school on quasicrystals in Balatonfüred, Hungary, on May 14, 1995.*

Let's start at the beginning.

I was born in Tel Aviv on January 24, 1941. During my infant years we lived in Tel Aviv, and then moved to the suburbs, first to Ramat Gan and then to Petach Tikva. My mother was born in Israel and her parents came to Israel in the second Aliyah, the second wave of immigration. Only a small fraction of the second Aliyah stayed in Israel and they were among them. Most of the rest went on to the United States. My mother's parents who were Zionists and socialists, disliked the Czar and the communists. In (then) Palestine, my maternal grandfather became quite a prominent figure and I looked up to him, but it was my grandmother whom I admired. My father was a newcomer from Russia in 1930, in the fifth Aliyah. He came via Poland where he prepared himself for coming to Israel by learning

*In part, this has appeared in *The Chemical Intelligencer* **1997**, *3*(4), 25–49.

agriculture. In Israel he was a laborer in the orchards and in construction. My parents were married in 1936, at which time my father started to work with my grandfather in his printing house. I never knew my grandparents on my father's side as they died in Russia years before I was born. My father passed away some years ago, but my mother is alive.

Only a fraction of my large family, the ones who came to Israel and the few who immigrated to the United States, survived the Holocaust and the Second World War. Everybody that survived lives now in Israel. Most of the members of my extended family are entrepreneurs and except me they are mostly independent and do not work for salary.

Do they realize that you are a celebrity in science?

Most of them. My mother realizes it to some extent. She is an educated lady, reads a lot and goes to every show in town (Tel Aviv). My wife is in academia, and she knows, understands, and appreciates my academic achievements, as I appreciate hers, and so do my children. My family has been with me on different occasions when I received awards, gave speeches, and associated with dignitaries and colleagues from around the world. They also took part in many of my trips around the world and lived part of their life in the United States.

What was your language at home?

Always Hebrew. Nobody in my home spoke English and nobody communicated in Yiddish or any other language. My father spoke Russian, but had nobody to speak to in this language. My grandfather contributed a lot to the Hebrew language and he was very insistent that we spoke proper Hebrew. I started to communicate more in English when I studied for my Ph.D. with David Brandon, who could speak only English during his first years in Israel.

Please tell us about your schooling.

I went to a primary school in Ramat Gan. This was after World War II, during and after the War of Independence, and during harsh economic conditions. When I was 14 years old, we moved further away from Tel Aviv, to Petach Tikva. This was the first time my parents bought a house and my mother still lives there. During my childhood years we lived in very small homes. I always longed for some privacy and when I built my

Quasicrystals (courtesy of Ágnes Csanády, Budapest).

own home, in later years, every one of our four children had a separate room in it.

I went to high school in Petach Tikva. At the time, mid-late-1950s, a large percentage of my peers did not attend high school. There were several disciplines in that high school: humanistic, biological, and realistic. I was a part of the realistic class which numbered about 15 to 20 students while most of the students chose the humanistic direction. Mine was a strong group, and we did relatively well also in our careers.

During my high school years I was involved in the youth movement Hashomer Hatzayir. This is a political Zionist-socialist movement and in the fifties it supported Russia and communism. It also defended Stalin's actions, which I could not accept. I was skeptical about many issues that came up there and needed satisfying explanations which seldom came. Other than that it was and I hope still is, a wonderful youth movement. In later years I recommended it to my children, but only one of them joined it. The movement gave us important values and shaped our characters through camaraderie, a strong feature of Israelis.

We had good leaders, people to look up to, although they were teenagers, just 2 to 3 years older than us, and I liked them a lot. The group, boys and girls, all from the same neighborhood, met a couple of times every week, in the evening, usually around campfire, singing and dancing. It was intensive and parts of the activity was very physical, and at times hard, like 5–6 days field trips in the desert, carrying all our supplies on our backs and learning to use as little water as possible, which today we know is dangerous practice. Although the activities were outdoors, and in the fields outside the neighborhood, we felt safe, and even the notion of safety problem did not come to our minds.

In addition to the physical activity in the youth movement we had paramilitary education in school which at times was quite harsh. We were running, for example, and if the first in line faced a barbed wire barrier, it was his job to jump and lie down on it so that everybody else could cross by stepping on his back. He then had to free himself and join the rest. We all did this and other exercises many times, but I do not suppose it could be done on today's modern razor-blade sharp barbed wires.

Can you single out a teacher who was a strong influence?

Yes, it was the principal of my high school. He was quite a character. He taught me probably only one or two classes, but he was prominent and influential. Later he became the principal of another school, for gifted children from all over the country, in Jerusalem. In later years, when I grew up I wanted to talk to him, compare notes and see him in the eyes of an adult, but I never did. Lately he died, the way I would have expected him to die. He had cancer and at a certain point he killed himself. I do not know why his personality effected me and in what way, but if I remember vividly one teacher, it is him.

Did he know that he was special for you?

No. I do not even know if he would have remembered me, although he might have. I know one thing though. When I was about to start my mandatory army service, he was asked to write about me. Years later, I happened to see the letter, and it was clear that the man saw in me what I could not understand myself at the time. He picked up correctly some important features of my character.

What were they?

A strong sense of justice and truth, investigative from a doubtful point of view.

What happened then?

In 1959 I went to serve in the army. My physical condition was not very good, as I had asthma in my youth, which disappeared when I got married later. I did the basic training and then was sent to serve in a unit of psychotechnic testers and interviewers. The group was hand picked for the job and our commanders made us feel very special. We took crash courses in psychology and interviewing, and worked very intensively in that profession till the end of my service, two and a half years later.

I then planned to start my undergraduate studies. I applied to study biology at the Hebrew University in Jerusalem and mechanical engineering at the Technion in Haifa. I wanted to become a mechanical engineer since I was a child. I wanted to be like one of my idols form a Jules Verne novel. He was an engineer who knew and could make everything, and I wanted to be like him. In my eyes, at the time, a mechanical engineer was the ultimate, and the place to study was the Technion. I was not accepted to the Hebrew University, but was accepted to the Technion. In 1962, I started my studies. Like in high school I was a good student but far from the top of the class. I graduated in 1966, a mechanical engineer with a bachelor's degree and the job market was dry. However, the Technion offered me to continue my studies for a Master's degree, and earn my living instructing students. That was an interesting proposition, and I was soon immersed in research on chromium diffusion coating of steels.

It was mainly metallography, X-ray crystallography, electron microscopy, diffusion, phase diagrams, and stress measurements. It took two years to complete my study, but in the meantime, in 1967, the first electron microscope arrived at the Technion. I was among the first few students to learn from Brandon, how to plan and perform experiments on the microscope and how to analyze the results. I worked first on fracture surfaces, by replica techniques, and when I finished my studies for the master's degree it was only natural that I continue to work with Brandon for my Ph.D. thesis. He did not suggest a subject, but at the time Nate Hoffman of Rockwell International was there, working on his Ph.D. thesis, and he recommended that I study titanium alloys that became important for aviation. I studied phase structure and microstructural defects in several commercial titanium alloys that were subjected to cyclic stress, and learned practical electron microscopy fairly well.

When I completed my Ph.D. thesis, I applied for a position at the Technion, but was told to first perform research abroad for several years. During 1972 I corresponded with about 100 universities and research institutes around the world, and ended up with two offers. I chose a National

Research Council scholarship to perform research at ARL, Air Force Research Laboratories at Wright Patterson Air Force Base near Dayton, Ohio. It was a good research laboratory, and I studied structural defects and properties of titanium aluminides. Two and a half years later, I still did not have a job offer in Israel, and decided to stay in the U.S. and take a permanent position with the Air Force. When the last papers had to be signed I received an offer from the Technion. In the summer of 1975 my family and I returned to Israel and I started as a lecturer at the Technion. It was in the Department of Metallurgy within Mechanical Engineering, which in later years became the independent Department of Materials Engineering.

How about your family?

I met my future wife, Zipora (Zippi) in the army when she joined my unit a year after me. We dated for three years and got married in 1964. During our first years together, she worked as a teacher and studied sociology and education at Haifa University. Our first daughter, Tamar was born in Haifa in 1967 and then Ruth, in 1969. During our years in Dayton, Ohio, Zippi completed her Master's degree at the University of Dayton and Eyla, our third daughter was born in 1974. In 1975 we returned to Israel and built the house in which we have lived since. Yoav, our son was born there in 1980. During our years in Maryland, starting 1981, Zippi studied for her Ph.D. at the American University in Washington, D.C., and upon our return to Israel, joined the faculty of the Department of Education at Haifa University. Two of my daughters are married now, our first granddaughter was born in 1995 and our first grandson in 1996. All my daughters study psychology-related subjects: Tamar and Ruth, for their Master's and Eyla for her bachelor's degree. Yoav is in high school.

Dan Shechtman with Ágnes Csanády at Balatonfüred, 1995 (photograph by I. Hargittai).

When people refer to you, I hear sometimes this expression "He is a character".

People sometimes view me from extreme perspectives. Some people like me a lot, others hate me a lot. Some people are afraid of me, and feel threatened, which I find especially strange. In our home the profession is psychology. Over the years I have learned to observe myself as well as other people. I try to understand why a few people find me threatening. Usually I am very flexible, mainly because I do not think that many matters are important. Sometimes, however, I express strong opinions about certain issues, which I consider important, and do not yield to peer pressure. On these occasions I can see why people can find me tough. The other matter is independence. I do not take anything from anybody and make it a point to owe nothing to somebody who is not a close friend, and then only if I can reciprocate. Giving, on the other hand is easy for me, but I become very stubborn when I feel that I am cheated out of something.

An emotion which I do not have, but find many times in other people, is jealousy. I am not jealous of anybody, but some people are jealous of me and act accordingly. I understand it logically, but I do not have the feeling. From my first days in my department, I felt bad vibrations. I was the youngest and did not have anything, I just started and had enemies. Now, that I help get young faculty started, it looks very strange to me.

What is your position at the Technion?

Since 1987, I am a full professor of materials engineering. I have the Philip Tobias Chair and recently I was appointed to be the director of the new Wolfson Center of Excellence in Interface Science. It is an independent entity, one of several such research projects at the Technion in which the members belong to their respective departments. The current eleven members collaborate to study different aspects of materials interfaces.

Tell us about the events leading to the discovery of quasicrystals.

In 1981 my family joined me for my first sabbatical at the National Bureau of Standards (NBS, now National Institute of Standards and Technology, NIST) in Washington, where eventually the discovery was made. It was John Cahn who suggested to me to come to NBS because I have developed a technique to study metallic powders by transmission electron microscopy, such as the ones studied at NBS at the time. My research at NBS was sponsored by the Defense Advanced Research Project Agency (DARPA, now ARPA). Jake Jacobson who sponsored my research told me specifically not to limit myself to the proposed plan but rather to expand in any direction I felt was interesting.

I started by studying rapidly solidified aluminum-iron alloys. I analyzed the phases present and the solidification patterns. I collaborated mainly with members of the metallurgy group, Bill Boettinger, Bob Shaefer and Frank Biancaniello. We wrote a series of papers together and understood rapid solidification better. It was in April 1982, half a year after I had arrived, that I discovered the icosahedral phase.

What was the background of the discovery?

In 1982, in the background was traditional crystallography. As far as I am concerned, modern crystallography started in 1912 with the commencement of X-ray crystallography. However, X-ray diffraction could not have been the tool to discover quasicrystals (QC). Since von Laue, all the crystals studied were ordered and periodic, and thus a paradigm has evolved that all crystals are periodic. Consequently you could see in textbooks statements such as that we should not expect the atomic lattice to have fivefold rotational axes, and that the allowed rotational symmetries are twofold, threefold, fourfold, and sixfold. There was perfect periodic order in crystals and in crystallography. The 14 Bravais lattices provided a significant tool for classification. The 230 space groups were there and the *International Tables of Crystallography* were the ultimate classification catalog for crystals. When a new crystal was found it was a straightforward matter to locate its place in the existing system. With time, crystallography became a mature science, there was nothing new in the classification tables and nobody expected a revolution. The discovery of 1982 and its publication in 1984 changed all that.

What was the sequence of events through that period of time?

First, let us pose this question: Why had quasiperiodic crystals not been found before 1982? Is it because it is difficult to make them? The answer is definitely not. In fact it is easy to make QC by solidification of molten alloys, deposition from the gas phase or by electrodeposition as well as by solid-state reactions, to name a few techniques. Maybe because QC are scarce or found only in exotic materials? Again the answer is not at all. There are about a hundred binary compounds known today in which quasicrystals appear. They may be based on aluminum or nickel or titanium, and so on. These substances are useful materials that regularly occur in the practice of materials scientists. Can QC be found in commercial alloys? Of course; they are there, too. So what has prevented QC to be observed and analyzed before? As a partial answer let us consider the sequence of events that led to the 1982 discovery.

At first, I was studying rapidly solidified aluminum-iron alloys, which we thought had some commercial future. Eventually, it turned out that although rapid solidification research resulted in several useful products, it did not develop into a widespread technology. This, however is not important for our story. In the aluminum-iron binary system there was one metastable phase Al_6Fe, which I studied. The equivalent Al_6Mn in the aluminum-manganese system is a stable phase, and I wanted to compare some crystallographic features of the two. We started therefore to produce a series of aluminum-manganese alloys with increasing amounts of Mn in them. Eventually I ran wild, from a practical point of view, since beyond several percents of manganese the rapidly solidified alloy becomes brittle and therefore useless. Among the alloy ribbons which I have prepared with Frank Biancaniello by melt spinning, there were alloys which contained over 25 weight percent manganese. On April 8, 1982, as I was studying by electron microscopy rapidly solidified aluminum alloy which contained 25% manganese, something very strange and unexpected happened. It is worthwhile to look at my TEM [transmission electron microscope] logbook records of that day. For plate number 1725 (Al-25% Mn) I wrote: "10 Fold???"

There were ten bright spots in the selected area diffraction pattern, equally spaced from the center and from one another. I counted them and repeated the count in the other direction and said to myself: "There is no such animal." In Hebrew: "Ein Chaya Kazo." I then walked out to the corridor to share it with somebody, but there was nobody there, so I returned to the microscope and in the next couple of hours performed a series of experiments. Most of the needed experiments were performed at that time. A few days later all my work was complete, and everything was ready for the announcement. Then it took two years to publish it.

Why was that?

Having the results, I started my inquiries at NBS about colleagues who would know anything about tenfold symmetry. In doing so I met a lot of ridicule. The sophisticated said, Danny, it must be twins, and I told them it was not and I had the evidence to prove that from my TEM data. The chief X-ray person said, Danny, please read this, and provided me with a textbook on X-ray crystallography. He told me that if I read the book I'd understand that what I was talking about was impossible. I knew the book, I had learned my X-rays. It sounds like an anecdote today, but when I was a student at the Technion, I had to prove, in a test, that fivefold symmetry is forbidden in crystals. Had I not proven it I would have probably failed the test.

All these symmetry rules were correct, of course, but only for periodic crystals. This fact was hardly mentioned though in the textbooks, as it was assumed that every crystal was periodic since during the years 1912 to 1982 nothing else was observed. There were several exceptions; incommensurate crystals started to shake the system, but then it was assumed that they were modulations of a periodic system, and the Janner group in Holland studied them extensively.

To put the history of this into perspective, I would like to add the following: I have discussed my fivefold diffraction patterns with many scientists. I even used diffraction reproductions as Xmas cards, and my DARPA sponsor had a copy of it on his wall. Yes, stumbling upon the icosahedral phase was luck, but from there on my results were known to a large number of scientists, and nobody came up with an explanation.

Had I known about Alan Mackay who showed earlier that Penrose tiling can produce patterns with fivefold symmetry when Fourier transformed, I would have probably had an answer at that early stage, but I did not know Alan or his results. I did know, however, about the Penrose tiles. Martin Gardner published them in 1977 in *Scientific American* as a mathematical game, and at the time I enjoyed the aesthetics of it. Other people made that connection at a later stage following our first publication, two years later. In mid-1982, John Cahn took my micrographs and diffraction patterns to MIT, to his previous Department of Materials Science, and showed them to several scientists. He came back saying that nobody knew what it was, but that several people thought it had to do with defects in a regular crystal. I told him at the time that my microscopy results proved that it was not a regular defected structure. It was monolithic with no boundaries to be seen anywhere within the crystal.

At that stage I stopped experimenting with these alloy, but every now and then I put them again into the transmission electron microscope to see whether there was anything else there, but there wasn't. In late 1983 I went back to the Technion and discussed my results with several colleagues. There was no interest in them except for one person, and that was very important.

Ilan Blech, who soon afterwards left academia to do business in California, was interested in my results. Ilan was our X-ray expert and the first scientist who believed in my microscopy results and their quality. Ilan and I looked for structural models that would, when Fourier transformed, give patterns identical to the TEM diffraction patterns. Icosahedral cardboard parts were built, which when connected properly, gave the required patterns with fivefold symmetry and the rest of the observed patterns. At a later stage this became known as the "icosahedral glass model". The model required that the icosahedra were joined by their edges or faces and did not change their spatial orientation.

I felt very good at that stage. There was at least one scientist in the world who was ready to stick his neck out with me and publish the results and the model. My collaboration with Ilan resulted in a paper which was sent to the *Journal of Applied Physics* in the summer of 1984. The way I wrote the paper was not fit for this journal. It was more metallurgical and we didn't focus on the discovery. All the information was there, but the paper contained also other information on phenomena which had to do with aluminum-manganese systems. *JAP* sent it back so quickly that I felt like on the tennis court when the ball bounces back into your face. The editor wrote me that this material was not suitable for the journal and it will not interest the physicists.

All this was still in the summer of 1984 and I was back at NBS and gave the paper to John Cahn. I asked him what was wrong with it. John was very busy, but when I finally got him to read the manuscript, I convinced him that this was real and that Ilan Blech and I had a model that worked. John suggested to call in Denis Gratias, a young and talented French mathematical crystallographer who was in the U.S. at the time. John then asked the following question, did Danny perform the experiment right? He said yes, I would have done the same. Next question, is there any other experiment that we should do? He said, no. So John suggested to publish a paper with only my TEM results, excluding the model. We wrote this second shorter article, sent it to *Phys. Rev. Letters* and it was rapidly accepted. Our paper was then published in November of 1984, and created a big wave in the community of physicists, unlike what the editor of the *Journal of Applied Physics* thought. However, this publication, was not made before David Nelson of Harvard had reviewed it. In this process, he showed it to Paul Steinhardt of the University of Pennsylvania who had worked, although not published, on a topic similar to Alan Mackay's results from previous years, to get a quasiperiodic Fourier transform of the Penrose

Dan Shechtman with Dov Levine at the Technion, 1996 (photograph by I. Hargittai).

tiles. He made the connection between the fivefold QC diffraction patterns and Mackay's findings on Penrose tiles and expanded it together with his student Dov Levine who is now Professor of Physics at the Technion. The right model, which explains the atomic positions of QC, lies, probably, somewhere in between the two models: that of Shechtman–Blech and that of Levine–Steinhardt.

What happened afterwards?

Many things happened. Starting in early 1985, I was very excited by the reaction of the scientific community and mainly that of the physicists. Frantic activity started almost immediately by several groups in the U.S. and in Europe, mainly in France. They started to make quasicrystals by rapid solidification and to study their properties. At the beginning we were a very small group of people who dealt with this but the group expanded very rapidly.

The group of the four who published the second paper, following the one by myself and Ilan Blech, stayed together for a short while only. Ilan Blech left soon and went into industry but John Cahn, Denis Gratias, and myself continued our collaboration. We decided to write a series of papers which we thought were interesting and useful. In January 1985 we all went to Paris to write the papers. At that time Richard Portier of Paris, who did high resolution electron microscopy, added an important result by performing high resolution TEM, when he showed that there were no boundaries in the Al-Mn QC electron micrographs. Also, almost immediately I received broad recognition in Europe.

A couple of years passed, though, before I started getting recognition in Israel. It did not come directly from the scientific community but from the media when the newspaper *Haaretz* printed a front-page article about the discovery on the basis of an interview one of their reporters did with me. Since then I've received good recognition from the Israeli scientific establishment. I have been invited to give talks all over Israel and around the world. There were three years in which I gave about 25 to 30 invited lectures every year at conferences and universities worldwide. I received several prizes, about one a year, some of them very prestigious, such as the prize of the American Physical Society for new materials, the Rothschild prize, and the Weizmann prize in Israel, and others.

The recognition came mostly from France, where they took the quasi-periodic crystal business really seriously, and spread later to other scientific centers around the world. If you look at applications for QC, France is clearly the leader. Not only did they find the first applications, but they also issued the first patents on quasiperiodic crystals. Jean-Marie Dubois, Christian Janot, and their colleagues were the first to look into the applications of QC.

Speaking about applications, it has been found, to the surprise of many who anticipated electronic properties for the first applications, that the tribological properties of certain quasiperiodic crystals are very useful. A process is being developed in France to produce kitchenware, which are coated with a quasiperiodic crystal. It outperforms Teflon as it is harder and does not scratch or peel off. Since these quasiperiodic intermetallic compounds have excellent high temperature, friction, and wear properties, I anticipate many more applications to follow.

Do you anticipate any income from these applications?

Probably not, unless I find new practical applications. I didn't claim any patents, and I couldn't have. When I discovered the first quasiperiodic crystal, I didn't find a use for it. You can't patent something that doesn't have a use. I didn't find such a use, but other scientists did, and they have the patents.

What research directions have grown out of your discovery?

From a very early stage it was clear that quasiperiodicity has proved to be a very interesting topic in several research fields. The first research directions surfaced during the first meeting on quasiperiodic crystals, organized in

Les Houches, France, in the winter of 1985. Later this meeting has come to be known as the first international QC conference. Following that meeting considerable branching took place in many directions, but mathematics and physics were in the forefront. Materials scientists constitute another group, which continues to develop quasiperiodic alloys. We now know about 100 binary and a much greater number of ternary combinations of metals which form quasiperiodic crystals. Another important group is the crystallographers, but they took their time to catch up with quasiperiodicity of crystals.

The traditional crystallographers rely primarily on X-ray diffraction to determine the structure of crystals and electron microscopy was never truly accepted by them as a crystallographic tool. The main reason for that is that electron microscopy is not a good quantitative tool. It is difficult or impossible to measure precisely crystalline length parameters by electron microscopy, but electron microscopy is an excellent qualitative tool and it can reveal phenomena that X-ray can't. One of the reasons why quasiperiodic crystals could not have been discovered by X-rays was their size. The first QC that I made by rapid solidification of metallic alloys were as small as 1 micrometer, thus preventing any possibility of getting single-crystal crystallographic information by X-ray diffraction. This however, did not pose any problem for electron microscopy. Now, of course, we can produce large quasicrystals, several centimeters in length. The uniqueness of quasicrystals was discovered and defined by electron microscopy, but X-rays have been instrumental in determining their precise parameters. Crystallographers seem to have waited for single-crystal work on quasiperiodic crystals and that became available by 1987.

In the 1987 Perth meeting of the International Union of Crystallography I had an opportunity to present QC to the community of crystallographers. There was general acceptance of QC and the Union established a new committee to cover this new area. One more area involved indirectly in quasi-periodicity is arts and architecture. Artists and architects find these structures, and especially tiling, intriguing and interesting. Several architect colleagues of mine have made use of quasiperiodic tiling of plane and space, and we may see more development there.

You seem to prefer the term quasiperiodic crystal to quasicrystal.

At least some of the materials that we call quasiperiodic crystals can be explained by quasiperiodic tiling of space. Others can be explained by the icosahedral glass model. The general term quasicrystal, coined by Dov Levine and Paul Steinhardt, is a nice popular term but it does not say scientifically what it is. In addition, the term quasicrystallographer, for example, is not acceptable; quasiperiodic crystallographer may sound better.

You left the field just as you initiated it.

I left the field following several years of activity but I didn't abandon it and it stayed in very good hands. I also found it difficult to find funding for research in this area, mainly because there was no immediate use for it at that time. I intend to return to the field and I think that I'll find it easier now to get support. During the last 5 years I was involved in another field of research, which deals with chemical vapor deposition of diamond. I study defects in diamonds as well as growth modes, and surface crystallography. It's also a wonderful field but, unlike the quasicrystals, it is limited and closed. I'm reaching the end of that field from the research point of view. The field is now in the hands of the manufacturers and application researchers.

If you are returning to quasiperiodic crystals you will probably get very good support and the best students. What are you planning to do with these resources?

You may be right qualitatively but not necessarily quantitatively in assessing the possibilities. I study structures and structural defects by electron microscopy and by other means. I'm especially interested in defects in quasiperiodic crystals. Over the past 10 years the field has developed in such a way that I can do work now which I couldn't have done at the beginning. We did not have good-quality crystals at that time for two reasons. First, we didn't have compositions that could grow into high-quality and large-size crystals, and we have them now. The other reason is that everything is better defined, more precise now than it was then. I'll of course, collaborate with people who grow these crystals, and have already started planning this collaboration. Properties of QC and their practical uses are also of interest to me and I may look in that direction for future activity.

I would like to ask you about your relationship with the late Linus Pauling? Even in what may have been his last interview he insisted that he did not believe in quasicrystals.

Linus Pauling heard about the discovery and contacted me in writing. He wanted some information which I sent him. Then he wanted more information and he sort of complained that I was not correct in the information that I had provided him with, which I was. Nevertheless, I repeated the work, performed the needed microscopy again, and sent him a short paper that I wrote just for him. He replied by saying that it was OK and what I did was fine, except, he did not agree with the interpretation.

Then at a certain stage I suggested to him that I'd come and visit him in Palo Alto, and show him the result. I went and gave a full lecture to an audience of one. He had many questions, which I answered but he was very negative and he did not believe in this. I showed him results, which for me were very conclusive. He said: "I don't know how you do that." If it were a student I would probably say: "OK, go and read a book if you don't know how to do that." But with Linus Pauling? This is the man who wrote the books. Anyway, as I was leaving, I asked him, "If you change your mind and if you ever agree with me, please publicize it, and let it be known."

Then we met several more time at conferences. It was a friendly meeting every time, and we invited each other to dinner. People were looking at us as if expecting a fist-fight but the conversation was always very pleasant. We also agreed on many things, like on the importance of Vitamin C, but never on quasicrystals. A couple of years later I was attending a big lecture by him at Stanford, organized by the American Chemical Society.

The topic of quasicrystals came up in his talk and he mentioned how bad it was. I was just sitting in the audience and nobody knew me. He was like a mixture of a politician and a priest. He had this quality to become a sweeping leader, enjoying the admiration of the crowd, with no questions asked. In his effort to explain the icosahedral phase as periodic, he presented a model of twinned crystal, which was very soon afterwards proven wrong by others. However, in this lecture he had the floor, of course. He talked about me by name, but he didn't know that I was there. At a certain point I turned to the man sitting next to me and said, Wow, he's wrong. And he said, What? and I repeated, Linus Pauling is wrong. And he shouted, WHAT??? as if he was going to hit me. It was a fanatic crowd.

In 1987 I met several of his close disciples in China. They came to the second international conference on quasicrystals. Each one of them separately and as a group said, Danny, we know you are right. And I said, hey, this is very important, I need this in writing, but they said: we can never put it in writing because it would kill our Linus, he trusts us, and we can't betray our old master. I felt very bad about it, I felt that science should not be done this way.

At one time Linus wrote me a letter suggesting to publish a paper together and settle the differences between us. I answered him with a letter saying the following: I'll be honored to write a paper with you but we have to agree on the principles first, the first principle being that quasicrystals exist, and they are not twinned crystals. He wrote me back and said that maybe it was too early to do this joint paper.

There is a saying by Max Planck, which in essence says that an important scientific innovation rarely makes its way by gradually winning over and converting its opponents. Rather, its opponents gradually die out, and the new generation learns the new ideas.

My experience is somewhat different. With the help of the first believers we have convinced a whole community some of whom were very strong non-believers, that quasiperiodic crystals are real. That took a lot of time and effort, but the attitude changed. The first person to collaborate with me and believe in me was Ilan Blech. He contributed a great deal to this. For a while I felt that we were two against the world. This was in the spring of 1984. Then, towards the end of 1984, John Cahn and Denis Gratias joined and that made a big difference.

Charles Townes in Stockholm, 2001 (photograph by I. Hargittai).

6
_____

CHARLES H. TOWNES

Charles H. Townes (b. 1915 in Greenville, South Carolina) is Professor of Physics in the Graduate School at the University of California at Berkeley. He was co-recipient of the Nobel Prize in Physics in 1964 together with Nicolay Gennadiyevich Basov (1922–2001) and Aleksandr Mikhailovich Prokhorov (1916–2002) "for fundamental work in the field of quantum electronics, which has led to the construction of oscillators and amplifiers based on the maser-laser principle."

He received his B.S. degree in physics and B.A. degree in modern languages, both from Furman University in 1935. He was granted an M.A. degree in physics by Duke University in 1936 and a Ph.D. in physics by the California Institute of Technology in 1939. He was a member of the technical staff of Bell Telephone Laboratories from 1933 to 1947. He served on the faculty of Columbia University between 1948 and 1961, chairing the Physics Department from 1952 to 1955. During the last two years of his tenure at Columbia, he was on leave being the Vice President and Director of Research of the Institute for Defense Analyses in Washington, D.C. Between 1961 and 1967, Dr. Townes was at the Massachusetts Institute of Technology, first as Provost and Professor of Physics, then as Institute Professor. He has been at the University of California at Berkeley since 1967.

Clarence and Jane Larson recorded a video interview with Charles Townes in his Berkeley office on March 21, 1984.* Professor Townes

_____

*"Larson Tapes" (see Preface).

kindly helped us revise the transcripts in the spring of 2004. First we communicate the Larson interview and then an interview of our own in February 2004.

Family background

I would like to start with a summary of my family history because one's early childhood experiences are of some importance to the general nature of what you do later and what your ideas and interest are. I was brought up on a small farm in South Carolina in the Piedmont region. My father was a lawyer, but as typical southerners all like farms, my father liked living on a farm. We had tenant farmers who did most of the work, but I did some of the work too and enjoyed that. I had three sisters and two brothers; I was the fourth in the family and since I was the youngest for some time, I learned a lot from my older brother and sisters. Two others came along later. My father Henry Townes was also born near Greenville, South Carolina, and was brought up on a farm. His father was a lawyer and also an editor. I think that my father, in other circumstances, could've become a scientist. He was very interested in science, and I don't regard law and science as all that different; the type of reasoning involved is frequently rather similar. My father enjoyed the law; it was the thing to do at that time. He put himself through school; the southerners were very poor when he was brought up, and there weren't many opportunities in science. When I came along, there were more opportunities and my family, both my father and mother, encouraged us to think about and be interested in natural history. I was interested in all kinds of natural history; my brother and I did a lot of field work and my father took us out on Sunday afternoons to some other farms he had in the outskirts of Greenville. My parents were quite interested in them, too.

As is typical with most southerners, we have long family traditions; southerners are interested in families. My ancestors all came over to the United States prior to the revolutionary war. Some of them came to New England; in fact, I am a descendant of Governor Bradford of the Plymouth Colony. The New Englanders and the southerners intermarried and knew each other and intermixed only before the Civil War. When the Civil War came along there was no longer any mixture and from then on my family would be completely southern. Before then, there was a lot of intermixture and I descended from a long line of congregational ministers as well as people who settled in the South. One of my ancestors was the editor of

Charles Townes, 1984 (during the conversation with Clarence and Jane Larson; photograph taken from the video recording).

a local newspaper in Charleston. He was an abolitionist and when the Civil War came along, he lost his paper. But until that time he was listened to and was fairly popular.

Studies

My parents both went to college in local Baptist colleges, Furman University in Greenville, and a nearby women's college where my mother went. It was a local scene. We knew everybody and my family lived there for many years and many of the people there went to the same colleges. On one hand, it was a very healthy environment because it was a friendly community, on the other hand, it was rather limited from the point of view of national connections and scientific connections. In fact, the science I could learn at Furman University was not modern research science. My first contact with real modern science came about through reading the technical articles of the *Bell System Technical Journals* in the local library. Bell Labs provided them free and I remember very well reading summaries about nuclear physics back in the 1930s. That was my first contact with modern physics.

I studied at Furman from 1931 to 1935. Nuclear physics was just being born during that period. I had very good teachers of great character, and intelligent, but they were simply not acquainted with modern research science. Some of them had Ph.D.s, but they did very little research. Some researched in biology and I enjoyed biology; I did a lot of field work. I collected for the museum and I got a little pay for that during the summers, and went to biological summer camps. I would perhaps have gone into biology except for the fact that my older brother was a biologist and he was so good that I got shied off from competing with him. But I learned a lot from my older brother; we were rivals in a certain sense, but we also did a lot of things together. That was very helpful to me; it was challenging and interesting. I liked physics because I liked mathematics, but mathematics was not so closely connected with the real world. It was interesting intellectually, and I liked something that was more closely connected with the real world and more quantitative and clear-cut than biology. I had some physics hobbies, electronics especially, but most of my hobbies were connected to natural history, collecting and observing things in biology, geology, and to some extent astronomy.

This was not so much of a consideration of how to make a living. I know that among Hungarian physicists, for example, it was common first to study chemical engineering. I did not expect to make a living in physics; the field was not even well known at that time and when I first decided that I wanted to major in physics, many of my friends did not know what physics was. I had to explain that it was somewhat like chemistry but more connected with electrical engineering. There were very few jobs in physics at that time. I hoped that I would find a job in a school where I could devote myself to teaching and also do research. That was my goal, but it was not at all clear at that time. The country was in a deep economic depression. But I liked physics and that was the way many scientists went into science; they did it because they liked it. The prospects of jobs were slim anyway in many fields.

To major in physics at Furman University, I had to take my last and fourth course in physics by myself because there were no other students to take four courses in physics and thus get a major. The professors didn't normally teach that many courses, so they simply gave me a book to read, which was G. E. M. Jauncey's *Modern Physics*. It's a very fine book, I enjoyed it. I worked the problems and reported every once in a while to the Professor of Physics what I was doing. That was my last course of physics and when I went to graduate school, I knew I had a lot to learn.

I actually finished physics in three years. I was young, I had skipped a grade in grade school and the southern schools at that time had only 11 grades rather than 12 that most other schools had. So I was two years ahead of my normal class when I entered Furman. When I completed my physics major, my parents felt that I was a little too young to leave for graduate school and I didn't disagree with them. I was not eager to break away, so I stayed another year and took a degree in modern languages.

I liked languages in general, and I also took Latin and Greek. So I took two bachelor's degrees, one in physics and one in modern languages.

I went to Duke University in 1935 for a year and I was taking some undergraduate courses at Duke. I went to Duke because they offered me a teaching assistantship, which was hard to get in those days; I'd applied to quite a variety of places and didn't get any offers from the bigger schools. But Duke was a reasonably good school and I stayed there a year. After that year I thought that I wanted to go to the very best place I could and again I applied everywhere, to four or five of the best schools in physics at the time. I was not offered any financial assistance so I worked very hard the following summer, saved up 500 dollars, and set out for Caltech. Caltech accepted me as a graduate student, but offered me no financial aid. At Caltech, I was again taking many undergraduate courses; they were big courses and stiff and I learned a great deal from them. After one semester there, I got a teaching assistantship and my 500 dollars lasted past that time and then on. I was adequately supported financially.

It was very fortunate that I went to Caltech because Caltech was probably distinctly the best place at that time for physics. It had a collection of people and a spirit which was really quite exceptional. Caltech has been an outstanding place for a long time, but at that period it was at its peak in relative standing in the country in physics, chemistry and biology. Robert Oppenheimer came down in the spring quarter and brought many of his students from Berkeley and I enjoyed meeting them and him; we used to do a lot of hiking with his students, together with the students of Caltech. I debated going into theoretical physics as opposed to experimental physics, but I liked laboratory work and at the time my eyes were giving me a little trouble; I was doing much night studying and I felt that I'd better mix theoretical work with laboratory work. So I did an experimental thesis under Smythe; this is the Smythe who wrote a book on electricity and magnetism. I learned an enormous amount from that book. He used me to try out his problems; he was just writing the book at the time and I worked all the problems and checked them all. I learned a great deal from that.

I think if one knows one field in physics very well, he finds that it applies over a surprisingly broad spectrum. Electricity and magnetism after all involves wave equations and their solutions and all kinds of things. Static and dynamic electricity combined covers so much of physics that I always found it enormously helpful. There is also overlap with optics and microwaves that have been my interest. I also enjoyed very much quantum mechanics by Professor Houston and Oppenheimer's rather more advanced lectures, which I found quite interesting. Caltech was a very rewarding and rich place; I was there only for three years and my goal was to finish up in three years. While I could support myself there, it was pretty skimpy living; by my last year I was skipping one meal a day to save money. I never felt that it was an enormous hardship because everybody was doing it; it didn't seem that much of an imposition, whereas today it would seem like an imposition to have to do that. But I was eager to get through and I got through in 1939.

Social behavior and independent thinking

I should also comment on a few other items in my early history. My parents were quite religiously oriented, as I am myself. They were also very insistent on doing what they felt was right in their judgment. They might have been considered somewhat asocial for that reason; they were not interested in society *per se*, they were interested in people, but they insisted on doing things they thought were right. Hence, I always felt it to be an honor to do something that you felt was right even if other people didn't agree with you. I think that viewpoint is also important in science. You have to be able to stick up for what you think is right; my parents never did it in an objectionable way; they were not the kind who would demonstrate in streets, they just quietly did the things they felt were the reasonable and right things to do. I was in a fundamentalist church of the Southern Baptist Church; my parents were not fundamentalists; they simply said that we don't agree with our minister on this, but the church is important, so they participated very fully in the religious activity.

There have been many times in my scientific career when some of my good colleagues didn't agree with me and insisted that I had to be wrong. I find it particularly useful and rewarding to do things that according to other people were probably not right as long as they were right. I could also make errors, of course. It's another important thing to be able to recognize when one is wrong. But if you look at a situation carefully and decide what is right, it's very important to stick with it. If one does things that everybody else agrees with and thinks are right, there's not much point in doing it, because all the other people are doing that. Advances are made that way, but on the other hand, the individual contribution may be greater if you don't do the things, which other people agree on and which are right. As for the obvious things to do, everybody is doing that.

In that sense I think it's more important to do those things with which everybody doesn't agree.

Career choices

Concerning my career, I always expected and hoped to be in some kind of teaching position. I very much wanted to go to a university where I could do research, but I didn't know if that would be possible. I had some plans; I hoped to get a National Research Council Fellowship and go to Princeton when I graduated. Recruiters from Bell Labs came along and my professors that I knew well, in particular Smythe and I. S. Bowen and W. V. Houston advised me to talk with the recruiters. So I talked with them although I wasn't terribly interested, because I hoped to go to a university. As it turned out, they offered me a job. That was in 1939. Jobs were very scarce then, and my faculty advisors told me that it was a very good job and I better take it. I knew, of course, that there'd been a good deal of fine work done at Bell Labs. I was acquainted in particular with C. J. Davisson's and L. Germer's work at that time. It obviously was a good place, so I finally acceded that maybe I better do that and I went to Bell Labs.

It was a very broadening experience for me, because I came in contact with engineering and with more different kinds of problems than I would have at a university. Bell Labs did a very generous thing when I went there. They sent me around three months each to four different departments for a year; that was the plan. I worked with the microwave group, doing microwave engineering, trying to invent new kinds of tubes. I worked with the vacuum tube group, electronic tubes, triodes, cold cathodes, and so on. I worked with the magnetic materials group, and then I was scheduled to go somewhere else, but I haven't gotten to find out what that was because I never made it there.

Radar and war efforts

I was suddenly called in to see Mervin Kelly who was the director of research, along with Dean Wooldridge who was my boss at the time. Dean had graduated from Caltech three years before me. I was informed by Kelly that beginning the next day I was to start working on radar bombing systems. World War II was facing us, we weren't in it yet, but the United States was trying to be helpful. Bell Labs in particular had collaborated with the British in trying to develop radar. I had not been at all involved in that although I knew it was going on, but the decision for the Labs was to get involved even more. I was simply assigned with Dean Wooldridge to start developing the first radar bombing navigation system. Now, systems like that were later developed at the MIT Radiation Laboratory, but I don't know of any other system that was started quite that soon. It came about as a result of work on potentiometers and analog computers, which had been started at Bell Telephone Laboratories. They had a system for guiding anti-aircraft guns, using potentiometers and analog computers. That seemed to be promising, successful, and they decided to try to do bombing and navigation devices using the same analog techniques.

I quickly had to start learning about radar and radar as a sensor. We developed the analog computers and used other sensors, for example for navigation, and put together a system which would allow an airplane to bomb at night or through clouds, and would allow the airplane to navigate and dodge anti-aircraft fire. We developed several systems during World War II, but none of them were used; basically they were too complicated — simpler bombing systems were used. The last one that I developed was put into the B-52 and used extensively after World War II. At least they were installed into airplanes, I don't know if they were used to drop bombs. From all that, I picked up a great deal about radar, microwave, and engineering techniques, electronics, and I was exposed for the first time to electronics techniques. Those were exceedingly valuable to my subsequent work.

Begin interest in microwave spectroscopy

It was during that time that I became interested in the microwave absorption of molecules and that gave rise to microwave spectroscopy. We started out with what was then a rather short wavelength, 10 centimeters. We developed a whole system and had it tested in the air, and it was within one year because everything had to move very fast. But even at that rapidity, by the time we were finished, the 3-centimeter radar was coming in. It was so much better in angular definition that we were told to develop a system for the 3-centimeter radar. We did that and tested that system as well; we did all the tests, dropping sand bombs on various places in the Gulf of Mexico. By the time we finished with the 3-centimeter system, everybody was moving to the 1.25-centimeter system.

We were now assigned the job of doing the 1.25-centimeter system. By that time I became a little annoyed that whatever we did seemed to be not quite the right thing and we had to go on and do something different. I was also concerned about the time scale. I was afraid that what we did would not turn out to be really useful for our war effort. Then I realized that the 1.25-centimeter waves could be absorbed by the water vapor. I looked at that hard and tried to convince my superiors, first in Bell Laboratories and then people in Washington. I explained to everybody I came into contact with that it was very likely to be absorbed so strongly that it would not be useful. Eventually, I was told by people fairly high up in Washington that the decision had already been made and we better go ahead although we cannot know how good this system would be. I was a young man at that time and my views wouldn't carry much weight anyway, but I was pretty sure that we were in for trouble, as we were as soon as the first radar was tried out. By then we were fighting in the Pacific where there was a lot of water vapor. This radar turned out to have very limited range and the project thus was cancelled.

However, this period turned out to be exceedingly useful for me. As a result of studying the water vapor, I came to realize that a very important branch of spectroscopy could be developed due to the interaction between free molecules and microwaves. In particular, what had not been apparently realized by most physicists and spectroscopists, was that the line width could be made very narrow. The lines are broadened only by pressure in the microwave region. By decreasing the pressure, the lines became narrow while they did not decrease in their intensity at the peak. It is a very important phenomenon — you pump out the gas pressure to be lower and lower, but the intensity of the peaks does not decrease; this was anti-intuitive, for it decreased in width only. The lines became much sharper and this made the spectroscopy all the more interesting because the transitions could be measured very precisely. The lines did not become weaker. That's what the theory said, and this was one of those occasions when many people differed with me. They felt that this just couldn't be true. I was proposing by that time at Bell Labs that I should be allowed to study this phenomenon — a new spectroscopy.

As soon as the war was over, one of the theorists looked at it and found that it looked as if I could be right. We had lots of parts that were almost free so it was possible to build equipment almost from what was considered to be junk. Microwave spectroscopy started mostly in commercial laboratories in three places independently, one in Oxford by Brebis Bleaney, another one at the Westinghouse Electric Corporation by

William Good, and the third by myself at Bell Labs. Soon afterwards, there was other work at RCA. The whole field started out of the wreckage of the radar program. Largely, most laboratories in the program were industrial laboratories where there was a basic physics program. The hardware was available and abundant and the techniques grew out of the radar program. So it was a fruitful period and I wanted very badly to get started on this field, but Bell Laboratories wanted very badly for me to stay in more engineering work and in particular to finish up the radar bombing systems. I did stay for six months after the war finishing up a radar bombing system. Then I switched as quickly as I could back into physics doing microwave spectroscopy.

I also wrote a memorandum trying to sell the idea to Bell Laboratories, which, in retrospect, was not so far wrong. It was about the possibility that spectral resonances of molecules or atoms or solids might become useful as signal elements as one moved towards the shorter wavelengths. This would be a reason for Bell Laboratories sponsoring that field, microwave spectroscopy. I didn't at that time foresee the possibility of amplification. In fact, my memorandum argued very plainly about the usefulness as passive circuit elements. I recognized that molecules could generate microwaves, but they would always be rather weak, limited by the black body radiation laws. That's the point where I was wrong, of course, as I later discovered. But concerning the passive circuit elements I was basically right. Microwave spectroscopy has developed rapidly to be a rich and important field for understanding molecular structure and to some extent liquid and solid-state structure.

After about three years, partly because I was always interested in going to a university and partly because my field seemed to be more of an academic field than an industrial one, when I got an offer from Columbia University in 1948, I moved there. I was eager to actively pursue microwave spectroscopy, and I didn't want to make a move that would make me inefficient in building up a lab. Columbia had the advantage that it too had been in the radar program. I. I. Rabi was there, who was a principal figure in the radar program and he initiated the Columbia Radiation Laboratory, which was basically a microwave laboratory for building magnetrons of particularly short wavelengths. So they had equipment and had other people in the general field of microwave physics. There was Willis Lamb and Polykarp Kusch and Rabi himself although he generally worked at somewhat longer wavelengths. They had the equipment — when I was invited there I thought that this would be a place where I could get started very fast. I had very good students and I worked happily at Columbia University for a number of years.

I worked on the development of microwave spectroscopy. Around 1955 or 1956 I completed a book on microwave spectroscopy and I felt that at that time for me that was a kind of closed chapter. After about 10 years of working in the field, I felt that most of the aspects of microwave spectroscopy that were of special interest to physicists were done. The book was a summary of the field waiting for others, like chemists, to take over. I felt the need to change my field occasionally and I thought that from now on I would do something different. It had become a sizable field with lots of momentum; there were then quite a few new aspects for physicists that had not been expected.

Microwave spectroscopy was successful partly because it recognized the use of a new frequency range that hadn't been previously used, the short microwaves. It became stronger as you went to shorter and shorter wavelengths. Many microwave spectroscopists were pushing toward shorter wavelengths; Gordy at Duke University, for example, who was working on using harmonics. I tried many different techniques to get to shorter wavelengths; I am talking about going down to below a centimeter and hopefully on down to a millimeter or even submillimeter wavelengths. That was the general goal. By the end of that period it was fairly easy to get down to half a centimeter, maybe to 3 millimeters, but then it became pretty difficult. Obviously, the spectroscopy became richer and better as you got to shorter wavelengths; most things gave stronger resonances, so it was very desirable to push on to shorter wavelengths — Cherenkov radiation was one possibility and I had a student do a thesis on that. It worked; the theory was OK, but it was awkward, difficult. Magnetron harmonics I tried, did some work on that — again we got down to a fairly short wavelength, but it was awkward and didn't seem to work well.

So I was looking hard for some ways of producing shorter wavelengths. Our work at Columbia was supported jointly by the three armed services, the Navy, the Air Force, and the Army, through the Signal Corps. They kept encouraging us to do applied things like making better magnetrons, but they were also open-minded for our doing other kinds of physics. I was never interested in building magnetrons, but we tried to build something, and I did want to produce shorter waves. The Navy people and in particular Paul Johnson of the Navy were interested in what we could do at shorter waves. He organized a committee at ONR [Office of Naval Research] with me being chairman to examine what were the most likely and best ways to get down to the millimeter region. He and I picked out the people, physicists and electrical engineers. These included John Pierce from Bell Labs, Marvin Chodorow from Stanford, John Strong who was an infrared man, and John Daunt who was a low temperature man. There were others as well, a wide selection of people in fields around the general area, who were real leaders of the field.

We tried to come to grips with what could be done. We were looking in the fields of low temperature, the infrared, and electronics. We tried to encourage people working in the field and we reviewed the entire Navy program and other suggestions that came along — there were some interesting ones. We considered all kinds of suggestions, for example, using ferromagnetic resonance as a slow wave structure and seeing how it interacted with electrons, but it didn't really look quite good enough. After I'd done this for a couple of years and we met a couple of times, I was feeling frustrated.

Developing masers and lasers

We had a meeting coming up in Washington; I checked into my hotel the night before; the meeting was on my mind. I woke up early, since I had small children at the time, I usually woke up early. I was rooming with Art Schawlow, who was still asleep, so I decided to leave and not wake him up and went into a nearby park, thinking about how we were going to run the meeting that day. I asked myself, "Why was that meeting?" We hadn't really made any remarkable progress and we didn't know what was needed. I went over the facts again. I knew that the resonators had to be very small. To make the resonators small and get the energy into them meant that they were overheated. Making the things very small while getting a lot of energy into them, that was the basic problem. We were all thinking of getting the energy from electron beams or something like this, but how can you get the energy in? They had to be small, very precise devises. I finally decided that as we get down to very small wavelengths, we just wouldn't be able to make these resonators. It dawned on me that we had to use some naturally occurring resonators and put those into the molecules. There is the ferromagnetic resonance in solids and other resonances in molecules and solids. However, I knew the usual argument that a collection of molecules would absorb more energy from a source than it would emit. I also knew, however, that it is possible to excite a molecule so that it would emit photons at the same frequency as the frequency of the stimulus. This process is the inverse of the absorption of radiation by a molecule in a lower energy state. The process cannot be a net gainer in thermal equilibrium according to the Second Law of thermodynamics.

I was probably sitting in the park for 45 minutes thinking about this when the revelation came. We need not be in thermal equilibrium! If the collection of molecules would consist of excited molecules, there would be no limit to the energy obtainable from such a collection and the greater the density of excited molecules, the more photons would the radiation wave going through them pick up and the stronger it would get. This would be stimulated emission. I even made some rough calculations in the park to estimate the number of molecules needed to maintain a self-sustained oscillation. Because of my experience with molecular beams at Columbia, it was obvious that the process should be accomplished in a molecular beam. You select a beam of molecules consisting of molecules in an upper state and send them into a cavity and stimulate them by radiation in the cavity. I knew the approximate intensity you could get from molecular beams and I saw that it would be possible though it might be marginal. It would be a more intense molecular beam than almost anybody had. It looked interesting and it might just work.

In any case at Columbia I was in the right place for trying out something like that. It was the big center of beam work at that time. It was started by Rabi, then Kusch, and I knew many of the students working in that field. Willis Lamb worked on theory and Norman Ramsey had been there and I knew his work very well. I was thoroughly familiar with the field. In addition, I had previously thought about stimulated emission and the possibility of doing some experiments with it, simply to show that stimulated emission occurred. Other people had too, in particular, John Trischka, who was a young postdoctoral man at Columbia. He thought about it and then talked with me about it, but thought it was too hard. I didn't personally believe in the feasibility of such an experiment and I'm not sure whether I told him this. I didn't much believe that there was a point in making such an experiment just to demonstrate it. In most of our work we had molecules in the upper and in the lower levels at the same time. The fact that absorption was as small as it was, was associated with the fact that there was stimulated emission from the upper state as well as stimulated absorption from the lower state. They almost cancelled. They cancelled to the precision of $h\nu/kT$, and this was maybe one part in a hundred or something like that. Any radio-absorption included such a demonstration of stimulated emission, but we knew this and I didn't see any point in doing an experiment just to confirm this. This was one thing that turned me away from making a demonstration experiment.

The whole field was fairly natural to me. I had thought about most of these ideas before, but I just never put them together. To get some intense radiation, to get an oscillator, the use of naturally occurring resonators and amplifiers was the right thing to do. I knew the field well enough to calculate the parameters such a system would have to have to make such an experiment possible.

We had our meeting and it was so tentative to my mind that it made me think about it some more. While I was generally quite ready to mention any new ideas I had to anybody, and I had brought up a number of new ideas already to the committee, I didn't mention this new idea at that time. I did talk with Art Schawlow, just casually, but immediately. A couple of weeks later I wrote it down in my notebook to record it as an idea. I had some experience at Bell Laboratories with patents and I knew I better have a record of it, and I had it witnessed by Art. That was in the spring of 1951. I wanted to do it, but I knew that it was chancy.

I decided to wait until I had a new student who wanted to do something new as a thesis. The student had to be strong to take a chance on trying this out. By the next Fall (I don't know the exact date), Jim Gordon, a very good student, joined me. He had done his undergraduate work at MIT with molecular beams. I explained the situation to him, I told him that I thought he could do some spectroscopy with this, at least, even if we didn't get it to oscillate, and that it would be interesting. He was interested in taking a chance on that. Then I hired a postdoc, Herb Zeiger, who had just taken his degree with I. I. Rabi and was also in the molecular beams field. I hired Herb with the grant which Carbide and Carbon had very generously given me, and I could hire a postdoctoral person every year to help me. Initially I had another student, George Dousmonis, a very young student, just starting to get interested in the general field of microwaves. For exercise, I let him do some calculations on this experiment.

There was an interesting coincidence, which is worth mentioning. Shortly before this idea occurred to me, I ran into a German physicist, Wolfgang Paul from Bonn. He had come over, and had just completed some experiments producing very intense molecular beams with a quadrupole focusing. That was a new way of doing molecular beams. The usual way was dipolar focusing. Paul's new approach allowed him to focus one state and defocus another state. Obviously that increased the intensity and I knew roughly what intensities he was getting. That too was important in stimulating me to believe that it was possible to do what I was thinking of doing. Otherwise, with the intensities which molecular beams normally achieved, it wouldn't have come close. With this technique, it appeared to me that one could get very close; it made it seem more likely that it would work.

So we started out with Paul's technique. My original idea was to try to get to very short wavelengths, at the millimeter range. I chose ammonia as the molecule; the first rotational state of ammonia is at about half a millimeter. This was the way I wrote it up in my original write-up and we did the initial calculations on this. However, I decided pretty soon that while this was the obvious way to go to try to get to short wavelengths, on the other hand it was a very hard experiment. We would do better to step back to a region where we had all our techniques already, we had the wave-guides and oscillators already, and it was the one-centimeter region. For this, we could use the ammonia inversion around 1 cm wavelength. This is why the first maser was built around 1 cm and not ½ mm. I didn't want to be too hard on the student to make him do an impossible job, and this seemed to be the right stepping stone, to start with wavelengths that we knew. Then we would move to shorter wavelengths. And, of course, it marked and produced the oscillator we named the maser, for *m*icrowave *a*mplification by *s*timulated *e*mission of *r*adiation.

I must say that when it worked, there were so many interesting things to do with it that for a long time I didn't get around to pushing very hard towards experiments with the short wavelengths. The experiment planned didn't give much power and I knew that we wouldn't get much power, it gave 10^{-8} watts at best. But 10^{-8} watts is a lot of power in spectroscopy. We recognized that it could help us make a very precise clock and we also recognized early in the game that it would provide us with an almost ideal amplifier. It was an essentially noise-free amplifier; I worked out its theory, and it was the most perfect amplifier one could get. So the amplification, the clock, and the spectroscopy made it quite interesting and it occupied me for a while.

Right at that period, in about 1955–1956, I was going on sabbatical. I was just finishing up a book and it was a turning point, so I asked the question, "What should I be doing next?" On my sabbatical, we spent 15 months away, 2 summers and the 9 months of the academic year. We went to Europe and then on to Japan. I taught in Paris and in Tokyo, and did a certain amount of traveling around. I wasn't going to decide what I wanted to do, I only wanted to explore a certain variety of things and figure out by the end of my sabbatical what I was going to do next. In Paris, I found one of my former students there, working in the same laboratory where I was going to work. He had just proven that the relaxation time for electron spins in semiconductors could be very long. He had a spin resonance, which was both sharp and had a very long relaxation time. If one could get an electron spin in an excited state, and have it stay there for a long time, obviously you could invert the population, then you would have an amplifier. I realized immediately that that was the right kind of thing and we should look for an amplifier, because it would be tunable. I worked on that fairly intensively for three months, and then I had to leave Paris.

Prior to that I played around with various ideas for measuring relativistic effects in radioastronomy. I have always been somewhat interested in astronomy. I did various things in astronomy. During that period I gave a talk at the International Astronomical Union. I was invited to give a talk on what microwave resonances might be looked for in astronomy. And I gave a talk about various molecular resonances that might be found. So I worked in that field, but it was only after I came to Berkeley in 1967 that I started an experiment search for that. At the time I didn't know if I would do something like that or something else in astronomy.

Well, I got interested in this amplifier possibility. At that time, I was supposed to leave for Tokyo, so I went to Tokyo and taught there. I was trying to work out just exactly how much noise there would be in a quantum amplifier of this type and whether it was possible to detect quantum fluctuations this way. I also did some other things in Tokyo, but I got caught up in that and talked to some of my Japanese colleagues, Shimoda and Takahashi in particular. Takahashi was more or less an applied mathematician. He was very helpful in working out a technique for handling the equations in the fluctuations, because of this skill in applied mathematics, but amusingly I first got onto handling the noise as I saw it in the amplifier by talking to a biologist friend.

I have sometimes talked to my biologist friends about this and they were rather pleased that a biological theory could contribute to the theory of physics. There had been a theory worked out by C. A. Coulson for population growth and population fluctuation. If you had bacterium with a certain probability to divide and multiply and a certain probability to be killed off, then the question is the fluctuation in population. He had worked this out and you can see the analogy now between the two states of a certain probability of generation of a photon, and a certain probability of the absorption of a photon — so what is the fluctuation in the number of photons? That was basically the noise problem. So I used some of Coulson's theory in this and Dr. Takahashi was very helpful in getting the equation solved in a very sophisticated way so we jointly wrote a paper on noise fluctuation, covering the basic situation and by the time I got back to the U.S. I was fairly well rapped up in finishing that up. I pretty well decided that really I ought to work further on masers. That was probably the most fruitful thing. It was not microwave spectroscopy *per se*, but masers in general were what I ought to do.

When the maser first worked, I realized that this was a brand new kind of device and it should have a name, but what should we call it? My first instinct was to derive a Latin or Greek word that would describe it — so I thought, how do you say stimulated emission and how do you say amplifier in Latin and Greek? I had a Greek student who helped me out with Greek and I knew Latin, but I couldn't find a word of any reasonable length that could be descriptive. I mentioned to my students one day that I should really have to switch over to some acronym — just an initial to something. It occurred to me that it is microwave amplification by stimulated emission radiation, so lets call it maser. It turned out to be a reasonable choice although one of my friends, Ed McMillan said: "I don't like that name, because a word with '-er' means it does something, it has to 'mase', and there is no word 'mase', so it is not an appropriate term." Etymologically he was, of course, right.

I was reminded by one of my students recently that he remembers how the word laser came up. Everybody began joking that this is microwave amplification, but there are other types of amplification — there is infrared that would be iraser; light, that would be laser; and X-ray — so you would have all kinds of words generated.

I often look back at the maser and the laser and think, why didn't someone invent this long ago? There was no component of the whole scheme that was really new. The resonator was known, the stimulated emission was known, spectroscopy was well known. If you look back you will find papers on all of these things. I think one of the most interesting early papers was in 1924, a paper by R. C. Tolman talked very clearly about the interaction of radiation. He talked about negative absorptions, saying that if you have more molecules in the upper state, you get negative absorption. It was a theoretical paper, it wasn't very clear, nor was it well understood at that time. Even some physicists didn't understand it, electrical engineers didn't know much about it because they generally were not trained in quantum mechanics.

Many of my electrical engineer friends were surprised that molecules could give off radiation coherently — they just haven't run into that. Nevertheless it was not unknown in the physics world, and many people had done theoretical work with it, so it was nothing new. The only thing that was new was just putting it all together to do this to generate or amplify radiation in a useful way. The one thing I never found anywhere else was the idea of using feedback with it. The feedback that enhanced the effect was not realized, I think, because stimulated emission was thought of by physicists and physicists were not that acquainted with electronic oscillators in the early days. Feedback was an electrical engineering idea and well known to electrical engineers, but not so common to the thinking of physicists.

The other thing that many people didn't realize was coherence. Coherence was known theoretically by physicists, worked out in a number of cases, but most physicists who knew about stimulated emission didn't realize that it was coherent. That is another thing that was missed in the early days.

Another thing that I had many arguments about was the coherence in an oscillator. Many physicists were stuck on the idea that you could not measure the frequency of the energy of a molecule more accurately than the uncertainty principle allowed. They said that if the molecule took a certain time to go through this cavity, then you only had a certain length of time to measure it. Hence, they argued, the frequency could not be determined better than $1/\Delta t$, where Δt was the time. That is a fundamental of the uncertainty principle and I had many arguments about that with some very distinguished physicists and one of them insisted that it couldn't do that.

In fact it was a collective thing. One is not measuring the energy of a single molecule, rather, the energy of lots of molecules and averaging it. Again to an electrical engineer it was probably more obvious, because if you take the feedback oscillator, even with a fairly broad resonance circuit, you get a pure, clean frequency generally. Even though the circuit is broad, when you put in feedback, it oscillates right on the peak. In a sense, you are measuring the center of frequency of that resonator very, very accurately. Basically it is the same thing with molecules.

No one ever worried about the fact that electrons spend only a short length of time in a triode amplifier, they knew the oscillators were pure. For an electrical engineer that was fairly natural, but for a physicist it wasn't, and I had many arguments about that with physicists. I remember talking to Niels Bohr and immediately he asked that question. I think I convinced him and he was at least kind enough to say that he thought I was right. Von Neumann, the Hungarian mathematician, at first didn't believe anything could be that narrow. Even after we had done the experiment he said that there was something fishy there. But then I happened to see him at a party, at a social event and he came back about 15 minutes later and he said, "I understand now, it can be right." The fact that it took him so long to understand is impressive. On the other hand, to some physicists, to I. I. Rabi, to W. Lamb, who are accustomed to dealing with molecular beams and with interaction with radiation, it was much more natural — they understood it immediately.

Those are some of the ideas that were missed. The final idea, with respect to the laser is that the laser oscillation frequency is so far removed from any frequency we had for oscillation before that most people just weren't thinking in those directions. In addition, the critical thing there was to find a way of isolating pure resonances in a multi-mode system. That is where the Fabry–Perot type resonator was very critical, and allowed the laser to work well.

I was very occupied exploiting the maser for some time. I wanted to build an amplifier to do radio-astronomy. I have linked these two things that I was somewhat interested in, the maser and radio-astronomy. Joe Giordmaine and Lee Alsop, my two graduate students, and I set out to build a good amplifier for radio-astronomy. When I returned back to Columbia from my sabbatical we first started on a rather complex material with a resonance in it, but very soon switched over to ruby, after we found out that ruby had very good resonances. By then amplifiers were very actively pursued by a number of groups. While I was on sabbatical in Paris, we had started to work on electron spin resonances in paramagnetic solid materials. We published a short paper about it. We were not able to make a system amplify very much, if at all. I only had three months there. I had come back to the U.S. during that period in Paris and made a point of talking to Bell Labs' people. I told them this idea of using spins in solid state, saying that I'll have three month there, I can try it, but I don't know if it will get done.

In the meantime, W. Strandberg at MIT had the idea of using electron spins, a device which he called a versitron. He gave a talk at MIT about it and Nico Bloembergen went to hear that talk. Bloembergen after that recognized that there is a better way of doing it using three levels unequally spaced in a paramagnetic material. He was very familiar with paramagnetic materials; it was part of his field. He realized that this would be a much neater way. He also had a copy of our paper, the work we had done in Paris. I had given some talks on it and published it. But he first realized the importance of paramagnetic materials from hearing the talk of Strandberg.

Bloembergen got onto the idea and told a Bell Labs' friend, H. E. D. Scovil: "I think I know a much better way of doing it and I am going to try to do it. It is much better that what has been suggested before." Scovil, I think, just overnight thought about it and said, what could Nico be thinking of — he must be wanting to use a three-level system. Sure enough he reinvented it, but he was decent enough to recognize that Bloembergen has been there first, so they got together. Scovil and Bell Laboratories built the first amplifier on that principle. Interaction between scientists is a very important part of scientific growth. Sharing of ideas, building on other people's ideas. So that produced the first demonstration at Bell Laboratories.

My students and I eventually swung over to using ruby and built a system, and I collaborated with people down at the Naval Research Laboratory, which was one of the foremost groups in radio-astronomy. We put the system on their antenna. C. H. Mayer in particular worked with us. We demonstrated that we got at least an order of magnitude improvement of signals from radio-astronomical sources. We were able to measure the temperature difference between the front and backside of the planet Venus and measure the temperature itself, which was something of a puzzle at that time. We did some astronomy and that was going well. I was continuing to develop what I think was the first amplifier to be used in astronomy or anything useful. But around that time I got caught up in other things, including the laser.

I think that most people recognized that the maser technique might be pushed to shorter wavelengths. Of course that has been my original idea, and I wasn't hesitant to talk about it. My original proposal was to get down to ½ mm, but it didn't seem to me to be worth a lot of work to do it, because it would simply be a demonstration. Maybe the paramagnetic resonances could be pushed on down with high enough magnetic fields and that perhaps would work. Basically I didn't feel that I had really the best idea at this point. I was waiting for an idea that looked more attractive, instead of just expanding what we already knew.

Of course, other people were involved, too. Bill Otting, who was in the Air Force Science Research and Development group came to me at one point and said that we really should push this down to the infrared, and they would like to sponsor that, would I be interested in doing something along these lines? I said that yes, I would be interested, but I am waiting to figure out the best way to do it, the right idea, and I don't think I want to do anything now. He then asked if I would be willing to write a paper about it, to encourage other people. I said that I am pretty busy right now, I don't think I have any great ideas to put down on paper, but obviously one can talk about it and show some ways of doing that. So I suggested some other people, but I don't think Bill found anybody to do it in the end.

Around the late summer of 1957, I felt that we really should think seriously about how to get down into the shorter wavelength range. I hadn't had any ideas that I felt were particularly good, but obviously it could be done. So how to do it? I should sit down and think about it rather than just go along with what I was doing. So, I simply sat down in my office to think about what to do and I started sketching out various ways of doing it. I realized that if you have a gas, the Doppler effect is such that the frequency shift is proportional to the frequency. I almost immediately recognized that the laws of radiation are such that it is as easy to get down to very short wavelengths in the optical region as it is to get down to submillimeter range. I was first interested in the submillimeter region. But the optical region in principle is just as easy, and we know the techniques in the optical region better. We know the resonances, everything is developed there, so why not just jump suddenly into the optical region or at least into the short infrared. So I began looking at that.

The problem was to get a resonator, which was sufficiently selective in modes. You could have a resonator that had a large wavelength. Also, the molecular container had to be fairly big so it contained many molecules or atoms, but you had to selectively pick out a mode if you wanted a clean oscillator. Now, I didn't initially find a very good one. I used a cavity with big holes in it, and put the energy in. It would damp out some of the modes and would probably oscillate in single mode momentarily. Or, if it is very stationary it will continue to oscillate in a single mode, but otherwise it is going to jump around between different modes. But that is not so uninteresting if it jumps around between modes, and still stays on the spectral lines and generates some energy. It may be a useful system, but it would be nice if we could control the modes. I was a consultant for Bell Telephone Laboratories at that time. Sid Millman, who got me into this, was very interested in the field, in radiation in general. He was a molecular beam man, and he convinced me to consult for Bell Labs on the basis of talking with him from time to time over eight months and going to Bell Labs occasionally. I could do the work that interested me right at Columbia, if I wished, consult with Bell Labs, and do something that interested Bell Labs.

So, I was in that mode of consulting with Bell Labs and I went out to talk with Art Schawlow. He was a postdoc who worked with me and had married my sister by then. Art said that he has been thinking about this question, too, and let's keep talking about it. It was during one of these conversations when Art suggested using a Fabry–Perot. It would get rid of most of the lateral modes of radiation. That was the real key, I think, to making a good laser. So we pooled our ideas. I was a consultant at Bell Labs. Most of my previous work was done back at Columbia, but since Art worked at Bell Labs, and since he was interested, I said, let's call this Bell Labs' work. We worked together, and I worked out first a theoretical demonstration that we could not only get rid of the lateral modes by picking the right geometry, we could also single out a longitudinal mode as well, by picking the right spacing. So there was a possibility of getting a clean single mode. We had to pick out the right gases and right materials, and we looked at solids, too, and got quite a variety of things together working out how the system should behave and how to build a system.

Certainly we both felt that such a system can be built. I was much surer of that than of the initial maser working. Nevertheless, I felt that it had to be very carefully planned. Otherwise it was a puzzle why someone hadn't run into one accidentally. People had been working with optics for a long time, and with discharges and excitation — if it is easy, somebody should have run into one, and obviously that hasn't happened. So we had to plan to make every step just right, to make sure the physics is right, and everything was under control. That is why we worked with alkaline atoms that were well understood, we knew everything that could be done. We calculated everything, and since all the variables were known, we could fix up conditions that were just right, and it had to work then. We recognized that solids might be good, we had solid-state resonances and some other resonances that might work, but it had to be planned very carefully, so that clearly it would work according to theory. So I started working on a system like that and I got a student, who was interested, and a younger but experienced physicist who came over from England also joined us after a while. He worked towards building a laser.

I generally worked on a fairly leisurely scale, on the graduate student scale of time — all the work I have done were connected with graduate student theses. We started in 1958, about the time Art Schawlow and I finished this paper, a few months after that the graduate student began working with me. The following late summer I was approached to go down to Washington. Well, I wasn't that eager to go to Washington, but I always had kind of a conscience to help out in national and public affairs. At that time I felt that there were not enough scientists in Washington. There needed to be more, the government needed more technical help. I think that was the general feeling in the scientific community. It would be nice if there was more scientific input in the government. In 1959, the missle gap story was talked about a great deal, Eisenhower had come in and the Sputnik was launched shortly before that. So there was an immediate push to get more scientific input in the government. I was urged to go down to be the Vice President of Research of the Institute of Defense Analysis, which was run by a group of universities.

I just felt I should do it, so in September of 1959 I moved down to Washington, which of course meant a hiatus in our efforts. But I came back on Saturdays to try to help my students; I had about 10 of them. One was working on the laser. That work continued, but we never quite got there on time. The first laser was of course by T. Maiman. He approached it in a different way, theory was not something he worried about very heavily. He just splashed a lot of energy on the ruby. It worked, so that was very exciting. The initial reports were not very clear, whether it was really lasing or not, but I talked with his people enough to be convinced that it was really working. Art Schawlow set one up and got spots on the wall, showing the beam was directional, so it was really a laser. That was the red ruby laser.

Many laboratories were skeptical about the laser, about picking out one single mode, or about the Fabry–Perot as a resonator. In fact, I rewrote part of that paper, because some people at Bell Laboratories doubted it. I tried to make it mathematically a little more complete in that respect. Sid Millman, who was very interested in the laser, had hired Ali Javan, who had an idea how to do this by collisional excitation and using helium/neon, which was a beautiful idea. Two other people worked with him to help him do it experimentally, Bill Bennett and Don Herriott, and of course Ali's helium/neon laser came along pretty soon after Maiman's system. It was a completely different kind of a system. Then another one of my students, Mirek Stevenson, who had gone to IBM, and Peter Sorokin, also at IBM, actually made the second and the third laser, which is not well known. They made a couple of other types of crystals lase, shortly after Maiman. I think they were uranium-doped crystals. Then Ali's laser came along pretty quickly, so by then lasers were coming on fairly rapidly. I was very interested, but also I was down in Washington doing this other job primarily.

The general interest in the field became terribly intense. When we first built the maser, we had lots of time; there was no competition. We showed everybody around the lab, told them what we were doing, people were mildly interested, by and large they did not recognize its significance. One of my friends, after we made the maser, said congratulations, I am glad you got it oscillating, but now I guess you can go back to some of the other things you were doing that are much more important. Some people did recognize its significance, e.g., Professor Feynman — he kept talking about its significance, and what a breakthrough it was.

By the time maser amplifiers came along, people were starting to get increasingly interested. There was more industrial interest — industrial companies did some very good work on them. Then, when Art Schawlow and I wrote this paper, showing the practical way of making a laser, interest became intense and it began to be almost a race to build the first laser. Then Maiman's laser came out, and soon lasers started popping up in a number of places, so it became a very intense activity, especially in industry. Initially it primarily came out of the academic world, but once people realized that this was something interesting, industry jumped in and contributed a great deal. You noticed that the lasers that were actually built started in the industry and not in academia — the ruby, the helium/neon, the carbon dioxide laser. I think only the chemical laser was started in academia, but other than that, industrial companies put a lot of money and a lot of talent into developing lasers. I didn't have to ever make a laser, unless I wanted to, because you could buy them and they were so good.

Lasers are very interesting and you do get a very high angular resolution of the beam, as well as great monochromaticity. The fact that we can handle light the way we used to handle electronics in the past has made it immensely flexible and it is a very important scientific and industrial tool. One of the fields that impressed me most that didn't occur to me initially as very important is civil engineering, where laying down a straight line became immensely popular. Right now one of the primary sources for the sale of lasers is to lay down a straight line and for farmers to plow their fields straight or to flatten out a rice field. Another area is heavy metal processing to cut or harden heavy metal pieces — the automobile industry uses it to harden surfaces. The laser provides a very refined high technology, capable of doing some very heavy work.

* * * * *

Almost 20 years after the Larson interview, on February 20, 2004, we recorded a conversation with Professor Townes in his office at Berkeley.

It's an obvious question to you: did you foresee the wealth of applications of the laser that we witness today when you began working with it?

Of course, I could not foresee all of those applications, but I foresaw some of them. For example, the Bell Telephone Laboratories thought that it was not going to be useful for communication, but I immediately

Charles Townes in his office at the blackboard, 2004 (photo by I. Hargittai).

recognized that it could be very useful for communication. I also recognized that it was a marriage between electronics and optics, so it was bound to have a good many applications. I was very interested in scientific applications and I could see a lot of scientific applications. Others I could not foresee. For example, one of the first medical applications was for detached retina. I even wrote a paper with a doctor who had approached me wondering whether the laser could have medical applications. So we wrote a paper together about it, but there is no mention of detached retina because I'd never heard of detached retina. I didn't know that such a thing existed.

Was the doctor an ophthalmologist?

No, he was just a general practitioner. We talked about some possible medical applications and we wrote about them, but we missed others. I probably foresaw more applications than anybody else. One of my friends even said that the laser was a great idea, but what good was it going to do us? When there is a new field there may be a lot of new applications that you can't foresee. The field has grown and the possibilities have changed in time.

How do you feel about the laser today?

Of course, I am very pleased that it has been so useful. It has made a great contribution to human life and will continue to do so. It has made a very big contribution to science. It has been a very useful scientific tool. You might even compare it to the screwdriver. The screwdriver is a useful tool that allows people to do things that they wouldn't be able to do without it. The laser as a tool has been crucial for about a dozen additional Nobel Prizes. It made it possible, for example, to achieve very low temperatures, much lower than ever before, down to a millionth of a degree above absolute zero. My primary aim was to contribute to science. I wanted to have a scientific tool that would enable us to make very accurate measurements at different wavelengths. Also, I did recognize normal industrial applications, such as cutting and welding, and in communication, which is perhaps the biggest industrial use at the moment.

Did you get a lot of citations to your papers?

I have never tried to count them. I have never bothered to check that. I have no idea.

I wondered about the Russians' contributions. Technologically Russia, that is, the Soviet Union, was not very much in the forefront, except in some fields. Was it a surprise to you that you shared the Nobel Prize with two Soviet scientists?

I was aware of their work and I knew them personally very well. They came over here to see my work. We had meetings together. One could argue whether they should've been given the Nobel Prize and maybe somebody else should've been given the Nobel Prize. For example, maybe Schawlow should have received it together with me. But I don't think it was all that unreasonable and the Nobel Committee is eager to see that the Nobel Prizes are spread out. That's fair enough. The Russians characteristically have been good in theory but not very good in experiment because they have been technologically somewhat behind us. But they had very good theorists. They can generate ideas and excellent theory. Basov and Prokhorov did have some good ideas. It's hard for me to know exactly how independent those were of what I was doing. It's difficult for me to know how much they knew of what I was doing. Some of our work was publicly available. Nevertheless, they certainly did some things. One thing that Prokhorov did, for example, was to suggest two parallel plates as a resonator. He did not develop it and he didn't quantify it, but he wrote a paper on trying to get into the far infrared using parallel plates. That was original. They certainly wrote about the idea of making a maser and I don't know how independent it was. So far as I know it was independent. But it was certainly after our work had been done and after our reports had appeared publicly.

Do you think that they could follow western literature closely?

I think they were pretty aware of what was going on. Our report had come out. It was not a normal scientific publication, but it was in some libraries in the United States. I don't know whether they had access to that. They were trying to build a maser and I heard them speak about it. The first time was at a meeting in England and interestingly, the British asked me to come over and talk about microwave spectroscopy. At that point I had just finished building a maser and made it work. So I wrote them and suggested that I might talk about this new oscillator, but they said, no, they weren't interested, they wanted me to talk about microwave spectroscopy. So that is what I did.

The British organizers also invited Basov and Prokhorov to come and speak, but typically for the Russians, they didn't tell anybody what they would speak about. They came over and, lo and behold, they spoke about trying to build something like what we had just done. After they talked about how they were trying to build it, I raised my hand, got up, and said that we had one already working. My impression was that they knew about our work, somewhat unofficially. I had a good time talking with them on the street. We walked along the streets in Cambridge and they talked much more freely on the street than they could in a public meeting. At that time the Russians were rather restricted. They were trying to use a molecular beam as I was, but they were focusing it with two plates rather than with four rods as I was doing. I explained to them that four rods gave much more intensity and that they should try that. So they went back and tried that and they got one working. After that, they used eight rods and that was still better, so they improved it a little bit. They were doing good work. The exact originality is something one could debate about, but there is nothing that one can prove clearly.

Did you visit their laboratory?

Yes, I visited their laboratory after I got the Nobel Prize. They wanted me to come to Russia and I agreed to go provided that I could see scientific laboratories. They arranged that and the next year, in 1965, I went over and they very generously showed me around. I went to many places where Americans had not been. There were still some places where they wouldn't allow me to go. There were cities where I could not go. But I had a good chance to look around.

Was it your impression that having better technological background and more money they might have achieved more?

At that time I felt that they had adequate equipment and adequate support. Of course, they were favored as compared with other places in Russia.

Microwave spectroscopy seems to have been falling out of favor in the laboratory, especially in chemical laboratories as a tool of molecular structure determination.

It has been falling out of favor except in astronomy. That's because we have been finding a lot of molecules in astronomical clouds. We do this by radio-spectroscopy or microwave spectroscopy. There has been some substantial work in our laboratory to confirm the frequencies of these molecules and find new molecules that might be in astronomical clouds. That has been a fairly active field. But just the general measurement of molecules by microwave spectroscopy, that has become much less popular.

You mentioned Schawlow, your brother-in-law, who might have been selected to share the Nobel Prize with you in 1964. He then received his prize in 1981. He died in 1999. Would you care to tell us something about him on the personal side?

He came to me as a young postdoc when I was at Columbia University. My wife introduced him to my younger sister who was then studying music in New York. They liked each other and soon became married. We were very good friends and he and I worked together on microwave spectroscopy. I would have tried to hire him as a professor at Columbia, but I was chairman of the department so I could not have hired my brother-in-law; it wouldn't seem appropriate. He went to Bell Labs and we continued to collaborate. We did a lot of work in the lab together before he went to Bell Labs. He never worked on the maser; he wasn't particularly interested in it, so my maser work was completely independent of Schawlow. Then three years later when I saw how to make an optical system, named the laser, I was consulting at Bell Labs. They asked me just to walk around and talk to people, and at that point Schawlow became very interested in the possibility of a laser and we decided to work on it together. His primary contribution there was a parallel plate system for resonator. I was troubled about the resonator, I didn't feel that I had a very good resonator, and he came up with the parallel plate solution. That was the right answer and we wrote a paper together. Another person who might have been involved in the 1964 prize was Bloembergen. He had invented the three-level maser and he had done some nice work on that. That would've been another possibility.

He then shared the Nobel Prize with Schawlow in 1981.

They were recognized for some other work in laser spectroscopy and non-linear optics with laser beams.

This year it will be 40 years of your Nobel Prize. Did you get used to being a Nobel laureate?

It gets worse and worse. You get called on for more and more things.

Isn't it fading away?

No. People want me to do things, things that ought to be done, but it takes time. It's a kind of a public service to show up and make a talk or to write something. There are all kinds of different things. Many people like to do such things and I also feel that I have to do them too. On the other hand, it's a distraction from my research.

I would like to ask you about the priority question in connection with the discovery of the laser. It seems to me complicated and I may have misunderstood the story. The Nobel laureates are not the ones that collect the money after related patents. Then, I just saw a book published by Nature *about great discoveries in the 20th century that were originally published in* Nature. *You wrote one of the entries about a pioneering paper on laser, but that pioneering paper was by someone else, Maiman.*

It was actually Maiman who made the first laser work.

After you had discovered it?

That's right. We wrote about how to do it and he built the first one. He made the first one work. *Nature* asked me to write about it and I did, because Maiman published his paper in *Nature*. The basic idea for maser and laser came out of university work, that is, from basic research. Everybody having to do with the early work on masers and lasers were at universities. After Schawlow and I wrote this paper about how the laser might be built, a lot of people tried to build it. I was asked to go down to Washington at that time to do some advising so I could not concentrate on building a laser. Furthermore, universities can't concentrate on a project as much as industry does. Once industry gets interested, they can put a lot of manpower into it. So all the first lasers were built in industry. Maiman worked at the Hughes Company and he built the first one; he had studied at a university under Professor Lamb in radio-spectroscopy; the second one was built at IBM, by one of my students and one of Bloembergen's students together. These were young people who had come from universities to industry and knew the field. The third laser, which was the important Helium-Neon laser, was built by Ali Javan, again one of my students at Bell Labs.

I didn't publish on the maser before one was completely working. When I had the idea, I wrote notes about it, I told people about it, and had a student working on it for a couple of years. People came to see it but nobody was interested in competing with us. They just didn't realize how important it was. I was quite open about it. Once we got it working, then everybody got excited, and it became very competitive.

When the idea of the laser came along, and I talked about it with Schawlow, we wrote a paper about it because we knew that it was a very hot field. Once we would say anything about it, people would try to do it and people even might try to publish a paper about it and beat us to publication. In addition it was being patented by Bell Labs. So we were very quiet about it for nine months while we were writing this paper. Nobody paid any attention to the possibility of building a laser at that time. People just didn't recognize that it was possible. We were very careful not to mention to anybody what we were working on and writing a paper about. This was confidential outside of Bell Labs because of the competition. We decided that we should write a theoretical paper first rather than trying to build one. Trying to build one would've involved others and substantial time, and people would have competed with us. I hadn't written a theoretical paper on the maser, I'd rather build one because it was a quiet field and nobody was interested in a competition. But for the laser, the situation had changed. After our paper had appeared, people immediately started trying to build a laser, and lots of different kinds were built.

Who wrote the patent for Bell Labs?

A Bell Labs lawyer, assisted by Schawlow and myself.

Would you care to comment on why someone else was collecting the money for the laser?

Sure. That's quite complex. There are additional things that one can add to a patent. Actually, I own the maser patent. That covered everything. That work was done at Columbia University, but they weren't interested in doing the patenting and let me do it myself; so, basically, they gave me the ownership. Patenting may be complex, it costs money and may be complicated to pursue. Columbia didn't want to be wrapped up in that. There is also the Research Corporation which collects patents from universities and then the income they put back into basic research. So I decided to give my maser patent to the Research Corporation with my getting 20% of the income and Jim Gordon who helped to build the maser getting 5%. The Research Corporation got 75% of the income. That patent was finished in 1959. Patents ran for 17 years at that time and then they ran out. I collected a moderate amount of money from the maser patent.

Then I patented the laser with Art Schawlow and that patent belongs to Bell Labs. Whereas my maser patent covered everything, the laser patent was specific about how to make a laser; in a way it was an improvement on the general maser patent. That was patented also fairly early, also in 1959, and it also ran for 17 years. Gordon Gould has made a good deal of money on a patent. He tried early to get a patent. It was turned down. Then he tried it again and he got some company that was willing to pursue it, but that also failed. Then he got another company which had a very good patent lawyer. They worked at it and they got a patent. By the time they got the patent, in 1979, it was 20 years after our patent, and it was also called an improvement patent. It was about certain specific things that were not in our original laser patent. These were minor changes but these minor changes were used in industry and that was what he collected on.

You might say that he had a laser patent but it was merely an improvement patent. Some of these minor changes were ridiculous, but his lawyers were very clever. For example, we patented an oscillator to make the laser beam. The oscillator amplifies the wave, the wave going back and forth, and getting amplified. Gould specifically patented amplification and we did not specifically patent amplification. Ordinarily, patent requirements are that anything that is obvious to a skilful person in the field can't be patented. I would've claimed that that was obvious, but his lawyers managed to put it over. There were a few other things that I've forgotten. There have been other improvement patents, for example, people have patented solid state lasers.

Would you have any general comment on patenting, about how scientists should approach patenting?

One thing is that it is useful to delay patenting if it's possible. If it's a new field, it takes time before it may get used.

Wouldn't it be advisable to come up gradually with innovations rather than in one stroke?

Yeah. I might have done that. I might have patented some additional ideas.

For patenting, it may not be the true pioneer who is in the most advantageous position.

That's true. It's frequently the case. If it's a brand new field, that is. If it's a going field, like automobiles, you patent minor things. Patenting is very tricky and a lot depends on the patent lawyers.

I would like to ask you about SDI, the Strategic Defense Initiative by President Reagan. As I understand, you were present in the White House, when it was announced.

I discussed this in my book *How the Laser Happened* [p. 166].

Did you have interactions with Edward Teller?

Yes, I have had substantial interactions with Teller. I believe he was sincere and well intentioned, but I disagreed strongly with him in a number of ways. He felt very strongly against communism and I believe took somewhat extreme positions partly because of this.

How should scientific advising operate?

I think it is important for scientists to be very objective in their advice to government and careful to not make public announcements on those subjects on which they are advising unless asked to do so by the advisees. This helps build trust. It is also important, but secondarily so, to be able to explain scientific reasoning and effects in simple ways and analogies so politicians and the public can understand as much as possible.

Is the impression correct that you were more involved with Republican than with Democratic administrations?

I don't believe that is correct. I first became involved under the Eisenhower administration, but then was active under the Kennedy and the Johnson administrations, then under the early Nixon period, after which there wasn't much science advice to the White House for a while. I became again involved under Reagan, but have not been much involved under Clinton or the Bushes, during which time there has been relatively little effective scientific contact with high government.

Quoting a book, I have shown you a letter Peter Kapitsa wrote to a Soviet leader, G. M. Malenkov, in 1950, in which Kapitsa brought up something that might be considered as a Soviet forerunner of SDI.[1]

His letter is striking. I can understand his desire and his search for a high-energy beam, but do not know just what he wanted to propose at that time — possibly high-energy particles.

Having read your book How the Laser Happened, *my impression was that you have had a very smooth life.*

What do you mean by smooth? Easy?

Easy may not be the right word, but things have gone invariably well for you. You didn't have to overcome adversities.

I was very fortunate with my parents and family, they were very helpful to me. My parents had a clear philosophy of life, they were religiously oriented, and I knew how to deal with my life. I could claim that I had many bad disappointments. Sure you run into problems in life.

What kind of problems? Some of my interviewees of your generation came from Jewish families, for example, and they had to overcome various hurdles.

Jewish people are very creative and part of the reason that they are creative is because they have to be different. They are different religiously and ethnically; people pick on them and they know how to stand up and be different. They accept that they are different. That's part of the reason that they have different ideas and new ideas, I think. My family told me that you do what you think is right. When I was youngish, I perceived how I may have to be different from many of my friends, for example in not drinking alcohol early and that sort of thing.

You said in your book that you didn't mind to be in the minority.

No, that's right. If I happen to be in the minority for doing what I think is right, that's fine, I accept that. In this, in being different, there may be something similar to being Jewish. But let me comment on some of the disasters, which turned out to be lucky for me. I went to a small university in South Carolina, which was not well known to the bigger universities.

So when I wanted to go to graduate school, I applied for scholarships or fellowships in many of the most famous universities in the East, like Princeton, Harvard, Cornell, and I got nothing. I got a scholarship at Duke University.

What may have been the reason?

My university just was not known. I had a good record, but that didn't mean anything to them. I couldn't get any strong recommendation from well-known people. So I went to Duke to get a Master's degree. But that was not an outstanding place for physics. When I graduated, I applied again to good universities and I got nothing.

Wasn't Duke well known at that time?

Duke was a little better known than my undergraduate university, but it was still not outstanding.

I first knew about Duke because of its excellence in microwave spectroscopy.

Gordy, a microwave spectroscopist, gave Duke a big push. Coming back to my story, again, I did not get anything. What I did was, I had saved five hundred dollars and I decided that I was going to go to the very best place there was and see how that worked out.

Paying your way.

Paying my way. I decided that Caltech was the best place at the time; Millikan was there and Oppenheimer was there, and Tolman, and others. So I went to Caltech with five hundred dollars; took a bus across the country; and after one semester Caltech gave me a teaching assistantship, so I was OK. But I had to go there and prove myself because they had no way of knowing whether I was any good or not. So I was very lucky to graduate from Caltech. It was a smallish place at that time and I loved the interactions with famous professors. It was very valuable to me. So my difficulty became a success.

How did it continue?

I wanted to be at a university. I wanted to do academic work, pure research. I graduated in 1939 and at that time the universities just weren't hiring

Charles Townes on a poster
in downtown Berkeley, 2004
(photograph by I. Hargittai).

anybody. I couldn't get a job. A Bell Labs guy came along to interview
people at Caltech. They hadn't hired people for a while and now they
wanted to hire people. So my professor said, why don't you go and talk
with them, so I talked with them, and they offered me a job. I wasn't
very interested in going to Bell Labs, that was industry. But my professor
said, you know, it's a job, you really ought to take it. I knew Bell Labs,
I knew that they did very good work and they had some famous people
there, so even though it was industry, I went to Bell Labs. But that was
a failure for me, a disappointment.

What did you do there?

I wanted to do basic physics and Bell Labs let me do basic physics for
about a year and then the war was coming along and I had to move to
radar engineering. Again, that was a terrible disappointment, but I learned
a lot. Much of my best ideas have come out of that experience. That's
the way life is and I have been very fortunate. I have been very lucky,
but I've had many failures that have become successes.

Charles H. Townes, 1970 (courtesy of the Lawrence Berkeley National Laboratory).

My impression was also that as you were visiting various laboratories in the U.S. and around the world, in Europe and Japan, you picked up ideas that other people may not have recognized as being as important as they were.

Well, having a lot of friends and talking with people is very important. Interchange between scientists is very important. Sometimes I discuss this as the sociology of science, the importance of interacting with people and picking up ideas. I talked with people in completely different fields and I found different things that could be very useful to me. It has worked very well for me and characteristically it's very useful in science.

Your science is not big science.

I don't like big science. I like to work myself and I like to work with my hands. I don't like to work with a big group of people or to use enormously big machines.

You think that small science still has possibilities today?

Oh, yes. My science still has possibilities. Much depends on the field. For high-energy physics, you need big machines, but that's one reason why I'm not working in high-energy physics. There are many things one can do in biology as small science. There are important things you can still do as small science in astronomy, building special instruments, for example.

Did you ever consider entering biology?

I have enjoyed biology. When I was a youngster, I used to do a lot of fieldwork, identifying insects and birds, animals and snakes and so on. My brother was two and a half years older than me and he was always so much better than me, and he liked biology too, and I sometimes say that was the reason that I felt that I shouldn't go into biology. I didn't want to compete with him. Then once I took physics, I liked physics. But I liked biology too and I was seriously considering going into biology. My brother went into biology and he did very well in biology. Physics at that time seemed to me more precise and quantitative than biology, which was largely descriptive. It has now become more quantitative. At this point, I would be very tempted to go into biology if I were not so busy with many other things.

Some other physicists have gone into biology.

Delbrück was one of the first ones I knew who did that. I knew Delbrück when he came to Caltech as a young postdoc. When physicists go into biology, they have to learn biology in order to do it deeply. If they do, they can do useful work. I have gone into astronomy, so I also changed fields from time to time.

Looking back on your career, can you single out some people who strongly impacted you?

My brother impacted me. We learned a lot of things together. We competed, but it was also stimulating. My father was also interested in scientific things, he could have been a good scientist; he was a lawyer instead. Then, my first physics professor at Furman University, Professor Cox was very good. He was not a research physicist, but he was very logical and careful and he set me up with physics.

Do you remember what turned you to science for the first time?

I always wanted to do science. The first thing turning me to science was wandering in the fields.

So it was not a book or a teacher?

No, no. It was the Universe. I liked the stars and I liked the trees. But my whole family was interested and open-minded.

I understand that you are religious. Do you separate your religion and your laboratory work?

I am religiously oriented and I think that many more scientists are religiously oriented than the public recognizes. They don't talk about it because when they talk about it, they get criticized, but more and more are becoming more open. The discussions about the interactions between religion and science become more public and I've given lectures on that subject. My own point of view is that religion and science are really very similar, much more similar than people recognize. Science involves assumptions or we can say faith. One of the basic assumptions in science is that this Universe is reliable and is controlled by fixed laws and these laws are reliable and we can trust them and so on. That's an assumption. We don't know it for certain and the whole thing might change tomorrow. We can't prove it wouldn't, but we have the faith that it doesn't. Ingrained in science is that the same is true every day. That's one extreme case of faith. Science and religion also both involve experiment, observations. We observe people, history, how society works. Take astronomy, for example. We don't play with the stars up there, we look at them, we watch them, which is observation. Religion observes, and from that we try to conclude what life is about. You make observations and you make conclusions and you try to use logic.

Much of the discussion these days involves the question of was there a Creator? Many scientists for a long time believed that the Universe never had a beginning and could not have a beginning. Einstein felt that and this is why he put in a cosmological constant because without this cosmological constant the stars would pull themselves together and the Universe would collapse. He put in the cosmological constant to keep the stars apart because it had to be always the same. Then Hubble found that the Universe was expanding and Einstein felt that he had made a mistake and threw away the cosmological constant. If something is expanding, it must have started from something smaller. Alternatively, some scientists have thought that new matter is being created all the time so instead of changing with expansion, it was always the same. Fred Hoyle pushed that a great deal and I talked with him about it. He gave lectures on this subject and I would point out to him that logically it could not be right. He admitted that it could not be right, but maintained that the Universe always had to be the same. Just had to be the same. That was his faith. Even after the Big Bang was discovered, he continued this interpretation, but finally, he had to give it up.

Everybody recognizes now that if the Big Bang was not the beginning, at least it was a unique moment in the past. In addition, scientists are becoming more and more convinced that this is a unique universe. There have to be very special physical laws that allow everything to come out just right and us to exist. It has to be very special. If you are not religious you have to assume that everything just happened by chance and there may be many other universes with their own physical laws. If you want to say that no, this wasn't planned in any way, then there may be billions of other universes and ours just happened to turn out this way. The assumption of there being many other universes is also a question of faith because it cannot be tested. Religion fits the observation that what we have is very special and has come out exactly or almost exactly as it should have. These are the kinds of discussions that are going on more and more.

My own assessment is that yes, there seems to have been a plan and I feel the presence of a God, or you might say a Spiritual Being if you don't want to call it God. There is something in the Universe beyond what science usually talks about. There is no reason why science shouldn't take this up and people could make tests. In fact, people do make tests, on the effectiveness of prayer, for example. We are making tests all the time because we are observing each other. We are seeing how people behave and trying to answer the question of what really makes a good life. That's my conclusion.

Don't you find it mind boggling the question if you believe, Where did the Creator come from?

It is mind boggling, of course. We can't visualize a beginning. In the beginning God created this Universe, but who created God? How did it begin? The beginning is always a problem. There are great problems in religion and there are great problems in science. People don't recognize how many uncertainties there are in science. For example, quantum mechanics and general relativity are not consistent with each other. We have known this for a long time but we believe both of them. It is a problem. In addition, the zero-point fluctuations, which quantum mechanics predicts, produce an enormous amount of mass, more mass than all the Universe or the energy of all the Universe, but it isn't here somehow. Yet we believe quantum mechanics although this is what it says. So there are a lot of inconsistencies within basic physics. Physicists are accustomed to these inconsistencies and we just kind of push them aside. We see inconsistencies in what we understand about religion too. My point of view is that we have to do the best we can and accept the inconsistencies, but let's make the most logical and sensible conclusion that we can as to what the realities are. That's what I try to do.

Do you belong to organized religion?

I go to church, it's a very liberal church, the First Congregational Church here in Berkeley. It's a Protestant church, it's rather open in its beliefs. Some of the things that people say I don't accept as being correct. On the other hand, the general atmosphere in the church, the general attitude of this church is good. I'm glad to be a part of it. Even though I don't believe everything that is said, there are a number of things in the Bible that can't be strictly correct, but as analogies or something like that, they are useful. I am not a fundamentalist and the fundamentalists make a mistake in believing every word in the Bible. The Bible in many cases provides similes and analogies which are very important, and how else can we express all those unimaginable spiritual things. People have to try to give examples and stories, but it doesn't mean that the stories are strictly correct. Humans have had the feeling for a long time that something spiritual is there.

Science has accelerated tremendously. I wonder if you would be willing to prognosticate? What is to be expected to come?

Changes. We are still in the middle of discoveries. This will continue and our point of view will change as well. For example, modern cosmology has changed our point of view very radically. Quantum mechanics has changed our point of view very radically. Science in the 19th century was deterministic, but quantum mechanics is not deterministic. Things are not predictable. This is a very radical change and science had to accept that. One must expect the same thing in religion, the idea of change in time. As I grew up, my ideas changed some, of course. I grew up as a Baptist. Our parents recognized that there were some things in the Bible that were not strictly correct. I talked with them about that and my parents did not mind my challenging some of the things. They were fairly liberal and they encouraged me to think openly. The Baptist Church itself tends to be somewhat conservative. When I was a boy, I was not a fundamentalist but I was closer to fundamentalism. Religion has been changing recently although the change is slower than in science. There have not recently been any radical new religious discoveries similar to what we have had in science.

I would like to ask you about your children.

We have four daughters. Our oldest daughter is in developmental psychology, the next one is a professor of neurophysiology, the third one is a professional musician and teacher, and the fourth one is a mechanical engineer interested in conserving energy. Religiously, my second daughter married a minister, my first daughter married a Jew and they go to a non-Jewish church, my third and fourth daughters are not so religious, they don't go to churches very frequently, but they respect religion.

When you received the Nobel Prize the prize money was rather low.

It was 63,000 dollars and I got half of it. The Russians got the other half. It was enough to pay our way there and back and I bought some gifts for my students to recognize them.

But that was not the most important thing about the Nobel Prize.

The prize was more important than the money. What I am most pleased about is the effectiveness of laser science and laser technology. It has been tremendously useful and many people feel that it makes me very important. You may be aware of the list of the most important people on which I am the 800th most important person in the last thousand years. Primarily I like to help society, I like to be useful. That's what life is about and that's my religious view too. I consider myself very fortunate.

Notes

1. The first two paragraphs from Peter Kapitsa's letter to G. M. Malenkov, on June 25, 1950 (quoted from Boag, J. W.; Rubinin, P. E.; Shoenberg, D., eds., *Kapitza in Cambridge and Moscow: Life and Letters of a Russian Physicist*. North-Holland, Amsterdam, 1990, p. 390):

    I am approaching you not just as one of the leaders of the Party but also because I have always greatly appreciated your interest in my work. I think that the significance of the question I am writing about justifies my giving you a detailed account.

During the war I was already thinking a lot about methods of defense against bombing raids behind the lines more effective than anti-aircraft fire or just crawling into bolt holes. Now that atomic bombs, jet aircraft and missiles have got into the arsenals, the question has assumed vastly greater importance. During the last four years I have devoted all my basic skills to the solution of this problem and I think I have now solved that part of the problem to which a scientist can contribute. The idea for the best possible method of protection is not new. It consists in creating a well-directed high-energy beam of such intensity that it would destroy practically instantaneously any object it struck. After two years work I have found a novel solution to this problem and, moreover, I have found that there are no fundamental obstacles in the way of realizing beams of the required intensity.

Arthur L. Schawlow, 1984 (during the conversation with Clarence and Jane Larson; photograph taken from the video recording).

7

ARTHUR L. SCHAWLOW

Arthur L. Schawlow (1921, Mount Vernon, New York – 1999, Stanford, California) was co-recipient of the Nobel Prize in Physics in 1981 together with Nicolas Bloembergen (b. 1920) "for their contribution to the development of laser spectroscopy". This was half of the 1981 physics prize. The other half was awarded to Kai M. Siegbahn (b. 1918) "for his contribution to the development of high-resolution electron spectroscopy". Schawlow won a scholarship in the Faculty of Arts of the University of Toronto, and he pursued his studies in physics. After war service, he continued his graduate studies at Toronto and then became a postdoctoral researcher at Columbia University under Charles Townes. After that, he worked as a physicist at Bell Telephone Laboratories between 1951 and 1961 and then as Professor of Physics at Stanford University until his retirement in 1991. He was a member of the National Academy of Sciences of the U.S.A., a fellow of the American Academy of Arts and Sciences, and other learned societies. He received the Arthur Schawlow medal of the Laser Institute of America (1982), the National Medal of Science (1991), and numerous other distinctions. Clarence and Jane Larson recorded a video interview with Arthur Schawlow on December 28, 1984, at Stanford University.* We are grateful to Charles Townes for having checked and corrected our transcripts.

I was born in Mount Vernon, New York, but my mother was from Canada. We moved to Canada when I was 3 years old and I grew up in Toronto.

---

*"Larson Tapes" (see Preface).

It was a nice place to live, a medium sized city, but a regional center, and it had all the cultural amenities that you could wish for. When I was 8 years old, I remember going downtown in the street carnival by myself and it was perfectly safe to do if you knew what you are doing. I went to school, like anybody else. I was rather clumsy with my hands and I think someone suggested to my mother that I should get a Mechano set, which is an English toy — the nearest thing in this country is known as the Erector set. It is a kit where you can screw together various strips and plates to make models of different things. There was also a Mechano magazine and I remember reading that for many years and learning about the achievements in engineering, bridge building, and all sorts of engineering advances and radio.

I was only 5 or 6 when I started playing with Mechano, but that was my earliest involvement with anything technical. My father was an insurance agent. He was born in Latvia, there were Germanic people living there in the Baltic States for centuries. He had a big family, some brothers went to the United States — he went to Germany to study electrical engineering at Darmstadt, but he arrived too late for the start of the term. So he went to New York to visit his brother and never returned to Europe, because of the very turbulent times. This happened around 1910. Later he met my mother in New York. I have a feeling that everybody meets in New York, because that is where I met my wife, too. He was strong in mathematics, but I don't think he would have been a good engineer, because he was not good with things. Anything that had to be fixed around the house, my mother did, using some very primitive way, but she would get it fixed. My father was very busy those days. As an insurance agent for Metropolitan Life, he dealt mainly with industrial insurance weekly premium, so every week he had to go and collect 25 cents or 10 cents on policies.

We lived in a working class district. People who lived around us were bus drivers or clerks. He had to work very hard and was often out in the evening — he had to go out when people were home. And I remember later helping him add up his accounts every week — long columns of figures. We didn't have an adding machine, so we would add the columns, we had to balance the collections, balance the changes in the books. At that time Metropolitan Life was the largest financial institution in the world. His position there was a fairly lowly one. He was a one-time assistant manager, but most of the time just an ordinary agent.

I went to school and did fairly well. I skipped a couple of classes of elementary school. Then I ran into a teacher who did not like me and said that I was stupid. So my parents consulted a psychologist — I had done very well up to that point — and he recommended that I move to the model school, which was attached to the teachers' college, the Toronto Normal Model School. I switched to that and I was several grades ahead of my age. The competition there was fairly demanding, so I never really felt that I was a genius, because I was up against people who could do comparatively well.

I was interested in lots of outside things, and let me mention radio again. I didn't have any money to do anything with it, but I was fascinated by radio. I remember that in the 1920s radio was a very exciting thing; in 1926, when I was about 5 years old we got our first radio, which was a battery-operated radio. We had to go to the hardware store and bring home an armful of batteries and it had one of those big horn loudspeakers. It was so exciting that the department store would have the broadcast of Santa Claus's ventures from the North Pole every night before he arrived at the store, before Christmas. All the kids in the block would come and listen to Santa Claus. Of course a couple of years later we got a batteryless radio. The newspapers had a column once a week on building radio sets; they got circuit diagrams and that sort of things. I did build a crystal radio set, somewhere in there, but it wasn't until the 1930s, when I was in my teens that I managed to scrape together enough money to build a short wave set with a tube.

I was a voracious reader and I used to go to the library in the summer with nothing else to do. I walked to the library to get several books then went back the next day to get several more. I read books on engineering; I read Alison Hawks's book on the pioneers of wireless, about DeForest and Fleming, and so on. I happen to think that DeForest's invention of the triode tube is the most important invention ever made, because it gave control at last in a way that we never had before and that led of course to all the computers and complicated circuitry. When transistors came along much later, people knew what to do with them, at least to begin with, because of the circuits developed for the tube. I wanted to be a radio amateur, but I was still an American citizen, so I could not get a license in Canada, but I had friends who could and I did manage to build a short wave radio set and listen to it.

I wanted to be an electrical radio engineer, but this was the deep dark Depression. We could not afford for me to go to a university if I had to pay the fees, which seem ridiculously small by modern standard, $125 a year and engineering was a little more. But that was only one of the two obstacles. The other obstacle was that I finished high school at the age of 16, but the Engineering School would not take anybody below the age of 17. So, I thought, a lot of people those days were taking a second year of the last grade of high school. We have 13 years of schooling in Ontario and they have a better chance of qualifying for a scholarship. Not knowing completely what to do, I thought I would try the scholarship exams for practice and to my surprise I found that I got a scholarship in mathematics and physics in 1937. Both my sister and I won scholarships to go to the University, she in English and I in Mathematics and Physics.

In those days we had the honors course system at the University of Toronto, somewhat modeled after what they had at the University of Chicago as you specialize right from the beginning. I chose Mathematics and Physics, then it would branch after two years and I could specialize in Mathematics or Physics or Physics and Chemistry or Astronomy. I of course switched to Physics, but it meant that your course work was pretty well prescribed and everybody in the class was pretty well qualified. It was very rigorous, it was hard work, but I did manage to come out close to the top in the first year and at the top of my class in the second and third years, so I was able to retain my scholarship. One of the amusing things was that it was very much specialized in mathematics and science; however, we had to take a few cultural courses and the University of Toronto was a federation of formally independent colleges. I enrolled in the Victorian College, which was affiliated with the United Church of Canada College. This was reasonable enough, because we attended the United Church of Canada, but it was also coincidental, because we knew nothing about the University and none of my family or friends have been there, so I just asked some of the teachers about the college. We actually took only one or two hours at the College and the rest was at the Physics Department or Mathematics Department.

I was sick and tired of writing essays on things when I had nothing to say, so I actually managed to get through four years of a good University without writing one single essay. Of course I wrote lab reports and since then I have written over 180 publications and have all of my life since then been writing publications and reports. I still feel very strongly that there are three rules for writing that I try to teach my students. First is the hardest one: have something to say; second: say it; and three: stop! I am amazed how some people can just write long reports without any substance. My daughters can do it — I can't. Not long ago, I was asked to give a seminar and was told that it should be of general interest, since they would like to publish the talk afterwards, so the title "Lasers and Civilization" was suggested to me. Well, I just could not do that.

I had some excellent instructors. Samuel Beckie was a Professor of Mathematics and he taught the introductory calculus course and later was Dean of the Faculty of Arts and then Chancellor of the University. He was an absolutely marvelous lecturer. I remember once he showed us how to do an integral by substitution, which is a very standard thing, but he had the class applauding. He could build up the tension — it was very fascinating. Then there was Professor John Soderly of the Physics Department, who taught introductory physics and he was a showman. Once a year he would give a liquid air lecture much in the tradition of the 19th century science lecturers. He did spectacular things with liquid air. For example, he would pour liquid oxygen on a loaf of bread and then set fire to it, the flames went to the roof of the lecture hall. The most spectacular was his goldfish experiment, where he took two goldfish and froze them in liquid air. One of them he would smash into little bits with a hammer, the other he would put back in the bowl of water and in a few minutes it would be swimming around again. But these things really didn't particularly inspire me. I was interested in physics and was really thrilled that we were learning this stuff. We didn't get too much of it in high school.

Our high school was a fairly new one; it was good, but not great. Some of the downtown ones had special advanced preparatory courses, but we, in the suburbs, didn't have anything like that. Like we had no calculus in high school and I was thrilled by two things when I went to the University. I learned some calculus and I learned to use the slide ruler. I went through all the regular courses and did all right in them. I had to work pretty hard at it, but it was interesting. I remember one experiment that I particularly enjoyed in our third year lab, when instead of having a prescribed laboratory experiment every day, they had one day when they turned us loose and Professor Soderly gave us a large balloon, and said, "See what you can find out about it." So we got a meter stick and a pressure gauge and measured the diameter of the balloon at various pressures, and so on. That was fun, but there was no way to tell if I could do research.

One other interesting thing was that when we started out in college, everybody in the class thought that they would end up teaching high school. This was an image we had in our mind, either mathematics or physics. That was a reasonable career. Since we were all not very wealthy, we had to think about how were we going to make a living afterwards. Then the war came along and Canada was at war in 1939. By February of 1941 they stopped all classes in physics and put us to work, teaching courses for the Army, Navy, and Air Force students, so they could learn the elements of physics and could operate a radar or a sonar or things like that.

I taught there until 1944; then the need for those courses ended and I worked the last year of the war in a radar factory. I had been very interested in microwaves, read a lot about it and they had a Klystron, which was a microwave generator, in the lab at the University, it must have been a very early Klystron. So I worked on designing microwave antennas and testing for the manufacturer at this radar factory. It wasn't a very important job. When the war ended in 1945 I came back to the University. I knew the people at the University very well, having been an undergraduate student there. The University of Toronto was in bad shape then. They had been strong in the 1920s. In the 1930s, they had given up all their research grants as an economy measure and were just limping along with almost nothing. Many of the best people have left, but there were a couple of good people and one of them was Professor Martin Crawford, whom I had known, and I had courses from him — he was not a great lecturer, but an inspiring person, because he would really discuss problems and speculate with the students. He didn't know all the answers. I worked with him and it was a very good experience. He suggested a good problem and let me work on it. It was hard to catch him, but I learned where and when he could be found, I got enough help from him. I was joined by another student a year later, Fred Kelly, and we needed several, because there was nothing there, everything had been torn apart. Anything valuable they had given to the war effort. There were a few interferometer plates from the 1920s that we were able to use and I built an atomic beam light source. I really wanted to do nuclear physics, which was one of my subjects, but they had no accelerator at Toronto and the nearest thing to it was to do optical hyperfine structure. Optical spectroscopy was considered pretty dead, old fashioned stuff then, but still if you could study nuclear properties, it worked well. Kelly and I built this atomic beam light source and then we were joined by another man, Matt Gray, who built a little spectrograph interferometer arrangement. We found our own problems to work on — I worked on hyperfine structure of silver, Kelly worked on magnesium and Gray on zinc. It was good experience — we were on our own, we had to think.

Then in 1949 I was looking for a job, and I had heard I. I. Rabi from Columbia University give a talk in Canada. There had been an association formed, the Canadian Association of Physicists; it was formed in the early post-war years, because they were afraid that physicists in industry would have to register as professional engineers to hold a job, and for that they would have to take other requirements. That might squeeze out physicists, who were indeed doing engineering jobs. So they formed this Canadian Association of Physicists and they held a meeting in Ottawa. We got a car and just drove there. It was dismal — a lot of talk about professional concerns and not about physics. Then Rabi came on and talked about the wonderful work that Lamb and Kusch have been doing recently there, which of course won the Nobel Prize a few years ago, and I thought I really wanted to go to Columbia University. So I wrote to Rabi and asked if there were any openings there. I wrote to other universities, too, and got several other offers, because there weren't that many physicists graduating yet.

Rabi wrote back and suggested that I apply for the Carbide and Carbon Chemicals Corporation postdoctoral fellowship to work on the applications of microwave spectroscopy to organic chemistry with somebody named Charles Townes, and I never heard of him. I wanted it so badly that even though I had no interest in organic chemistry, I applied for it and got it. I found that Charlie Townes was a wonderful person and working with him really was a marvelous experience. At Columbia, Rabi had a very stimulating atmosphere; he had very high standards, but he could also be very cruel, because he would not stand for anything less than the best. I'll never forget how thrilled I was when he came back from Japan and I was struggling with the usual things experimentalists do, trying to find leaks in the apparatus and he popped his head in the door of my lab and said, "Well, what have you discovered?" Well, I hadn't discovered anything, I never even thought I could discover anything.

There were no less than eight future Nobel Prize-winners at Columbia when I went there. Powell was visiting there and he got the prize just a few months later. Born was there and Townes and Kusch and Lamb were on the staff and Fitch was a student. Everybody who was anybody visited — I remember Pauli visited and toured the lab. It was very exciting coming from Toronto, which was not in the center of things at that time. It is much better now; it is much more in the mainstream. I was interested in microwaves and appreciated Charlie's interest in shorter wavelengths.

When I had been a student, I thought vaguely, well gee, I have all these molecules — you can't build resonators at short wavelengths but they exist. I really didn't know enough quantum mechanics to think of how to use them and that was really Charlie's great advance in showing how to use stimulated emission to amplify and put them in a resonator. It was a very good time, we started a book on microwave spectroscopy, which didn't get finished until late 1954, and it was published in 1955. After I was there for two years I had to get a job and went to Bell Laboratories in New Jersey. I would come in every Saturday and work on this book. It was an amazing experience, because Charlie Townes has the most amazing ability to concentrate. He was chairman of the Department by then and a student would come in and ask him about a course or something like that. He gave the student his full attention, but as soon as the student left he would get back to where he was. He could keep things firmly in his mind and move from one to another. He is a very stimulating person, very good with students — he had a large number of students. I try to handle my students the same way he did — let each of them develop in his own style. This is the only way to do it. I met his younger sister, Aurelia, who came to New York to study singing and we got engaged, and got married in May of 1951. So I had to find a job in the spring of 1951.

A strange thing happened. Sidney Millman, a Columbia Ph.D. was a recruiter for Bell Labs; they sent their senior staff around to different universities to keep contact with the professors. It seems that John Bardeen, who is one of the inventors of the transistor, had switched his interest to superconductivity and he wanted to have someone to do experiments on superconductivity. Nowadays, you would have dozens of Ph.D.s in every specialty with superconductivity, but I never worked with solid state or low temperature, yet they hired me to work for Bardeen to do experiments in superconductivity. Well, it was a little worse than that, because by the time I got there in the fall, he had left — he decided to go to Illinois. I thought that superconductivity is honest physics, it isn't just doing engineering, so I tried to do something by myself and there wasn't really anybody doing anything similar at all. Warren Matthias was working on superconducting materials, but he wasn't really involved or interested in the phenomenon. Harold Lewis came along a little later, he was interested in the theory and we had some good interactions. It was rather difficult to learn about superconductivity and I felt I was rather isolated. The only people who were interested in this thing were somebody in England and somebody in Russia. It was a very small field.

I may mention something here about Charlie's maser. He tells a story, which I don't know whether it is true or not, but he tends to repeat it about the invention of the maser and the important part that I played in it. He said that in the spring of 1951 we both went to the meeting of the American Physical Society in Washington, D.C., which I am sure happened, but the rest I don't remember. He says that we shared a room at the Franklin Park Hotel and I can't recall that at all, but it was only a few weeks before I got married, so I was distracted. He claims that since I was still a bachelor and used to working late at night and getting up late in the morning, I was still asleep when he woke up — he had small children. Not wanting to disturb me, he got dressed and went outside to sit on a bench in Franklin Park. It was a beautiful spring morning, he started thinking about problems and that is when he invented the maser. See, if I had woken up, there would not have been any maser. I did witness his notebook a few weeks later and I do remember him talking about it, but I was already scheduled to go to Bell Laboratories and work on superconductivity, so I didn't work on the masers at all. I still wonder whether it wasn't somebody else sharing a room with him. I never was particularly fond of rooming with somebody else, but maybe I did, I don't know. I also had forgotten that I witnessed his notebook until he showed me a copy of the page, but I didn't hear that story until 1959, being told just before the first Quantum Mechanics Conference.

One thing that is worth pointing out is that he did not publish the idea of the maser right away, and he may have told you about it, but the reason, as it is worth recounting, is that right after the Second World War a lot of people were rebuilding their laboratories and they didn't have any equipment, so they published articles about what they were going to do. People even joked that *Physical Review* should be called *Physical Preview*. In 1950 it wasn't the thing to do to say what you were going to do. If you were going to do something, then do it and then tell us about it. Although he talked very openly about it, took visitors to the lab, put it in unclassified progress reports that were in libraries, he didn't publish an official paper saying here is how you would make a maser. Fortunately, he did give a talk in Japan and someone there wrote down an account of it and published it, but that was the only record of it until he got the thing working in 1954. Meanwhile the Russians

Basov and Prokhorov had some of the same ideas and Weber at Maryland had some ideas, too. None of them had as much as Townes already had, but they got it into print quickly, so when later we were working on the idea of the laser, we were very aware of this history, and in fact there were many more people working in the field, so we published our theoretical proposal rather than build one first. Townes shared the Nobel Prize with the two Russians and they did do the work independently as far as we could tell.

So I went to Bell Laboratories and did the work on superconductivity, but just on the side I attended a conference and heard about nuclear quadrupole resonance. Ralph Livingston gave a talk on that and it looked so easy that I thought I could not resist trying that. It was very simple, I just threw together a one-tube oscillator and put the sample in the coil. They had a vibrating capacitor for frequency modulation and listened on earphones until they found the signal; the pitch of the node would change as you scan. You get some amplitude modulation as you scan the vibrating capacitor, as you reach the resonance you get a sharper sort of tone. So we did some nuclear quadrupolar resonance on a few compounds, we did temperature dependence, but this was on the sideline as I was setting up to do some work on superconductivity. We found a method to show the intermediate state of superconductors by sprinkling not an iron compound, but a niobium compound, which would move out of the region where there was magnetic field and so it would indicate where the magnetic field was penetrating through the sample. I worked on that for a few years, but became a little disgusted, because it became apparent that actual details of the pattern depended on the microstructure of the material, on imperfections; where it would penetrate would be determined by the imperfections of the material. We did one experiment in conjunction with Hume of Westinghouse, who gave us some arc-welded rhenium samples. These behaved like soft superconductors with fine grain intermediate state pattern and then we could work them by just filing them and then they worked as hard super-conductors, showing that introducing defects changed them from apparently a soft superconductor to a hard superconductor.

Then I also did some work on the penetration depth of magnetic fields into superconductors. I developed a rather cute method, where I wrapped a coil closely around a rod of tin and I measured the resonance frequency of an oscillator, which was connected to this coil by a capacitor and as the radio frequency magnetic field penetrated deeper into the metal then the frequency of the oscillator would shift and as you go down to low temperature, the field would be pushed out of the metal and the volume in the coil would decrease, so the conducting would decrease and the resonance frequency would shift.

Our penetration depth experiments were done just about the time when Bardeen, Cooper, and Schrieffer developed their theory with the annihilation gap. We found a departure from the earlier theory that was consistent with the energy gap, so it was very timely, and it did give us some confirmation of the BCS theory, although there were lots of other confirmations. I also did an experiment where I tried to measure the penetration of the magnetic field through a cylindrical film. I had a little pick-up coil inside and I saw spikes in the output of the coil at near where the transition temperature comes through and jumps. Well, I really missed something there, because that was evidence of flux quantization. I had heard the word, but it seemed that it was associated with defects in the sample so I just sort of pushed it aside, but I think I was very close to discovering flux quantization several years before Fermi, Liefer and Niebauer. I feel that I was really stupid for missing good things that I should have seen. I like to tell students that I am really stupid for missing things right under my nose, and the only thing that saves me is that nearly everybody else is pretty stupid, too. They should realize that there are still a lot of simple and beautiful things lying out there, still waiting to be discovered that just all of us have overlooked. Every couple of years when somebody publishes these, you think, "My God, how could I have missed that?"

So I was working on superconductivity. I had one technician there, a very good technician, and that was all. We used to have tea in the afternoon with the theoretical physicists. It was a very small group at Bell Laboratories, also a very excellent group. When I came there, the solid state group had about ten physicists or so. I used to wonder that time, which of these people, obviously all of them very bright, would be very well known, say, ten or fifteen years from then. And essentially, all of those who stayed in physics were. It included two members of the National Academy of Sciences and Walter Brattain, who got a Nobel Prize, and Phil Anderson, with another Nobel Prize. It was a very stimulating group of people. Bell Lab policy was to cover a lot of fields. They said that the purpose of having people do research was not to get inventions, like the transistor, although that is nice, but it took an enormous amount of money to develop. Rather, they should have the best possible window on the world of technology, because if they could save a cent on 100 million new telephones, that would be a lot of money. You couldn't do that by having people sitting in a library just reading magazines. You had to have people working in the fields, who could talk to the other leaders, and know what is really going on.

I think they saved a lot of money. One illustration of it was that once they had a meeting whether they should do some work on superconducting logic devices, cryotrons, as they were called those days, and they decided not to. That was a very good decision at that time. IBM threw a lot of people into that, built a big group, explored the thing, and abandoned it. So by having a few people, you could really understand what you could do and what you couldn't do, and you were able to make a sensible decision. One time Hal Lewis and I went to the boss, Stanley Morgan, and said should we think of some devices that you can build with superconductors, should you write down some of these ideas for possible patenting? He asked, "Does it require liquid helium?" We said yes, so he said don't bother then. That was kind of rough, but it was a good sensible decision. Anyway, Bell Laboratories in their basic research covered a lot of fields with one or two people in each; they didn't have big groups. Other people, from the outside that look at Bell Laboratory publications, didn't understand that. It took me about 5 years to understand how the place really works. I was pretty lonely, I sort of worked by myself, but I realize now how

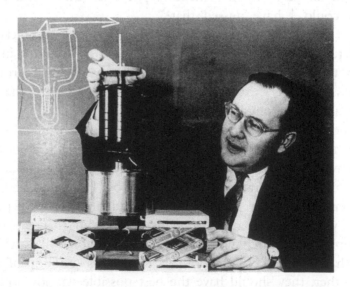

Arthur Schawlow with an experiment (courtesy of Charles Townes, Berkeley, California).

it works. Somebody gets an idea for an experiment, let's say Mr. A. He then goes to Mr. B and Mr. C, who could provide him with samples for this, then takes them to Mr. D and Mr. E who make measurements on them and then to the theorist Mr. F, and then you have a paper with all these people combined. Other people said oh, Bell Labs put a big group on this project, but the fact is, by the time it is published they are probably not even speaking to each other anymore. But there are a lot of lonely people who are free and willing to drop what they are doing for a while, to work on an interesting idea, so they can cover a lot of different things very effectively with a relatively small number of people.

In 1957 I was working on superconductivity and I had not worked on masers at all. After the maser operated, the boss came and asked if I wanted to get back into that, and I said no. I would stay with what I was doing, because I couldn't see anything very fundamental, but by the summer of 1957 I remember thinking that the original idea of the maser was to try to produce a wavelength shorter than that of what you could get with vacuum tubes. I was very much interested in that problem, having been familiar with the whole history of radiowaves, how they got to shorter and shorter wavelengths, and their practical uses. I started thinking vaguely that maybe you could use ions or crystals to get into the far infrared, but I didn't really have any ideas. Then in the early fall of 1957, Charlie Townes was consulting with Bell Laboratories. I hadn't been talking to him much, but we had lunch one day and he said he was thinking about whether he could jump over most of the infrared region and go up to the near visible region. He had a scheme in his head and thought that he might be able to do it. So, we decided to work together and examine the problems and see if we could find solutions. This was just a spare time activity; I would do this at odd moments. Nobody was looking over my shoulder in the lab, my main business there was to get on with my experiments in superconductivity. So I looked at his valiant scheme and gave him reasons why I didn't think it could work and I suggested why don't we look at some of the optical ions, like sodium or potassium, because there is more information about them than about most other elements, particularly about the transition probabilities, they were not very well known.

I picked potassium to concentrate on for a silly reason. Namely, the only piece of equipment in my lab was a wavelength spectrometer as a visible device you could look through and measure wavelength and I bought that back in the superconductivity days to try to measure the thickness of thin films by interferometry. Potassium has the peculiar property that the two lines in the visible spectrum, the first two resonance lines, are in the visible, one is in the red, the other is violet, while the other alkalines have one in the UV. So I picked potassium. I worked through the calculations of the energy levels. Charlie had the maser formula and calculated that you should be able to get with a reasonable pumping power enough excited atoms to get laser amplification or optical maser. Then we had to think about a resonator and Charlie felt that some kind of a box would be fine. Although there would be a lot of modes in the visible, there are so many ways you can pack in the wavelengths, still a few of them would have higher gain and would pick themselves out. Martin Peter, who had come there a year or so before from MIT, and worked on microwave spectroscopy and had worked for his thesis with multimode resonators, kept saying, you had to find a way to pick out a mode. So, prodded by Peter, we had worked together on some superconductivity, I thought about it, and I had taught at Bell Labs, they had a course for their engineers, and they asked me to give lectures on solid state, which seems a little ironic, because I was absolutely self-taught in that. In that course I became very familiar with the Debye theory of specific heat, in which you cover up the number of vibrations in a solid by thinking of them as waves of different wave-lengths — long ones and short ones, which are going in different directions, and the atoms would have a fairly narrow resonance, so they would only respond to a narrow range of wavelength, but you have all these different directions. Then it occurred to me that if you take this big box, the resonator, and throw away everything except for two little pieces facing each other, then only the waves that went straight back and forth would stay in there, so it would be a good resonator for those and not for anything else. It would pick out things within an angle, which was roughly the solid angle.

I mentioned this to Charlie and he said that it was better than that, because the waves would go back and forth many times, being amplified in between and you would get good directionality, so this should be a way to pick out one mode. So we satisfied ourselves that we could get enough excited atoms and we knew how we could pick out a mode. Remembering that earlier when he stopped to build one, he had been partly anticipated, we started writing a paper on this, so we worked on this during the spring of 1958, and submitted it in the summer of 1958. We circulated copies to our colleagues at Bell Laboratories, and they didn't understand this mode selection thing enough to believe it. Somebody said that there would be longitudinal modes. True perhaps, but I didn't understand it any deeper than that, to tell you the truth, but I was convinced that our picture was the one that had to be. So Charlie put in a little more about the diffraction facts and they let us publish it. The patent department didn't think it was worth patenting, it was just another kind of maser, but he persuaded them, so they did apply for a patent. That was my first patent. When I went to Bell Labs, they gave me $1 for all patent rights, so I didn't get anything for all that, but I was not upset — they had been supporting me for seven years. I had several other patents, but none of them amounted to anything. Most patents never get used. So we published this article and I think I sent it off the same day I sent off an article on superconductivity to *Physical Review* or at least very close to the same day. I still didn't try to build one; Charlie had some students working on it.

Being at Bell Laboratories I had the feeling that you can do anything in a gas, anything is more calculable in a gas, but we mentioned solids in the paper. I only thought about them vaguely, and wrote one paragraph that said solids. You don't usually have a lamp that produces just the same wavelength, but you have an even better solution, because there are often broad bands. I had in mind ruby that people were beginning to use for masers, but I didn't really know anything about it. I thought anything you can do in a gas, you can do better in a solid. So I thought I would like to drop superconductivity and start learning more about solids in the hope of finding some optical maser material. All I had to do was go to the boss and tell him that I would like to drop superconductivity and work on this and he said fine, that was all I had to do. He arranged for funding, any amount of money I wanted, I could get. I had never bought any fancy equipment, in fact when I first went to Bell Labs in 1951, you could hardly buy any equipment, they had a strange situation, the lab was set up as a non-profit corporation, subsidiary of American Telephone and Western Electric. If they added capital equipment that would be equivalent to making a profit, they did not want you to do that. You could only buy capital equipment if you could junk some old equipment. So when it came to something new, it was very hard to get capital equipment. In fact when I bought that wavelength spectrometer and wanted to buy a camera attachment for it my boss asked me if I really needed it and I said that I wasn't sure, to which he replied "You better wait until you are sure," and I never got it. I was sorry of course later.

In about 1956, after I had been there for five years, they suddenly realized that there was another way; if the equipment was bought for one specific experiment that was not adding to their capitalization. As they loosened the purchasing, Varian magnets started sprouting all across the halls. Although I didn't get one, I still operated rather frugally, from habit I guess, but when I got into this laser stuff, I thought that it might be important and I wanted a high resolution spectrograph, since we had shown that the gain was inversely proportional to the line width, we had to have narrow lines. I also bought the most expensive oscilloscope that I could find, which was a dual beam device from Techtronics, and these both turned out to be very wise choices.

I did try to build a laser, and started working on the spectrum of ions and crystals of ruby and some people down the hall who were working on microwave masers had a drawer full of rubies, synthetic rubies, you could borrow some so I started looking at the spectrum and I wondered about it. I found that there was a thesis at Johns Hopkins by a student, where he had studied the spectrum of ruby. It was a little bit mysterious, because there was a theory at that time why it should be a very simple spectrum, it should only be the two doublet lines in the red, yet there were a whole lot of other lines, so-called satellite lines or neighbor lines in there. Nobody knew where the lines came from. The student had measured a lot of these lines, but had no idea what they were; in fact they were discovered around 1905. They had been measured and studied various times, but nobody had any idea what they were caused by. Of course the early scientists had no theory where any of these lines came from. I thought, maybe they are caused by the crystal vibrations and if we can understand these lines, maybe we could understand the crystal vibrations better.

We had crystal growers, who grew us some crystals of gallium oxide, which is like chromium oxide, or like aluminum oxide. The technician, who had no formal training to speak of, but was very observant, noticed that the spectral lines were different in different samples; they were dependent on the concentration. Once I realized that, it was immediately obvious that these could come from pairs of chromium ions, which could have their energy levels split by exchange interaction. So we discovered the *para* lines and did some work and tried to analyze them. We had a lot of fun. We applied stress to the crystals and split the lines in magnesium oxide. We did chromium magnesium oxide, then displaced them in ruby, which helped us to understand why the lines were so broad at low temperature when you do the strains. When you strengthen them, you can move the lines appreciably. I also realized that we could make a four-level laser out of these *para* lines. All the atoms of ruby are in the ground level to begin with. No empty levels are nearby, but in these *para* ions the levels are split often by several hundred wavenumbers, so there would be levels that could be thermally empty at low temperature. These would be good final levels for the laser, you wouldn't have to get more than half the atoms excited before you get laser action. Again, I tried to get the Bell Laboratory people to patent that, but they said it was not worth the patenting, because we got ruby masers and this is just a different concentration of ruby and wavelength.

I did mention it in talks, but just casually mentioned that our line of ruby is not suitable for laser action. Again as I said I really outsmarted myself, because that was the one that Maiman used. I was not quantitative, just qualitative. Even if I had been quantitative, I would have come up with the wrong answer, because several people have measured the fluorescence efficiency of ruby and they said that it was between one and ten percent and if it were that low, you would never be able to get enough excited atoms to get laser action. In fact, I already knew that it was wrong, because we had done some work on measuring the fluorescence lifetime and we found that it depended on the size of the sample, we were getting trapping of resonance radiation. In sapphire, which had about a part per million chromium ions, we got a lifetime of 4 ms at low temperature. In a chunk of pink ruby we would get a lifetime of up to 12–14 ms. I remember one wonderful day when I had Frank Bersani, a student at Johns Hopkins work with me and Darwin Wood of the Chemistry Department, and we suspected this trapping, so we started cutting pieces up. We reduced the lifetime from about 14 to 11 ms, but then we realized that even in the thin slab there was still some trapping, so we ground the stuff up into fine powder and I got the lifetime down to 5–6 ms. Then we realized that the grains of powder could still see each other and light could be omitted from one, but absorbed in the other. So finally when we embedded the powder in some black pitch cube, then we got the 4 ms.

So we knew that radiation would be trapped, and we published that, and that should have told us that the fluorescence efficiency could be high and I should have done the calculations. I have to leave it to Maiman's credit that he did the calculations on the efficiency and what pumping light was needed and he showed that it worked. I had already said that dark ruby could work, and have also stated specifically at a conference that the structure of a solid phase optical maser would be especially simple. Essentially it would be a rod, with the ends polished flat and parallel and coated to reflect radiation and reflect the light and the sides left open to the mid-pumping radiation. So when I saw this picture in the newspaper with Maiman with a flashlight on a rod with the ends flat and parallel, it was exactly what we were talking about. But I tried dark ruby, and I actually got a rod of it, but I only had a small flash lamp of about 25 W/s. I didn't have any determination to buy a big one — if I had it, it would have worked, because after Maiman published his results, and I worked with some others, we verified the properties predicted for the laser, coherence and directionality, which Maiman hadn't at that point. Then I thought, maybe I better go back and try the dark ruby. I asked Al Clogston who was my boss, a very good boss, do you think I should try that? He said that you owe it to yourself. We did and it worked, and again, it was the same rod, and again I felt really stupid; I should have done that a year before, but you can't win them all.

It was very exciting when we got our first ruby laser after Maiman has built his, but we knew what to do. There were two other groups at Bell Labs building it in the Physics Section and in Materials. I wasn't involved, but it just looked like so much fun that I couldn't resist and they needed a good spectrograph, so they came down and worked with my spectrograph and I got involved.

Of course we were arguing what should we do next. The thing was very fragile, had metal coatings and it would blast off after a few shots. The ruby rods we had, we just adopted from masers, they were very poor quality. We had a flash lamp, just like the one that was shown in the photograph of Maiman. Later on I heard that it wasn't the one he used. I heard two stories about it. One of them said that all the ones they actually used were broken, so they could not be used in the picture. Maiman later said that the photographer thought that this one would look better. Whatever it was, he was vague and said they used a crystal of centimeter dimension. That was obviously GE FT 524, you could see the rods, so we built lasers using the same flash lamp. This lamp was rated at 4000 V, 400 μF at 3200 J. Anyway, we tried this miserable rod and it did not glaze at 4000 V. So, knowing about the threshold effect, we raised the voltage and at 4200 V, it worked, so it was very worthwhile to be rough with the apparatus. A few years later, when I came to Stanford, one of my students, John Holzrichter was building an early flashlight pumped dye-laser, and it didn't work. So I told him, why don't you overrate the lamp a bit and see what happens, the worst you could do is blow out the lamp. He did, and it worked, so it was a very useful experiment. Sometimes with the laser, since it has a threshold, when you are below it, it is hard to know, but when you reach it, it is all clear.

We did think very crudely — I like to improvise — with the flash lamp, we just put cardboard around it, but still a lot of light would fill the room when you flash that thing, so we didn't see the beam, if there was a beam. Maiman had suggested that there would not be a beam, because of the reflections of the sidewalls of the crystal. However, he had shown that there was a sharp increase in the output. Well, we thought that is easy, we'll just leave the sides of the rod rough ground, and not polished. So we set it up, but we were busy with measuring other things. Finally one night I just could not sleep, I had to know whether that thing was giving a beam or not. We just directed it into a camera with some Polaroid film and we saw that we got a small spot, which showed us that it was directional, as I predicted it, with about half a degree of divergence. We didn't see this beam, we didn't know if you could see this beam, but about a week later two amusing things happened. The other group got theirs going and they got a beam also, and their rod was polished and we didn't understand that at first. Later on it turned out to be a rather sophisticated thing that there is a kind of focusing effect from the pumping light by the sidewalls of the rod, so the intensity is greater on the axis than near the walls, so the material is still absorbing near the walls and the reflection from the walls doesn't occur when it is amplifying at the center. Later that led us to invent another device, the CLAD-rod, where you have a clear outer section and a ruby core. We hadn't actually seen the beam, because of so much stray light, but later on we boxed in ours and we could see the spot. We hadn't known whether you could, because it only lasts half a millisecond or less, and it was also very deep in the rod, and our eye is 200 times less sensitive than when it is in the green, but it spots very easily in the visible.

You remember, I told you that I bought the most expensive oscilloscope I could find in the catalog — and it turned out to be wonderfully usable. The technician, who had wonderful instincts, asked if there are any signs of relaxation oscillation. So we looked carefully with the oscilloscope at the light output, and there seemed to be a little bit of spikiness on it. But we had this dual beam oscilloscope, and we could set one beam to look at a small portion of the pulse, delayed by the right amount from the other, and we could see microsecond pulses within the millisecond bursts, a series of short spikes. So having the right equipment we were able to discover this spiking phenomenon, which occurs in ruby lasers.

We wanted to check the coherence and we tried to get two slits cut at the end of the mirrors. We finally got the one slit diffraction pattern. We had a definite cutoff; we were going to submit this paper by a particular date, whatever result we had, because we thought it was a competitive field. It turned out it wasn't that competitive at all. We got the single slit diffraction and the next day we got the two-slit diffraction, showing that the light was coherent, directional, and also that it was monochromatic — all the things that we predicted. That was a lot of fun, all these exciting things happening every day.

Then I started getting offers from universities — eight different universities approached me within the next year and Stanford made me an offer that I couldn't refuse — I decided that it was a good place to come. This was partly because it was a good university, partly because I had an autistic son, who was born in 1956, and was 5 years old in 1961. New Jersey didn't have a Medical School and there wasn't any Health Department or any school or anything. Bob Hofstadter had an autistic daughter, Colleen. His wife and others had set up a program for people like that and that was a big inducement to come out to Stanford. Also I had a feeling that I wanted to leave Bell, because there I worked with just one technician. If I had a big group, I would have tried to build a laser, but I didn't ask for one and nobody offered one. But here, I would have students. I really agreed with the Bell Laboratories policy that you shouldn't have physicists working for other physicists, because if they are creative people, they want to be independent, but students need you for a while and you can get really first-rate people, who are young and inexperienced and it works out. It has been wonderful working with students, some of them who are already very distinguished in their own field.

This was a small department, so I was again by myself, which was fine. People were asking how in the world I could compete with Bell Laboratories, but I wasn't going to compete with Bell Labs, I was going to do something different. I really had the feeling that any idea I had, and I had so many ideas, if anybody wanted to take them and work on them that is fine. We didn't really work much on lasers. We did do some laser work in the 1960s, but we mostly worked on the questions of solid state physics related to lasers. We even got into magnetism, and discovered spin-wave side bands and anti-ferromagnetic materials. We did do some work on high-powered lasers. John Emmett was a student, he is now the Director of the Laser Program at Livermore. John Holzrichter, who worked with him is one of the top people in lasers; they work on laser fusions and isotope separation. Emmett is a great machine builder. Having him as a student was like holding a tiger by the tail. Very nice guy, but he knew so much more about building high-powered lasers. In fact he was the world's leading expert on flash lamps when he was a student. I managed to get him a grant to go to a summer school that Charlie Townes organized on lasers in 1963. Charlie was so impressed by what Emmett knew that he had him give a lecture and had him write it up for publication. I could not get him to stop building equipment and do something with it. Years later he told me that he understood that, he liked to build apparatus, he did not want to finish, he was having too much fun, he knew that if he took measurements, he would have to get out. We did some work on pulse ruby lasers in the 1960s, but mostly we did solid state physics, spectroscopic problems related to lasers.

It was very stimulating. I had up to a dozen wonderful students. For a while it was just myself, for a while we had an Assistant Professor working with us, but mostly just me and the students, occasionally a postdoc or a visitor. The students would help each other, the more senior ones would help the younger ones, there was great interaction. All along Leonard Schiff, the chairman, said that you should get a younger man working with you, if you would like, we could always find a place for one, but I didn't see anybody who I really wanted. We just had a few people here — just like at Bell Labs, each running their independent programs, all very good. Then in 1970 I got a letter from a colleague at Heidelberg, whom I met at a conference. He said that he had this young man, who was getting his Ph.D., could I hire him as a postdoctoral associate? I wrote back saying I didn't have any money, but he answered asking that if they got him a fellowship would I take him. So I said, "Oh, all right." So, Haensch arrived and within a few weeks you could see that he was absolutely brilliant. I found some money to supplement his rather meager fellowship, and two years later we made him Assistant Professor. A year later we gave him tenure, a year after that he became full professor, because he had offers from Harvard, Yale, and Heidelberg, which was his home town.

This marked a big turning point in two ways. One was that about that time tunable dye-lasers came along — the early lasers, we couldn't tune them. We actually did some work on photochemistry using a ruby laser that was tuned to different wavelengths to separate isotopes, but we tuned the ruby by changing its temperature. The only substitute we could find that coincided with the ruby, since we couldn't tune much, was bromine, and this was a really bad choice, because bromine always undergoes a chain reaction once you start. We tried to show that we could initiate a reaction selectively, but it didn't end up selective. So I dropped photochemistry, partly because I couldn't get students interested, partly because I became aware of the dangers of isotope separation. I still have a horrible fear that somebody will find a simple way to separate uranium isotopes, which will lead to the proliferation of bomb materials even faster than it is occurring now. I just leave isotope separation to government laboratories, where they can protect their secrets — I hope.

In fact, Charlie and I didn't really think about the application of lasers, we didn't need to. We knew the history that shorter wavelengths would be useful. Vaguely we thought it might be used for spectroscopy and because

Arthur Schawlow with a laser (courtesy of Charles Townes, Berkeley, California).

of the interaction with Union Carbide, Carbide and Carbon Chemicals, we thought of isotope separation in photochemistry. But we didn't know what the properties would be, and I think it is very important that we did not have a specific application in mind, because that would have imposed additional requirements and might have made it impossible. You do what you can do and later you can extend it.

Anyway, we got tunable lasers, the dye-laser was developed at IBM and by others in Germany, then people at AVCO showed that you can pump nitrogen into the laser, which then they would sell commercially. We managed to get some money and bought a nitrogen dye-laser; just as Haensch came, we got that. It wasn't very narrow band, but Haensch found ways to make it narrow band, so we could do some high resolution spectroscopy — you could tune the wavelengths that we wanted. Also, he discovered the saturation method of getting rid of Doppler broadening. We just dropped all our work on solids and went on to work on atoms. Particularly, he worked on hydrogen, and I worked on other atoms and molecules, and found a way of simplifying the spectrum, which sort of worked itself to the Nobel Prize. Oh, they clearly had in mind the fact that I was one of the co-inventors of laser, they mentioned that. The director of the Nobel Foundation told me that it had been a close thing in 1964. They told me that they couldn't divide it between too many subjects. Siegbahn got it for electron spectroscopy and Bloembergen and I for laser spectroscopy. Bloembergen worked on non-linear optics, although he did invent the solid-state maser; they had to have some unification. We spent 10–15 very interesting years working on laser spectroscopy and things like simplifying very complicated spectra.

Of course all the time you worry — you have the chemists looking over your shoulder. In 1970, I remember giving a talk entitled "Is spectroscopy dead?", which was a fashionable view since physicists didn't work much on it anymore. When Felix Bloch asked me what I meant by "dead", I answered, "Turned over to the chemists." That is what happened to microwave spectroscopy and nuclear magnetic resonance. When Carl Djerassi asked Felix Bloch to give a talk on the discovery of nuclear induction at the Chemistry Department, he also told him, "Just tell us about the beginning, the rest we know much better than you," which is true. Nowadays, chemists know more about molecules and often can build equipment better suited for spectroscopy. Still, there are some very important fundamental physics problems left for physicists with very simple atoms and unraveling structures of more complicated ones. It is also important to realize that people have to be able to change the focus of their research, like I did after I got my Ph.D. or after I went to Bell Laboratories. It is not good if a young scientist repeats the same thing for the rest of his life that he did for his Ph.D. studies.

Physics and chemistry are very closely related and there are things that a chemist knows, but I don't know. I remember Charlie Townes told me once that he had a conversation with a very distinguished chemist in Southern California. He said that aren't you worried about the physicists in this field being tough competition? The chemist answered, "Oh, no, there are molecules that the physicists never heard of." But it is good that we can talk some common language. Still, physics has the position of dealing with the basic laws of structure of matter and energy and the way things are put together. I think the task of physics is not just to understand the proton or the hydrogen atom, but to understand the Universe. I think that there is a very rich field working gradually to greater complexity, not to jump right to the DNA molecule, but to understand the atoms and the molecules. There are always some surprises. Even in sodium, which you think of having one electron outside its core shell, the d level in the fine structure is always inverted. The reason is now understood, it is the polarization of the core electrons, which is enough to invert the fine structure. I see my field as exploring problems of slightly more complexity than the hydrogen atom, so that we can deeply understand the fundamental principles, as well as developing new techniques, with which we had a lot of fun.

Lasers have become indispensable tools in physics, chemistry, and biology. They have done a lot of nice little things, but I expect that sooner or later there will be some rather major breakthrough, which has not occurred yet, through the use of lasers to do some wonderful things. There are still some fundamental discoveries to be made, at least by my standards. Every year they have a panel of Nobel Prize-winners and the two questions asked is, do you believe in scientific intuition? And, what is it and what field will be given the Nobel Prize in the year 2000? Well, I certainly cannot answer that one at all, because I believe that the most important and interesting discoveries in physics will always be the unexpected ones. The expected ones are always discounted. We know some broad problems, but a new twist always gives you a new framework for looking at things. For example, the discovery of the neutron was really quite unexpected.

You need to develop good instincts to have a feel for what is a worthwhile field; you never know what is around the corner. For example, one of the uses of the laser is the surgery of the retina of the eye to prevent retinal detachment. Well, neither Charlie Townes nor I have ever heard of a detached retina and had no idea that you could use the laser for that. If we were trying to help prevent blindness, we wouldn't have been fooling around with stimulated emission of light from atoms.

Leon N Cooper, 2002 (photograph by I. Hargittai).

8

LEON N COOPER

Leon N Cooper (b. 1930 in New York City) is Thomas J. Watson, Sr., Professor of Science, Professor in the Departments of Neuroscience and Physics, and Director of the Brain Science Program and the Institute for Brain and Neural Systems at Brown University. John Bardeen (1908–1991), Leon N Cooper, and J. Robert Schrieffer (1931) shared the Nobel Prize for Physics in 1972 "for their jointly developed theory of superconductivity, usually called the BCS-theory". Leon Cooper received his degrees (A.B. 1951, A.M. 1953, and Ph.D. 1954) from Columbia University. After short employments at the Institute for Advanced Study in Princeton, the University of Illinois, and Ohio State University, he has been at Brown University since 1958. He is a member of the National Academy of Sciences of the U.S.A., the American Academy of Arts and Sciences, the American Philosophical Society, and other learned societies, and has been awarded a series of distinctions and honorary doctorates. We recorded our conversation in Dr. Cooper's office at Brown University in Providence, Rhode Island, on February 4, 2002.

Is the BCS-theory still valid?

It's very valid. If I may say so, the BCS theory is one of the great theoretical achievements of the 20th century. In addition to its immediate applicability to superconductivity, including high-temperature superconductivity, to nuclei, helium-3, neutron stars, concepts, such as the notions of broken symmetry, that originated in our theory are at the heart of the most important developments in particle theory.

What are the Cooper pairs?

Why ask me? [Dr. Cooper is heartily laughing.] It's sometimes difficult to explain something you did yourself. But I'll try. It's a pair of electrons, both near the Fermi surface, often (but not always) moving in opposite directions with opposite spin. If there is an attractive interaction between them, they form a kind of atom. What makes this atom special is that it is at the Fermi surface where there are so many other electrons. Suppose you had two fermions that attracted each other in empty space; they might form an atom, but they would have available to them all the fermion states, from zero up. But in the metal, other electrons occupy the lower states. So the wave function for electron pairs can't use states deep inside the Fermi sphere; it uses only states near the Fermi surface. That makes it a special kind of an entity; it gives the pair wave function its coherence, and its size. A normal atom is usually about 10^{-8} centimeters; the size of a Cooper pair may be as much as 10^{-4} centimeters in a typical metal. It spreads out over the metal and there is a very large number of other pairs that are intermingled within one another, satisfying the Pauli principle. When people focus on Cooper pairs they don't sufficiently appreciate that in addition to the idea of pairs and the idea, contributed by Bob Schrieffer, of the wave function with all of the pairs intertwining with each other, our theory involved some very complex calculations. We obtained many results, such as the coherence effects, that were new and surprising. People tend not to appreciate how original and how difficult (using the techniques we then had available) those calculations were.

I have read an anecdote about John Bardeen. He was known to be a quiet person, and a colleague remembered meeting him one day in the hallway of the physics building of the University of Illinois. The colleague sensed that Bardeen had something to say, but it took some time before he spoke up: "Well, I think we've explained superconductivity."[1] Such anecdotes may have contributed to the notion that it was the result of a brainstorm rather than a result of tedious calculations in addition to brilliant ideas.

The calculations were complex but not tedious. It was January 1957 when John agreed that we had the key idea that would distinguish the superconducting from the normal state. But then to forge the initial ideas into a full theory required intense, extremely complex, and highly original calculations.

Although you were only one year older than Schrieffer, you were a postdoc and he was a graduate student. Were you the two who did the calculations?

We all did the calculations, but divided the major responsibilities. Bob was primarily responsible for thermodynamic properties, John took on transport and non-equilibrium properties, and I calculated electro-dynamic properties. But we all were heavily involved in every aspect of the problem since results in one area usually influenced results in the others. Colleagues at the University of Illinois were very excited about our progress; Hebel and Slichter were doing the nuclear spin relaxation experiment that provided an early test to our theory. Charlie Slichter became so expert and so involved that he could check our calculations and find errors. I think we did several years of calculations in about three months.

So you were aware of the fact that something big was in your hands.

Sure. We knew what the magnitude of the problem was. First there was a period when I knew I had a great idea, but nobody believed me. Once we began seriously to work on it all together, things fell into place so beautifully that there was no doubt about it anymore. We certainly did not have any doubts.

When was the moment it occurred to you that you might get the Nobel Prize?

We knew it was Nobel Prize caliber from the very beginning.

Did you think that you might not receive it because Bardeen had already received one?

Well, that's another matter. That's why we had to wait 15 years. It was a terrible period. There is a big difference between being nominated, realizing that one is on the list, being regarded as a potential laureate, and actually winning the prize. However things finally worked out. The Nobel committee made the right decision to award the prize to the three of us. It was a joint effort.

I may be blunt in my questions.

That's OK, I'll be equally blunt in my answers. As the French say, there are no indiscreet questions; but there are indiscreet answers.

You remember the cartoon, which is a decoration in the Faculty Club of Rockefeller University, that shows some scientists discussing the fame of Prometheus: "Sure, he discovered fire, but what has he done since?"

Sounds like my father, who asked me, "What are you going to do next?"

What would you single out from your works since the BCS-theory?

I have made reasonably important contributions both to neural networks and the study of the physiological basis for learning and memory storage. We have one of the interesting theories in neurophysiology, a theory of synaptic plasticity, called BCM-theory,[2] B being Elie Bienenstock and M being Paul Munro. I seem always to be surrounded by brilliant colleagues. People have asked me if BCM is as important as BCS; probably it is not, but it's still very important. I'm not the one who should say, but it's a ground-breaking theoretical structure in neuroscience, one of the first. We have worked out the theory; we have suggested experiments; there are experiments that have been done and are in agreement with our theory. The postulates of the theory have been checked experimentally, as well as many of its consequences.

What kind of experiment would that be?

One of the underlying postulates of the theory is that if incoming (pre-synaptic) cell firing is correlated in time with post-synaptic cell firing, synaptic modification occurs. (Synapses are specialized sites of interneuronal contact.) This is a Hebbian postulate. Our specific postulate is that if the post-synaptic cell fires weakly, the synapse is decreased in strength. If it fires strongly, that synapse is increased in strength. The crossover point between decrease and increase, we postulate, should move, depending on cell activity. This has become known as the BCM modification threshold. The decreasing part is now called long-term depression (LTD) and the increasing part long-term potentiation (LTP). Long-term potentiation had been previously seen, but the decreasing part was theoretically required. One of my colleagues, Mark Bear, who is an experimentalist, believed the theory sufficiently to invest a great deal of effort and found this decreasing part — now called long-term depression. The experimental protocols he devised have been repeated in laboratories throughout the world.

Where did he find it?

In slices of the brain, the hippocampus.

Was it human brain?

That's an interesting question. It was initially found in rat brain, but it has since been found in many species: rats, cats, mice, and in human beings. It has been found in young animals and old animals and it has been found in different parts of the brain, such as hippocampus and visual cortex. So it appears to be a fairly general mechanism. This is one kind of experiment that has been suggested; in addition, Mark has seen the moving threshold, experimentally. This has also been confirmed in various other laboratories. These experiments confirm the postulates of the theory. You can check a theory by experimentally checking its postulates. (For the Newtonian theory of gravitation, the equivalent would be directly measuring the inverse square gravitational force.) In addition, you can check the consequences of the theory (for Newtonian theory, elliptic orbits, etc.). There are various subtle consequences of our theory. In addition to our ability to explain a variety of previously-obtained experimental results, there are unexpected predictions. This is getting complicated. Are you sure you want me to go on? For example, there is a rearing condition called monocular deprivation in which an animal is raised in a normal visual environment with one eye closed. It is known that in this rearing condition the connections between the neurons coming from the closed eye to visual cortex, very rapidly will go to zero. In a sense, the animal will become blind in the closed eye; and this happens much more rapidly than if you close both eyes. This has very important clinical consequences, as was pointed out by Hubel and Wiesel. For example, when a child has pink eye — as I understand it, not being a medical person — the recommendation is, close both eyes or leave both open but never close only one eye.

Theoretically we can show that if you leave one eye open and you close the other, the rate at which the closed eye is disconnected from the cortical cells increases with the amount of noise coming into the closed eye. It is a counterintuitive prediction. How could you check this experimentally? By using special drugs or by closing the eye with a black patch versus a translucent patch, one can control the noise coming in from the closed eye. The prediction is that the eye with the translucent patch would lose its cortical connectivity more rapidly than the one with the black patch. That experiment was done here in Mark's laboratory and it confirmed our prediction — possibly a first in neuroscience: to construct a theory of such a complicated system and to be able to compare it with experiment. The whole field has evolved very rapidly. We are moving now into questions concerning the cellular and molecular basis that underlie learning, memory storage and memory consolidation. That's one of the things I have been doing since 1972. Perhaps it's not the moral equivalent of fire — but it keeps us warm.

Can this be in any way related to the genetic studies of memory and learning that Seymour Benzer at Caltech and others in other places are doing on Drosophila?

It is related although not directly for the moment. For example, the CREB activation sequence is more of an overall switch that governs downstream gene activity that in neurons finally leads to memory consolidation.

What turned you to science originally?

I think children want recognition. You find something that you're good at and you keep working at it. I was good at it; I liked it and kept working at it.

When did it happen?

I might have been 10 years old; I had a little laboratory. I was interested in things like photography, chemistry and electromagnets. I didn't know what I was doing but it was fun doing it.

Was it in the family?

No, and, honestly, I don't quite remember how it happened.

Your mentors?

I had some wonderful and encouraging teachers in Junior High School and especially in the Bronx High School of Science. My thesis advisor was Robert Serber who had worked at Los Alamos; then I was a postdoctoral fellow with John Bardeen. Bardeen had written to Frank Yang; he was looking for a young theoretician who knew the techniques of quantum field theory and who might be interested in working on superconductivity. Frank asked me and I said, I would try it. This was the time I was at the Institute for Advanced Study. Bardeen stopped by the Institute to interview me. I told him that I didn't know much about superconductivity, but he said that he could teach me what I needed to know about it. That's how I got to the University of Illinois.

How did you happen to be at Brown?

I came at the invitation of Bob Morse, who was doing the very important experiment on ultrasonic attenuation in superconductors (also an early confirmation of our theory); he was at Brown, and I wanted to come back to New England.

Have you transformed yourself from a physicist to a biologist?

You don't really transform yourself; I know a lot about the area of neuroscience, in which I'm working, but that doesn't mean that I've lost my interest in physics. I am the Director of the Institute for Brain and Neural Systems and the Brain Science Program; we have about a hundred people involved in brain research: applied mathematicians, neuroscientists, physicists, and others. It's a very active and powerful program. Brown created one of the first departments of neuroscience. We've done work in neural networks, image recognition, and on the biological basis of learning and memory storage. We have a nuclear magnetic resonance imaging center.

Vartan Gregorian, a former Brown president, used to say that we do more with less, but he added: "We're running out of less."

You have been involved with companies.

There is Nestor, which is involved in applying neural-network systems to various commercial applications. I recently participated in founding a new company called Sention which is involved in creating compounds that would aid memory consolidation and thus could aid in various age or disease related memory disorders.

How did the Nobel Prize change your life?

The Nobel Prize opens many possibilities, among them the possibility of making a fool of yourself.

My impression is that you are a very careful person. Is there anything that you are passionate about?

I won't make a list. Of course I feel strongly about many issues, but to paraphrase Warren Buffet, "People think we know what we're talking about, so we should be careful."

Is there one issue you are willing to talk about carefully?

I feel strongly that stem cell research should be done without any political interference. It's a shame that the issue even comes up. It's a shame since the divisiveness is over what is really a religious issue. The United States was founded on the notion that you cannot get people to agree on religious matters; you either accept different peoples' beliefs or else we burn each other at the stake.

The intrusion of a person's religious ideas about such questions as when life begins, which is not a scientific question, such an intrusion into public policy or scientific affairs is damaging. When life begins is one of those non-issues people get ferociously excited about. The scientific answer, in my opinion, is that we have somewhat of a continuum — consider for example the distinction between living and non-living (salt, prions, viruses, bacteria, etc.) — where shall we put the boundary? Or consider the distinction between organic and inorganic materials, a much made distinction prior to the 20th century — a distinction preserved now almost exclusively in academic chemistry departments. A possible best answer is the punch line of an old joke: life begins (a) at conception, (b) when the fetus is viable, or (c) when the children go away to college and the dog dies. It's ludicrous to listen to politicians or ethicists reciting pompous opinions about which line of stem cell research is most promising when the super experts are not sure. It is absolutely clear that we should be exploring all possible avenues. Success is by no means assured, but to close off such promising directions for the cure of horrible human ailments, in my opinion, is criminal.

But what about the argument that public money should not be used in such research?

Public money is used for all sorts of things. My opinion is that if a person feels, as a religious matter, that stem cell research is immoral, they are free not to use the results of this research. Abortion is a similar issue. People opposed to abortion are free to follow their own beliefs, but they shouldn't impose their religious beliefs on other people with other religions. My tax money is being used for purposes with which I don't agree. For example, I don't agree with missile defense, but the IRS would not look on kindly if I deducted my share from my income taxes. In any case, without public (NIH) funding, stem cell research is crippled, and the most likely result is that people will suffer and die unnecessarily.

You don't agree with missile defense because it's not going to work or because it is not useful even if it works?

Technically it seems highly problematic to me. If it had a chance to work, I would support it.

What do you read?

Lately I don't read too much besides scientific articles and reviews, but when I do read, I read everything. As a teenager I loved Kafka especially and read essentially continuously.

Do you write?

Not as much as I would like to. Apart from scientific papers, I would like to write some essays about science. When I was a teenager, I wrote all sorts of things; later I even wrote a novel. It was made into a script; my agent tried to sell it in Hollywood. Apparently, someone liked the idea and appropriated it. It seems to have metamorphosed into a movie with Steve Martin called All of Me.

What would be the message of your essays about science?

It would be about what kind of knowledge science is and how it relates to other knowledge. How it fits into what human beings can possibly know?

We met in Stockholm last December [2001] at the Nobel Prize Centennial. There were about a hundred or more other scientists Nobel laureates. Did you feel a special atmosphere among them?

They are a special group of people. One thing I had not realized before and dawned on me during those days that many of them took great gambles with their careers, worked out things that other people considered insoluble. Obviously very talented people, and generally a bit adventurous.

Leon Cooper at the Nobel Prize Centennial celebrations in Stockholm, 2001 (photograph by I. Hargittai).

References

1. Lubkin, G. B. *Physics Today* April **1992**, 23.
2. Bienenstock, E. L.; Cooper, L. N.; Munro, P. W. "Theory for the development of neuron selectivity: Orientation specificity and binocular interaction in the visual cortex." *J. Neuroscience* **1982**, *2*, 32–48.

Alexei A. Abrikosov, 2004 (photograph by I. Hargittai).

9

ALEXEI A. ABRIKOSOV

A lexei A. Abrikosov (b. 1928 in Moscow) is Distinguished Argonne Scientist at the Argonne National Laboratory in Argonne, Illinois. He was co-recipient of the Nobel Prize in Physics in 2003 "for pioneering work on the theory of superconductivity and superfluidity" together with Vitaly L. Ginzburg of Moscow and Anthony J. Leggett of Urbana, Illinois.

He received his Diploma in Physics (equivalent to the M.Sc. degree) from Moscow State University in 1948 and his Candidate of Science degree (Ph.D. equivalent) from the Institute of Physical Problems in 1951. Lev D. Landau was his scientific advisor. He stayed on at the same Institute after graduation. In 1955, he defended his higher doctorate and received the Doctor of Science degree. Abrikosov became head of the Condensed Matter Theory Department of what is today the L. D. Landau Institute in Moscow. Between 1988 and 1991, he was director of the High Pressure Physics Institute in Troits, Moscow District. In 1964, he was elected corresponding member of the Soviet (now Russian) Academy of Sciences and in 1987, he became full member. In 1966, Abrikosov, V. L. Ginzburg, and L. P. Gor'kov were awarded the Lenin Prize "for the theory of superconductivity in strong magnetic fields".

Abrikosov moved to the United States in 1991 and has been at Argonne ever since. In 2000, he was elected to the National Academy of Sciences of the U.S.A. and in 2001, to be foreign member of the Royal Society (London). He has several honorary doctorates and other awards.

We recorded our conversation in the Abrikosovs' home in Lemont, Illinois, about an hour and a half drive southeast of downtown Chicago, on January 31, 2004. In the first part of our conversation, we discussed at length his discoveries, and in this account, first I summarize his Nobel Prize-winning discovery. Then follows the second part of the conversation, which is quoted verbatim.

* * * * * * * * * *

Superconductivity is the absence of electrical resistance at low temperatures. It was discovered in 1911, but it took time to understand what it really is and what the physical phenomena are, which lead to this effect. The phenomenon was understood by Bardeen, Cooper, and Schrieffer (BCS) in 1957, but even before, people understood many things about superconductors. In 1950, Ginzburg and Landau published a theory of superconductivity, which was based on Landau's theory of second-order phase transitions, which he constructed in 1937. It was a relatively simple but at the same time general theory, which could describe phase transitions of the second order in many different substances. The Ginzburg–Landau theory made many useful predictions, but they required experimental verification.

One of Abrikosov's colleagues, the late Nikolay Zavaritskii was an experimental physicist at the Institute of Physical Problems where they both worked at that time. They had known each other from their university studies and they always discussed Zavaritskii's experiments. At some point, Zavaritskii started to do experiments for checking the predictions of the Ginzburg–Landau theory. Zavaritskii's scientific advisor at the Institute, Alexander Shalnikov had done similar experiments several years before. However, at the time of Shalnikov's experiments, there was no theory with whose predictions he could have compared his measurements. Zavaritskii was a Ph.D. student of Shalnikov and now everything was together to make such comparisons. Zavaritskii found that the Ginzburg–Landau theory described his experiments brilliantly.

These experiments were done on thin films, which Zavaritskii prepared by evaporating a metal on a glass substrate. Everybody was satisfied, but Shalnikov was not. He was a perfectionist and said something like this, look, I am not satisfied because we do not know how well you prepared your film. You evaporate a metallic wire onto a glass substrate. The metal atoms reach the substrate, but the substrate is warm, it is at room temperature and therefore the metal atoms are free to move around and they probably form some micro-crystallites. Thus – Shalnikov continued – instead of having a uniform film, you have an assembly of small crystallites. Therefore your film is poorly characterized.

Shalnikov proposed a way to resolve this problem. He suggested keeping the glass substrate at liquid helium temperature. In this case the atoms reaching the glass substrate will stick to the substrate and will not agglomerate into crystallites, and a smooth and uniform film will form. Shalnikov also warned that the film should not be heated until the measurements have been completed. Although this was not easy to do, Zavaritskii managed, and he made all the measurements for such low temperature films. The results astonished everybody because the measurements did not fit the Ginzburg–Landau theory at all.

Abrikosov and Zavaritskii started to discuss what actually has happened. Of course, one could always say that the theory was wrong, but the theory was beautiful by itself, and it explained other properties correctly. These experiments and discussions prompted Abrikosov to work out a new theory, and the superconductors for which he worked it out he called superconductors of the second group. Eventually, they became known as Type II superconductors. It turned out that the Type II superconductors were the widespread kind and what was considered Type I, seldom occur.

After the Bardeen–Cooper–Schrieffer theory of superconductivity was published, another colleague of Abrikosov's, L. Gor'kov showed that the Ginzburg–Landau theory was a limiting case of the Bardeen–Cooper–Schrieffer theory. In other words, the Ginzburg–Landau theory qualitatively contained everything that was in the BCS theory, but the BCS theory was a more general approach. The equivalence of the BCS theory and the Ginzburg–Landau theory — under certain conditions — was understood later.

Abrikosov and Zavaritskii published their papers separately in the same issue of the journal of the Soviet Academy of Sciences — Zavaritskii his experimental data [*Dokl. Akad. Nauk SSSR* **1952**, *86*, 501] and Abrikosov his theory [*Dokl. Akad. Nauk SSSR* **1952**, *86*, 489].

Abrikosov stresses: often a discovery comes in an unimportant form. But then the significance of the discovery becomes more and more recognized. At the beginning, the Type II superconductors were considered to be something very exotic. It referred to a film that was prepared in a very special way. At that time it was impossible to predict that all superconductors discovered since 1916 were Type II superconductors. Today, people even ask why we should distinguish between Type II and Type I because all superconductors are Type II. The fact, however, remains that their understanding started with Zavaritskii's very artificial thin films.

Following this initial success, Abrikosov went on and made further important theoretical discoveries, including the regular structures of what is called today the Abrikosov vortex lattice. Some of his theoretical findings gave the interpretation for experimental results that another Russian physicist had produced in 1938. His name was Lev V. Shubnikov, a

Ukrainian scientist who had studied with a Dutch professor by the name W. J. de Haas. In his Nobel lecture, Abrikosov paid special tribute to Shubnikov whose career was cut short when the KGB, the Soviet secret police, arrested him and falsely accused him of organizing an anti-Soviet strike. Shubnikov was executed by the KGB.

Here we continue with the actual conversation with Alex Abrikosov. This is what he had to say when I asked him whether it was difficult to have his results accepted:

First I showed my results to Landau. He disagreed with my theory completely, with the idea of the vortices, and so on. If I had insisted on my results, I might have convinced him, however, at that time, we were busy with quantum electrodynamics. There were many developments there, which started at the end of the 1940s and continued into the 1950s, and about which we were very much excited. Landau and his group were universal theorists; they were interested in many things simultaneously. Before we could become members of his group, we had to pass a system of tests, which was called the Landau Minimum. It was a set of nine tests in different fields of theoretical physics and two in mathematics.

How many were you in his group at any given time?

It was a complicated organization. He had a group at the Institute of Physical Problems, which is now the Kapitsa Institute. That group was not very large, at that time it consisted of E. M. Lifshitz, I. M. Khalatnikov,

Lev Landau (1908–1968), late 1940s–early 1950s (from Abrikosov, A. A. *Academician L. D. Landau: Short Biography and Review of his Scientific Works.* Nauka, Moscow, 1965, in Russian).

and myself, in addition to Landau himself. Later that group increased when L. P. Gor'kov and some others joined us. Landau then had his former students in different other institutions, at the Kurchatov Institute, at the Lebedev Institute, and elsewhere. He had associates in many places. These people came to him and they attended his seminars. Those who lived in Moscow, they attended these seminars regularly, every week. All of them participated in the seminars and this participation meant the following: there was a list of speakers and everybody had to give a talk when his turn came. The speaker had to tell about some works about which he read in the journals. In some cases, they reported on their own works. I was a kind of secretary of the seminar and I had to come to Landau with the journals that he was interested in, and above all the *Physical Review* and he marked the papers that he wanted to hear about. Then his associates came to me, looked up the papers and chose their assignments.

How many people attended the seminars typically?

The nucleus was about 15 people, but others could come and sit in; Landau did not object to their presence. Preparing the reports was a heavy duty. Landau was very critical and if a person did not prepare well his talk, meaning mainly that he did not fully understand the paper which he was supposed to talk about, then Landau was furious. Although I was the secretary of the seminar, I also had to prepare reports. Once I was reporting on a paper that was especially difficult for me because I was not familiar with the field, and it meant a tremendous effort for me, but I remembered that paper for the whole of my life. I have benefitted from the ideas of that paper and accomplished much more in that topic than the original author did.

You became a corresponding member of the Soviet Academy of Sciences very early; you were 36 years old at the time of your election, which was unusual.

It was early and it was good.

Previously you said that Landau disagreed with your theory about the vortices. How did you finally convince him as I presume you did?

I will tell you the story. Landau was interested in rotating helium. Landau created the first theory of superfluidity of liquid helium and correctly predicted many of its properties. He was interested in everything concerning liquid

Murray Gell-Mann (Nobel Prize in Physics, 1969) and Lev Landau (Nobel Prize in Physics, 1962) in Moscow, 1956 (from Abrikosov, A. A. *Academician L. D. Landau: Short Biography and Review of his Scientific Works.* Nauka, Moscow, 1965, in Russian).

helium. But rotating liquid helium was a problem. On the one hand, he understood that it had to rotate if you rotate the vessel where it is contained. On the other hand, there could not be microscopic rotation and as a consequence, there could not be vortices; there was a condition of the absence of vorticity. But in the center, there would be a vortex. Landau and Lifshitz published a paper in which they described concentric cylindrical layers of superfluid helium, which rotate with different velocities. The center is not rotating at all.

Then we learned about a theory by Richard Feynman, which considered quantum vortices. Landau was so interested in Feynman's theory that he did something exceptional. He went to the library, found Feynman's paper, and read it. He usually did not like to read other people's papers. The seminar existed partly just because of that. After having read Feynman's paper he came to us and told us that Feynman was right and we were wrong. I asked him immediately why he believed it when Feynman wrote about it and did not believe it when I did the same? I took my manuscript out of my drawer and showed him. Landau understood that I had shown him the same at least two years before Feynman did it. Landau now agreed with me. Of course, Feynman did not do exactly what I had done; I worked on a theory of superconductivity and Feynman worked on liquid helium although liquid helium might be considered to be a superconductor with some extreme characteristics. In any case, I published my paper two years after Feynman's paper and it contained some more findings too, but

Lev Landau and Niels Bohr (1885–1962, Nobel Prize in Physics, 1922) at Moscow State University, 1961 (from Abrikosov, A. A. *Academician L. D. Landau: Short Biography and Review of his Scientific Works.* Nauka, Moscow, 1965, in Russian).

Feynman's paper helped me in convincing Landau about the validity of my claims.

Landau was known to be very sharp. Why did he not see the value of your theory?

Landau had a very special mind. Because he understood things much deeper than others, some things were not so easy for him to understand. He often noticed contradictions, which the original author did not even think about.

In 1962, he had a tragic accident.

After his accident, he became a different person.

Did you have any communication with him after that?

By then I had become quite independent.

When did you become independent in research?

He always encouraged his students to develop their independence. People talk about the Landau School, but on average, he had one student per three years. Landau never proposed any topics for research. His students had to find their research projects for themselves. After I passed the Landau Minimum, he told me to look for a topic for my own research. I asked him, how? He told me to read the journals, attend the seminars, and what is most important, discuss with experimentalists. I have done that all my life. I have read the papers, listened to people and listened to them very attentively; I have developed long ears.

Did your work suffer from the isolation from the West?

Maybe not so much, but, of course, my work suffered from that, of course, it would have been better if I could have come to the West and talk with people.

When did you go for the first time?

It was to India at the end of 1965. I gave some lectures on my work and the visit was useful.

Did you teach in Moscow?

Even some additional work that you might not be eager to do may turn out to be useful. For example, I never wanted to teach. However, at one point there was an acute need for my teaching at Moscow University. I lectured there on condensed matter theory. I hated that because it took away valuable time from my work. Eventually, however, I became a good speaker and a good teacher. I had the same experience with looking for my own research topic because at first it was very hard. Then, I enjoyed my independence. Landau's motivation in not proposing research topics for his students was not entirely out of his desire for them to become independent. It was also greediness. When he had a problem, he wanted to work on it himself.

Greedy in a good sense?

I do not know whether it was good or bad, but he wanted to do his work himself. He said to us many times that the most important thing in a research project is the initial idea after which everything is merely sophisticated technical details.

I have read that he had participated in the Soviet atomic bomb project and then he quit at some point.

He quit at the first possible occasion. He quit after Stalin's death.

Did you participate?

No, I did not. He wanted to include me and I did not object. However, the KGB did not agree for me to be included in that program.

Were you Jewish?

My mother was Jewish, but Landau was entirely Jewish and so was Lifshitz and so was Khalatnikov, and they participated in the program. I was only half-Jewish.

So this was not the reason.

No, it was not. Only much later did I learn about the reason. My father had a younger brother who was a diplomat and when the Revolution happened, he was serving at the Imperial Russian Embassy in Japan. He never returned to the Soviet Union; he remained in Japan, first as a member of the staff, later he became the acting ambassador. When Japan recognized the Soviet government, he had to leave his post and he became a private citizen and stayed in Japan until the end of World War II. Then he moved to the United States and did not live for a long time. However, he left behind a manuscript, which was eventually published under the title, *Revelations of a Russian Diplomat* by Dmitrii Abrikosov. It took quite a while before it was published, after some historian researcher discovered it in the Library of Columbia University. I have read it, it is a fascinating book. Although the KGB did not know about the book, they knew that I had an uncle abroad. I did not know about his existence, but they did. We were living in the Soviet Union and my father never told anybody about his brother. This was the reason why the KGB did not let me participate in the program. In hindsight, it was lucky for me because I probably would not have done this work if I had gone into the bomb program. I consider myself a very lucky person although, seemingly it was not always so.

Alex and Svetlana Abrikosov in their home in Lemont, Illinois, 2004 (photograph by Magdolna Hargittai).

You had a brilliant career in the Soviet Union.

I had a good career. It might have been even better if I had not married a French woman in 1970. It was my second marriage. I had divorced my first wife after twenty years of marriage. This French woman was the wife of my colleague in France, but we first met in India. My second marriage lasted for seven years, we lived in Moscow, and we have one son. Her father was Vietnamese and her mother was French, but they lived in France. Her father was a college professor but in the family he was a dictator; his three daughters rebelled. My former wife was the youngest and she married very young and they had two children. She was very gifted in languages; first we communicated in English, then she learned Russian fast. Even in her first marriage she did not feel independent enough and to enhance her independence might have been her motivation to marry me. However, in Russia, she depended on me entirely, not because I would suppress her in any way but because as a foreigner she was completely helpless there. She became depressed and on one occasion when she went to visit her relatives in France, she did not return, and we got a divorce. Eventually I married my present wife. We have been married for 28 years and we have a daughter who went to college here, graduated from medical school, and is now a surgical resident in Savannah, Georgia.

How do you compare your positions in Russia and in the United States?

I work at Argonne National Laboratory, I am Distinguished Argonne Scientist, the only one in such a position. In Russia, I became full member of the Russian Academy of Sciences in 1987 and at that time I was director of the Institute of High Pressure Physics. This was exceptional because I was not a party member.

Why were you not a party member?

I can tell you, but it takes some time. When I was young, I was a member of the Communist Youth Organization as everybody was and without which nobody could get into higher education. My assignment was to talk with people at election times to make sure that everybody went to vote; I was as it was called at that time an "agitator". I was good at it and was promoted to so-called "propagandist". I had to teach others — grownups — Stalin's biography, which to a large extent was also the history of the Soviet Communist Party. To facilitate learning, I found some books for my "pupils" to read and at our study sessions we would discuss these books. When the regional party people came to inspect our progress, they were terribly pleased with what I was doing. They decided to further elevate me and I was appointed deputy party secretary for propaganda in spite of the fact that I was not even a party member. They wanted me to become a party member, of course, but at that point I sought my father's advice.

My father was a very well-known scientist, full member of the Academy of Sciences, Vice President of the Academy of Medicinal Sciences; what made him most famous was that he performed the post-mortem autopsy of Lenin and later Stalin. He had all possible awards and his name was known to almost everybody. At the age of 65, he became a party member. I described him my situation, and asked for his advice. I told him that I would probably like to join the party, it was the ruling party, and of course, our psychology was that this system would last forever. If we did not join it, it would be left to some nasty people. I thought that it was our duty to join it. My father said, "Do you understand that joining the party means that you would become a soldier of the party?" I was prepared for that, I told him. Then he said, "You probably do not know what it really means. Do you know how I resigned from my position as director of the Institute of Morphology, which I created myself?" I told him that he probably felt old and sick. He said, "It was nothing like that. One day I was called to the regional party committee and I was asked whether I knew that all nationalities in the Soviet Union were equal. Of course, I knew. Then they told me that half of the co-workers in my Institute were Jewish whereas in the whole population the Jews amounted to a few percent only. They told me that this situation was unjust to the other nationalities and that I should correct the situation."

When did this happen?

A few years before Stalin's death. By then he was a party member. My father never openly revolted, but he did not take any action. When he was called the second time, he explained to the party people that he examined the files of all his collaborators and found no excuse to let any of them go. They were very good and they had to be because he had personally hired them in the first place. The party people told my father that if he could not take action, they would do it themselves. My father went home and wrote his resignation. My father told me that I must be prepared for such things if I became a party member. I immediately understood that I would not become one.

Do you think if he had not been a party member the party would have not tried to direct him to such an action?

I don't know.

Why, do you think, he joined the party?

Because he was a soft person and if they pressured him, he succumbed and joined it. Many people were in the party without ideological devotion. Besides, the party was not anti-Semitic initially; it became so later.

So you did not join the party.

I told them that when I am doing something I am doing it with zest. If I joined the party, I would make a very successful career in the party even if I had no intention of doing so. That would mean that eventually I would have to give up science. I felt — I told them — that I was at a branching point and it was at this point that I had to stop.

Was Landau a party member?

No.

How did it happen that you left Russia?

After Brezhnev died, his successors died rapidly and Gorbachev came. He initiated perestroika — reconstruction — and political changes started to happen. I always read the newspapers attentively and could read between the lines. I understood that under the pretext of perestroika, they were destroying the socialist economy without replacing it with something. I anticipated that the first victim of the situation would be basic science. I was trying to create some industrial connections for our institute, the Institute of High Pressure Physics of which I was director. Such connections might help us earn money especially by producing diamond instrumentation. We actually started such activities but when we wanted to legalize them, there was no way to do so; at that time private enterprise was still frowned upon officially. It dawned on me that we had no chance to survive. I understood that there was no future for us and we had to emigrate. Some of my friends were already abroad, mostly in America, and had satisfactory positions. In 1991, I accepted the offer of the Argonne National Laboratory and came here.

Did you not feel sorry to give up the prestige and positions that you enjoyed in Russia?

My reputation was always based on my science and not on my position.

In my conversation with Philip Anderson, we talked a little about the Soviet Union and he told me that by 1982, the Soviet Union was already a paper tiger. What do you think about this?

The Soviet Union was a paper tiger in the sense of the economy. We could see that the variety of consumer goods decreased. There were only two kinds of bread and two kinds of sausage. One was for 2.20 and the other was for 2.90 and that was all. Anderson saw the situation from the outside and what could be seen from the outside was how the space program deteriorated. The earlier Soviet advantage had disappeared and the Americans took over.

The collapse of the Soviet Union had been predicted since the 1920s, but prior to its collapse, nobody had predicted it.

The real collapse was due to Gorbachev's perestroika.

Don't you think that the arms race also contributed to the collapse?

Of course, it did.

In this sense, the United States was pressuring the Soviet Union to spending more and more on defense.

Yes.

Don't you think that in this sense the Strategic Defense Initiative — Star Wars as it was known popularly — whether it would have worked or not, also contributed to the collapse of the Soviet Union?

Yes.

So Edward Teller played a role in this.

He did but he was more important in creating the hydrogen bomb in the United States. His main argument was that the Russians would create their own hydrogen bomb anyway. By the way, the Soviet scientists had better ideas than Teller about how to build the hydrogen bomb. Vitaly Ginzburg was the person who proposed to use LiH, which made it simpler and cheaper and more effective than the American bomb.

I met Teller once and had a long conversation with him. I was very impressed by him. He was smart and friendly; it was a surprise for me because he had always been portrayed as a hawk.

Coming back to your personal history...

I understood that there was no future for us in Russia. The situation was volatile in the country. On the other hand, I was over 60 years old and it was not trivial to get a good job in America at such an age. Again, I found myself at a branching point because I knew that my associates at my institute would try to prevent me from leaving if the situation further deteriorated. It was not that they would hold me by my shirt but by moral pressure. Incidentally, after I married that French woman, for many years I was not permitted to travel abroad. At the end of the 1980s I got permission for the first time after many years. I got permission because I went directly to the KGB and asked them to give me permission. When I could travel again, on one of my visits to the United States, I called some reliable friends and asked them to find a position for me. One of them told me that it was a difficult task because I was in a high position in Russia and such high positions were very rare in the United States. I was determined to come only if I had a job.

Most people when they immigrated, came first and tried to find a job afterwards. I did not want to do that. Because there was no news for some time, I thought that it just did not work out. Then, there was an international conference on low temperature physics in Brighton, England, and one of the people whom I had asked, told me that he had a position for me. That was the position at the Argonne National Laboratory. It worked in the following way. I went by invitation to Venezuela for one month. I used that trip to come to Chicago and spent a week at Argonne. I did not know anybody in the Laboratory at that time. I talked with many people in different labs, which was not difficult for me because I was always interested in other people's work. As a result of this visit, both sides came to the conclusion that we liked each other. I returned to Moscow, but had made an agreement that if they write me a letter inviting me for a one-month visit that would mean that I have a permanent job at Argonne. The letter came, we arrived, and only upon arrival I learned that Argonne had arranged with the Department of Energy for a Distinguished Scientist position for me. They had never had such a position before and afterwards they decided that there would be only one such position at Argonne at a given time.

Do you drive to the Lab?

I do. I drove back in Russia too. I got my first driver's license when I was 19 years old, exactly the same year when my present wife was born.

Have you been back to Russia since?

Never. I have travelled everywhere else.

Any plans to visit Russia?

No. I do not want to. Most of the good physicists are here. I used to be a member of a close group of friends together with my first wife. When I divorced her, these friends stayed friends with her. We did not become enemies, but the intimate friendship was gone for me. There are no close friends for me in Russia whom I would miss.

The President of Russia, did he send you a telegram on the occasion of your Nobel Prize?

He sent me a telegram, he congratulated me, and I got an invitation from the Russian Embassy to attend a reception in Moscow.

Not in Washington?

No, in Moscow. In Washington, I attended a reception by George W. Bush. The Russian Embassy invited me to go to Moscow for a reception by President Putin. I responded and declined saying that I was very busy, which was true. The first day after the announcement of the Nobel Prize I could not even eat because I could not leave the phone. I got thousands of e-mails, which I had to answer.

You knew about your nominations.

Of course, I knew about them, and I said to myself that if it won't happen this year, it will never happen.

Why?

There were two things. One was that I knew that this year several Russians who are working in Scandinavia — and there are quite a few of them — were trying to promote my nomination. Therefore, I thought that this was something different from what had happened before. Previously Russia was rather isolated and countries like Sweden and other Scandinavian countries and other countries, like Japan, were not eager to accept Russians in permanent positions. But their attitude has changed. There are many Russians now all over the place. During my recent travels I met many, many more than I had expected. The professors who work in Scandinavia exert a great influence on the Nobel Committee so this worked in my favor. I have powerful enemies but I have many more good friends. I had done a lot for them. For example, in St. Petersburg, practically everybody who had got the Doctor of Science degree — the second degree in Russia, higher than the Ph.D. — had me as a reviewer. Many of the people from St. Petersburg ended up in Scandinavia. So that was one thing.

There was also something else. Never in my life, before the announcements did I get any warning. This year I got a letter from the Nobel Committee for Physics, which said that I was nominated for the Nobel Prize. Afterwards, I discussed this with the chairman of the Committee and he told me that this was absolutely impossible and that it should not have happened.

So how did it happen?

I know the person who sent me the letter, but I can't tell you his name. Somebody in position, a person whom I did not know before.

It took 50 years from your original discovery.

My paper about the Type II superconductivity was published in 1952. From that time it was 50 years indeed. However, we should think about different things as well and we should see them in their complexity. My work on vortices was published in 1957. In the same year, the Bardeen–Cooper–Schrieffer work was published, the microscopic theory of superconductivity, which was recognized immediately. It was something that people were waiting eagerly. Therefore, they were awarded the Nobel Prize relatively soon, in 1972. However, the Nobel Committee cannot award prizes for the same topic very close to each other. As they were waiting further, my work was becoming rather old. It became realistic to expect the award at the end of the 1970s. I knew I had nominations. Nevertheless, the Nobel Committee thought that one prize for the theory of superconductivity was given already. Of course, other topics kept coming up, and so on. In the meantime our work had become yet older and older. However, in the 1980s, there was a renewed interest in the vortices that could not have been predicted. Many labs in the world started working on related topics. Accordingly, the number of nominations kept increasing. They came from all over the world. I knew about them. I also knew that there were no nominations from Scandinavia before, but this year this changed dramatically.

Is your life changing now?

I travel more although I try to limit it because I want to continue my work.

When you met with Ginzburg, the two of you, what did you talk about?

Nothing special. We talked about our lectures so as not to overlap too much.

His waiting period was even longer.

Yes, because their work was in 1950. Most of his life he was afraid that I would get the Nobel Prize and he would not. There was much more rumor about my work than about his.

Was it because it was difficult to delineate his work from Landau's work?

That may be, because Landau had already received his Nobel Prize after his accident. There have been important omissions. For example, for quantum electrodynamics, Feynman, Tomonaga, and Schwinger got the Nobel Prize [in 1965], but Dyson did not get it.

Ginzburg could have been easily included with Landau in 1962.

No, because Landau got it for a different discovery. Landau got it — as the citation said — for theories for condensed matter, especially liquid helium. This meant the whole concept of quasi particles, theory of superfluidity, and Ginzburg had nothing to do with them.

So you are ascribing the 2003 Nobel Prize in Physics to the renewed interest in your works.

It was a kind of renaissance, starting in the 1980s and it kept increasing. In every issue in *Physical Review* you will find something about vortices.

Where did Leggett come into the picture?

Leggett was lucky in the following sense. I now understand how these nominations work. I know that there were nominations also for Gor'kov and for Andreev for this year. Gor'kov's role was that he showed that the Ginzburg–Landau equation is the limiting case of the BCS theory. Andreev's work was independent, it was also based on the BCS theory; he predicted the so-called Andreev reflection, which became popular recently. It was old work but interest in it increased recently; something similar happened to it as was the case with my work. Apparently, the Nobel Committee could not decide between Gor'kov and Andreev and they chopped off both names. They still had a third slot and they used in for Leggett who was also nominated.

May I ask you about religion?

I am not religious. My parents were not religious either. How could they be? My father was originally orthodox Christian. His first wife was a Polish woman and he also converted to become Roman Catholic. However, Russian people are not very religious. The new government now is trying to re-introduce religion. Putin is a hypocrite. On the one hand, he is proud of his service in the KGB, which killed millions of people and on the other hand, he now says that he is religious.

When I lived in Russia I did not agree with much of the politics of the government; I did not like it. I liked much more what America did. Even in those days I was more American than Russian.

And today?

I was asked in Stockholm by a Russian journalist whether I considered myself American or Russian. My answer was that it was a wrong question because science is international. If the question is about my citizenship, I am an American citizen.

But also a Russian citizen.

Yeah, but I never use it.

Do you have a hero or heroes?

My heroes have changed with age. Now my hero is Niels Bohr for the following reason. He continued working in physics until he was very old, almost until his death, and he published good work. There is an illness in old age: on the one hand, the abilities of a person decrease, on the other hand his self-esteem increases, he may think that what he can do now is only great work. Of course, this is not the case and such people usually do not produce anything. Bohr had a different attitude. He worked for fun, not fame, and with that attitude, he could work any time. I am doing exactly the same.

You seem to be vigorous and youthful. Can this have something to do with the fact that here you are not the big and revered academician and director, but a member of an egalitarian community?

Of course it helps.

For the Americans, past glory means very little.

It depends. They revere the British royalty. When somebody is knighted in Great Britain and is addressed as Sir, the Americans also address him Sir.

Do you have any message?

I always try to convey that science is such an interesting thing; it is interesting in itself; its interest is independent of the awards, titles, and the money it can earn. For any young man in science there comes a time for a hard choice. He may follow fashion either in methodology or in the subject of investigation or both and his papers will be easily accepted by journals, but he will never get the Nobel Prize. The other path is much more difficult. This is to stay independent of fashion; and go independent.

I would like to add that I was never a dissident; I respected the people who were like Sakharov, Orlov, and others; I knew them. I consider that their cause was a lost cause. They could never reach their goals. In any case, I had nothing to do with that because I wanted to do science and it was incompatible with what they were trying to do. Nonetheless, the authorities were always suspicious of me, particularly when I married a French woman. They found me unmanageable.

I was independent. My mother was clever and able, but she was also very hard, even dictatorial, and for this reason, from my very childhood, I wanted independence. I learned to fight for my independence. At the same time, I was smart and I did not do foolish things. In science, I could have a choice and Landau encouraged it, so I chose my own way. Think thoroughly about various things and choose your own way. For a career, publications, it is harder, but it is also more enjoyable.

Did you have any difficulties because your mother was Jewish?

No. But I was never considered to be a Jew; I was always considered to be a Russian. The approach to this question in Israel and the Soviet Union was the opposite of each other. I had a Russian surname, my paternal name (from my father's first name) was Russian, and my father was Russian, so I was Russian. You had to choose the right parents, I used to say. If you had a Jewish father and Russian mother in the Soviet Union, you were considered a Jew. In Israel, it goes the opposite way.

The only time I had problems was when at first I was not accepted to the Institute of Physical Problems after I got my Ph.D. I did my Ph.D. in this Institute, which is now the Kapitsa Institute, and I wanted to stay there after graduation. The representative of the KGB at this Institute — and every institute had a resident representative of the KGB — said that I could not be accepted to that Institute because my mother was Jewish. In this case, I was just lucky. It is very important to be lucky and I was always lucky. At the time when I was trying to stay in the Institute, Choibalsan, the Mongolian communist leader died, and he died in Moscow. There was a medical communication about his death in the newspapers, which was signed by several leading doctors and among them there was my mother who was the head of the pathology department of the Kremlin Hospital. She was a former student of my father.

What was her name?

Her name was Fanny Davidovna Wul'ff.

Did she use this name?

No, she used her married name, Abrikosova, Fanny Davidovna Abrikosova, but Fanny Davidovna gave away her being Jewish. The KGB man was impressed. If she was trusted to carry out the autopsy of Choibalsan, it may be all right to let her son work in the Institute for Physical Problems. So I was accepted. You may not have expected such a story, but I trust you will want to include it in the interview.

Of course, I will.

You remember Shalnikov from the beginning of our conversation; he was a very nice person and he had a very good sense of humor. He told me, "You should hang a portrait of Choibalsan in the coffin on your wall."

Luis W. Alvarez, 1968 (detail from a photograph courtesy of the Lawrence Berkeley National Laboratory).

10

LUIS W. ALVAREZ

L uis W. Alvarez (1911–1988) received his degrees from the University of Chicago: B.Sc. in 1932, M.Sc. in 1934, and Ph.D. in 1936. He joined the Radiation Laboratory of the University of California at Berkeley, rose to be Professor and stayed there. The only interruptions were during World War II, when he worked at the Radiation Laboratory of MIT (1940–1943), at the Metallurgical Laboratory of the University of Chicago (1943–1944), and at the Los Alamos Laboratory of the Manhattan District (1944–1945).

Of his many distinctions, he was a member of the National Academy of Sciences of the U.S.A., the National Academy of Engineering, the American Academy of Arts and Sciences, the American Philosophical Society, and the American Physical Society (of which he was President in 1969). He received the Medal of Merit, the National Medal of Science, and many other awards. He was awarded the Nobel Prize in Physics in 1968. The unusually long citation read, "for his decisive contributions to elementary particle physics, in particular the discovery of a large number of resonance states, made possible through his development of the technique of using hydrogen bubble chamber and data analysis."

Alvarez published his autobiography shortly before his death, *Alvarez: Adventures of a Physicist*, Basic Books, Inc., Publishers, New York, 1987.

Clarence and Jane Larson recorded a conversation with Luis W. Alvarez on March 13, 1984, in Berkeley.*

---

* "Larson Tapes" (see Preface). In part, this has appeared in *The Chemical Intelligencer* **1999**, 5(1), 43–49.

Origins

I was born in San Francisco, right across the bay in 1911. My father was a physiologist who used to work in San Francisco downtown as a clinician in the afternoon to make a living. As his fees went up as he got more and more proficient, he spent less and less time in the office and more in the laboratory. My introduction to science was through my father's laboratory. I found that his biological work was of no interest at all to me, but I did enjoy the electrical equipment that he had; I learned how to run the Wheatstone bridge and to make all kinds of electrical measurements when I was in high school.

Dad was invited to join the Mayo Clinic in 1926, so at age fourteen I moved to Minnesota in the middle of the winter. It was cold and I'd never been in snow or ice before, so it was quite a shock. After dad had done a few years of medical research, physiological research, the Depression came along and the Mayo Clinic could not afford such a big staff in the laboratory, so he went back to clinical work. When he retired, he found some measure of fame as syndicated medical columnist. I'm always asked if I'm related to Dr. Walter Alvarez and my answer is, yes, he is my son. I'm working actively with my son who is a geologist by the name of Dr. Walter Alvarez.

Schooling

I took chemistry and physics in high school, and I liked them very much. When I went to the University of Chicago, I majored in chemistry because I'd never heard that there was such a profession as a physicist. I'd heard of lots of great achievements of chemists. Even later, when I used to go to cocktail parties, after I had my doctor's degree and people would ask me about what I did, I would always say that I was a chemist, because it was too complicated to explain what a physicist was and what he did. In college, I did discover physics and I wasn't a very good chemical student. I enjoyed physics tremendously so I switched over and became a physicist. My first work was with Arthur Compton. He won the Nobel Prize for the discovery of the Compton Effect in X-rays. He'd just changed fields and was working with cosmic rays. He suggested an experiment to find the electrical sign of the particles that make up the cosmic rays. I built the apparatus and took it down to Mexico City where I made the measurements and we published a paper together in about 1933. We surprised everybody by showing that the cosmic rays were positively charged. Then I did a few

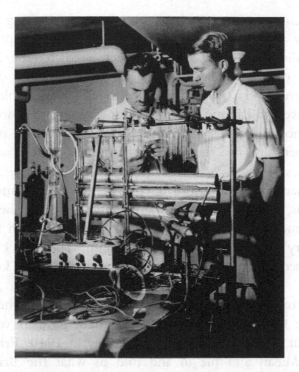

Graduate student Luis Alvarez with Arthur Compton (Nobel Prize in Physics in 1927 for the Compton Effect) in 1933 (courtesy of the Lawrence Berkeley National Laboratory).

other things that didn't amount to much. I had the fortunate circumstance that my sister was Ernest Lawrence's secretary. Ernest offered me a job here in Berkeley in the laboratory and I've been here ever since except for five years during the war. I've always told people that I got my job here because of my sister but I've probably kept it because of myself. I don't think I would've lasted for almost fifty years on the basis of my sister's recommendation.

First Discoveries

The first day I got here, Ernest said to me, "I just got the money to build a big cyclotron, I want you to design a magnet for it." I told him, "Professor Lawrence, I have no experience designing magnets; I don't know anything about it." He said, "You'll learn" and I did. I made lots of model magnets and eventually came up with a design of what turned out to be the 60-inch cyclotron, the largest one at the time in the world. Then I got into nuclear physics and spent four years or so, full time doing a lot of interesting things. Probably the first useful thing I did was to discover a new mode of radioactivity called K electron capture. This was a mode of radioactive decay that had been predicted by Enrico Fermi. Then I found out about the radioactivity of hydrogen-3 and helium-3. At that time everybody believed that hydrogen-3 was stable and helium-3 was radioactive and I found that it was just the other way around. Tritium is, of course, now a famous material and it is hydrogen-3; it is frequently in the newspapers, and it is the material that people used to make hydrogen bombs.

Mark Oliphant and Rutherford discovered hydrogen-3 and helium-3 as high-speed ions, but they didn't know what happened to them afterwards. This is like somebody discovering the alpha particle, which is quite a different discovery from the discovery of helium. They are closely related, but if you discovered the alpha particle, you still don't know anything about helium.

One of the other important things I did was probably when Felix Bloch and I made the first measurement of the magnetic moment of the neutron; that took about a year in 1939. Then one day in 1940, Ernest Lawrence called Ed McMillan and me in and told us what the British scientists were doing in the defense work. He told us about radar; it was the first time that we heard about that. He said that the United States was setting up a laboratory in Cambridge, Massachusetts on radar, in particular working with the newly-invented cavity magnetron that was invented in Mark Oliphant's laboratory by J. T. Randall and H. A. Boot. That, in effect, made microwave radar possible. At that time magnetrons typically gave 2 or 3 watts, and all of a sudden, Randall and Boot had 50,000 watts in their very first attempt and inside of a year they were up to a million watts. It was an enormous improvement in technology that made radar possible. It was then that I left for MIT to work on radar.

Ground Control Approach

While working at MIT, one of the most significant events was when one day I was watching the first automatic radar that could follow an airplane. Before that, if you wanted to follow an airplane, you had to turn some cranks and watch the needles. This was a system in which the radar antenna automatically kept pointing at the airplane, following it no matter where the airplane went. An arrow on screen could lock at the plane and follow it. It had several mechanisms the first of which I had ever seen. Suddenly it occurred to me that if you could locate an airplane well enough to shoot it down, with three coordinates, that is, two angles and a distance, then you ought to be able to feed that same kind of information to the pilot, and help him when he was up in getting him back down on the ground. That was the basic concept of GCA or Ground Control Approach. In those days, if anybody had an idea, and people thought it was a good idea, then you could start immediately working on it. You didn't write proposals, have proposals and that sort of stuff. So within a day or so, I was working on the GCA. I put together a team of people. We thought we could use this so-called gun-laying radar system in directing the pilot. The difficulty was that when the plane came close to the ground, half the time it would look down into the runway and see the reflection of the airplane above the runway. The runway was a good mirror. That was a kind of a useless thing if you are going to track somebody down, you wouldn't want to tell him, you're 25 feet down in the ground, you want to fly up to get onto the runway. So we had to go back to the old drawing board and I invented a new kind of antenna that made it possible to distinguish between the airplane itself and its reflection in the ground.

All during the war and even after the war, during the Berlin Airlift, GCA was the only way that people had for landing aircraft in bad weather. Nowadays we use a thing called the ILS, landing system, but it wasn't available until several years after the war. At the time of the Berlin Airlift [in 1948], it was already in development, but they couldn't use it because it had trouble from the reflections from buildings. The runway coming into Templehof Airport in Berlin was between a couple of apartment buildings and GCA did its job just fine, and brought in all the coal and food.

Now commercial airplanes carry radar as well but it's only for the weather. There is then the respondent system in which the pulse from the radar on the ground triggers up an instrument on the airplane called the transponder that sends back a much louder signal and in fact identifies the airplane and tells its altitude. So the radar has been enormously improved. I have the fundamental patent on radar transponders. It has run out long ago.

Manhattan Project

Robert Oppenheimer had tried for a long time to get me to leave the radar business and come out to Los Alamos. I finally finished up the three big jobs that I did at MIT and I talked only about one of them; I had two others, radar systems that I essentially invented and developed. I went over to England in the spring of 1943 and demonstrated GCA to the RAF for about three months. We landed all kinds of airplanes; every kind of airplane that the RAF had and we met every kind of pilot from sergeant to air chief marshal. We demonstrated the virtue of our system for all kinds of weather.

Then I came back and I had planned to go to Los Alamos, but I got a cable from Robert Oppenheimer when I was still in England saying, would I mind working with Enrico Fermi for a while. Enrico wanted Emilio Segrè to leave Los Alamos and come to Chicago and work with him, but Robert Oppenheimer got me traded and I went to Chicago instead, and worked with Fermi for six months, which is hardly a fate worse than death for a nuclear physicist. It was certainly a delightful experience for me and a whole new way of doing physics. I had lunch with Fermi just about every day for six months.

I felt that I was completely disconnected from the war. I had been living with guys who were fighting the war and having all my meals, in England, with RAF pilots who would go out at night flying over Germany, drop bombs and very often not come back and would not be there on the next day, just dead. I really thought that there was a war going on and I didn't feel happy just playing around in nuclear physics. It was clear to me that I should go to Los Alamos and I did go.

In Los Alamos, my main job was something that made possible the development of implosion bombs. Plutonium had originally been proposed to be shot at the uranium, but the discovery of spontaneous fission made that impossible. Then the implosion method, which had been first proposed by Seth Neddermeyer, was the only way that plutonium could be detonated. There were three things wrong with it, three problems that had to be solved; one of which had to do with the simultaneity of detonation of electric detonators. At that time the best simultaneity that one could get out of detonators was about one millisecond. Those were the so-called seismic detonators that were used in oil prospecting business. The standard detonators had about two milliseconds. We needed about a tenth or a hundredth of a microsecond. So an enormous improvement had to be made and I suggested a way to do that and my young colleague, Larry Johnston implemented it. The first time he tried it he got a microsecond. It was a solid breakthrough and I worked on it for many months during the war.

The discovery of spontaneous fission of plutonium changed the lab upside down; I've never seen such a sudden drastic change in any program in my life. Joe Kennedy was in charge of the chemistry division there and his main job, as far as I can tell, was to purify plutonium to get rid of the light elements like lithium and boron and beryllium so that the alpha particles from plutonium would not likely meet neutrons. But when you find that you are living in a rainstorm of neutrons, that job suddenly disappeared. So they just dismantled that whole section of the chemistry department and turned it into things having to do with high explosives. The detonators were one of the things that made it possible to create such explosive weapons that had never been described in the open before to the best of my knowledge.

When this was solved, it was in April 1945, I went to Robert Oppenheimer and told him that I was through with my old job and I'd like to have a new one that would get me overseas. I had been overseas in the European theater with radar and I thought it would be interesting to get out in the Pacific. So he said that as a matter of fact we had a job that would just fit you. Normally, when you develop a new weapon, like a new bomb or a new rifle shell, you take it out to Aberdeen Proving Grounds [in Maryland] and test it and test it until you learn all about it, how it shoots and how much energy it releases. For the atomic bomb we had to take the proving grounds over enemy territory, and make the test when the bombs are dropped in combat. So he said, you figure out some way to measure the yield of the bombs that we drop on Japan.

So I figured out a way to do that using the acoustic method. We very quickly designed and built pressure-measuring devices, which would be dropped out of an accompanying airplane on parachutes, which would then stay essentially put, dropping slowly as the two airplanes made approximately 180 degree turns to "get the hell out of the place". The acoustic sensor pressure-measuring device in the parachute gauges would then radio the signals back to the airplanes where they would be recorded and reduced. We recorded pressure versus time curves and using theory and knowing the distances and the altitudes, we calculated the pressure. The difficulty with this is that nobody paid a speck of attention to our measurements because before we had a chance to reduce our measurements, President Truman announced that the yield of the bomb was 20 thousand tons of TNT. That was one of the projected yields. He didn't know that, he just thought that that was the number and he released that. For 25 years that was the "standard" yield of the Hiroshima bomb. People at Los Alamos could make those numbers agree with what they measured in Hiroshima, but the intensity of the burning and various other indicators of pressure that they had made it look like that it was somewhat less. Somebody remembered that we made some measurements and I started getting letters from Los Alamos, saying whether they could see my records. I had our records in my personal files so I made Xerox copies of them and I sent them to Los Alamos and they analyzed them and said, it was more like 13 kilotons and that is now the accepted number and is now being used in place of the old standard 20 kilotons as the Hiroshima number.

Message to the Emperor

Harold Agnew and I flew over Hiroshima but neither of us flew over Nagasaki. But the same pressure gauges were in the airplane that accompanied the bomb-dropping airplane to Nagasaki. Larry Johnston whom I mentioned earlier and a couple of our sergeants from the Los Alamos SET group were in the plane and dropped the gauges and got the measurements, so we got good measurements over Nagasaki as well. I didn't go along and Harold didn't go along. But the night before, I did get this idea that it would be interesting to get a message to the Japanese High Command and so I sat down and wrote a letter out long hand to my friend Dr. Sagane who was at the University of Tokyo and who had spent

Four future presidents of the American Physical Society, left to right: Luis Alvarez, Robert Oppenheimer, William Fowler, and Robert Serber in 1938 (courtesy of the Lawrence Berkeley National Laboratory).

a year and a half or so in Berkeley before the war. I knew him quite well. I addressed it to Dr. Ryokichi Sagane from three of your former colleagues at the Berkeley Radiation Laboratory. I had enlisted the support of two of my friends, Bob Serber and Phil Morrison. None of us signed our names but at least there were three of us instead of just one. I wrote the thing out by hand with two carbon copies and they approved it and we put them in envelopes and taped them on to the pressure gauges and they were dropped out over Nagasaki.

I have actually seen the report of the Naval officer who opened those envelopes, who probed around in the pressure gauges. I always thought he must have had a lot of courage because the newspaper reports, in this country at least, said that bombs were dropped on parachutes. So he must have thought that there was a good chance that he was probing around with an atomic bomb. There was never any mention of any instruments coming down with the parachutes.

The interesting thing is that I have in my files at home the letter that went down on Nagasaki. Dr. Sagane sent it to me after the war. I sent him a full set of *Physical Reviews* for the whole war period. I had to do that surreptitiously because General MacArthur didn't want any intercourse between the Americans and the Japanese in the field of physics. He destroyed the 16-inch cyclotron. I got the journals to Sagane by circuitous route and as a favor in return, Sagane sent me the letter. In fact, he had already given it to Arthur Compton's brother, Wilson Compton who was then President of Washington State College. At my suggestion, Sagane wrote President Compton a letter asking if he would give this letter to me, which President Compton did at half time during a football game when Washington State College was playing here and I had an appointment to meet Wilson Compton. So I have the letter in my files and I also have the pressure gauge.

I know there is one copy of the letter in the museum in Hiroshima. There was an article by Lowell Thomas in the *Saturday Evening Post* one time about it. It was entitled, "Under Separate Cover One Atomic Bomb". The other thing that probably is important is that I learned that that letter did get to the High Command essentially immediately. They took it seriously. The fact is that they offered to surrender the next day. Whether that had anything to do with our letter, I don't know, but I like to think it did.

Incidentally, talking about letters in that period, I had another interesting letter. On the way back from Hiroshima to Tinian, I wrote a

Headquarters
Atomic bomb command
August 9, 1945.

To: Prof. R. Sagane.

From: Three of your former scientific colleagues during your stay in the United States.

We are sending this as a personal message to urge that you use your influence as a reputable nuclear physicist, to convince the Japanese General Staff of the terrible consequences which will be suffered by your people if you continue in this war.

You have known for several years that an atomic bomb could be built if a nation were willing to pay the enormous cost of preparing the necessary material. Now that you have seen that we have constructed the production plants, there can be no doubt in your mind that all the output of these factories, working 24 hours a day, will be exploded on your homeland.

Within the space of three weeks, we have proof-fired one bomb in the American desert, exploded one in Hiroshima, and fired the third this morning.

We implore you to confirm these facts to your leaders, and to do your utmost to stop the destruction and waste of life which can only result in the total annihilation of all your cities, if continued. As scientists, we deplore the use to which a beautiful discovery has been put, but we can assure you that unless Japan surrenders at once, this rain of atomic bombs will increase manyfold in fury.

Facsimile of the Alvarez letter to Dr. Sagane (courtesy of Clarence and Jane Larson).

long letter to my son, who at that time was five years old, telling him about my experiences, what it was like to go into combat for the first time in an old plane. It's a long letter and it has also been reproduced a few times, but it's a fairly personal letter.

Linear Proton Accelerator

Chronologically the next thing I did when I came back to Berkeley, was to design and lead a team that built the first proton linear accelerator. Ernest Lawrence cast me and Ed McMillan and Glenn Seaborg who had all come back to the laboratory; we'd all been his young kids, so to speak, when we ran away. We had all done important things by ourselves and he knew that we would get offers for good jobs from other places, so he made sure that if we came back, we would have a chance to do whatever we wanted to do, we knew that he would support us. I told him that I wanted to build a proton linear accelerator which nobody had ever done before. He said, fine, and backed me to the hilt, and it was built very quickly. It was running in 1947, in less than two years from the time we started. On our team I enlisted the aid of "Pief" [Wolfgang] Panofsky, who is now a very famous particle physicist, who had already signed up to go to the Bell Laboratories. If I hadn't asked him to come and join the laboratory in Berkeley, he would've probably been one of the co-discoverers of the transistor. He was about the smartest person I met during the war, but he was completely unknown, because he was not either in Los Alamos or at the Radiation Laboratory or any of the other big laboratories. He worked on a little project together with his father-in-law. I happened to know him because he was the one who built the microphones we used in our pressure measuring gauges. As soon as Ernest told me that I could build a proton linear accelerator, and that I could hire any five people I wanted, I told him that the first guy I would hire was Pief Panofsky, the smartest guy I met during the war. Ernest had never heard of him. The most important thing I did after the war was to get Pief out of the Bell Labs.

It was no great physics that came out of the proton linear accelerator. The reason they built linear accelerators was that if you plot the cost of an accelerator as a function of its energy, the cost goes up as the cube or the square for a magnetic machine but it only goes up linearly for the linear accelerator. If you plot this on log/log paper, you always get to a higher cost for the magnetic accelerator than for the linear accelerator. There was a difficulty in that the magnetic guys kept changing their design. The linear accelerator is now largely used as an injector into magnetic machines although there is a very large, almost a billion volt linear proton accelerator at Los Alamos. The linear proton accelerator turned out to be a very useful machine, but it was not the world beater that we thought it was going to be.

Another interesting thing there — and it sounds very conceited to say — but we went ahead not knowing how to build a proton linear accelerator and had an operating one just under two years. Then, two or three years later, one day Ernest came in and said that, "By the way, Luis, did I ever tell you that I gave your linear accelerator to USC?" The USC [University of Southern California] people came out, learned how to run the thing, brought it down to Los Angeles, and it took them over three years to get it running again. So Pief and I and a few others patted our shoulders after that experience.

Bubble Chamber Experiments

Don Glaser invented the bubble chamber and he used hydrocarbons in it, ether, for example. When I first heard about it in 1953, I immediately decided that if we were going to make any use of the bubble chambers, we had to use liquid hydrogen because in liquid hydrogen all its target nuclei are protons. In hydrocarbon bubble chambers the target nuclei consist of twelve protons and neutrons, making a big blob, which is a very unattractive thing to a nuclear physicist or a particle physicist. As I said, I learned about the invention of the bubble chamber in 1953 in Washington at the Physical Society meeting from Don Glaser. I told my two young friends who were with me in Washington that we were going right home and build liquid hydrogen bubble chambers and that is going to be the greatest detecting device that they had ever seen. We did do that and very quickly we went from a 1-inch chamber — where we saw the first tracks in hydrogen in my group — to larger chambers. A young chap named John Wood was the one who built the first bubble chamber with liquid hydrogen that showed tracks.

Then we built a 2-inch bubble chamber. Probably, the most important discovery in our group at this period was the fact that you didn't have to have smooth-glass walls to make the bubble chamber work. Don Glaser pointed out — and he was very emphatic on the point — that you had to have very smooth walls because you had to be able to reduce the pressure on the liquid and not have bubbles form at the walls. When we first started building what was then called dirty bubble chambers that started automatically boiling at the walls because there were casketed seals between the glass windows and the metal bodies. We didn't like the name dirty chambers, but pretty soon all the chambers that anyone was building were dirty chambers. The clean chambers never amounted to anything; they never did any physics.

Luis Alvarez at the Berkeley Laboratory in 1966 (courtesy of the Lawrence Berkeley National Laboratory).

There were probably two most important discoveries in my group. One was that you could use liquid hydrogen and see tracks. The other was that you could use metal casketed to glass windows and that permitted us to go to very large chambers. We made a 4-inch chamber that worked quite well and then the 10-inch chamber was the first one that was designed on a drawing board by an engineer. All the others before had been just hacked out in the machine shop without drawings at all, maybe just a sketch. The 10-inch chamber did some really beautiful physics down at the betatron.

Before that was running we had the plans for the 72-inch. That's an interesting story, how that came about. I went to Ernest Lawrence and told him that we wanted to build a real big bubble chamber and he asked me how big. I said, we'd got planned 72 inches long, 20 inches wide, and 15 inches deep. He asked me, how big is the biggest chamber now working? I said, 4 inches in diameter. He said, That's a big extrapolation to my taste. I said that we were building this 10-inch chamber and if that works, and I am pretty sure it will, then a 72-inch is guaranteed to work because our extension system is such that the 72-inch can be thought as being a whole lot of 10-inch chambers expanding outward, several of them on top of each other and two side by side. I said, "If we cannot make the 10-inch work then we'll give the money back to the AEC [Atomic Energy Commission]. If we can make it work, we'll be well on our way to make the 72-inch chamber work." So he said, in his characteristic way, "I don't believe in the 72-inch chamber, Luis, but I believe in you. I'll go to Washington and I'll work to get that money for you." I learned extrapolation from Ernest Lawrence, but he became a little more faint-hearted as he got older. Shortly after that we went to Washington together. One morning we called on three of the Atomic Energy commissioners, Johnny von Neumann, Lewis Strauss, and Will Libby. That afternoon they said, "We had a meeting this afternoon and voted you the money." My young friends cannot believe that now because they know about writing proposals, peer review, and all that sort of stuff. They can't believe that in one day, less than one day, talking in the morning and going to a cocktail party in the afternoon, we had the money.

Of course, then we had the money and we had to make the thing work. Nobody else got into the business until our chamber was working. I think a lot of them expected us to fall flat on our face. We had a big chance too, but we did make it work. As soon as the 72-inch chamber was working, then Brookhaven started an 80-inch chamber, which took 5 years to come on, and CERN started a 2-meter chamber, which is very nearly the same size. It took them 6 years. So we had a field to ourselves for several years.

The Nobel Citation

The Nobel citation mentioned the discovery of a lot of new particle resonances, which came out early in the 15-inch chamber and a lot more in the 72-inch chamber. The citation went on to say that these were accomplished through the development of the liquid hydrogen bubble chamber and data analysis techniques. One of the things we had to do with the bubble chambers was to make data analysis equipment that would keep up with the enormous number of nuclear events in the machine in the chambers. Before that time, we used cloud chambers; the density of a cloud chamber gas may be 1 percent or less than the material in the bubble chamber. So there weren't that many things happening and they could take all day or half a day to analyze an event. But in my original write-up on the 72-inch chamber, I said that if we hadn't developed in time to use this data analysis technology, which we did develop, then the bubble chamber would simply have been an expensive toy, because in one day, it produced enough interesting events to keep the whole cloud chamber fraternity world-wide busy analyzing them. Actually, they would've never gotten all analyzed. So we did all this work of automatic measurement of pions, automatic track following, which came out of my radar experience and using computers to create the tracks in three dimension and to provide stereo pairs. Then we had to analyze those afterwards. It was a very big development to get the data analysis techniques going. The Nobel citation recognized that too, not just the development of the chambers.

Cosmic-Raying the Pyramids

I did some work with cosmic rays using high-altitude balloons; nothing world-shaking came out of it, but it was a lot of fun. Then I got the idea to X-ray the pyramids of Egypt using cosmic rays and spent a few years on that. It was the Second Pyramid, the one right next to Cheops's Pyramid, which was built by the son of Cheops. Cheops's Pyramid has three big chambers under the ground level. The pyramids that were built just before that had two chambers in them and the ones before that had one chamber, so with the number of chambers increasing in number, I was convinced that Chephren's Pyramid, built by the son of Cheops would have three or four chambers in it. As each pyramid builder built his pyramid, he got a little more clever in hiding the chambers from grave robbers. The original pyramids all had their entrances in the middle of the north face. If you knew that, all you had to do was to get a battering-ram, get some people with metal spikes and start drilling in on the center and you'd run right into the chambers.

Mamun was the caliph of Cairo in the ninth century A.D. and he got the idea that it would be nice to find out where the Pharaoh was buried because there was probably a lot of gold in there. He set his tunneling team to work on the middle of the north face and told them to exit south. They dug about a hundred feet and if there hadn't been an unusual accident, they would've dug six hundred feet and come out of the far side of the pyramid. But after they'd gone a hundred feet, they heard a noise off to their left. What they had done was dislodge a block from the roof of the long descending passage way that went down to the basement. That block went tumbling down and they could hear it; the rock is a good transmitter of sound. When they reported this to the caliph, he said, there's got to be something over the left, so turn your tunnel to the left. They turned sharply to the left and they ran into the descending passage way. The descending passage way intersected the ascending passage way. At that point there was the block covering up that junction and that dropped down to the bottom of the pyramid. Then they went around some big granite blocks and went up the ascending passage way and went into the Grand Gallery and into the King's chamber and into the Queen's chamber, which was just below that. That's how it was found. Had it not been for that lucky accident, they would've gone right to the south face and said, sorry fellows, there are no chambers in there.

My theory was that now that people had stopped being allowed to trough ground in pyramids that probably there were three or four chambers up in Chephren's Pyramid. Nobody had found them and we would find them using cosmic rays. It would be like taking candy away from a baby because we had these fancy tools, namely, cosmic rays, which would go through the pyramid right to the other side. I enlisted the aid of some Egyptologists, some physicists, and we got money from the Atomic Energy Commission. When Glenn Seaborg, the head of the Atomic Energy Commission, and an old friend, was coming back to Washington from a meeting in Japan through Egypt, I'd intercepted him and had shown him the pyramids, and he gave us the money. We set up our equipment in a room underneath the Second Pyramid. So we proved that the Second Pyramid was solid. Lots of people said to me, Luis, I hear you didn't find any chambers in the pyramid. Then I say, no, it's not that we didn't find any, rather, that there weren't any. It's quite a different thing. Most people didn't find any chambers, but that doesn't prove anything. We found that there weren't any chambers there. This was the only time that cosmic rays were used for a practical application.

How Did the Dinosaurs Disappear

For the last five years I have been working intensely with my son Walter, who is a geologist, and with Frank Asaro and Helen Michel, who are nuclear chemists. We have been combining our various expertise in a solution of a problem of what killed off most of the life on Earth 65 million years ago. We now have a theory that's believed by almost everybody; there are two or three holdouts, but there are a few people who don't believe quantum mechanics and there are a few people who don't believe plate tectonics.

The impact theory explains everything that I know. The idea is that a large chunk of extraterrestrial material, either a comet or an asteroid, hit the Earth 65 million years ago, about 10 kilometers in diameter, came in about 25 kilometers per second and threw up an enormous cloud of dust. The dust went outside the atmosphere; it was transported around the globe worldwide by ballistic orbits. It fell down through the atmosphere and in falling down made day into night, stopped photosynthesis, and we are now beginning to understand how the killing was done.

The original proposal was that the darkness would've stopped photosynthesis, cut off the food chains, killing all the animals that ate plants and killing the animals that ate the flesh because there wouldn't be any animals to eat. It's holding up very well and it's getting stronger every day. I just heard today that a very good young paleontologist has correlated the kinds of animals that went out with whether or not they ate live food or dead food. The correlation is good. Now the paleontologists really believe this, and they'd kind of dug their heels in for several years, but now that they believe it, they've all jumped in and can explain everything by this theory. It's something like what happened in plate tectonics. Wagener got the idea that the continents drifted around; Africa and South America are seen to fit neatly together and he said they pulled apart. Everybody said, you're out of your mind. Paleontologists said, they had all kinds of evidence that this was not so. But once there was good geophysical evidence for it, then the paleontologists came in and wrapped it up. They invoked all kinds of evidence to prove that it was right.

So the original theory that the extinction of the dinosaurs 65 million years ago was triggered by an impact of an asteroid or a comet, that theory is now almost universally accepted. Just two or three months ago, a new feature was injected, which is very exciting to everybody. That is that two people at the University of Chicago, David Raup and Jack Sepkoski have found that extinctions in the last 250 million years are periodic with a period of about 26 million years. So the paleontologists have gone from not believing that any extinction was due to impacts of extraterrestrial bodies, now believing that all of them are and they come at regular intervals of 26 or 28 million years. We had a meeting here at Berkeley 10 days ago in which all the players in this new round, which we might call the periodic comet shower theory, they were all here.

In addition to the theory of periodic extinctions, my son Walter and one of his friends have shown that the craters on Earth, which are made by impact, were also periodic, with that same period. That ties the craters to the extinctions with the same period; something nobody would've guessed five years ago. In fact, when somebody first suggested it a couple of years ago, everybody thought that he was out of his mind. Now, most everybody at this meeting believed it because the evidence is overwhelming. Another one of my young friends has come up with an explanation for this. That is that the Sun is not a single star the way most everyone has assumed but is part of a double star. In the past it's been strange to find that the Sun was a single star because most stars are parts of multiple systems, more than half of them are. Now that it's known that the Sun has a companion, which is on a 26 or 28 million year orbit, it goes out about 2.5 million light years. When that companion comes back close to the Sun, it can inject a shower of comets, about a million comets in a sudden pulse, right into the Earth's orbit, so some of those would crash into the Earth, making craters and the craters making the dust in the sky, causing cold and darkness and that makes the extinctions. All tie together beautifully. Now the big question is, where is that star? It's got to be out there. I've spent an amount of time in the last two months searching through catalogs of stars, looking for that star. A lot of people are working on it and I'm convinced that within one year we'll know where that star is.

People will be able to see it and say that that guy has been going around with our Sun for the last few billion years and that's what allowed us to be here. There wouldn't be any people on Earth if the major extinction 65 million years ago hadn't been triggered by a comet shower. We've now given up the idea of single asteroids for multiple comets. We always said we couldn't tell what kind of a chunk of rock it was; in our theory it was either a comet or an asteroid. I've said several times in lectures that I doubt if anybody will ever be able to tell whether it was an asteroid or a comet. Now, sure we know, it was a comet, one comet or more than one comet. Eventually we'll find the star and people will say that that's the guy that made it possible for us to be here. Had that star had not sent in the comets to wipe out the dinosaurs, we wouldn't be here. The mammals could not live in an environment where the dinosaurs ran the world. There was no way for the mammals to develop; they were there with the dinosaurs, but they were only about the size of rats. Immediately after the dinosaurs had disappeared, the mammals grew in size, and grew in complexity and abilities and they radiated out (the word used by the paleontologists) into all the little niches of the environment, and we are the results of that. The dinosaurs had to be cleaned up first and that's what the comets did, and the comets were triggered by the solar companion. They gave it a name,

Nemesis, even before it was observed. It may look like finding a needle in a haystack, but it has been so for me for the last 25 years, finding needles in haystacks, taking millions of bubble chamber pictures and finding one event, we discovered one particle; that's been our business. That doesn't scare me a bit.

Parting Thoughts

My main thought about such interviews is that people much sooner should be interviewed. Science is a young man's game. It always bothers me when you take pictures of somebody who is 72 years old like I am and say that he is going to be a role model for people so that they all know how science is done. I have a picture of Einstein when he was really young and that is the Einstein who was doing relativity. When he got old and white haired, then he didn't do science anymore. Then people used to interview him and take pictures of him and make statues of him. I would say that that's the guy who used to be Einstein. Young men in their forties should be interviewed when they're really doing the most important science they are ever going to do. I have the good fortune of still doing good science at an advanced age, but most of the people at this age had stopped doing science many years ago. I'm not being critical of them. I'm always outraged when I go to the Academy of Sciences and see that statue of Einstein; that's the guy who used to be Einstein. I can't think of anything that turns people off from science more than thinking that you have to look like Einstein to be a good scientist. Einstein was the greatest scientist this century has seen, but the pictures you usually see of him are not of that particular person; it is of the person he then developed into.

William H. Pickering, 2004 (photograph by I. Hargittai).

11

WILLIAM H. PICKERING

William H. Pickering (1910, Wellington, New Zealand – 2004, Pasadena, California) was Professor Emeritus of Electrical Engineering of the California Institute of Technology (Caltech). He entered Caltech in 1929 and received his B.S. degree in 1933, M.S. degree in 1934, and his Ph.D. in Physics in 1936. The same year he joined the Caltech faculty and became full professor of electrical engineering in 1946. He was director of the Jet Propulsion Laboratory (JPL) between 1954 and 1976. Dr. Pickering was a member of the National Academy of Sciences of the U.S.A., a founding member of the National Academy of Engineering and held other memberships in various learned societies. His numerous American and international awards included the Columbus Gold Medal (Italy), the Guggenheim Medal (AIAA), the Distinguished Service Medal (NASA), the Edison Medal (IEEE), the National Medal of Science (U.S.A.), Honorary Knight Commander of the British Empire, the Japan Prize, and he was an honorary member of the Order of New Zealand.

Theodore von Kármán (1881–1963) was a Hungarian-born American aeronautical engineer, sometimes called the "father of modern aerodynamics".

We had our conversation with Dr. Pickering at the Athenaeum Club of Caltech on February 4, 2004. A few weeks later we were saddened to hear about his death.

You were at Caltech already at the time when Theodore von Kármán started the Jet Propulsion Laboratory. Kármán was one of that distinguished group of Hungarian scientists who made their mark on the 20th century.

This was a von Kármán joke, which I heard from him more than once and it went something like this. "If you ever wondered why all of these eminent scientists came from Hungary, as a matter of fact, we are all men from Mars and when we decided where to land on Earth where we would not be noticed, we decided that Hungary was the place to start from."

Did you have personal contacts with him?

Yes. My contacts with him were primarily in JPL. He came to Caltech in 1928 or thereabouts as head of the Aeronautics Department. There is a building around here, the aeronautics lab, which was built at that time for him. The Aeronautics Department in the late 1920s and 1930s was a very good department; a lot of good people went through it, and had a very close relationship with the aircraft industry. The way JPL got started was that a young graduate student in the late 1930s, Frank Malina, decided that he wanted to do his thesis on rockets. He went to his professor, Professor Clark Millikan and Clark said, "You're out of your mind, this is ridiculous, there are all kinds of good aircraft problems to work on, and here you want to work on rockets." Anyway, Malina went to von Kármán who was the head of the Department, and von Kármán encouraged him to do his thesis on rockets, so he did. The initial thrust of Malina's work was just trying to learn how to build a rocket and what to use for fuels, and generally the operations of the potential engines. By the early 1940s, they had done that and they were working on both solid and liquid propellants, and they demonstrated the jet-assisted takeoff of aircraft, first with the solid fuel rockets and then with the liquid fuel rockets.

When Malina left the Laboratory after World War II, in about 1946 or thereabouts, he went over to Paris to join von Kármán who was then setting up the Advisory Group for Aeronautical Research and Development, AGARD for NATO. Malina worked for von Kármán in Paris on getting it going. Malina was also close to von Kármán when the International Astronautics Federation was set up, which was an association of people who were interested in space exploration. Malina then got interested in art over there in Paris, especially in kinetic art, and he set up the *Leonardo* magazine.

Von Kármán was very active in World War II.

He was very close to General Henry H. Arnold, who was head of the Army Air Corps. They worked very close and Arnold encouraged him.

Left to right: Dr. William H. Pickering, former JPL Director, Dr. Theodore von Kármán, JPL co-founder and Dr. Frank J. Malina, co-founder, and first director of JPL (courtesy of NASA).

After the war, they set up the Scientific Advisory Board for the U.S. Air Force. That was before von Kármán went over to France.

Did he ever come back?

He made some short visits back here, but he was never here for very long. At the end of World War II he was invited to go over to Europe and see what the Germans were up to, particularly in aeronautics and rocketry. I also went over on one of the trips and on that particular trip the British set up a launching of a V-2 rocket so that people could see the actual procedure. They were careful to choose a site to make sure that the rocket would not land in England. Von Kármán and I went to see that launching.

Did it work?

It worked very well. It went up and disappeared over the horizon somewhere. On that trip, incidentally, after we had spent some time in Europe, we went to Japan, but von Kármán didn't feel himself up to the trip, so he left us. But we went to Japan and made a stopover in China as well. It was very interesting for me that wherever we went in China and Japan and other places, a bunch of Kármán students would turn up. He had students all over the world. One of his best students was a Chinese student who worked with him during World War II and some time afterwards. Then this Chinese student got into a big argument with the U.S. authorities because he wanted to go back to China to visit his elderly father. There were two groups in the U.S. government: one thought that he knew too much and the other just wanted to get rid of him. They argued for a long time and finally they let him go. When he left he said that he would never come back. He went over to China and ended up running the Chinese rocket program. I think he even got involved in their nuclear program too.

Were the students you met in different places his former students in the U.S.?

Most of them but not all of them because he had had students before that as well.

What did he know that other people did not?

I suppose it was the understanding of the mathematics of aerodynamics.

What was his most important discovery or innovation?

In summary, it was aerodynamics, but what in particular, I don't know. He had an understanding of the whole field. There is also a textbook that he and another Caltech man put out and which is the standard book on aerodynamics. As far as rocketry was concerned, he did not have any particular input.

He had made important contributions to building up German aviation. Did he ever express regrets about that?

No. But his main contribution was that he had been the teacher of those who then had established the capabilities of air forces almost anywhere in the world, not only in Germany but also in China and in Japan.

Did you have any experience of him as a political person, as someone who was interested in ideology?

I can't think of him as a particularly political person.

Like Edward Teller who was very political?

Von Kármán was not. On the other hand, he did have a very good relationship with General Arnold, which was vital not only for what he did at JPL, but also for the whole future development of our Air Force. This Scientific Advisory Board was charged with forecasting what to expect of the Air Force in ten years and in twenty years, and so on, and they did some very good work. Kármán did get along not only with Arnold but also with the other generals of the Air Force.

By the time he died, was he still actively involved in the work here?

No. After he had left for Paris, his main involvement was with NATO.

But he got the first National Medal of Science from President Kennedy.

It was for past accomplishments, for his support for the Air Force.

Did he become a real American?

That's a good question. I don't know.

What was your impression?

I never thought about it, but I'm pretty sure he did not become an American. I think he stayed European but I can't really know.

How did you get started?

I was born in New Zealand in 1910. I came over as a student in 1928 and I got my doctor's degree in 1936 here at Caltech. During the period 1938–1942, I worked with Robert Millikan and H. V. Neher. Millikan was the first American-born Nobel laureate who received his prize in physics in 1923 for his work on the elementary charge of electricity. With Millikan and Neher, I studied the absorption properties of primary cosmic rays.

Did you do war-related work?

I was investigating Japanese balloon warfare techniques. They released these balloons from Japan; the balloons drifted across the Pacific Ocean to the United States where they dropped firebombs.

What happened after the war?

I became a Professor of Electrical Engineering over here. This had nothing to do in the early days with JPL. There were rocket people and mechanical engineers and chemical engineers. What happened was that as they developed the skills, they found more and more need for good instrumentation, high-speed instrumentation, so forth. So they came to me because I was doing electronics and electrical engineering. They persuaded me to spend some time over there in the Lab to help them out. So I did for a while and by 1954, when Malina's successor left, DuBridge invited me to become director of JPL.

You were director of JPL ...

From 1954 to 1976.

Kármán was still a member of JPL.

Yes. After the Lab started, more and more money was brought in and it became a big operation and Malina was in charge of it. Physically, JPL was removed from the Caltech campus because while still back on the campus, they had a rocket explosion. JPL is now about 8 miles from the Caltech campus, up against the hills. There are now four thousand people. The physical plant up there was built and paid for by the government. Then the government gave a contract to Caltech to provide the staff and to operate it. All the employees up there are employees of Caltech. The work which is done up there is worked out mutually between the head office of NASA in Washington and the people here, and the work is funded by the money given to Caltech.

It sounds like a civilian operation rather than Air Force.

It is a civilian operation.

Didn't Kármán work directly for the Air Force?

He did when he was setting up the Scientific Advisory Board. That had a direct relationship with the Air Force. During World War II he got funds which went into the Lab from the Air Force. In that sense he worked for the Air Force. The first government-sponsored project up there in the Lab was to work out the jet-assisted takeoff to help aircraft make a short run. It was obviously of great interest to the military because in many places it was impossible to have long runways. At the beginning of World War II, they came to the Lab and asked Kármán to make a lot of those rockets. The problem was that Caltech as a university didn't want to get involved in just producing and manufacturing rockets. This is why Kármán and a few other people in the Lab set up a private corporation called the Aerojet Corporation, which is still in existence out here. Then Aerojet became the source of rockets for the military as a commercial operation. But the Lab continued and at the end of the war Kármán wanted to continue it because he felt that the research program that the Lab was involved in was not appropriate for a university. There was then another thing that happened during the war. As the German V-2 system began to be understood, the artillery part of the Army came to the Lab and wanted to support developing a ballistic missile by the Lab. By the end of World War II, the support went from the Air Force support to the Army Ordnance support and after the war it became some Air Force support but mostly the Army Ordnance support. All this, however, went through this process, through Caltech.

When was the Air Force established independent of the Army?

At the end of World War II or a little after.

Was there a rivalry?

There was a tremendous rivalry between the Army, Navy, and Air Force. When it was decided to develop intercontinental ballistic missiles, the question came up whether it should be done by the Army or by the Air Force? The Army General, who was in charge of the ballistic missiles was an artillery man. Long-range artillery was always a function of the Army and the intercontinental ballistic missiles could be considered to be long-range artillery, so he thought that the Army should do it. But the Air Force had a lot of political clout and they got the intercontinental ballistic missiles. The Army was given the intermediate ballistic missile program (IMBM). I don't think we actually produced any significant amount of the IMBM. I remember being at meetings where the three services were represented by generals, and they sounded like the ambassadors of three warring nations.

The person who was the main organizer of university scientists for the war efforts was Vannevar Bush. He made this assertion that if you wanted to deliver warheads at intercontinental ranges, you had to do it with an airplane, and an unmanned airplane at that. This is because, he said, rockets would not be appropriate for guiding the warheads to their target. This is what then determined U.S. policies in the late 1940s and early 1950s. It was only in the early 1950s that people suddenly realized that the Russians were working like mad on intercontinental rockets. We have also realized that rockets could be made much more sophisticated than previously thought, and they could be guided accurately. Another factor in favor of rockets was that the nuclear weapons had become much lighter than the original very heavy bombs. So the U.S. started the rocket program in the early 1950s and this was further strengthened by the space program. We were suggested to make a small scientific satellite. Eisenhower supported the program, but he wanted to keep it separate from the military program. He didn't want anything to interfere with the military program, which was developing the new weapons. The whole program of building the small scientific satellite was then started from scratch, and the program was given to the Naval Research Laboratory. They had to design their rocket for the satellite program from the ground up. It was only after the Soviets had put up their successful satellite, followed a month later by a second one, that the Army was told that it could go ahead and use a military rocket for the first stage. This military rocket had been designed by the Germans who had been brought over to this country. It was a sort of an upgraded V-2. As soon as the long-range rocket for nuclear warheads got established, the problem between the services came up, not only between the Army and the Air Force, but also both of them and the Navy. The Navy wanted to get into the act too. They wanted the ship-launched device. First they started work together with the Army on the intermediate-range missiles, but then the Navy decided to do it alone, and they developed the Polaris rocket for submarines, which turned out to be very successful.

You directed JPL for more than 20 years…

When I took over, we worked on the short-range ballistic missiles for the Army. We developed two systems, the first one was called the Corporal and the second the Sergeant. The Corporal was about a hundred-mile range missile; it was shorter range than the V-2, but more or less equivalent to that except it was radio-guided. The Sergeant was a solid propellant rocket of about the same performance, but inertially guided. The Lab had the good fortune to be assigned to work both in liquid and solid propellants and radio and inertial guidance. So we had quite a variety of experience to put into the space program when that came along.

The developments in this country were pretty fast. The Soviets launched in October and then in November of 1957. Then in December of the same year, the official government satellite program that was called the Vanguard had a demonstration firing from the Cape, which was a complete disaster. The rocket only went up a few feet and fell back down and bore a fire.

That was done with complete publicity, all the media were there; it was an awful mess.

Actually, in November the Army had been told to go ahead and come up with a backup to the Vanguard. Von Braun's people were in Alabama and they provided the first stage, a beefed up V-2 and we at the Lab put the thing together, and provided a three-stage solid propellant rocket put on top. We also built the satellite itself to provide communications. Most of the new stuff was done out here and very quickly, the whole thing within about three months. We were given the go-ahead in November and we launched at the end of January 1958, and that was a successful launch. That was followed up by the Congress deciding that the U.S. should have an ongoing scientific civilian satellite program, independent of what the military wanted to do. The result was eventually the National Aeronautic and Space Agency (NASA), in June 1958. This is very fast for a legislative body to do anything. NASA officially started on the first of October and about the first of December the top NASA people came out to see me and talk about having JPL transferred into NASA. First I was concerned and DuBridge was concerned; he was president of Caltech at the time. But soon DuBridge and I were delighted because we were much happier having the university run a civilian space program than a military program. They asked me which part of their program we would like to have and I preferred the deep space program, which meant through the Solar System and they gave it to me. That was then our assignment, to develop a deep space scientific program for NASA.

We were also proud that we participated in developing a re-entry test vehicle that permitted missile warheads and other payloads to traverse Earth's atmosphere following long-range ballistic flight. This technique was eventually applied in the Mercury, Gemini, and Apollo programs. JPL participated also in numerous other high-profile and very successful projects.

Are you excited about recent developments, about the Mars program?

Oh, yes. It's very nice to see it. I get continually fascinated when we have some problems out there in millions of kilometers. The troubles have to be immediately analyzed and the people in the Lab can do the analysis promptly and solve the problem.

What do you do nowadays?

I am working on my memoirs.

William A. Fowler, 1974 (courtesy of California Institute of Technology).

12

WILLIAM A. FOWLER

William A. Fowler (1911–1995) received half of the Nobel Prize in Physics in 1983 "for his theoretical and experimental studies of the nuclear reactions of importance in the formation of the chemical elements in the universe". The other half of the physics Nobel Prize that year was awarded to Subramanyan Chandrasekhar (1910–1995) "for his theoretical studies of the physical processes of importance to the structure and evolution of the stars". William Fowler studied at the Ohio State University in Columbus, Ohio, and earned a degree in Engineering Physics. He went to graduate school at the California Institute of Technology (Caltech) in Pasadena where he prepared his doctoral thesis on "Radioactive Elements of Low Atomic Number" under the supervision of Charles Lauritsen. He stayed at Caltech and worked at its W. K. Kellogg Radiation Laboratory until his death. In 1954–1955 he spent a sabbatical year in Cambridge, England, and established a lasting co-operation with Fred Hoyle, and Geoffrey and Margaret Burbidge. They continued their joint work at the Kellogg Lab and published a seminal paper "Synthesis of the Elements in Stars". An independent study at about the same time by A. G. W. Cameron resulted in similar ideas.

Dr. Fowler was elected to the National Academy of Sciences of the U.S.A. in 1956; he received the National Medal of Science in 1974, and he was the first recipient of the William A. Fowler Award for Excellence in Physics, Ohio Section of the American Physical Society in 1986. These are only a few of the many awards and distinctions he received.

Clarence and Jane Larson made a video recording at the Kellogg Laboratory of the California Institute of Technology in Pasadena, California, in 1985.* We prepared and slightly edited the transcripts of this recording and Drs. Charles A. Barnes and Bradley Filippone of the Kellogg Lab kindly checked the text for us in January–February 2004. We are especially grateful to Dr. Barnes for helping us acquire illustrations for this interview.

Childhood, Family Background

I was born in Pittsburgh, Pennsylvania, in 1911. My father was transferred to Lima, Ohio, when I was 2 years old, and my recollections of Pittsburgh come from the fact that all my grandparents were there, and uncles and aunts and cousins were there. Every summer, on my father's holidays, we got on the train and went back to Pittsburgh. I spent two weeks every year in Pittsburgh during my youth. Pittsburgh was a big city and it had a baseball team and a football team. I am still a Pittsburgh Pirates fan; the Pittsburgh Pirates are in the National Baseball League; and I am still a Pittsburgh Steelers fan; the Steelers are in the American Football Conference of the National Football League. On our moving to Lima, it was a great change. Lima, Ohio, is now and was then a small community with a population of some 40 thousand people and it was surrounded by a farming community. But it was a very interesting town; it was a railroad center. The main line of the old Pennsylvania railroad from Chicago to Pittsburgh and to New York went through Lima. The main line of what was called The Nickel Plate from Cleveland down to Saint Louis went through Lima, and a very important branch line of the Baltimore–Ohio line from Cincinnati to Detroit went through Lima. Furthermore, Lima had the Lima Locomotive Works, which built steam locomotives when I was a boy. It continued to do so up till almost 1960. My younger brother worked for the Lima Locomotive Works as a design engineer all his professional life.

So I was surrounded by steam. Our home was only about half a mile from the big switchyards of the Pennsylvania Railroad just outside of Lima. Much against my father's wishes, I spent a considerable period of my youth in the switchyards at Lima. In those days, the rules were not so strict and the switch engine drivers would see this young fellow standing alongside the rails and they would invite me up on the foot plate. They

---

*"Larson Tapes" (see Preface).

taught me how to drive a steam locomotive, which is not as simple as just handling the break; there is the so-called Johnson's bar which determines when the steam is let into the pistons, and so on. I was 6 or 7 when I drove my first locomotive. If I'd been older, I doubt if the engine drivers would've paid any attention to me. I became fascinated with steam locomotives and have maintained that interest all my life.

There is a fascination with steam, which has carried over in my scientific career. A steam locomotive has a fascination because you can see how it operates. You can see the big wheels, you can see the big driveway, you can see the motions of the piston, and all that sort of things. In a diesel train engine you can't see anything and it doesn't have that fascination. My interest in science and engineering started with the steam locomotives of the Pennsylvania Railroad in Lima, Ohio. It continued when I went to high school and very soon it became very clear to me that there was something different in engineering than driving a steam locomotive and that there was something else beside engineering, which my teachers called science.

Education

In particular, my teachers in physics and in chemistry in high school had an enormous influence on me and they really got me started in my interest in science. After I graduated from Lima Central, I decided to go into engineering and — for a strange reason — I went into ceramic engineering of all things although ceramics is a very important industry in Ohio. I had written an essay, sponsored by the American Chemical Society of those days for high school students, on some subject in chemistry. I wrote an essay on the production of Portland cement. I won third prize, which was four hundred bucks; that was a handsome prize in 1929. It came in handy when I went off to college the next fall.

When I went to Ohio State [University in Columbus, Ohio], I enrolled in ceramic engineering, but very soon I became acquainted with physics. All freshman engineers — as is common in many institutions — took the same courses and it included first year physics for engineers and mainly physics laboratories. I was just fascinated by the physics laboratory and that soon led to a decision to go into physics rather than into ceramic engineering. This was made possible in part because just at that time the Physics Department at Ohio State under Alpheus Smith decided to establish a physics option in engineering, so they established engineering physics.

In fact, there I met Leonard Schiff, who later became quite well known in physics. His book on quantum mechanics was <u>the</u> book and he was head of the Department of Physics at Stanford before his death. Leonard and I were two of the very few who opted to take engineering physics. He was quite young; only 15 then; we were freshmen at Ohio State when this happened and he was a child prodigy. We got to know each other. His family was quite wealthy and his grandmother was very orthodox and the family lived in Columbus. Leonard would invite me to his home for dinner and for the first time in my life I had exotic orthodox Jewish food. I was raised in a small country town in Lima so it was a revelation for me how different life could be. Leonard and I remained good friends all during his lifetime. He went to MIT after graduating from Ohio State in engineering physics and I went to Caltech.

Then, I stayed here as a postdoctoral fellow and he came out to Berkeley as a postdoc to J. Robert Oppenheimer, Oppy as we all called him. I'll get to that a little later on. At Ohio State, once I got to the end of physics, I found out that taking engineering physics was really up my alley because in addition to taking physics courses, I took quite a few courses in engineering, especially in electronics and in electrical engineering. I took laboratory courses in engineering with big electrical machinery. I remember that one of the professors, Johnny Byrne permitted me to work in the brand new electronics laboratory that the Electrical Engineering Department had just built. He gave me his keys so I could work there on weekends and at nights and I remember spending a whole term studying the characteristics of a pentode, the five-electrode vacuum tube, which was <u>the</u> thing in those days. I learned a lot that way. And I shouldn't stress perhaps but it also meant that I didn't have to take courses in the humanities and social sciences, which probably I should've done. I emphasize this because it was in the engineering laboratories — which were much more modern than the physics laboratories — that I really got some training in experimental physics that stood me in good stead when I came here to Caltech. The fact that I was able to work in the electrical engineering laboratory at Ohio State as an undergraduate shaped more than anything my decision when I came to Caltech to go into experimental physics in contrast to theoretical physics.

I stayed in experimental physics for the first 30 years of my career. Finally I got to the point where I felt that I've got to stop grabbing away in the lab and start doing some thinking about what I've been doing all these 30 years. Quite frankly, during the last 20 years I've been to a certain extent a theoretical physicist in a rather limited sense in elementary particle physics; certain theoreticians call it phenomenologist because they translate what's done in the high-energy laboratory into the basis for doing high-powered physics on the experimental work. That's been mainly what I've done in the last 20 years, translating the work done in our laboratory and elsewhere on nuclear physics into essentially the rates at which nuclear processes take place in stars. That's again getting ahead of the story.

There were several people at Ohio State who influenced me quite a bit; Alpheus Smith was very kind to me. He was a rather gruff man, he was called Bulldog Smith because of the heavy jaws he had. Many people were scared to death of him, but I got to know him and he gave me a lot of privileges and in particular he arranged for me to work with Willard Bennett who was then a staff member at the Physics Department. Bennett was the first one to really introduce me to laboratory physics where you didn't know the answer. He was doing research himself and he let me do an undergraduate thesis on magnetic focusing of electron beams. That was just a whole new world for me because in the electrical engineering labs you were essentially doing things for which the answers were known. When I measured the characteristics of a pentode, there were textbooks you could go to and read, which told you what the correct answer was.

Bennett put me to work on a problem for which the answer was not known. Of course, that's the key to the fascination that all of us have in scientific research where we can work on something for which the answer is not known and we can find it out for ourselves. Bennett got me started in that regard and I've always felt a great debt to him. He also helped me a great deal with my experiment. He built most of the equipment; there was some quite complicated glass blowing involved and the tube in which we produced the electron beam was a long glass tube, which he had to blow and shape. I did some of the things and then I made the measurements and he helped me. He had spent some time here at Caltech as a National Research Fellow and he strongly recommended that when I finished my undergraduate work at Ohio State I go to Caltech.

Caltech and the Kellogg Lab

In fact, I had always had some interest in coming to Caltech. I had tried to come to Caltech as a transfer student during my junior year at Columbus, but Caltech charged tuition whereas Ohio State did not. I was a resident of Ohio so I went to Ohio State free; I paid a laboratory fee, some small amount a year, but there was no tuition. Caltech told me that I had to pay 300 dollars a year and that was a lot of money for those days. I had been attracted by Caltech because in 1921 Millikan had become the head of Caltech and everybody knew about Millikan in those days. Our high school textbook in physics was written by Millikan and someone else. Millikan was the first native-born Nobel Prize-winner. Michelson had won it previously, but he was born in Europe, so Millikan was very famous. When he went to Caltech, that made Caltech rather famous too. To make a long story short, I applied for a graduate fellowship at Caltech and to my great joy, I received a telegram from the great man himself, Robert A. Millikan, saying that yes, indeed, I was admitted to the graduate school and I had a graduate fellowship which would provide me with room, board, and tuition, but no cash.

I had to stumble to raise enough money to pay for the train ride to California. It was pretty tough because my father, who had left school at 13, had proudly declared that his children would go to college. He had helped me in so far as he could in my undergraduate work at Ohio State although I had to do quite a bit of work on my own. I served meals, washed dishes for the Phi Sigma Sigma Sorority and stoked the furnace in the wintertime and I worked at the Central Market in Columbus on Saturday, starting late Friday night, all day Saturday, to make some money. But when I went to my father at my graduation from Ohio State, he expected me to come home, get a job in Lima, and help with my younger brother and younger sister through college. When I told him that I was going to graduate school he just about fainted; he had never heard of graduate school. He thought that his boy was goofing off. But he finally realized and he helped me again as much as he could. He got me a job that summer at the local YMCA before I came out to Caltech after graduation. Thus I earned some money to pay for my train fare and it all worked out. In 1933, I got out here to Caltech where I have been ever since. I came here as a graduate student and now I have been here for 52 years and around the campus I am known as the oldest graduate student.

Charlie, Charles Townes, was a graduate student when I was a postdoc from 1936 to 1939. I got my Ph.D. between 1933 and 1936. I knew him very well and Charlie played an important role in my research. When I came here in 1933, I had a predilection to go into the photoelectric effect because Millikan had measured the value of Planck's constant that way. But out here I found that most of the excitement had to do in the experimental work in nuclear physics. 1932 is considered to be the great year in nuclear physics. Harold Urey discovered deuterium; Carl Anderson discovered the positron; Chadwick discovered the neutron; and Cockcroft and Walton showed that you could disintegrate nuclei by accelerating particles to energies of the order of a million volts or so, or even less. It was also at that time that Lawrence started work with the cyclotron.

The man I went to work for and the man who had the greatest influence on my life was Charles Christian Lauritsen who had been here since the middle 1920s. In fact, Lauritsen built this laboratory, the W. K. Kellogg Radiation Laboratory. He designed it and it was finished in 1931, just two years before I came. I have worked in this laboratory essentially since its foundation. It was Lauritsen who had the dream and built the lab and Millikan got the money from the corn flakes king, W. K. Kellogg, to pay for the construction of the laboratory. It was the same Lauritsen who had developed the Lauritsen electroscope as a very sensitive detector — given its simplicity — for all types of radiation, gamma rays and neutrons in particular. That was the instrument that we used for much of the detection of the radiation produced in nuclear reactions in the early days. Charlie was an incredible person; he was not only a physicist, he was a radio-engineer and he was an architect, before he came to the United States from Denmark; and he was also an accomplished musician. In those days, he had a Friday

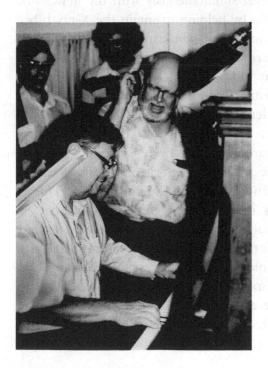

An informal moment at a party following the usual Friday evening nuclear astrophysics seminar in the Kellogg Radiation Laboratory. Following a long tradition established in the Laboratory by Professor C. C. Lauritsen in 1931, Professor Fowler leads a large group of graduate students in song. At the piano keyboard is Professor Charles Barnes, a Laboratory colleague of Fowler's from 1953 until Fowler's death in 1995 (courtesy of California Institute of Technology).

night seminar for all of his graduate students and after the seminar we would go to his home to drink beer and to sing. His son, Tommy Lauritsen, who later on joined us, was an undergraduate in those days and he played the piano and Charlie played the violin. All the graduate students had to sing. Mainly what we sang was Carl Michael Bellman, who was a famous poet and musician of 17th century Sweden; he wrote a great number of songs. Most of the songs were drinking songs and that made all that much the better. I must say that one of the most enjoyable parts of my visit to Stockholm two years ago was the fact that I was able to sing along with the Swedish students, sing Bellman with them; I at least knew some of the words of Bellman's drinking songs.

Charlie had an enormous influence on my life; he guided my graduate research and at the same time, Robert Oppenheimer played a significant role. Charlie and I were doing research in what was then this new science of nuclear physics when Charlie found out that Cockcroft and Walton could disintegrate nuclei with accelerated energies less than one million volts. Lauritsen changed one of his X-ray tubes, which was a part of an ion accelerator, and started doing nuclear physics. He'd built the X-ray tubes because he wanted to do X-ray physics and cancer therapy. Mr. Kellogg supported the research because of the cancer therapy. Research in X-rays and cancer therapy lasted up to the war simultaneously with the new work in nuclear physics. I got my fellowship by helping maintain the X-ray tube, which was located in this building. The doctors who carried out the research in cancer therapy were all located here and we graduate students who had fellowships kept the million-volt X-ray tubes running so that in the morning the doctors could treat cancer patients. Then in the afternoon, the tube was free for Charlie's graduate students who were still working in X-ray physics.

I was in nuclear physics with Charlie and my thesis was on radioactive elements of low atomic number. We studied the positron decay of radioactive nuclei like carbon-11, nitrogen-13, oxygen-15, fluorine-17. This is where Robert Oppenheimer comes in. He had a joint appointment in those days between Berkeley and Caltech. In those days Berkeley started early in the fall — some time in August — and then their term finished early in the spring, so he was able to come down — after school had finished at Berkeley — to Caltech for our spring quarter. He was essentially the theoretical advisor for my Ph.D. research. Charlie and I found that the beta decay energy of the series of radioactive nuclei increased quite uniformly. This energy was proportional to the electrostatic energies in these positively-charged nuclei; carbon-11 has charge 6, nitrogen-13 charge 7, oxygen-15 charge 8, fluorine-17 charge 9, and the energies increased uniformly. Robert pointed out to us that this was just the electrostatic energy, which would be the difference between the so-called mirror nuclei and to make a long story short, he pointed out that the nuclear forces were charge symmetric. You had to have the same force between two protons as between two neutrons if you exclude the fact that the protons are charged and also have an electrostatic energy. Take that away and the forces were the same; so we discovered something fundamental about nuclear forces, but I wouldn't have known what it meant if Robert hadn't told us what was going on.

It was very characteristic of the situation in those days. Charlie was a great man; he had a deep knowledge of physics, but he was primarily an experimentalist. For him the fact that Oppenheimer could come down during the spring quarters meant that his students could get some theoretical advice. So it was a great thing. Another man who played an important role was Richard Tolman who was Professor of Chemistry there, but he was essentially a physicist. Tolman and Lauritsen and I were very close. When I was working as a graduate student with Charlie, quite frequently Robert and Richard Tolman would pop up in the lab and sit down and just watch while we were working. Charlie would let me take the readings and I never forget the discussions they had, ranging from everything to the coming world war and to basic problems in physics.

Then I must go back to mention the role that Charlie Townes played. Shortly after I finished my thesis on the radioactive nuclei of low atomic number, Lauritsen and I decided to study interactions of the isotopes of carbon and nitrogen with protons. One of the isotopes of carbon is a rare one, carbon-13. In normal carbon, most carbon is carbon-12, it's 99 percent. Carbon-13, the heavier isotope is only about 1 percent so it's very rare. If you take an ordinary target of carbon and try to see what comes from carbon-13, that's swamped by what comes from the more abundant carbon-12. Charlie Townes was then working with W. R. Smythe as graduate student in mass spectroscopy and was separating the isotopes of the light elements. When I wanted to bombard carbon-13 with protons to study what happened, I went to Charlie Townes and he supplied me with enriched samples, which he was producing as part of his thesis research under Professor Smythe. So Charlie Townes helped me with my postdoctoral research.

Charlie [Lauritsen] and I bombarded carbon-12, carbon-13, nitrogen-14, nitrogen-15 with protons and that culminated when all of a sudden, in 1939, Hans Bethe announced that the interactions of protons and those isotopes are part of the carbon–nitrogen [CN] cycle by which energy is generated in many stars. That was just an awakening; we suddenly realized that what we were doing in the laboratory was studying processes that occurred in stars and generated the energy with which stars shine. I was hooked, Charlie was hooked, and we decided to concentrate on the aspects of nuclear physics, which had to do with energy generation in stars. We came eventually to call this field nuclear astrophysics. I emphasize that it's a benign application of nuclear physics. As we all know, there are many applications of nuclear physics, but they are not all benign. It has always been a paradoxical situation for me, but I tend to look on the sunny side of things and in this case I use the word sunny rather literally, not by the carbon–nitrogen cycle but by other processes which Hans Bethe also suggested, called the proton–proton chain.

Helping the War Efforts

The war came along just about that time and eventually this laboratory was transformed into a center for the development of rocket ordnance. We built small rockets primarily for the United States Navy and we helped the Navy establish the Naval Ordnance Test Station at China Lake. Toward the end of the war, when it became clear that the atomic bomb was going to work after the Trinity test, and even before Trinity, we transferred all the rocket ordnance work to the Navy and began producing non-nuclear components of the atomic bombs for Los Alamos. Charlie and I — mainly because of our connection with Robert Oppenheimer who was the director at Los Alamos — we began to spend a great deal of time at Los Alamos at the same time as we were phasing out our rocket work.

We tried to find out what we could do to help the efforts at Los Alamos. For example, for the plutonium weapon, which was housed in an enormous bomb casing — the Fat Man — the ballistics, that is, the trajectory that such a bomb would take when it's launched from an airplane, just wasn't known at all. With a number of dummy bomb cases, which we called pumpkins, which had to be built, and we built hundreds of them, and they put the center of gravity in the same place, and we would modify the structure according to what Los Alamos wanted. We would take these up to Wendover, Utah, where the Air Force had a test range — I never went up there — and they dropped these things, untangling what the ballistics were going to be. This is the sort of things that we did.

Finally, the war ended and, fortunately, successfully, but with consequences that remain until the present time. That's again a great problem for all of us who took part; we felt we had to do what we did, but there were lingering doubts whether the development of the atomic weapons was the right way to go. I personally think it was, but I can see why many people feel that perhaps there were alternatives that might have succeeded. My own recollection is that at the time we were convinced that it had to be done; we were scared to death of Hitler and the Germans and we were scared to death of the Japanese. So it happened and we had to live with it.

Low-energy Nuclear Physics

When the war ended, Charlie and I and Tommy, his son — who had done his Ph.D. just before the war with his father just as I had, and joined the faculty after the war — we set about getting rid of all the defense work and re-establishing Kellogg as a nuclear laboratory. We made a quite different decision than Ernie Lawrence made, not to say that what we did was right or wrong; I think that all decisions worked out quite well. Ernie decided to go to higher and higher energy nuclear physics, which has resulted in things like the Fermi Lab where the energies are in the billions of volts range and now they are going to be in the terravolt range, and the CERN laboratory in Geneva. But the Lauritsens and I decided to stay in low-energy nuclear physics because that's where the applications in astrophysics occurred. Contrary to the way most other people were going, we developed low-voltage electrostatic accelerators, following most closely the work of Ray Herb, the pressurized electrostatic accelerators, which was an improvement that Herb in Wisconsin made over the original designs, the open-air designs of the van de Graaff generator. We followed Herb's direction and worked for very high resolution in the energy of the beams produced by our electrostatic accelerators and for very high sensitivity in the detection scheme. Our first accelerator was powered with alternating voltage because it was powered from an alternating current transformer that had been built to test the insulation of the transmission lines. When Charlie changed from producing X-rays with alternating voltage, he made his tubes into positive ion accelerators and that meant that you had a beam which had all energies from zero up to a maximum. Actually, the focusing was such that the beam was mostly at the higher energy, but it had an energy resolution maybe of 20 or 30 percent, not of the tenth of a percent which you could get from a home-built van de Graaff accelerator.

So Tommy, Lauritsen, and I built the first one and we built it before the war, in 1938–1939. When the war came along, it was moved into a corner and after the war we had to put it back into operation. That machine — as I remember — had a maximum voltage of about 1.8 million volts and we built another one for still lower energies and higher currents and still higher resolution that operated up to 700 kilovolts, and then we built a still larger one that operated to 3.5 million volts. With those three machines we were able to maintain a very extensive program partly in pure nuclear physics but primarily devoted to studying those nuclear processes, which the theorists told us occurred in stars. As you go lower in energies in bombardment, the probability of a nuclear reaction gets smaller and smaller. That's due to the fact that the interacting nuclei, the protons, which are the nuclei of hydrogen and the nuclei of carbon-12, they are both positively charged and the like charges repel. So you have to give the proton quite a bit of energy to get into the carbon nucleus, where it can fuse and then break up in that case into the emission of gamma radiation, with a transition from carbon-12 to the nitrogen-13 nucleus. In classical mechanics this can't even happen; in quantum mechanics, there is a penetration permitted through what's called the Coulomb or electrostatic barrier.

As you go to lower energies, the probability of that penetration gets smaller and smaller, in fact, it's an exponential decrease. We had to have higher and higher currents at lower and lower energies and that's what we were concentrating on. You can't design one machine to cover the whole range of energies and that's why we built three of them. Eventually, in the 1960s, to jump ahead a bit, we obtained a commercially-built tandem electrostatic accelerator, which operated up to 7.0 million volts, and it is still in operation in the Sloan Laboratory adjacent to the Kellogg Laboratory here. Then, just a few years ago, we found that all of our old equipment that we had built was just becoming obsolete and was costing more to maintain than was worth it. So we got a 3.5 million volt machine, a high-current tandem, from Ray Herb's outfit in Wisconsin, the second one that we hadn't built, and it's our work horse now. It delivers very high currents down to very low voltages, and has still higher resolution. Along with the advantages of the van de Graaff accelerator in producing higher resolution beams, you have to go beyond that, you have to have big magnetic analyzers to put the beam through, so that you can further improve the resolution by a feedback mechanism. If the beam going through the magnet with its proper setting begins to move off, feedback tells the accelerator to correct that. In addition you have to have very thin targets, you have to have very sensitive detectors, and, of course, right after the war, and up to the present time, the detectors have been complicated electronic devices.

Well, as much as I love these gadgets, recent developments, especially with the microchips, have made possible what we were doing in the lab and we've tried to stay abreast of that. We have one of the best-equipped labs in that regard in the country although we operate at a considerably smaller scale than the big laboratories, like the one that Lawrence built at Berkeley, the Fermi Lab, SLAC at Stanford, and CERN at Geneva. But we have a staff of 5 full professors and quite a few graduate students and quite a few postdoctoral fellows. We have an operation that involves about 50 people and we are still primarily concerned with studying those nuclear processes, which, we think, take place in the stars.

Modeling the Production of the Chemical Elements

Going back, after the war we did get into the study of the carbon–nitrogen cycle, which we had been working on before we'd learned from Bethe that it was important in stars. We improved our measurements and we went on to do other things that were directed toward problems in nuclear astrophysics. One of the very popular theories after the war was that due to George Gamow, which suggested that all of the elements had been produced in the early high-temperature high-density stage of the Universe, which we call the Big Bang. Even before the war, Hans Staub and William Stephens here in Kellogg had confirmed the fact that there is a mass gap in the periodic table at mass 5. There is no stable nucleus at mass 5, where by 5 we mean the mass that's roughly 5 times the mass of a proton. There are two radioactive nuclei, helium-5 and lithium-5, but they both break up in a millionth of a micro-microsecond. As fast as you make them they disintegrate. Staub and Stephens had shown it as well as other people.

Then after the war, Alvin Tollestrup, who was one of our early graduate students after the war, Lauritsen, and I showed that the same thing happened at mass 8. There the nucleus that would be stable would be an isotope of beryllium, beryllium-8. The geochemists had long suspected that beryllium-8 was not stable because the stable form of beryllium in nature is beryllium-9. After the war we looked at beryllium-8 and found that it broke up into 2 helium nuclei of mass 4. As fast as you made it, it again broke up in thousandths of a micro-microsecond as the lifetime. So there are gaps at 5 and 8; and Gamow's theory was that all of the elements and their isotopes had been made by successive neutron capture with the emission of gamma radiation. The neutron also has roughly a mass of one. With neutrons you have to have every number; you start with hydrogen, you go to deuterium, you go to helium-3, then helium-4, but when you go to 5, it breaks down immediately to neutron plus helium-4. Even if you get around that, by the time you get to 8, it breaks right back down to two nuclei of helium-4. We played a role in this laboratory in convincing Gamow that his scheme would not work beyond mass 4.

The ultimate solution again turned out to involve stars. After the fact was discovered that there were no mass 5 and no mass 8, it was suggested that two helium-4 nuclei were put together in the first instance of the Universe, the so-called Big Bang. They stuck together and then, with another helium-4, they made carbon-12; you make mass 8 out of two helium-4 or alpha particles, and they do stick together briefly. If the density and temperature are high enough, another helium-4 will join in and form carbon-12. However, that won't work in the Big Bang because after helium-4 is made, the Big Bang is an expanding Universe and when something is expanding, the temperature and density are dropping. So after helium-4 was made, the temperature and density continued to drop, so you did not have enough collisions to form carbon-12.

It was Edwin Salpeter at Cornell who first suggested, on the other hand, that this process could occur in red giant stars. In a star, quite contrary to what happens in the Big Bang, after it had gone through what we call the main sequence stage, in which it converts hydrogen into helium by nuclear processes and kicks off the energy by which the stars shine, the gravity of the star takes over. So, it's not expanding like the Universe. There is increased density and higher energy and the three helium-4 nuclei fuse into carbon-12, as Salpeter pointed out, which is stable. That came about because when Salpeter came here in 1951, he found out that Tollestrup and Lauritsen and I had shown that beryllium-8 was only unstable by something like 90 kilo-electron-volts. At that point he realized that it could stick together long enough in the center of the red giant stars. That's when Fred Hoyle comes in.

I've worked ever since 1953 with Fred Hoyle and I must say right at the outset that the grand scenario of element formation in stars is due to Fred Hoyle just like the grand scenario of energy generation in stars is due to Hans Bethe. Hoyle took Salpeter's suggestion quite seriously and worked on exactly how this fusion of helium took place. He was a theoretical astrophysicist, working on stellar structure. He came to the conclusion when he compared the theoretical calculations on red giant stars with the observational material that the astronomers had collected about the red giant stars, that the red giant stars had to ignite the helium fusion into carbon at a lower temperature than Salpeter's theory gave. So Hoyle modified the theory, and he did it by saying that when a pair of heliums are temporarily stuck together, and another helium was coming along, there was a resonance of their interaction in that process, which would enhance the rate beyond what Salpeter had calculated without the resonance.

I know that's highly technical but Hoyle immediately realized that a resonance in this process required there to be an excited state in the carbon-12 nucleus at an energy that he could calculate fairly precisely, just from the astrophysics. He calculated something like 7.58 million electron volts, and he was convinced that this excited state had to be there. There had been some observations even before the war that maybe there was such an excited state, but after the war, improved work at MIT had looked for this excited state and hadn't found it. Tommy Lauritsen, who kept the books on all excited states, erased it. When Hoyle came to the Lab in 1953, and talked to us about this, the two Lauritsens and I told him to go away, that there was no state there, and that we were busy, and to please stop bothering us. But Ward Whaling, now Professor Ward Whaling in the Laboratory listened to Hoyle and he and his group found this state in carbon and they found it to be almost exactly where Hoyle said it was. They got 7.56 instead of 7.58; and that made a believer out of me.

Bethe had gotten us fascinated with energy generation in stars. Hoyle was proposing elements in the stars and there was now a new reason for working in nuclear astrophysics. That was all in 1953 and Tommy and Charlie and I jumped into the problem after Whaling had shown that this state exists. We, along with a graduate student, had shown that such a state could be formed by putting three helium-4 nuclei together. They had to have some special properties; they had to have the correct spin and parity in our terminology.

Then I got a Fulbright Scholarship to go to England to work more with Fred Hoyle in 1954 and Fred introduced me to Geoff Burbidge and Margaret Burbidge. Then, in 1955, the Burbidges came back here to Kellogg with me. Geoff got a Carnegie Fellowship at the Mount Wilson Observatory and Margaret had a postdoctoral fellowship here in Kellogg. One thing led to another, primarily some work that Hans Suess and Harold Urey did, which unraveled many of the apparent complexities in the abundances of the elements and their isotopes. Once we were aware of what Suess and

William Fowler with three of his frequent co-authors at a conference in honor of his 70th birthday, held at Caltech in 1981. From left: Professors Margaret Burbidge, Fowler, Sir Fred Hoyle, and Geoffrey Burbidge (courtesy of California Institute of Technology).

Urey had done, then we came forward with a general theory along the lines of Hoyle's grand scenario. We posed nuclear processes which could produce all of the elements all the way up in the periodic table to the radioactive elements thorium and uranium.

I must say that at about the same time, in 1957, this was independently done by A. G. W. Cameron who was then in Canada. He is now Professor of Astrophysics at Harvard. It's very important to know that Cameron independently did what the Burbidges and I did. When I look back on how furiously and hard the four of us had worked on all these problems after we got an inkling of how it would go from Suess and Urey, I realize that Cameron did the same thing single-handedly, it was fascinating.

We are Stardust

While I'm passing out credits, I think it's very important for me to stress that the award to me I regard as an award to this Laboratory, to the memory of Charlie and Tommy Lauritsen and to my colleagues here who really have done most of the experimental work over the years. I've already mentioned Ward Whaling, and in addition Charles Barnes and Ralph Kavanagh, both full professors and Brad Filippone who is a vigorous young Assistant Professor. They carry on the experimental work. The theoretical work, which was started by Oppenheimer, was in the early days carried on by Bob Christy, who is now Professor Emeritus of Theoretical Physics at Caltech, and at the present time by Professor Steve Koonin here in Kellogg. I have to emphasize that if it hadn't been for the work of so many people, all of the staff members I mentioned and legions of graduate students and postdocs, the field of nuclear astrophysics could not possibly have been recognized in the way it was. I have to emphasize too that it's true that our Laboratory has been a leader in the field, but there were many other laboratories all over the world that have contributed to the field. It's a very active field both experimentally and theoretically in other places.

It's still an extremely exciting field and Filippone and Barnes are studying the next process that occurs in red giant stars after the carbon is produced. Carbon-12 may fuse with another helium-4 and that makes oxygen, the main isotope of oxygen, oxygen-16. The ratio of the rate of that process measured in our Laboratory to the rate of the production of carbon-12 determines the ratio of how much carbon to oxygen is made in the red giant stars. That's where our carbon and oxygen are made. That's important to us because we are mostly carbon and oxygen. We are 65 percent oxygen, 18 percent carbon and we are still about 10 percent of the primordial hydrogen in the water. But carbon and oxygen are important to us and because they are made in stars, I sometimes tell my audiences that we're all, literally and truly, a little bit of stardust.

Vera C. Rubin, 2000 (photograph by Magdolna Hargittai).

13

VERA C. RUBIN

Vera C. Rubin (b. 1928 in Philadelphia) is Senior Fellow at the Department of Terrestrial Magnetism of the Carnegie Institution in Washington, D.C. She received her B.A. degree at Vassar College (1948), her Master's at Cornell University (1951), and her Ph.D. at Georgetown University (1954). She stayed on at Georgetown University until 1965, when she moved to the Department of Terrestrial Magnetism of the Carnegie Institution of Washington. She is most famous for her results, in the 1970s, indicating that most of our Universe is dark matter. She is a member of the National Academy of Sciences of the U.S.A. (1981), the American Academy of Arts and Sciences (1982), the American Philosophical Society (1995), and the Pontifical Academy of Sciences (1996). She received the U.S. National Medal of Science (1993); the Gold Medal of the Royal Astronomical Society (London) (1996 — the second woman; Caroline Herschel was the first in 1828); the Weizmann Women and Science Award (1996), the Peter Gruber International Cosmology Prize (2002), the Bruce Medal of the Astronomical Society of the Pacific (2003), and the Watson Medal of the U.S. National Academy of Sciences (2004). She has been awarded numerous honorary degrees, from Harvard, Yale, and Smith College, among others. We recorded our conversation in her office at the Department of Terrestrial Magnetism on May 16, 2000.* There was a follow-up to our conversation in writing in May, 2004.

---

*Magdolna Hargittai conducted the interview.

First I would like to ask you about your family background.

My father came to the U.S. with his family as a child. He was the first in his family to go to college, and became an electrical engineer, but he was interested in many things, including mathematics. My mother did not have a college degree, but she had studied singing, played the piano, and had many interests. Our extended families were very close, and my older sister Ruth and I grew up in an intellectual atmosphere. We were encouraged to follow our interests. My sister is a lawyer, ultimately a judge; she also lives in D.C. and is my closest friend.

What made you interested in astronomy?

Looking at the sky, looking at the stars as an eleven year old. I had a bed that was under the window and I could see the sky and I just got more interested in watching the stars than in going to sleep. When it was time to go to college, there were not too many places where a girl could study astronomy. I went to Vassar College, which is a woman's college. I knew that a famous woman astronomer, Maria Mitchell, was the first professor of astronomy there when the college opened in 1865. So I knew that there was a place where a woman could study astronomy. I needed a scholarship and they gave me one. I met my husband-to-be while I was in college; he was a graduate student at Cornell, so I joined him at Cornell and got my Master's degree there. Then he finished his Ph.D. and we moved to Washington. I entered Georgetown University as a Ph.D. student in astronomy, although I wrote my thesis under George Gamow, who was at George Washington University.

Could we please go a little slower? What was your Master's thesis about? I have heard that it already created quite a stir.

There were 109 galaxies whose radial velocities had been measured by 1940, and these had been used to establish that the galaxies were moving away from us, that is, the universe was expanding. Any observer, anywhere in the universe, would see the galaxies moving away from her. But I used the observed velocities to ask a different question. If we remove the average expansion from each velocity (based on the distance that we would place the galaxy from its apparent brightness), were there patterns in the residual velocities that remained? Did the galaxies show patterns that indicate that over large distances, groups of galaxies are moving en masse. I applied the mathematics of a rotation to these velocities, to see if there was a large-scale rotation in the universe. That was because I knew the mathematics involved in such a rotation, but I knew of no models to describe a large-scale motion.

In spite of all the criticism you've got for your results, I understand that they actually helped Gerard de Vaucouleurs to establish the local supercluster.

Yes, that is so. I had plotted all of the residual motions on a globe, and looked for an "equator" on which to base the geometry of their large-scale motions. Some years later, de Vaucouleurs used about the same coordinate system, defining it as the "equator" of the local system of galaxies. He was one of few astronomers to take my early work seriously, and that made a great bond between us.

Was it easy to get into graduate school? I can imagine that in the 1950s some colleges may not even have welcomed women into their grad school?

There was no problem. Unlike the important physics department at Cornell, the Astronomy Department was tiny — 2 faculty — and undistinguished. When I arrived I was advised by the Chairman to go find something else to study. He said the world didn't need more astronomers. I ignored that advice.

You mentioned that you did your doctoral thesis with George Gamow. What was that about?

Gamow had heard of my Master's research. When we moved to Washington, my husband went to work at the Johns Hopkins Applied Physics Lab, and worked with Ralph Alpher and Bob Herman. Ralph, who had earlier written his thesis under Gamow, and Bob Herman were still working with Gamow on problems of the early universe. Through this connection, Gamow called me on occasion to discuss galaxies. Ultimately, his question "Is there a scale length in the distribution of galaxies?" seemed to be a good problem for a Ph.D. thesis. Even though I was enrolled at Georgetown University (the only local college which offered a degree in Astronomy) and Gamow was a professor at George Washington University, I made arrangements to write my thesis under his direction.

University of Michigan 1953 summer school for young astronomers. George Gamow is in front row, 4th from left. To his left is Walter Baade. Vera is behind and between them, Bob is in Hawaiian shirt to her right. Owen Gingerich is to Gamow's right. Geoffrey Burbidge is leftmost in photo, with Allan Sandage to his left, and Margaret Burbidge to his left. Don Osterbrock is at left of top row, with his wife Irene to his left (courtesy of Vera Rubin).

What was it like working with George Gamow? What kind of a person was he?

He was very pleasant, very amusing, liked scientific games and jokes, but was not really interested in the scientific details of an analysis. Except for very general guidance, he was not involved in the details of my calculations, but he was enormously interested in the implications of the results: patterns exist in galaxy distribution, and galaxies are strongly clustered.

His initial suggestion for the Big Bang was not well received. How did he take this?

He was not disturbed. He enjoyed throwing out many ideas, important and imaginative ideas, about the universe. And he had great delight in those that survived. He was very generous, and liked the interactions and the social aspects of science. He started almost every talk with amusing scientific demonstrations. He liked to have fun, especially with science.

Could it be that the poor reception turned him away from astrophysics and prompted him to become interested in molecular biology?

I do not think that this had anything at all to do with it. Gamow was a physicist, but interested in, and curious about, all of science. He understood the important implications of what was about to happen in biology, and he was versatile enough to have important ideas in this field too. He went wherever his curiosity led. I think I am correct in stating that he was the first to devise a helix (but a single, not a double helix) for the DNA structure, noting that the 4 nucleotides would pair to form 20 amino acids. He made an early model of the DNA helix, formed a RNA tie club of 20 (male) leaders in the then baby field of DNA structure, sent each a tie with one of the 20 structures, ties and tie pins which he had specially made. I think he understood from the start that the DNA was a code that expressed the "language" of biology.

How did he react to the discovery of the remnant heat by Penzias and Wilson? How did he feel that Penzias and Wilson were universally recognized and his contribution was not so much appreciated?

Gamow left Washington in 1956, 2 years after my Ph.D. degree. He was in Berkeley in 1954 when I defended my Ph.D. thesis in Washington. After that, I saw him at meetings or occasional visits before his death in 1968. I had only a few discussions with him about many of these matters. The last time I spent any time with him was at a meeting on Relativistic Astrophysics (the "Texas" meeting that was that year held in New York). I am certain that he would have been delighted with the discovery of the remnant radiation by Penzias and Wilson. It proved that the cosmology of Alpher, Herman, and Gamow was very close to the cosmology inferred from the Penzias and Wilson observations.

How do you feel about the fact that they received the Nobel Prize and Gamow practically never got any recognition? How do you evaluate Gamow's general importance in science? How would you react to what Penzias said when he placed Gamow above Galileo?

How do "I" feel? Well, let me say the following. My husband Bob and I remained very close friends with Bob and Helen Herman, and continue to see Helen now, some years after Bob's death. I know that he and Ralph Alpher believed that their work was never properly recognized, and were sometimes very bitter about this. I feel sorry for scientists who believe that their work was not given due credit, but I don't think that Gamow thought that his work had been neglected. It is recognized that important ideas in nuclear physics came from Gamow. His joy came from interactions with scientists, with dreaming up ideas that might be correct. He had a distinguished career, and he was recognized for his many innovative ideas. He did get recognition. He was an active member who enjoyed his activities with the U.S. National Academy of Sciences. (I do not know the quote about Galileo).

Coming back to your life, already at Cornell you had great names as teachers, such as Hans Bethe and Richard Feynman. Did they make a difference for you?

It is always inspiring to study under great scientists and great teachers. I had minimal interactions with Feynman or Bethe, but Bob and I did become friends with Phil and Phylis Morrison, a friendship we have enjoyed

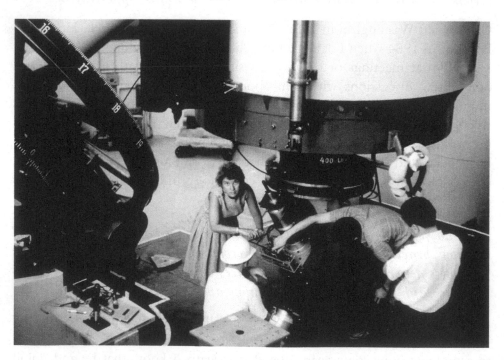

Vera Rubin at the 60-inch telescope of the Ohio State and Ohio Wesleyan Universities at Lowell Observatory, Flagstaff, Arizona, 1965 (photograph by Bob Rubin).

over many years. It was Phil I asked about the dark matter in the very early stages of trying to understand it. Phil said, "The vacuum has energy." I'm still trying to learn what that means — I think this is what we now call "dark energy".

Please, tell me something about your most important achievements.

Well, I guess, I have two. For many years I've studied the ways that stars orbit in galaxies. I am an observer, I go to the telescope. When I arrived at the Department of Terrestrial Magnetism, Dr. Kent Ford, a staff member, had just built an image tube spectrograph, an instrument to get spectra of faint objects at a telescope. I wanted to work on a project of interest, but on ideas not actively pursued by other astronomers, so that I could work at my own pace.

I had long been interested in the outer parts of spiral galaxies, which were then little studied. It was known that near the center of a galaxy, the stars orbit with high velocities. By analogy with planets in the solar system, we expected that for stars farther out from the nucleus, the stellar orbital velocities would be slower. However, what we saw in virtually every case (and by now there have been thousands of galaxies studied), is that the orbital velocities of the distant stars and gas are just as high as they are for those near the center. The best explanation is that the bright matter is responding to the gravitational attraction of matter that we cannot see. The distribution of this dark matter is very different from the distribution of bright matter. The bright matter is highly concentrated in the center, and then falls off rapidly with increasing nuclear distance. The distribution of the dark matter is less near the nucleus, but it becomes more significant with increasing nuclear distance; it falls off much more slowly, and extends much farther than the bright matter. It composes about 95 percent of the galaxy mass. Thus, the distribution of bright matter in a galaxy is not a good indicator of the distribution of matter.

This leads to the conclusion that almost all of the matter in spiral galaxies (and also in most of the few studied elliptical galaxies) is not radiating; it is invisible. Although Fritz Zwicky over 50 years earlier had said that something strange was going on in clusters of galaxies and he thought that it must be matter that we're not seeing, at that time no one knew what to do with that observation. So it was just ignored. It wasn't that it wasn't believed; no one knew what it meant. Once rotation curves started being observed all the way out and we saw that they, too, indicated dark matter, the community was really very rapid in accepting the possibility and believing that there must be matter, which does not radiate. When we look at a galaxy we're seeing the distribution of light but we aren't seeing the distribution of matter.

Does this mean that dark matter does not radiate or just that it does not radiate in the electromagnetic spectrum?

We think it means that this is the kind of matter that does not radiate at all. Our current cosmological theories place a limit on the amount of "normal" matter that exists, that is, atoms and subatomic particles. And this limit is less than the amount of matter required by the observations. So the remainder must be an exotic form of "matter" — neutrinos for example. But now we have a constraint on the neutrino mass, and it looks like that is not enough. Particle physicists think that their next generation of accelerators will give us the answer. We'll see (rather "know"; we don't "see" dark matter) in a few years.

There is still another possibility, which is much less likely, but I am surprised we haven't been able to rule it out yet: the possibility that Newtonian gravitational theory does not hold over distances as great as galaxies. That would be a shock, truly a revolution! It is easy enough to write down equations to describe what we see but that's not enough because we know that Newton's laws work in some domain and relativity works in a different domain and whatever changes are made, the theory still has to reduce to both of them in the domains where they are valid. Thus you have to invent a new cosmology, which is a daunting task. But some scientists are attempting this.

We have not yet covered all your important contributions…

OK. The dark matter was one. The other one would be my studies of large-scale motions of galaxies in the universe. This was really a return to the problem I worked on for my 1950 master's thesis, as I described above. That scientific paper describing that work was rejected by both *The Astrophysical Journal* and *The Astronomical Journal*, so I never got it published. But I did give a paper at the AAS so the work at least got out in abstract form. Then in the 1970s I returned to this work.

I took a sample of about 100 galaxies, all farther from us than the Virgo Cluster, the local supercluster of galaxies of which we are an outlying member. The galaxies were distributed at about the same distance from us all over the sky, like a shell. I tried to use galaxies of all the same type, of the same true brightness. I asked the question whether the Hubble expansion was the same, that is, isotropic around the sky. It turned out that it was not. On one half of the sky, the velocities were systematically smaller than the velocities on the other half. We interpreted this not as a distortion of the smooth Hubble flow, but as a relatively large motion of our galaxy. In the region of the sky we were moving toward, galaxies have apparently smaller motions; in the opposite direction, they have apparently larger motions. The astronomical community named this result the Rubin–Ford effect. At that time, the mid 1970s, it was believed that our galaxy could not be moving so fast because earlier observations of the residual radiation from the big bang had indicated that our galaxy did not have a large motion. Within a few years, however, observations of the cosmic background radiation improved in accuracy enough to detect a galaxy motion of about the same amount as we had detected, but in a different direction. So for a decade or so, the subject was full of confusion, as different studies returned different results. But within a few years, numerous larger observational studies were initiated, and the subject of large-scale motions of galaxies in the universe became an active one for astronomers. There was an important meeting at Lake Balaton, in Hungary, to discuss such studies. It has been fun to watch this subject become an important part of cosmology, after the early disbelief of my 1950 MA work.

What other research of yours would you like to mention?

I can't neglect a very special galaxy, NGC 4550. This is a rather unremarkable galaxy in the Virgo cluster of galaxies, until you study the motion of its stars. In its single disk, some stars orbit clockwise about the center, some counterclockwise. No collisions: stars beyond the nuclear regions are very very far apart. Newton's laws of motion relate the orbital velocity and the distance of the star to the interior mass causing the motion. But the velocity enters the equation as the square, so either a positive or a negative velocity will satisfy the equation. But it was a shock to see it on the spectra, and it took about two years before I believed that I understood what I was seeing. No one yet knows how to make such an object, for if two disk galaxies merge, the energy transferred to the stars will puff up and destroy the disk.

There have been different periods in human history when especially important astronomical discoveries have been made; the time of the Greeks, Kepler, Galileo, the Copernican revolutions. Do you think that we also live in such revolutionary times concerning astronomy?

In some sense we certainly are because we've discovered many things that were not known before. This is mostly due to our technological advances. Being able to see gases in the radio frequency range; hot plasmas in the X-rays; different temperature regimes radiate at different frequencies and so we've just learned many things that weren't known before. At the same time, there are still many major things that we still don't understand. I think that every couple of years or maybe even more often we are going to learn something very, very new, and important.

Looking at different statistics we see that the proportion of women decreases rapidly as you go from the undergraduate level to higher. What's the problem?

You can write books just on this subject! It is perceived as a women's problem and I believe it will never be solved until it is perceived as society's problem or an academic problem. So I am really more pessimistic now than I was 50 years ago. Then, with so many women entering college, it looked like a gradual evolution would take place. Yet in the U.S. the number of women full professors in science departments of the top 50 or so colleges and universities, is 6% (or was in 1998) and that's really outrageously low! Twenty years ago, some 20% or more (depending upon field) of all Ph.D. degrees went to women — these women should now be professors, but they are not. In science, women now get more college degrees than men. And even in science Ph.D. degrees, numbers of women are now about equal to men. So I think the real problem lies in academia. Some women get a Ph.D. degree in physics and never have studied physics under a woman.

There are still meetings in science, too many, where the speakers are all, or almost all men. Organizing committees are all men — without women on committees, on faculty, it is too easy for men to "ask their friends". That's why there must be women on faculty, on committees, in Academies of Science, to see that the injustices do not continue to propagate. I still hope that this change will come about easily. There are now overwhelmingly large numbers of brilliant women receiving Ph.D. degrees in science. Science faculties will have to work hard to NOT hire them. So we'll see.

Is it only the men who are to be blamed for this? Isn't it possible that women are not interested enough in science or are not brave enough to stay?

I don't think it is bravery. If you are bright, and you are not welcome, you will have the brains to go elsewhere. Too many women enter college thinking that they want to be scientists, but they do not survive. In many cases this is due to a lack of welcome, a lack of support, and the poor treatment by their male professors or colleagues. They are ignored, they are not listened to. If the community of scientists was more welcoming, many more women would succeed. I often visit universities because I like spending time with students and I always talk to the women. At universities with large graduate science departments the women can name the professors they are warned not to attempt to work with. Although the colleges deny that such exist, the women know that they would never succeed with these male faculty. Someone called this the Bluebeard syndrome. If you go in his laboratory you will never come out alive. So the universities are often a large part of the problem; I think it unfair to blame the women. Being a graduate student is hard, being a scientist is hard, but if you add more impediments along the way, the likelihood that women won't survive

Vera Rubin measuring spectra in her office, the Department of Terrestrial Magnetism, Carnegie Institution of Washington, early 1970s (courtesy of the CIW).

is very large. Women who are determined enough surely can survive, but even less tough women should be welcome into the fields of science. Science will lose too many of the most brilliant minds, if it is not more welcoming. That's the tragedy.

Do you have children?

We have 4 children. They are all Ph.D. scientists, all wonderful people. I think they had thought that their parents were having a lot of fun, so they got interested in doing science themselves.

How did you manage?

Well. I muddled through. I just did whatever I had to do. I wanted children, I still adore children. I think the greatest miracle is to watch a child grow up. I had one child before I got my master's degree. I had two children while I was working for my Ph.D. I put them to sleep at 7 and I did all my work from 7 p.m. to 2 a.m. It was very hard. But by the time I came to Washington, to do my Ph.D. I had done a lot of physics at Cornell, so my course work was relatively easy. Astronomy at Georgetown was only a graduate department, so all the courses were given at night and I went to school at night. I did all my calculating at home. I had two children come to my Ph.D. graduation. Having the third one and even the fourth didn't change things very much. I stayed on the Georgetown faculty for about ten years. When I came to Carnegie, I asked for a part-time (2/3) job. I went home at 3 every day.

You must have a very supportive husband.

Oh, yes. He was more than supportive. When I would say that I could not possibly go to a meeting he would tell me I should go. He was encouraging; more than just supportive.

What is his profession?

He was trained as a physical chemist, he was a student of Debye's at Cornell; he worked most of his life as a physicist, a mathematical physicist. Then about a dozen years ago he switched to biology, so now he is doing the same kind of mathematical modeling, random walk modeling, but now he does it in biological topics and not in physics. And he loves it; I think he is enjoying it more than he was enjoying his physics.

Today many young women think that they would rather first establish their career and then have children. Do you think that's a good idea?

I don't know. I don't know how to give advice on that subject. I know that's what they think and I am sorry they do. I think they have to do what they think they have to do. Maybe it's harder today than it was when I was young. It's a very personal decision.

These days we have the strange situation that the very few high-positioned women are overwhelmed by committee memberships, speaking engagements, and such. Don't you think it will hinder their creative work in science?

Perhaps. But I don't know what the answer is. It is related to the question about having children. I think you have to decide what your priorities are, and then do the best you can. I think that the science has changed. There are many more astronomers, so many more competitors. Science

Vera Rubin and astronomer friends, in the garden of Judith Perry, astronomer, Cambridge University, England. From right: astronomers Dr. Dennis Sciama, Dr. Margaret Burbidge, and Vera (courtesy of Vera Rubin).

advances much more rapidly. There are many more committees, many more books, many more sites on the web. I was at Georgetown University for ten years and I did what I think was a very good work. Most of the time I did not even have a mailbox; nobody wrote me a letter. I was totally isolated; there were not many astronomers around. I read journals, I went to meetings, I wrote papers and published them, people were interested in my work. It was at an international summer school in Leiden, The Netherlands in 1960 (my parents were at our home, baby-sitting our daughter and our 2-month-old son; Bob's parents in Florida were baby-sitting our other two boys) that I realized that the other students and faculty knew about work I did not; there was an active preprint system so astronomers sent out copies of their papers many months before they were published. I was amazed, but I got into the system.

Until that time, I wasn't called upon to do committee work. In this and other ways, I really had an unconventional career. I did not study at places that typically train astronomers. I think this was wonderful for my career. I had ten years to learn how to be an astronomer, how to be a good astronomer. The Kitt Peak National Observatory was coming online just during that time so I started observing first with the 36 inch, which was the first telescope there, then with the 84 inch, then with the 4 meter. Very few astronomers, male or female, have the privilege of having years like that. Unfortunately, now the community decides when you are 22 whether you are brilliant or not. And if they don't notice you then you may not have the opportunities. I am now besieged with things to do but I feel that I am old enough and somebody has to do these things. But I am not as happy internally. If I could do what I want, I would still do science every day, all day. There are now weeks when I don't get to do any science.

What was the greatest challenge in your life?

Finding good care for my children. Nothing compared to that.

Who are your heroes?

Maria Mitchell, the early American astronomer. Margaret Burbidge. She was one of the few women astronomers when I was a student. My husband has been an enormous influence both professionally and with the family. For many years he was the person I went to for scientific help and support. The Burbidges also had an enormous influence on me. We spent the academic year 1963–1964 in La Jolla. My husband had a senior NSF fellowship and one of the reasons he picked La Jolla was that I could work with the Burbidges. The year (actually at half-time so that I could be home with the children) gave me enough courage to walk into the Department of Terrestrial Magnetism and ask for a job. The year in La Jolla was the first time that I really worked with astronomers and they took me seriously. That's when I decided that I was really an astronomer.

Do you mind to answer questions about religion?

Not at all. I am Jewish (she is laughing). I am laughing because I have on my desk, you won't believe this, the following material. I have gotten involved with the Hebrew Union College, which, although they have been training women to become rabbis since 1975, still do not have a woman on their faculty. Now I find myself involved with trying to get women to their faculty. Until now my activities on behalf of women have been in science, and now this is something different. Of course, this is not religion, but organized religion.

Are you religious?

Am I religious? Judaism to me is kind of a moral code and on that level I surely am religious. I really believe that people should be good and do

With Magdolna Hargittai during the interview (photograph by I. Hargittai).

the best they can. If you are asking me if I believe in God, then, I guess, the answer is no. I believe that the Universe evolves, and galaxies and stars and atoms and molecules and planets and animals and humans are what comes. I am not religious in the sense that I believe there is a deity who is thinking about what we all are doing and who is directing large-scale activities.

How do your children relate to religion?

That's interesting. I would say that as a group they are probably more religious than I am. My three sons have all married Jewish women. Judy, our daughter, married someone who was not Jewish. They have a daughter and they were divorced after 10 or 12 years. For the last 10 years Judy has been living with a woman who has converted to Judaism. Religion is a very important part of their lives. One of our granddaughters is in graduate school. She is very religious, modern orthodox, and her aim in life is to make the Orthodox Jewish religion more environmentally friendly.

Since you brought this up, how do you relate to your daughter's way of life?

Easily. Probably more easily than I would have thought. Her partner is a very interesting woman. My daughter was apparently very unhappy in her marriage and we are delighted to see her so happy now. She has a wonderful daughter who is now a very charming 20-year-old. I'd rather see her happy than unhappy. She is also an astronomer, the only other astronomer in the family.

Have you ever had a joint research work with her? Have you ever published together?

Judy also studies galaxies, mostly, but with a radio telescope, in the millimeter spectral range, so she studies gas in galaxies. We have several times published papers together. She would occasionally ask me to get an image or a spectrum of a galaxy she was studying, and I would ask her to observe in the millimeter band a galaxy I was studying, and we would combine the results in a paper. It was fun but not a big deal — we each have our own interests. When she was in high school, I hired her one summer to pick out (from thousands of photographic charts of the sky), galaxies with very special similar appearances, relating to their spiral structure. These are the galaxies I observed for the study of large-scale motions in the universe.

What do your sons do?

The oldest and youngest sons, David and Allan, are both geologists. David works at the U.S. Geological Survey branch on the campus of the University of California in Santa Cruz, where he is also an adjunct professor. He is a sedimentologist who studies, among other things, how rocks form. About a month ago (April 2004), he was called to Washington by NASA, to be the "expert" on a Mars rover press conference. He showed pictures of rocks and sand along the Grand Canyon with the same arcs and feature as the Mars rocks, implying that they had formed in standing water. Several times a year Dave rafts down the Grand Canyon for his research in sediments and rock formation.

Allan is a professor at Princeton University, where he studies earthquakes and volcanoes and dikes in rocks. His research has taken him to Iceland and Hawaii. I think our many visits to the U.S. West during summer vacations turned both sons into geologists. We still have lots of rocks in the attic, and in their former bedrooms. Every time Allan's young children come, they leave with a few rocks. They don't know that they are doing us a favor.

Our middle son Karl is a mathematician, about to move from Stanford to the University of California at Irvine. He has done distinguished work in number theory, and received the Cole Prize in number theory of the American Mathematical Society, given every five years to a young number theorist. The entire family has a hard time understanding his work.

Grandchildren? How often do you see them?

We have 5, the youngest aged 4 and 3/4 (according to him) to 26. Two are out of college (one in grad school, one about to enter grad school), one graduates from college in 2005, one in high school, and one to enter kindergarten. The three oldest are all interested in environmental and ecological studies; all enjoy the outdoors, and are lovely, interesting people. We see them generally a few times a year, often vacation together (or in subsets) in the summer. A few years ago we all spent 10 days on a walking trip in Costa Rica. Only the oldest grandson was missing. He was in India on a college exchange program.

In 2 weeks, our California son and daughter-in-law are coming for a visit, with their daughter Ramona, our oldest granddaughter. Our other grand-daughter, Laura, will join us, for she will have finished the college term, and have a week or so until she flies to California to start a summer job to

The Rubins, July, 1998, celebrating the 50th wedding anniversary of Vera and Bob (center), Taos, New Mexico. Standing top row right: oldest son geologist David, next to wife Michell with son Zan behind and daughter Ramona bottom row right. Standing left: astronomer daughter Judy Young, partner Gene Stewart, and Judy's daughter Laura Young. Sitting right: mathematician son Karl, and Alice Silverburg. Sitting left: youngest son geologist Allan, wife Donna and son Eli in front. Youngest grandchild (Donna and Allan's son, now 5) was 1 year when picture was taken (courtesy of Vera Rubin).

clean and make repairs along the Pacific Ridge Trail, which extends from Canada to Mexico. There are often visits like this. For all the grandchildren, coming to my office and arranging the pencils/pens/rulers/junk in my top desk drawer has been a treat on their visits to Washington (at least when they were young).

What are you doing these days?

Still studying motions of stars in galaxies. Not pretty spirals, but the small irregular types that have been very little studied. They may hold surprises too. They are fainter and consequently more difficult to study, but once again bigger telescopes are available.

You have achieved a great deal. If you look back, what gives you the greatest pleasure to remember? Again, looking back, is there anything, you would do differently today?

I would definitely get a secretary, early in my life. I've never had one. I'm swamped with paper, both at home and at work.

My greatest pleasure has come from combining the roles of wife/parent/astronomer. None would have given as much joy alone. I love science because I have an unending curiosity about how the Universe works, and I could not be happy living on Earth and not trying to learn more. For me, it is the daily internal satisfactions that make a life in science so wonderful. Cold dark nights at a telescope have been among the greatest treasures of my life. Now they are less often, less cold, but equally treasured.

Neta A. Bahcall, 2000 (photograph by Magdolna Hargittai).

14

Neta A. Bahcall

Neta A. Bahcall (b. 1942 in Tel Aviv, Israel) is Professor at the Department of Astrophysical Sciences, Princeton University. She received her B.S. degree in physics/mathematics from the Hebrew University, Jerusalem in 1963, the M.S. degree in physics from the Weizmann Institute of Science in Rehovot in 1965, and her Ph.D. from Tel Aviv University in 1970. Professor William A. Fowler of Caltech (where she worked on her Ph.D. thesis) was her thesis advisor. She has served as an astronomer in various positions at the Space Telescope Science Institute between 1983 and 1989. She has been associated with Princeton University since 1971, from 1989 as Professor. She is a member of the National Academy of Sciences of the U.S.A. (1997), and has been active in many national committees. Our conversation took place in her office at Princeton University on April 25, 2000.*

What made you interested in astronomy?

I started in physics, not in astronomy. I grew up in Israel and I had very good teachers in the sciences in high school. I enjoyed the way they introduced us into mathematics and physics; I enjoyed the topics, and it came very easy to me. Thus I decided to continue these fields in college. I graduated from college in mathematics and physics and originally I thought that I would go back and teach in high school. But I enjoyed physics

---

*Magdolna Hargittai conducted the interview.

so much that I did not want to stop, and stayed for a master's degree. I received my master's in physics and went on to start my Ph.D. studies in physics at the Weizmann Institute. I worked in nuclear physics, which was an exciting and rapidly-moving field. Just when I started my Ph.D., I met my husband; he came to Israel for a visit from the United States. Eventually we were married and moved to the United States. He had a faculty position at Caltech and we both went there. Caltech had an excellent department of astrophysics and astronomy; Israel at that time did not have any. I did my Ph.D. work at Caltech, connecting what I did earlier in nuclear physics with astrophysics. The field is called nuclear astrophysics, addressing the question of how the stars shine; that is, what the nuclear reactions are in the stars. I was lucky to have a wonderful professor, Professor Fowler, who was a Nobel laureate. He died a few years ago.

While at Caltech, I talked a lot with the astronomers, including Maarten Schmidt, Fritz Zwicky, Wal Sargent, and other famous astronomers. Since then I have directed my work to astronomy. The same thing happened to my husband. Many scientists in our generation started in physics and then continued in astronomy.

Looking at your vita you did all your studies at exceptionally good places. Was it just by accident, or did you have good advisors?

That is a good question. When I got started in Israel, I was accepted at the Hebrew University because of my good grades and record. Then I moved to the U.S. because I married John. I was very fortunate that John, my husband, was at Caltech and that Professor Fowler took me as a student. He was an excellent advisor. Caltech was a turning point for me towards astronomy. Then Princeton, again, was a wonderful place to come to.

You wrote a paper in Science *about the anti-gravity force and the cosmic triangle. Please tell us something about it.*

Last year I was asked by *Science* magazine to write a review article on the state of cosmology of our Universe. I have been working in that field for a while trying to understand the structure of the Universe, how this structure formed after the Big Bang, how it is evolving with time, what the mass density of the Universe is, and how it can explain the structure we see today. Specifically, in the past few years I have been working on trying to understand what the mass density of the Universe is — that is how much matter there is in the Universe. *Science* magazine asked me and Dr. Perlmutter, who was leading the supernovae project that discovered the anti-gravity force, to write a review article. We introduced in this paper "The Cosmic Triangle" as a summary of the cosmological results.

At this time it seems that the cosmological picture fits together very nicely. It is not clear whether or not this picture will continue to hold, but currently the data still point to the same cosmological model. What we tried to answer was: what do we know about the Universe and its mass density? Does it have the critical density, that is the amount of matter needed eventually to stop the expansion of the Universe? We know that the Universe is expanding but depending on how much matter exists in the Universe, gravity will pull it back and, if the Universe has the critical density or more, then it will eventually stop expanding and may collapse. Other questions were: Is there anything else in the Universe, besides matter? For example, is there any dark energy, the anti-gravity force, which does not clump with matter? What is the curvature of space; is space curved or flat?

We answered these questions based on observations. These three questions (what is the mass density of the Universe, how much dark energy exists — if any, and what is the curvature of space) relate to each other by Einstein's equations of general relativity. The answers to the first two questions tell us what the curvature of space is because the curvature is determined by the sum of the mass density and any other energy density in the Universe. What we found was that the Universe is lightweight; it has less than the critical density. The data also showed that there is an anti-gravity force, a dark energy existing in the vacuum that works against gravity and increases the expansion rate of the Universe. Thus the Universe not only expands but does it faster and faster. These two together also tell us that the Universe is flat, that is, there is no curvature; the sum of the mass density and the dark energy density in the Universe appears to be equal to the critical density, which, in turn, indicates a flat Universe. We summarized all this in what we called the Cosmic Triangle. The triangle is based on these three specific questions I mentioned: the mass, the energy, and the curvature of the Universe.

It is hard to imagine the three-dimensional Universe as "flat". That would suggest something two-dimensional ...

You are right, it is difficult to imagine or explain it because we live in a three-dimensional world. It is easier to explain it with a two-dimensional analogy: it used to be believed that the Earth was flat, so that if you went along two parallel lines you would never meet. Of course, now we know that the Earth's surface is curved and so if we start walking along two "parallel" lines that go to the North Pole, eventually we will meet. That is a two-dimensional analogy of the three-dimensional space. If space was curved like the surface of the Earth, two lines would be curved just as two lines towards the North Pole on Earth. Another way of describing it, which is based on Einstein's theory of general relativity, is that matter density causes space to curve; space will be curved because of gravity.

So then which one is it, flat or curved?

On relatively small astronomical scales, where we see clusters and galaxies, space is curved by their gravity. Thus, there is curvature, locally, in many places in the Universe. There is even a little bit of curved space around us as well because we are a little bit of matter so we curve the space around us; but it is a tiny effect. If we go to large galaxies and clusters of galaxies, where there is much matter and strong gravity, we can actually measure how they curve light from objects behind them. We can observe this effect through gravitational lensing. However, when we talk about the Universe as a whole, it is a somewhat different picture; here we look at the entire space of the Universe. Current observations suggest that the geometry of the Universe is flat, which means zero curvature.

Is this related to what Einstein called the "cosmological constant" and what finally he called his "greatest blunder"?

Exactly. That is why this is so interesting. Still, we all find it hard to believe that the cosmological constant exists; it is not natural. In fact, we don't know if the dark energy (if its existence is indeed confirmed by additional observations) is a cosmological constant, or not. Constant means that it is constant over time; and this we don't yet know. The observations suggest the existence of anti-gravity dark energy, the same or similar to the cosmological constant (which may or may not be constant) suggested by Einstein around 1917 in his models of the Universe. This came out of his mathematics. In his model the Universe was not static, it was expanding or collapsing. And at that time we did not yet know that the Universe was in fact expanding; the observational facts about the expansion came a few years later. Einstein predicted that the Universe was expanding but then he himself could not quite believe it and he introduced an additional term into the equations to balance the force of gravity and make the Universe static. That was the cosmological constant; some kind of energy in the vacuum where there is no matter; energy that prevents the Universe from collapsing. When Einstein found out a few years later that the Universe is in fact expanding, he threw out this constant and said that it was the biggest blunder of his life. The question is, of course, are we now introducing back this biggest blunder or is it real? It is still a big question. There are two independent observations that suggest that this is real. One is the following: we can measure objects at large distances, supernovae, that are believed to be standard candles, that is, we know exactly how much light they emit intrinsically. If we know how much light they emit, and can measure how much light we receive from them on the Earth, then we can determine exactly how far away they are. By measuring the distances to these "candles" and their redshift, we can determine how fast the Universe is expanding. If there has been nothing else in the Universe but matter, we would expect that the expansion of the Universe has been slowing down because matter pulls things together and slows down the expansion. If there was no matter in the Universe, it would just keep expanding at a constant speed. If there is matter, the Universe will slow down.

But what the scientists found from the supernovae observations is just the opposite. They found that the Universe, rather than slowing down, is expanding faster and faster with time. There is no way that this could happen if there was only matter in the Universe. The only way for the Universe to expand faster and faster is if there is some "anti-gravity" force pushing it out. It is as if you throw a ball up in the air: it goes up, slows down, and then falls back due to gravity. But if you find that instead of dropping back eventually, it flies away faster and faster from you, you know that there must be something else that is pushing it away. This is why the observations suggest the existence of the anti-gravity force, some kind of dark energy, like a cosmological constant.

The third point in the cosmic triangle addresses the question of how much matter is there in the Universe; in other words, what is the mass of the Universe. This is one of the topics I have been working on. What we found from different methods, such as the structure of the Universe, clusters of galaxies, and so on, is that the mass density of the Universe is very low; the Universe is lightweight. There is not enough matter to stop the expansion of the Universe. Therefore, it will keep expanding forever. We found that the Universe has only about 20% of the critical density (the density that would be needed to stop the expansion). Currently there are many different observations confirming this result, but when we first started, quite a while ago, many people did not believe it. At that time we did not yet know about the possible existence of the dark energy, or about the flatness of the Universe. Of course, many people expected the Universe to be flat based on theoretical grounds but this was shown observationally only recently.

We established the mass of the Universe by different methods, all of them show that it is approximately 20% of the critical density. We simply don't see more mass around us; it is just not there.

As I understand, much of the matter in the Universe cannot be seen. What is this dark matter?

This is probably the most fundamental question in astrophysics and cosmology: trying to understand what our Universe is made of. Most of the matter in the Universe is dark. Even if the total amount of matter is only 20% of the critical density, most of it is dark — we cannot see it, unlike the luminous objects (stars, galaxies) that we see in the dark sky. The biggest question is what is this matter. We don't know the answer to that.

So what can it be? One possibility is that the dark matter is just normal matter that does not shine like the stars. It could be planets or very faint stars that do not shine strongly enough to be noticed. It could be rocks. It could be black holes. These examples are all baryons, which are normal matter that we just don't see because of the above reasons. But some of the dark matter is expected to be nonbaryonic — some exotic matter that relate to particle physics.

When you say we do not see, do you mean only the visible spectrum or also the other regions in the electromagnetic spectrum, that we can still measure with instruments?

That is a very good question. Most scientists refer to "dark matter" as the matter that cannot be seen at any wavelength. Certainly, if it can be seen in the visible or infrared or ultraviolet or X-ray or any other region, it should be counted as "luminous", not dark. Here is an example, a system on which I work quite a lot: a big cluster of galaxies. When we add up the matter in all the galaxies that we can see in the cluster, their mass adds up to only a small fraction of the total cluster mass; there is still a lot of matter in that cluster unaccounted for. During the last thirty years it was discovered that there is a lot of hot gas in these clusters and the hot gas weighs even more than all the galaxies. We observe the hot gas in X-rays. Earlier this counted as dark matter but now that we detect the hot gas in X-rays it is not dark matter any longer. So while earlier we said that the dark matter makes up about 95% of all the matter, now it is much less, say, 80%, because we can observe some of it.

We can say that about 20% of matter in the Universe can be accounted for and observed as luminous galaxies and gas. But the other 80%, which is dark and unknown, can be inferred only from its observed gravitational effects.

Coming back to the question about the dark matter, some of it could be planets and stars that do not shine. Some of it can be the so-called "MACHOs", which stands for Massive Compact Halo Objects. There has been a huge effort to determine if the dark matter could be such compact objects (dark "stars" or planets). Could such compact objects make up the mass of the halo of our galaxy? These compact objects could be any objects that do not shine, not only planets and dim stars but also black holes, for example. Black holes are collapsed objects that take up only a very small volume. However, these objects do not appear to make up the entire halo, maybe only a small fraction of it, so this is still an open question.

There could be some other, more exotic components of this dark matter, that are not baryonic, that is not made up of the normal particles that we know, neutrons and protons. Some of it could be neutrinos, which move very fast. They have a very small mass, so we do not think that they can make up much of the unseen dark matter, but they are part of it. The current standard model is that some of the dark matter is made up of so-called cold matter. This fits the astronomical observations well. But such cold particles have not yet been detected and this is therefore still a big puzzle. Cold particles do not move fast; this is why they are called cold. They have enough mass but do not move fast and thus help gravity to collapse the early fluctuations in the universe to form objects such as galaxies; they provide the seeds of structure formation, eventually forming the structure we see today. There are many experiments these days trying to find the dark matter. Theoretical physics has various predictions for these particles, but none of them have yet been detected. We expect that the new accelerator experiments now underway will indeed detect these particles.

Cosmology and astronomy is probably one of the oldest scientific disciplines; it was already practiced in ancient times; then there was the exciting period when Kepler, Copernicus and Galileo worked. Do you think that we live, perhaps, in another such exciting era for this field?

Oh, yes, this is an enormously exciting time for cosmology. Recent observations brought up totally new ideas and discoveries, things that we have not seriously thought about before, such as the accelerating Universe and the dark energy, as well as large-scale structure, microwave background fluctuations, and the flatness of the Universe. These are truly revolutionary times. The next questions are: What is the dark matter? What causes the dark energy? These are amazing and interesting questions for physics. When we find the answers, it will teach us something fundamentally new about physics and about the nature of the Universe. I believe that the next few years will tell us the answers to at least the observational questions — how much mass and dark energy exist. They will not yet tell us why the dark energy exists, where it comes from, or what the dark matter is. But it will tell us if these exist, and how much. We all feel that a revolution is on its way.

Einstein said that one of the most incomprehensible things about Nature is that it is comprehensible ...

That is exactly true. Often when I talk to the public and they ask: don't you feel, studying the cosmos, that we are so small and insignificant? I always say that it makes me feel exactly the opposite; it makes me feel so incredible that we sit here, people, on this tiny planet around this small star in one galaxy amongst billions of galaxies and we try to figure out this unbelievable question of the entire Universe, or understand DNA, or particle physics. It makes me feel that the human mind and innovation is spectacular — being able to answer these questions.

What do you think about extraterrestrial intelligence?

It is very likely that it does exist. As the Copernican principle has shown, we are not in the center of the Universe, not in the center of our solar system, or in the center of our galaxy, we are not unique in any way as far as the Universe is concerned. It is just a matter of appropriate physical conditions that our species, our intelligence or life in general could develop. It is quite clear that it has to exist in other places. It cannot be that Earth is the only planet in the Universe where intelligent life has developed. Of course, it does not have to be the same type of life form as we are, it depends entirely on the particular conditions. It takes time for intelligent life to develop, but much less time than the age of the Universe.

Can your observations shed light to the question of how the Universe started?

We try to find the answer to the question: what happened after the Big Bang? How did the particles form, how did matter form, how did the structures we see today form? To go all the way to the Big Bang itself, is much harder because we don't know what the physics and the conditions were. We don't know if there are other Universes, for example. Our work only concerns the area where we can make observations within our Universe.

You often mention the "structure of the Universe". What is it, how is the Universe built up?

The skeleton of the Universe is mainly the large-scale distribution of galaxies. The galaxies are distributed in numerous concentrations, filaments, and clusters. In between the galaxies, we think that there is not much matter; just some gas, floating stars or black holes, and mostly empty space. It is the galaxies that create the skeleton of the Universe, they ARE the Universe. Then, of course, in the galaxies, themselves, we have the stars and the planets. How did all this form? We try to visualize it in computer simulations. After the Big Bang, first there was a uniform distribution of radiation and then matter. In the uniform distribution some random fluctuations existed. Even if they are very tiny, gravity starts pulling on the fluctuations and after billions of years they form the galaxies and the large-scale structure that we see today. With our computer simulations, we can follow this and see how it happened. We create tiny fluctuations, about 10^{-5} of the mean density of the Universe at that early time, and then let gravity operate. We stop the simulations after 15 billion years — the Hubble time of the Universe. We then compare what was formed in the computer with the structure observed in the Universe. It is remarkable how similar they are.

There is this huge project of mapping the whole universe. Are you involved with that?

Yes, I am. We are imaging the northern sky in five colors. This map of the sky will have about a hundred million objects. The computer will automatically identify all the galaxies, all the stars. Then we'll observe the spectrum of each of the brightest one million galaxies and the computer will tell us how far they are from us based on their spectra (by the Doppler-shifted spectral lines). This way we will have a three-dimensional map of the Northern Universe. This will be the first time that we will be able to study in detail the structure of the Universe. The project started a couple of years ago. In our early data we have already discovered the most distant known objects in the Universe and the coolest known stars (which are between stars and planets). This is a very ambitious and exciting project and will be done in about five years. But already from the early data we can learn a tremendous amount of new information about the Universe, its large-scale structure, its galaxies, quasars, and stars.

This project reminds me of another very ambitious one, the human genome project.

Yes, that is even bigger.

I would like to talk a little about you. What was your family background?

I was born and brought up in Israel. My father came from Romania and then from Vienna just before the Holocaust in Europe. He was a lawyer. My mother came from Russia and she was a head nurse at the Hadassah hospital. There were no scientists in our family and I was an only child. My father wanted me to study law and be a lawyer and work in his office. I did that during summer holidays when I was in high school and it was good, but I did not particularly want to become a lawyer. I liked the sciences in school. First I wanted to become a medical doctor but I was discouraged to do that; at that time it was very difficult to become a medical doctor in Israel. I did not know much about astronomy at that time but I enjoyed physics and math and studied these subjects.

Did you serve in the Army in Israel?

No. It was allowed to start with your studies without going to the Army. Then immediately after my studies I was married.

Generally speaking, in most of the western countries while at under-graduate level there are about equal number of women and men, as you go higher and higher, and eventually reach professorships or member-ships of national academies, the number of women is only a few percent of those of men.

In Israel I never thought of discrimination against women. In the United States I heard and learned more about it. However, at the highest levels, in science and other fields, there aren't more women in Israel than here. When I lived in Israel as a young student I have not felt any discrimination.

I was elected to the U.S. National Academy of Sciences a few years ago and the number of women there is very low. This low fraction of women at the highest levels is not necessarily an overt discrimination; people do not say, "We will not select this person because she is a woman." When a selection or promotion for full professorships comes up, most of the people on the committees are men. They seem to know their other fellow men better then they know the women in the field and thus vote for them; I think this is one important reason for the low number of women scientists at the high levels. People tend to think of and select people who are more like themselves.

Does not it depend partly on the women themselves?

Some women decide not to follow a career in the sciences for various reasons. It is demanding and competitive and some see the struggle other women go through and are discouraged. Sometimes they may feel that they will end up sacrificing their family life and they don't want to do that. Some women try to stay and don't make it; this, of course, happens to men too.

Nowadays we often see that departments go out of their way to hire highly qualified women. This is very important because unless you increase the number of women, the trend will not change. And I am happy to see that the trend is changing — slowly but surely.

Do you have children?

Yes, we have three children.

How did you manage?

I am asked this question frequently by young women. I think that this is a cultural thing; I never asked myself in Israel whether I could do a job and have a family at the same time; it was always obvious to me that I would have both. My first son was born when I was still a graduate student, my second son was born just when I finished my Ph.D., and my daughter was born when I was a postdoc here, in Princeton. I did it in the same way that you do any other job while you have kids. Of

The Bahcall family, 2003; from left to right: Dan, Neta, John, Safi, and Orli (courtesy of John Bahcall).

course, you have to have some help, a daycare, or someone helping with the children. You also need a supportive husband who helps when needed. But it never occurred to me that I could not have a family because of my profession as a scientist.

Are you religious?

I am not very religious, but am very Jewish. My husband is also Jewish and we raised our children Jewish.

Do you believe in God?

I combine the science that I do with the religion's question about God in the sense that all the laws of physics that created the Universe and the enormous amount of beauty in the Universe represent the connection to God.

You said earlier that your husband is also an astronomer ...

Yes. His research is in somewhat different areas. He has worked in many different fields. Since we met, over 35 years ago, he has been working with solar neutrinos, neutrinos that come from the Sun, and tries to answer the fundamental question of why the Sun shines. He is working to confirm the nuclear reactions going on in the center of the Sun. He did the theoretical work on how to detect and interpret the neutrinos that come all the way from the center of the Sun and thus confirm the nuclear reactions that power the Sun and make it shine. He has been involved with experiments testing to see if the neutrinos have mass. He also worked with the Hubble telescope, with quasars, and with dark matter in the galaxy. Our fields are different: I work in cosmology and he works in other areas.

Do you talk shop at home?

Oh, yes, we talk a lot of astronomy at home; we enjoy that a great deal. It is wonderful to come home in the evening and talk about my interests with someone who really understands it. Many people ask, isn't that too much, to come home and talk about the same work that I have been busy doing all day; I think this is great. If you like what you do, you enjoy talking about it any time.

How did your children take it?

It probably influenced them somewhat. All three of them started out in science although by now moved away some. Our oldest son received his Ph.D. in theoretical physics at Stanford in condensed matter physics. He was a postdoc and had a faculty position but decided to move towards biology — an exciting and rapidly-developing field, and he is currently a CEO of a new and successful biomedical company in Boston. Our second son just received his Ph.D. in cognitive science. He was studying vision and the brain. Our daughter completed her B.A. degree in molecular biology at MIT, was a Marshall fellow at Oxford, and is now completing her Ph.D. in biology and epidemiology in London.

Who are your heroes?

I have heroes in different categories. I have heroes in my field, in astrophysics; these go back to the very early times up to current times. From current times, Vera Rubin, a well-known astronomer, is a wonderful role model for women astronomers; an excellent scientist and a wonderful person. Margaret Burbidge is another astronomer, who has been an outstanding role model for women in astronomy. Other role models in science, for their work, their dedication, and enthusiasm include my advisor, Professor Fowler; Professor Zwicky from my early days at Caltech; and Professors Lyman Spitzer and Jeremiah Ostriker from my early days at Princeton. In Israel, the people who started the country, including the first president, Weizmann, who also started the Weizmann Institute; Ben-Gurion, Golda Meir, and Itzhak Rabin.

What was the greatest challenge in your life?

I would say that the two greatest challenges in my life are my family and my work; these are what I find most important and most enjoyable.

Many women who want to have a career feel that it is better to establish themselves first in their field and delay having families. You did not follow that path.

No, although I see that people do that more now. This is a personal choice and you should do what is best for you. I think it is good to have children at an early age. I think you can do both, you just have to find the right balance. But, again, this has to be one's own choice.

Have your children ever suffered from you being a working woman? Have they ever resented it?

Not that I know of. None of them has told me so. They were always very supportive. It happened often that I told them, maybe I should not go to a particular meeting but they always said, no, you should go, we are fine.

If you started your career today, woulsd you do the same thing?

Oh, yes, it is a wonderful field; it is more like a hobby than a real job for me. It seems that we are playing and having fun in trying to figure out puzzles of our Universe that are very exciting. We are very fortunate.

* * * * * * * * *

Added in 2004: Beautiful observations in astronomy in the last four years, including mapping the early fluctuations in the Cosmic Microwave Background radiation, mapping the galaxy distribution in the Universe on large scales, with high precision, using large surveys such as the Sloan Digital

Sky Survey, and observations of distant supernovae, have nicely confirmed our earlier work (including the "Cosmic Triangle") that the Universe is lightweight (with ~25% of the critical mass-density), is flat, and is accelerating (i.e., contains dark energy). This is an amazing Universe. I am proud and humbled to have participated in some of these important discoveries about our Universe.

Rudolf E. Peierls (courtesy of The Rudolf Peierls Centre for Theoretical Physics, University of Oxford).

15

RUDOLF E. PEIERLS

S ir Rudolf E. Peierls (1907 Berlin, Germany–1995 Oxford, England) was one of the pioneers in 20th century physics with his main contributions in solid state physics, quantum mechanics, and nuclear physics. He was also one of those scientists driven out of Nazi Germany who took an active part in the Allied war effort. He and Otto Frisch estimated the energy released by fission and the critical mass of uranium-235 needed for a fission bomb, and alerted the British government to initiate a nuclear program. Peierls participated actively in this program, and when it moved to the United States, he joined the Manhattan Project. He was awarded the Royal Medal and the Copley Medal of the Royal Society (London) and was knighted in 1986.

Clarence and Jane Larson recorded a conversation with Rudolf Peierls on April 25, 1989 and our narrative is based on this recording.*

Rudolf Peierls was born in a suburb of Berlin, where his father was the director of a large AEG factory. He became interested in technical and scientific matters early on, mainly by reading but he was also growing up in a technological environment. He wanted to become an engineer but was talked out of it because he wore glasses and he was considered to be clumsy with his hands. He then decided to do the next best thing and become a physicist. Before the start of his university studies, he apprenticed

---

*"Larson Tapes" (see Preface). In part, this has appeared in *The Chemical Intelligencer* **2000**, *6*(2), 54–57.

for six months in a telephone factory, which gave him useful practical experience.

He entered the University of Berlin in 1925. Since the experimental physics laboratories were crowded, he focused on mathematics and theoretical physics and got hooked on them. During his first semesters, he attended a course by Max Planck, who was a poor lecturer, just reading from one of his books. The students knew that Planck was famous but had no idea what he was famous for. Peierls got his first notion of the excitements that were going on in physics from the lectures of Walther Bothe, the famous nuclear physicist, who gave a course on X-ray physics. Such new terms as the Bohr orbits and the K shell popped up in his lectures; these had not been part of high school physics. After two semesters in Berlin, Peierls continued in Munich where Sommerfeld was the great teacher of physics. At some point during the three semesters Peierls spent in Munich, Sommerfeld gave him some papers by Dirac and Jordan on transformation theory and asked him to give a seminar. Sommerfeld told him that he had not understood it and asked Peierls to explain it in his seminar. It was quite a challenge for the young student. Peierls found the papers very interesting and gave two talks on them rather than one. Hans Bethe was one of his fellow students, one year his senior, and they became life-time friends. When Sommerfeld left for a sabbatical, he sent Peierls to continue his studies with Heisenberg in Leipzig. Felix Bloch was also there, working on his thesis on the electrons in metals, and Heisenberg suggested to Peierls to apply Bloch's results to the Hall effect. There was a paradox in that while some metals behaved in the expected way and the transverse voltage was what would be expected from the deflection of electrons by the magnetic field, some metals showed the Hall effect with the wrong sign. It was called the anomalous Hall effect. Peierls explained the anomaly by Bloch's band theory by introducing the concept of what is today called holes. The holes, that is the absence of electrons, behave as positive charges and their effect is the opposite to that of the electrons. This gave Peierls his first paper at the age of 22. Looking back, Peierls reflected that almost any problem originating from the in-adequacies of the old physics could lead relatively easily to new results.

After two semesters Heisenberg went on a sabbatical, and Peierls went to study with Pauli in Zurich. He is grateful to the sabbatical system, which gave him this opportunity to have a unique combination of teachers. Pauli was not that easy to get used to but that was compensated for by his profound thinking in theoretical physics. Peierls wrote his Ph.D. thesis on the heat conduction in non-metals due to the lattice vibrations. Debye had an elegant and simplified theory on this subject that was typical of him, but Peierls found it in this particular case too simple, and not quite correct. Even Pauli had done something incorrect on this subject, the only case to Peierls's knowledge in which Pauli had done anything erroneous. Pauli had suspected, though, that there was a problem, and this is why he had suggested to Peierls to look into it. There were some interesting results, such as that the thermal conductivity of a crystal at very low temperatures should grow exponentially as the temperature goes down, which was not verified until the 1950s.

Peierls defended his Ph.D. in Leipzig but stayed in Zurich as Pauli's assistant for three years and continued working mainly on solid-state problems. When Peierls had explained the anomalous Hall effect, he had applied Bloch's theory of the electron bands in metals, which was formulated for very tightly-bound electrons. For real metals this was not a very good approximation. Peierls then considered the other extreme, that is, electrons moving in a very weak potential — nearly free electrons in the limiting case — and found that when the potential was gradually increased, the behavior of the electrons was very much the same so his previous conclusions were proved to be correct. Peierls worked out his theory for the one-dimensional case and Léon Brillouin (1889–1969) in Paris extended it later to two and three dimensions, hence the term, Brillouin zones. Peierls was interested in the applications of quantum mechanics, and at that time it was still possible to read everything that appeared in one's field.

In 1930, Peierls went to a physics meeting to Odessa in the Soviet Union, which proved to be of great interest to him not only from a physics perspective. He met there a recent physics graduate, and they spent a lot of time together. Six months later, she and Peierls got married when he went to Leningrad for another brief visit. She did not continue in physics in a direct way but made a great contribution by educating and looking after many young physicists.

In 1931, Peierls visited Copenhagen and the great Niels Bohr for the first time, to be followed by many other visits. He found himself in an embarrassing situation on the occasion of one of his first visits. He had written a paper jointly with Lev Landau who had visited Zurich. In their paper, they claimed that an extension of the uncertainty principle was needed when it came to relativistic situations. The manuscript was ready when they met the next time in Copenhagen, and they showed it to Bohr, who did not agree with them. They had a long argument and Bohr finally accepted what they claimed. Following this encounter, Landau and Peierls were doubtful whether they should acknowledge Bohr's advice and they felt they could not do so without Bohr's permission. Usually people acknowledge somebody in a paper, it gives the impression of support, and in this case the interaction was basically a disagreement. When Peierls asked Bohr about this, Bohr misunderstood and he explained it to Peierls that he only wanted to help them but if they had any doubt about that, he would refuse them to let them mention his name at all. Eventually the problem was straightened out and Landau and Peierls mentioned Bohr in the acknowledgments.

In 1932, Peierls won a Rockefeller traveling fellowship for one year and decided to split it between Rome with Fermi and Cambridge. By the time Peierls got to Rome, the neutron had been discovered and Fermi was preparing to use this new tool for future experiments about which Peierls would learn only later, when he had left Fermi. While Peierls was in Rome the situation changed drastically in Germany. Peierls had been offered a prestigious appointment in Hamburg and he had decided to accept it. The Peierls, however, found the conditions in Germany intolerable by the end of 1932, that is, even before Hitler came to power, and Peierls decided to complete the Rockefeller fellowship and went on to Cambridge. Peierls had had contact with Dirac, and in Cambridge he also saw Rutherford, who impressed him a great deal. When the six months in Cambridge was over, Peierls was hoping for a lectureship at Manchester. However, the University of Manchester was being attacked for appointing foreign scientists at a time when many British scientists were unemployed. When Michael Polanyi was appointed to the chair of physical chemistry some older chemists had protested, and so they could not do anything for Peierls. For the time being, Peierls was helped by a grant that had been set up to provide assistance to German refugee scientists, and the Peierls stayed in Manchester for two years. Bethe was there too for the first of these two years (1933–1934) and they worked together, first on solid-state problems, concerning alloys and superlattices and their thermodynamics. William Bragg had worked out a crude theory for these systems but Bethe and Peierls were not satisfied with it and they improved on it.

Bethe and Peierls became interested in nuclear physics when Chadwick, the discoverer of the neutron, challenged them to work out a theory for a phenomenon he had already observed, though he did not tell Bethe and Peierls that he had. It was the disintegration of the neutron by gamma rays. However, they met the challenge with success. Around this time, Fermi came out with his theory of beta decay, and they worked on that too and published two letters in *Nature* on some further developments. One of their predictions was that the neutrino would never be observed, but this later proved to be incorrect.

In 1935, Peierls was offered a job in Cambridge at the Mond Laboratory of low temperature physics that had been set up for Peter Kapitsa. By then Kapitsa, who had been working in Cambridge but was a Soviet citizen, had been detained on one of his visits to Moscow. The Royal Society shipped his equipment to Moscow in 1934 to enable him to continue his research and the money received for it was used to duplicate the instruments so that the Mond Laboratory could also continue its operations in Cambridge. The money that would have been Kapitsa's salary made it possible to establish two research fellowships at the Mond Laboratory one of which was given to Peierls. Ferromagnetism and alloys were two of Peierls's topics of research. After two years, in 1937, the University of Birmingham decided to create a new professorship in applied mathematics, which meant theoretical physics. Mark Oliphant had been appointed to the physics chair in Birmingham and he persuaded the University to create the applied mathematics chair. Within two years of Peierls's move to Birmingham, World War II started. Oliphant was working on the radar and wanted to involve Peierls, but since Peierls was still a German citizen, and thus, technically, an enemy alien, the Royal Navy refused to grant him permission to work on radar. However, Peierls was eager to contribute to the war effort and signed up for the local fire brigade.

In the meantime, fission was discovered. First, the presence of secondary neutrons was shown by Joliot, Halban and Kowarsky in Paris, which prompted people to speculate on the chain reaction. Peierls saw a paper by the French theoretician, Francis Perrin who had developed a theory of the chain reaction based on a rather crude approximation. Peierls first thought that the theory's prediction of a critical mass, that is, such a sharp change in behavior, was the result of the crudeness of the approximation. Soon, however, he understood the correctness of the critical-mass concept. He then made important calculations in this connection, but because of the possible application for the development of a bomb, he hesitated to publish his findings. There was a debate at that time about imposing secrecy on the scientists involved in nuclear physics in the West, but Peierls was not connected with them. He himself came to the conclusion that secrecy might be needed.

Then Otto Frisch came for a visit to Birmingham, and together they saw a paper by Niels Bohr which argued that it was impossible to make a bomb from natural uranium, which was correct. Peierls felt relieved, and decided to publish his paper on the critical mass. Then one day Frisch brought up the possibility of separating uranium-235 and having a sufficiently large amount of it for the bomb. When they calculated the critical mass and it was in pounds rather than in tons, the possibility of the bomb loomed over them once again. They were afraid that Nazi Germany would make such a bomb and were convinced that the Western powers should make one before the Germans did. They did not underestimate the difficulties and costs of isotope separation. Frisch said that even if building a plant to carry out the isotope separation were to cost as much as a battleship, it would be worth doing it. That proved to be a vast understatement, of course. Isotope separation had been done on laboratory scale, but producing the large quantities of uranium-235 that would be needed for the bomb would have seemed like science fiction until they realized its importance. Frisch and Peierls wrote a report to the British authorities. Then their activities diverged. Frisch first worked in Birmingham and then joined Chadwick in Liverpool to do experimental work, eventually doing investigations on isotope separation with the thermal diffusion method in the gas phase. It didn't go anywhere. It turned out that the only gaseous uranium compound that was considered at that time, uranium hexafluoride, has a zero coefficient of thermal diffusion.

Peierls was working intensely on the theoretical problems of isotope separation and was building up a small group in which his first collaborator was Klaus Fuchs, who later became infamous for being a Soviet spy and passing the atomic secrets to the Soviet Union. Finally, it became clear that the project was too big for Britain in wartime, and many of the scientists involved moved to the United States. Peierls himself spent six months in New York working on the design of isotope separation by gaseous diffusion. Then he spent the next 18 months in Los Alamos. He was also interested in the nuclear aspects of the bomb. Teller had been asked to look into the hydrodynamics of the implosion, but he was more interested in longer-range projects, notably the hydrogen bomb, and Peierls took over this problem. Fuchs accompanied Peierls to Los Alamos.

Peierls stayed in Los Alamos throughout the war and stayed on a little after it ended to close down the operations and then he returned to England. In Birmingham, he had to start everything from scratch. He continued research in nuclear physics, involving graduate students and research fellows, and also in field theory, and he returned to some solid-state problems. While he was writing a book on solid-state physics, he realized that when you have a linear chain of atoms, if it is metallic, in other words, for example, it has one electron per atom, the regular arrangement of the chain is not stable, and the chain would have a tendency to separate into pairs, to dimerize. To Peierls, it was a surprising finding and appeared to be generally valid, but he did not think it too important because it concerned a one-dimensional case. It became just one paragraph in the book, but later it became known as the Peierls transformation. People realized that crystals very often can be considered to consist of long chains of atoms so Peierls's findings found broad applications. The title of the book is *The Quantum Theory of Solids*. It appeared in 1955. Almost at the same time, he decided to publish another book, a popular book on theoretical physics that would not use any mathematics and would be devoid of jargon. Since there was a great interest in atomic energy after World War II, Peierls had been asked to give many lectures. He found that the questions from the audience always concerned basic physics. The book was called *The Laws of Nature*. It sold well and was translated into many languages.

In 1963, after having spent 26 years in Birmingham, Peierls was invited to join the University of Oxford, where he stayed until 1974, when he retired. In 1967, he spent a sabbatical in Seattle, Washington, where he was appointed a Professor of Physics, half-time, until he reached the university's retirement age in 1977.

Speaking about the future developments in physics, as of the time of the Larson interview in 1989, Peierls offered some guarded predictions. He anticipated that nuclear physics would become less important than before. It had started as a front-line area of research and it was not that anymore. In the utilization of nuclear power, the problems were of a technological nature rather than problems of nuclear physics. The field of elementary particles was wide open, but Peierls predicted that the difficulty of achieving higher and higher energies in the accelerators and more and more expensive detectors would pose the limitations to this field. He said that there are, of course, cosmic rays of very high energy but the number of particles is very small. He also commented that, with the prevalence of concerted research efforts performed by large teams, it must be depressing for young experimentalists not to be able to do any individual research.

Emilio G. Segrè, 1954 (courtesy of the Lawrence Berkeley National Laboratory).

16

EMILIO G. SEGRÈ

Emilio Gino Segrè (1905–1989) went to school in Tivoli and Rome and started his university studies in Rome in 1922. He was Enrico Fermi's (Nobel Prize in Physics 1938) first doctoral student and received his Ph.D. degree in 1928. He served in the Italian Army in 1928–1929. He started working at the University of Rome in 1929. He did post-doctoral studies with Otto Stern (Nobel Prize in Physics in 1944 for 1943) in Hamburg and with Pieter Zeeman (Nobel Prize in Physics 1902) in Amsterdam. He was at the University of Rome in 1932–1936 and at the University of Palermo in 1936–1938. He emigrated to the United States in 1938 and joined the University of California where he taught and did research for the rest of his career except for 1943–1946, when he was a group leader in the Manhattan Project in Los Alamos. Emilio Segrè was awarded the Nobel Prize in Physics in 1959, jointly with Owen Chamberlain "for their discovery of the antiproton". Here we communicate edited excerpts from Emilio Segrè's narrative from a video recording by Clarence and Jane Larson at the University of California, Berkeley, on March 20, 1984.*

I was born near Rome, in Tivoli, an archeological and artistic place where tourists visiting Rome often go. My father had a paper mill and we were well-to-do. I went to school and was also tutored privately. I had two uncles, one a jurist and another a geologist, both of whom were members

---

*"Larson Tapes" (see Preface). In part, this has appeared in *The Chemical Intelligencer* **2000**, 6(4), 54–57.

of the Academy of Science. It was an intellectual surrounding. As soon as I learned to read and write at the age 5 or 6, I got interested in physics. There were books at home, physics books with illustrations. There was even a book with experiments for children. I started doing these experiments and I still have my note book where I described them. An example is a description of an experiment, dated March 7, 1912, about producing the colors of the rainbow by letting the sunlight pass through a pitcher full of water and through a prism. I have made drawings of my experiments too. I also had toys for mechanics, a small induction coil, and other physical instruments.

We moved to Rome in 1917 when I was 12 years old, and I no longer had the facilities that I had in Tivoli. In Rome I went to school and learned Latin and Greek and other conventional things. I continued reading physics books and I memorized parts of them. When one of my uncles gave me a physics book, he inscribed a dedication to me that he hoped that physics would also be used for peace. That was at the time of World War I.

I entered the University of Rome as an engineering student in 1922, the same time Mussolini came to power. I knew modern physics and the physics in Rome didn't appeal to me at all. In the first year, excellent mathematicians taught us mathematics. When the courses became more professional, they also became less interesting. By a tremendous stroke of luck, something like winning the Irish Sweepstakes, Enrico Fermi came to Rome in 1927 and was looking for students. When he offered to teach me physics, I realized at once that this was a unique opportunity and I jumped at his offer. I left engineering and moved to physics. I took a doctorate in physics, which corresponded more or less to a Master's degree in America.

For a couple of years Fermi was teaching us; our group included Amaldi, Rasetti, Majorana. He gave us private lessons twice or three times a week. Fermi would later give these courses many times, first at Columbia, and then at Chicago. He was rather ruthless in choosing his students. He would devote an infinite amount of painstaking effort to his students, and he didn't want his efforts wasted. You find all kinds of famous people among Fermi's former students.

After graduation, I became an assistant to Professor Corbino who was the director of the Physics Institute. He was also a Senator of the Kingdom of Italy and a sort of political protector for all of us. We liked him very much and he did a lot for us. He had a great aspiration to make physics important in Italy and enlisted Fermi in his efforts. Corbino was middle aged and no longer active in research when I became his assistant. I soon started working on my own projects and Fermi, who had first suggested topics to me, encouraged me to find my own direction. I started doing spectroscopic work. For some of the experiments I went to Pieter Zeeman in Holland. He then helped me to obtain a Rockefeller Fellowship to go to Otto Stern's laboratory in Hamburg. Around 1932 I returned to Rome.

In the spring of 1934, we started the neutron work, using radon and beryllium for neutron source. We had no idea then that beryllium was poisonous. At that time the hospitals in Rome regularly received a radon supply, and we had an arrangement through the office of public health to use part of that supply. I participated in this work with Fermi and in October 1934 we discovered the slow neutrons. It was a discovery made in one day, in part it was chance, and then Fermi understood what happened. In the morning, we didn't know about slow neutrons, in the evening the paper was written and sent. During the previous week we were observing things that we could not explain. An object would be activated on a table but not on another table. Then we put a piece of paraffin between the source and the object. We immediately saw a great increase of radioactivity. It was a reproducible effect but we didn't understand it. We all went to lunch and took our siesta and came back at three o'clock and Fermi said, "I know. The neutrons slow down by collision in the paraffin and become more effective." We understood immediately that the neutrons would slow down by collision but we didn't understand why the neutrons of less energy should be more effective than those of greater energy. The next day we told Corbino about our discovery and he predicted big industrial applications for this effect, and we filed a patent. Frantic activities followed in which we systematically tried all available elements. I went around Rome with a basket, visiting chemical laboratories, collecting all the elements I could find there.

Even before the discovery of slow neutrons we did a lot of experiments with uranium. We observed various mysterious things and we also made some mistakes. We then left for a while after we had discovered the slow neutrons. Hahn and Meitner were also working on uranium in Germany and they confirmed our observations and they were much better radio-chemists than we were. In the meantime I won a chair in Palermo in

1936 and I left Rome. It was a big promotion but it meant a departure from the research in Rome. At the same time terrible things were happening in European politics. Hitler was firmly in power in Germany and we were fast moving towards World War II. There was tremendous turmoil and stress. Rasetti didn't want to stay in Italy and left for America.

In Palermo, I discovered the first artificial element, technetium in a material that I had received on a visit in Berkeley in the summer of 1936. I worked together with Perrier who was Professor of Mineralogy in Palermo. I was very lucky to be able to do something new in Palermo, which was a place that had been in disarray for years.

In 1938 I came to Berkeley in order to study short-lived isotopes of the element 43 because in Palermo I could study only the long-lived ones because the samples were coming to Palermo by ship. Within weeks of my arrival, using the techniques worked out in Palermo, Seaborg and I found a short-lived isotope of technetium, the one that is now medically important. While I was in Berkeley, in Italy Mussolini promulgated anti-Semitic laws, I lost my job in Italy, and remained in Berkeley. Fermi was also preparing to emigrate. His wife was Jewish and he didn't want to stay in Italy where his children couldn't go to school. He came to America in early 1939 and went to Columbia University.

At Berkeley first I worked a lot on radiochemical things and we discovered more new elements, such as astatin. In January of 1939 we heard of fission. It was an electrifying discovery, which also explained some of the uranium mysteries. I had told Abelson during my first visit to Berkeley to look into uranium because of these mysteries. With the strong sources at Berkeley and with their good chemistry it should have hit somebody on the nose that a chain reaction could be done. After the fission discovery many people did calculations and it became clear that something very important was there. When it became known that uranium-235 was a fissionable material, the problem of isotope separation came up. I did not believe in the feasibility of isotope separation, and I was wrong, but not by a large margin. In late 1940 I visited Fermi and he said that he had a hunch that element 94 might be a slow neutron fissioner. This is now known as plutonium, and I tried to prepare a sample of it to see whether it was fissionable or not. By the spring of 1941 we had enough plutonium to prepare a thin layer to do some chemistry on it. We had one microgram and we found that it was fissionable, similar to uranium-235. That opened new horizons be-cause if we could build a reactor, we could obtain plutonium and it could be extracted chemically, bypassing isotope separation. This would provide another nuclear fuel whether for a bomb or for producing power. Pearl Harbor gave a strong push to these efforts, although, even before, they were not much less intense than they could possibly have been. You can't start such things on a big scale; you need time for growth. By the time the scientific underpinnings were in place, there was sufficient momentum of the efforts.

In the beginning, many of the Americans were working on radar. For a while the atomic business was left to foreigners, people like Wigner, Fermi, myself, Teller, Bethe, and others. Then it became necessary to build up a big organization; to determine the critical mass of the pile, separating the isotopes, and other problems were gradually leading to a big project in 1942. By the summer of 1942, I had Chamberlain and Wiegand and other people with me, and we started making physical measurements that were necessary if we wanted to make a bomb. The work was very much compartmentalized. Ours was experimental work. Another group — Oppenheimer, Serber, Teller, Konopinsky, and others — were the theoreticians; they were doing calculations. To increase the efficiency of the project, it was decided to bring these efforts together in Los Alamos. I was one of the founders of Los Alamos, although I was not among those who were choosing the site. I was invited to go there at the very beginning. I remember when Fermi gave me a paper to read. He locked me in his office and left. It

Emilio Segrè standing between Otto Frisch and Enrico Fermi during a physics meeting in Basel, 1949 (photograph by and courtesy of Ingmar Bergström, Stockholm).

was necessary because he didn't want me to leave the paper around in case I wanted to go to the toilet. I have never been in a similar situation ever since. When I read the report I was all shaken. I saw the nuclear pile at the beginning of 1943 when I came from Los Alamos to Chicago to use it for some experiment.

We had to study the spontaneous fission of plutonium and we found that it was plutonium-240 rather than plutonium-239 that made the spontaneous fission. Then came the idea of implosion. Neddermeyer had the idea first, but he was not a good organizer and the implementation, which had to be done in a short time, fell on other people. The scientific part was done by von Neumann, Taylor, Fermi, and Bethe, by hydrodynamicists. The measurements were made by Kistiakowsky, Rossi, and Staub. After that came the test in which my group was also involved.

After the war, I came back to Berkeley, and for a few months I worked on old projects concerning radioactivity that I had started before the war. One of the things that I enjoyed very much was to change the half-life of radioactive substances by chemical means. When the new accelerator went into operation, we wanted to use it. For several years we studied nucleon–nucleon collisions. The proton–proton and proton–neutron collisions were considered a central problem in nuclear physics at that time. It turned

Enrico Fermi, Hendrik Kramers, and Emilio Segrè during a physics meeting in Basel, 1949 (photograph by and courtesy of Ingmar Bergström, Stockholm).

out that it was not as important as one thought. The hope was to uncover the nuclear forces in these collision experiments. We did some very good work, which at the time set the standard. But it was not really the right problem because it was too complicated for that time as the nucleons are not simply nucleons; they had the quarks inside, and with the available energies we could not separate the different interactions.

Then, as soon as we could be above the threshold of the anticipated antiproton, we tried to find it. This was in 1954–1955. There were good reasons for the antiprotons to be there, yet we could not be sure that they should exist. That was a question that had to be settled. The essential problem was this: you produced a few antiprotons with a tremendous background, 50,000 different particles and one antiproton. You had to identify it and had to be sure that it was what you thought you saw. You had a very short time, less than a millionth of a second, to distinguish the antiproton from all the other particles. We built essentially a mass spectrograph. We measured the momentum, velocity, and energy. Momentum and velocity would have been enough as you had to measure two of these three

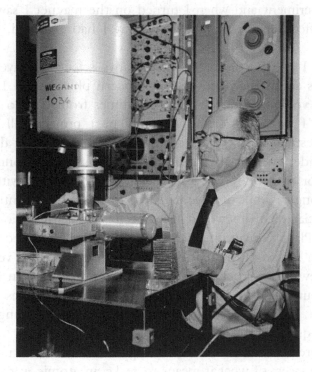

Clyde Wiegand in Berkeley, 1969 (courtesy of the Lawrence Berkeley National Laboratory).

quantities, in theory, but in practice, we had to measure all three to eliminate the consequences of the background. Once we had seen the electronic signature of the antiproton we carried out all kinds of tests to verify that what we had seen was the antiproton. Later, we worked on antineutrons. Eventually other accelerators became more powerful than ours and we could not compete anymore. Like at this moment CERN has such a machine that nobody else can compete with.

Chamberlain and I got the Nobel Prize in 1959 for the discovery of the antiproton, but the work was done by Chamberlain, I, Wiegand, and Ypsilantis. Ypsilantis had just gotten his Ph.D., and Wiegand had been my student and my collaborator for a long time. Chamberlain had also been my student. We went to Stockholm, of course, and it was a tremendous experience but not as tremendous as the discovery itself. It's nice to receive a prize but to make a discovery is very, very thrilling. I did my very first discovery, not a very big one, around 1930. I was a student and I had figured out the origin of certain spectroscopic lines. Nobody knew what they were. I figured out that they had to be a consequence of quadrupole radiation and had to originate from the Zeeman effect of a certain kind. I did the experiment and when I turned on the magnet I saw the pattern of Zeeman effect, appearing exactly the way I had expected, and that was a great thrill.

Nowadays I am too old to do any creative science. I have written two books. One is about the discoveries in modern physics, and I just finished a companion volume, which is about the times from Galileo to Hertz, to about 1895. I don't yet know what my next project will be.

It is interesting to speculate as follows: if a particular discovery had not been made by that particular person, how much longer would it have taken for somebody else to make it. For example, without Planck, how much longer would it have taken to discover the quantum? Without Einstein, special relativity would have been found within a year or two. But without Planck, quantum theory may have been delayed by five years or even more. It was uncharted territory and it was really a very surprising discovery. Newtonian mechanics probably would have been found rather rapidly without Newton. Huygens was near it and so was Leibniz. He was very near. Of all the things the one that was the strangest of all to me was the quantum.

As to my present concerns, I am unhappy about the scarce understanding, even at high places, of what it means to make an atomic war. People don't seem to believe that that would be the end of it all. It is a scary situation: how things accelerate, how things can happen at shorter and shorter notice in an increasingly automated way. On the other hand, I am not scared of atomic energy. On the one hand it is ridiculous what they do to develop fears about atomic energy and, on the other hand, there is the lack of fear about weapons. It is like treating a little pimple and cancer in the same way.

Harold Agnew (courtesy of Harold Agnew).

17

HAROLD AGNEW

Harold Agnew (b. 1921 in Denver, Colorado) is presently adjunct professor at the University of California, San Diego. He received a B.A. in chemistry from the University of Denver in 1942. He joined Enrico Fermi's research group at Chicago in 1942. First, he was sent to Columbia University and then moved with Fermi back to Chicago and participated in the construction of the atomic pile under the west stands of Stagg Field. He was a witness at the initiation of the first controlled nuclear chain reaction on December 2, 1942. Following this event he moved to Los Alamos in 1943. On August 6, 1945, he flew with the 509th Composite Group to Hiroshima with Luis Alvarez and measured, from the air, the yield of the first atomic bomb over the target.

In 1946, he returned to Chicago to complete his graduate studies and received a Ph.D. in 1949 under Fermi's direction. Then he went back to Los Alamos and started work in the Physics Division, eventually becoming the Weapons Division leader (1964–1970). In 1970 he became the third director of the Los Alamos Scientific Laboratory. In 1979 he retired and became president of General Atomics from which he retired in 1983. He was scientific advisor to SACEUR at NATO (1961–1964), a member of the President's Science Advisory Committee (1965–1973), and a White House science councillor (1982–1989). He was chairman of the General Advisory Committee of the Arms Control and Disarmament Agency (1974–1978).

He also had a political career being a New Mexico state senator from 1955 to 1961 when he resigned to join NATO. Harold Agnew is a member of the National Academy of Sciences of the U.S.A. and a member of the National Academy of Engineering of the U.S.A. He has received many recognitions for his service including the E. O. Lawrence Award in 1966 and the Enrico Fermi Award of the Department of Energy in 1978.

Dr. Agnew and his wife, Beverly, visited Budapest in August 2003 and we used this occasion to record a conversation with him just at the time of the 58th anniversary of the first atomic bombs dropped in August 1945 over Japan. It was natural to ask him about his thoughts on this anniversary.

No regrets. We did the right thing. I still don't like what the Japanese did. We killed more people with fire bombs, we killed more people in Tokyo, we would've killed many more without the atomic bombs. So it was an easy out for the Emperor because otherwise the military would never quit. We would've lost a lot of people, and I don't really care about how many they would've lost but I do care about how many we would've lost.

There was an interesting point you made earlier that even a lot of American POWs were probably saved by the atomic bombs.

Yes. Because their practice was to get rid of the prisoners. They worried about them rebelling. The survival rate of the American POWs in Japan was very low; the Japanese did not follow the rules of war.

You also mentioned leaflets.

After Hiroshima, we dropped leaflets.

Not before.

Not before.

Where did you drop the leaflets?

All over. Over Nagasaki, but not only over Nagasaki. The leaflet said that we've got this new kind of weapon, we've destroyed Hiroshima, and that you should surrender. Alvarez also wrote a personal letter on canisters.

You were Director of Los Alamos between 1970 and 1979. By then the second weapons lab in Livermore had long been established (in 1950). How did you feel about it?

The team with the instrument with which they measured the yield of the Hiroshima bomb. Standing, from left: Harold Agnew and Luis Alvarez; squatting, from left: L. Johnston and B. Walman. Tinian Island, 1945 (courtesy of Harold Agnew).

It's been good. They've been good competition, good people did good work. At the time we didn't like the idea because we thought it was a reflection on what we were doing. It took Livermore 10 years before they got something successful in the stockpile. It took them 10 years; they had lots of failures. The reason was that they wouldn't build on what we had done. They wanted to do everything on their own. Today the two places cooperate very well. Both are run by the University of California, but maybe not for long. There is a Congressman who wants the contracts to be competed. So that everybody can bid to run the laboratories. Maybe the University of California won't compete, and why should they? They don't make any money; they do it as a service. It may also be that if they would compete, they wouldn't win. Some big, politically-connected outfit, like Martin Marietta or Battel Institute would win.

But there is no question that the two places should continue.

I think so. Also, there is no question in my mind that the two places should be under one management. They were always under the same management, the University of California. There was always competition between the labs, but sharing of facilities, sharing of people. This would be very hard under different main managements.

As I understand, Edward Teller was instrumental in initiating the second laboratory.

I think so, although their first director, Herb York says, "No." According to him it was E. O. Lawrence.

It's interesting because Teller told me that he was very proud of having initiated it.

I know and in my opinion, he was a leading person to initiate it because of his interest in the hydrogen bomb. He didn't think Los Alamos was working hard enough on it. I think he had more to do with starting Livermore, but York says it was Lawrence. It's called the Lawrence Livermore Laboratory.

Lawrence and Teller were politically close.

I don't know. I never met Lawrence; I didn't know him.

But you knew Teller.

I knew him very well.

What hurt Teller? He became very unpopular.

The Oppenheimer Affair. Only that.

But Teller was not alone with his opinion.

Of course, not. And if you read his testimony, I don't think his testimony was all that bad.

Edward Teller and Harold Agnew in Los Alamos in the early 1970s (photograph by Bill Regan, courtesy of Harold Agnew).

He formulated it in a clever way, but he practically said that he would not trust ...

That's right. He would feel more confident with somebody else. But he did say that Oppenheimer did a good job as Los Alamos director. I don't think he initiated the business against Oppenheimer. That was done by Lawrence and his people and the Air Force.

So why did he suffer alone and not the others?

There weren't any other leading scientists who testified. He was the only one and the people who turned against him were people from Caltech and Berkeley, who were maybe not so fond of Lawrence. I don't know. I don't understand why everybody turned against Teller. There were lot of hero-worshippers of Oppenheimer. Edward was the only real scientist who testified.

Lawrence and Seaborg and others agreed with Teller, as far as I know, but they were quiet. They were clever. Christie, who was to become Provost of Caltech was very unhappy with Teller and so was Bacher. On the other hand, Hans Bethe never attacked Teller. Bethe wrote a review of Teller's *Memoirs* and it was very complimentary.

Did you like the book?

I liked it, it's wonderful. There are a few errors and I pointed them out to Edward. He said, "If I have a second edition, I'll consider correcting them."

I would like to tell you about my impression because I am curious of your reaction. I always thought that Teller was a very determined person who always knew what he wanted to do. In this book he came across for me as one often hesitating about his actions and asking for other people's opinion before making up his mind. In a way he was seeking justification or approval from others.

I didn't sense that. No. I asked him if he kept a diary and he said, "No." He just remembered everything. One of the reviews, which I did not agree with, said, "Edward has a fantastic memory; he even remembers things that never happened." The things that I know about were factual. He got a few little dates wrong, who was where and at what time, but that doesn't matter. I enjoyed reading it, it's well written, it's easy to read, for me.

It's a funny thing. After the Oppenheimer Affair, since that time, Edward was essentially not allowed to come to Los Alamos.

Not allowed?

Not invited. He was just *persona non grata*. He never came. When I became director, I invited him. I did not care about politics and I did not think he did so badly on Oppie. I liked Oppie; my wife was Oppie's secretary for a while. We knew him very well. But Oppie lied. He lied.

About what?

About his affair with this Chevalier guy or whatever it was. He didn't tell the truth.

Did you know that at the time?

Of course, not. We learned about it afterwards. If someone asks you to tell them secrets, the rule is that you go to the government and tell that this guy is doing that. Oppie didn't do that. He broke the rules. That's what they nailed him on. The book *Brotherhood of the Bomb* [by Gregg Herken] is about that. It's about Oppenheimer, Lawrence, and Teller. They followed Oppenheimer all the time because of his Communist connections. They were suspicious of him all the time. I even knew a man of the FBI who was detailed to follow him when he was in Washington.

So I invited Edward to Los Alamos to spend the summer there. It was good for Los Alamos because for the young people to get to talk to Edward was a very stimulating thing. He profited too and Livermore profited because we had started a laser program. We weren't doing much but Edward saw that as a program that Livermore could take and run with. As a result they have got now this big multi-megawatt laser program. That started from Edward coming to Los Alamos in the early 1970s and talking to a man named Keith Boyer who was working on our laser program.

For completeness, I may add that Edward can sometimes make you very angry; he just can.

I would like to ask you about your work as a physicist, prior to your directorship of Los Alamos.

Most of the stuff I did was just measuring cross sections. That was in the Manhattan Project. In graduate school, I was a student of Fermi and did work on beta-decay. However, none of this work was especially remarkable. The most important thing I did was conceiving the idea of what was originally called the Permissive Action Link of weapons, PAL. This is the black box that prevents anybody unauthorized from using a nuclear weapon. That was my idea and I got the first one made. Then I went to the Congress and told them that we should do this. I went to Kennedy's science advisor and told him that we should do this. That's why I got sent to NATO because that's where we were going to implement it. The military were violently against it for two reasons. One, they said, "You don't trust us," and second, "What if we don't get the code word?" My answer always was, "If you don't get the code word, you're not supposed to use it."

Harold Agnew and President Gerald Ford (courtesy of Harold Agnew).

Did you share this idea with the Russians?

No, they learned about it. It was important for everybody. When Pakistan and India got their bombs, our government was dumb. We always had sanctions. What they should've said, "OK, you guys, you're nuclear powers now, Mr. President of Pakistan, Mr. President of India, you don't have the bomb, some lieutenant has the bomb physically, you better make sure that he can't use it. You need this technology; it's not classified; we should make it available to you." When the Chinese first started their nuclear program, they wanted this information and they came and asked us, but we said, "No." So they went to the Russians and got help from them. This is why I'm saying that our government is dumb. In any case, that's one program that I started, I got an award for it, and it's physics.

Also, measuring the yield of the atom bomb with Alvarez was another example. Some of the instruments were mine. There is the so-called Agnew box, which was used for calibration.

What was the atmosphere like in Los Alamos when you were building the bombs? Did you sense a feeling of urgency?

We had a feeling of urgency all the time, but I don't think we cared either way. You could think that having so many Europeans among us, they would've been unhappy using the bombs in Europe.

Wouldn't they be happy if the bomb would end the war?

I don't know. The European war was different than the Japanese war. We had not invaded the main Japanese islands. We were working these individual islands. We've been bombing, bombing, bombing Japan, fire bombing.

With so many Jewish scientists in Los Alamos, don't you think they would have preferred bombing Auschwitz and Germany?

Not Auschwitz because that's where the Jewish people were.

They were being killed.

You still hoped that maybe somebody would survive when the war would be over.

How about Germany?

Maybe, but this never came up. Our real worry was who was going to get the bomb first? I remember Fermi was always worried because all these people had studied in Germany. They all knew that the center for nuclear physics at that time was mostly in Germany, with a little bit in Britain. Their worry was that the Germans had the talent, but we didn't know what was going on. I think it was Fermi who got the Norwegian heavy water plants bombed.

Do you think if the bomb had been ready before the war was over in the European theater, it wouldn't have been used in Europe?

No, I was with the military when we went to Hiroshima. Before that we practiced a lot with the airplane people with our gauges. We learned how to drop them. Their orders were, be prepared to use the bomb in Europe or Japan. They would be told when. They were to be ready to use it in either theater. But the bomb was not ready in time for Europe.

So it would've been used.

I don't know.

It might've been.

Might have been. But I don't know what the target would've been. Maybe on Rommel's army; it was a different type of thing; I don't know. I only know that they were told to be prepared to use the bomb in either the European theater or in the Asian theater. I don't know whether they would've had a base for using it in Europe. We'd built a base on Tinian for us. We had our own base [that is, for the atomic bomb project]. The runways were built there. In Europe, they would've had to build some facilities for using the bomb. The plane was so low to the ground and the bomb was so fat, we had to have a loading pit with a hydraulic lift to come up. You couldn't get the bomb under the airplane when the airplane was on the ground. You needed special facilities. They were never built in England, I don't think they were. We would've flown out of England, I'm sure.

You were associated with the project, among the very first participants.

I had a bachelor's degree in chemistry. Before Los Alamos, we were in Chicago. Beverly was the secretary of the Head of the Metallurgical Lab.

Oppie wanted her to come to Los Alamos. They didn't want me; they wanted her. I had gotten bad radiation exposure in Chicago from my radium-beryllium sources. I had to stay away, but I was there when we brought in the first pile with Fermi, and I helped build it. I was there on December 2, 1942, when it was turned on and Wigner brought the bottle of Chianti to celebrate.

Did you get a sip of it?

No. I was doing an experiment in the same building down the hall. I came to watch it; I was there before lunch, then we went to lunch and I was there after lunch. When everything was finished, Fermi said, "ZIP in," and George Weil put the control rod in. Everybody cheered and I went back to my lab. It was then that Wigner brought up the bottle. At least everybody knows I was there when the pile went critical.

What did you do after the war [World War II]?

We went back to Chicago. Fermi took me on and I got my degree in 1949. Then, I went back to Los Alamos in 1949. I stayed there until 1961 when I went to NATO in Paris because of this command and control thing. I served as advisor to Norstad and Lemnitzer for three years and I came back in 1964. I became head of the weapons division in Los Alamos and in 1970, I became director. I was the third director of Los Alamos. The first was Oppenheimer and the second was [Norris] Bradbury. I had a rule, you take a job for no less than 5 years and no more than 10. So I retired very young, at 58.

What did you do afterwards?

I went to California and became President of General Atomics. It was a good company; they were trying to build high-temperature gas-cooled reactors. I'm still on their Board and they are now famous as makers of the Predator, the unmanned air vehicle, the no-pilot airplane for surveillance. I stayed as president for four years. I have always had a relationship with the University of California as Adjunct Professor. I used to give lectures there. After 1984 I really retired and have served on government committees; I was on Ronald Reagan's Presidential Science Advisors' Committee. I had run the General Advisory Committee to the Arms Control Agency under the Nixon Administration. I still serve on a few committees and a couple of boards.

I presume then that you are a Republican.

No, I'm a Democrat. Edward is a Republican. I was once a State Senator in New Mexico and served two terms but then I decided to do it no more. I don't vote party, I vote persons. I'm a registered Democrat but it's not terribly important unless you don't have a mind of your own. They're all crooks. They come in with good ideas and work hard but the people bribe them and make them crooks.

Did you take time off from your job?

I did and without pay. As a State Senator, I received five dollars a day. That's why they become crooks. Most of them are lawyers and insurance people and they still have their businesses. People give them business for favors on legislation.

There was this terrible arms race between the United States and the Soviet Union. This is over now but the world doesn't seem to be much better off.

At least they didn't fight each other because each one was so big. Today, there are the terrorists and I don't understand where they get their explosives from.

Are you an optimist?

I've lived this long, so I guess I am.

********** * * * * * * * * * *

Harold Agnew on Enrico Fermi

When I subsequently asked Dr. Agnew to tell me more about his interactions with Enrico Fermi, he sent me the account reproduced below.

In January 1942 I went to the University of Chicago to join the Manhattan Project. I was immediately sent to Columbia University to work with Enrico Fermi. When I first met him the only unusual thing that I noticed was that all of his pants pockets had zippers, all four of them. At the time he was conducting experiments using a large pile of graphite. The structure was entirely encapsulated with a sheet metal cover and was evacuated using mechanical vacuum pumps. The pile had a radium-beryllium neutron source at its center and we measured the slowing down of the neutrons using indium foils, which were activated by the source's neutrons. We would insert the foils at different levels in the pile for a specific time, then remove them and run about 100 ft to the counting room where there was a set of Geiger counters. We did this hour after hour for about 10 hours each day. Fermi not only directed the work but also actually took on a shift the same as the rest of us. Inserting the foils, running to the counting room with the activated foils and then taking the data. He was one of us.

This always distinguished Fermi. He clearly was a genius but acted with no pretentiousness. He was a very unassuming person. He had a wonderful sense of humor. The array of counters in their lead shields all had names, taken from the Winnie the Pooh books. They were named Pooh, Pigglet, Heffelump, etc. For non-nuclear safety reasons he decided to move the experiments to Chicago and we started to build CP-1, the first man-made chain reaction. One day a several ton load of graphite blocks was delivered around 4 p.m. We had to unload the truck so along with the rest of us Fermi took off his coat and pitched in and helped unload the truck. This was Fermi. He not only supplied the brains at Chicago but when needed also supplied the brawn.

Chicago is cold in the winter and people went ice-skating there near the University. Fermi had never ice-skated and decided he would. We all went to the rink, got Fermi a pair of skates, and after a few falls Fermi caught on, and before the end of our first session was skating as well as anyone else. He was an excellent athlete and loved to compete. He liked to play tennis especially. Later on when I returned to Chicago as a graduate student we used to play tennis during the lunch hour. This required checking out a net and setting it up on the court. The professor and student took turns with this task. Fermi was a very regular person, not at all impressed with his position. The only sport at which he was a failure was in fly-fishing for trout. Segrè who was a very good fly-fisherman never let Fermi forget that at this sport he was no good.

In 1946, after the war, housing was very scarce in Chicago. I was unable to find a place for my family to live. Fermi who had a fairly large house suggested that my wife, small daughter, and me come live with them. His wife Laura wanted to visit her sisters in Italy and when she was gone my wife Beverly could run the house and do the cooking, etc., for Fermi and his children Giulio and Nella. We did this for almost three months until I found a place for us to live. Being part of the family for three months was a wonderful experience. Fermi preferred non-spicy food and

Harold and Beverly Agnew in Budapest, 2004 (photograph by I. Hargittai).

always diluted the red wine we had for dinner half with water. We stayed on for a month or so after his wife Laura returned.

One evening she told Fermi that she had gone to the local appliance store and put her name on a waiting list for a General Electric dishwasher. [After the war appliances were scarce and one had to sign up on waiting lists for appliances, car, etc.] I was astounded. Fermi had been the major consultant for General Electric who were building reactors at Hanford for the production of plutonium. I said, "Enrico, you know the president of General Electric. Just tell him you want a dishwasher and he will send you one tomorrow." Fermi thought for a second and said, "No, that wouldn't be fair for others, we will wait our turn in line." This was classic Fermi.

Fermi liked to swim. Sometimes after work his team of which I was a member would go to Lake Michigan. On one day he decided we would swim across a little bay. I had been a varsity swimmer in high school so thought I was pretty good. But after about 15 minutes in the choppy cold water of Lake Michigan I was falling behind. Fermi who swam with what I would call a "dog paddle" style swam back to me and asked if I was OK. I said I thought so but clearly my Australian crawl swimming style wasn't best for choppy Lake Michigan. I barely made it to the other side of the bay and with difficulty climbed up the sea wall and sat down. Fermi said, "Meet you back where we started" and plunged back in and swam back to our starting point. I had difficulty just walking back.

Fermi was known by his colleagues as the "Pope". This made it all very clear that he was the supreme authority on all matters. He held this position in all of our minds as an accepted fact. No big deal. Just an accepted realization that he really knew more than the rest of us, or anyone else involved in our scientific work. Fermi especially liked young people. He, in his position, entertained a lot but preferred to have young people. The top floor of his Chicago house had a large room in which he would invite students to come and square dance; I usually did the calling and a good time was had by all. He and Laura had these parties about once a month. When he had dinner parties for his peers he always said, "We need to dilute 'so-and-so' and 'so-and-so' with some young people." The "so-and-so" were too stuffy.

Chicago had an open enrollment system for graduate studies but required a 3-day written examination to decide one's future. Choices were, flunk and out, pass with a Master's degree and out, or pass with option for going on for a doctorate, if you could find a faculty sponsor. I was terrified about taking the exam because I felt my peers were much smarter than me. [Subsequently 4 of my classmates have received a Nobel Prize in Physics, and they were not all the really smart ones.] The tests were given so that those scoring the written results had no idea as to whose papers they were grading. I kept putting off taking the test but Laura Fermi kept urging me to do so.

I went to Fermi and asked what he suggested I read. He said he had no idea because he didn't read much. I asked how he always knew what was going on. He said people came and told him and explained things to him. Then he said — which amazed me, that there were people who said they immediately understood things — but he wasn't one of those. He said it took him a long time to understand what people were explaining to him but many times he realized that they really didn't understand what they were describing to him but he did. He also volunteered that one who was very quick to say he understood even before the person finished was Oppenheimer, but a lot of the time Oppenheimer really didn't understand the technical information the way Fermi understood it. He told me that if you really understood [Fermi's way of understanding] about ten things in physics you could know almost everything.

I had been getting a week's lecture on Brillouin zones, which I never understood, and asked him about it. He went to a small blackboard and in less than 5 minutes developed the whole theory and at the time I thought I understood it. But as was with most of Fermi's lectures, they were so clear and so simple that you really thought you understood all. But when you tried to repeat it afterwards on your own, you became lost. Very much like eating Chinese food, you end up very full and satisfied but shortly very empty and hungry.

Of all his colleagues of his vintage, Fermi's favorite for his intellectual ability was Edward Teller. He told me this and years later Laura Fermi and his daughter confirmed this when I raised the question. Among his young people I believe Fermi thought Dick Garwin was the brightest and I also believe this even to this day.

This is just a short snapshot of my interaction with Fermi. There are many other stories such as how he saved our nuclear weapon program when he came up with the idea that plutonium from Hanford would be different than that produced in a cyclotron. He had Segrè confirm his worry.

Clarence E. Larson, 1998 (photograph by I. Hargittai).

18

CLARENCE E. LARSON

Clarence E. Larson (1909–1999) was an energy consultant in Washington, D.C., when we recorded a conversation with him on April 24, 1998, in the Larsons' home in Bethesda, Maryland.* He graduated from the University of Minnesota majoring in chemistry in 1932 and received his Ph.D. from the University of California at Berkeley in 1936. His career included various assignments in the nuclear program during World War II and after; he was an isotope separation scientist; director of Oak Ridge National Laboratory; president of Union Carbide nuclear division (until 1969); and Commissioner of the United States Atomic Energy Commission, 1969–1974. He was a member of the National Academy of Engineering and other learned societies. From 1984 until his death, he was President of Pioneers of Science and Technology Historical Association. He and his wife, Jane, recorded videotapes of conversations with over 60 significant figures of science and technology of their time.

I would like to ask you first about your background and education.

I was born in 1909. Originally I'm from Northern Minnesota, from Cloquet, about twenty miles west of Duluth. Cloquet is a French name because it was first a French settlement. There was nothing unusual in my childhood except when I was nine years old there was a big forest fire which burned down the whole town of about 10,000 people. We became refugees

*In part, this has appeared in *The Chemical Intelligencer* 1999, 5(1), 43–49.

and fled to Duluth where they took care of us until they could build us temporary housing in about two months. Fortunately, the lumber industry was around the river so the town was able to start up again.

My father was a railroad engineer. He had emigrated from Norway in 1880. I went to a regular school and didn't remember any outstanding teachers but they were adequate. After graduating from school I worked for two years to earn some money for college. I was very interested in science in general and at the age of 12, I made a radio transmitter and receiver. That was before radio broadcasting. I was also interested in chemistry.

When I went to college at the University of Minnesota, I followed the regular curriculum for chemistry majors. I was especially interested in physical chemistry. During my undergraduate years I read an article on Balmer and his spectral lines which interested me a lot. It was in the early 1930s and Kolthoff was there at Minnesota in analytical chemistry and I remember Heyrovsky's visit there. I also took some courses in chemical engineering. I graduated in 1932.

For graduate work I went to California. I got my Ph.D. from the University of California at Berkeley. That was the golden years for science at Berkeley. There was a lot of work on radioactivity. E. O. Lawrence invented the cyclotron. Oppenheimer was there; he had a chair at both Berkeley and Caltech. The discovery of carbon-14 also happened there at that time. My supervisor was Dr. Greenberg. He is not particularly known for any of these big discoveries, but he was a very competent man and a very good teacher.

In my graduate studies I did some work in electrochemistry and published a couple of papers. I also happened to do some work on radioactivity and published a small paper. I had a friend at the agricultural department and he wanted to follow the metabolism of calcium and phosphorus in the egg cycle. It was nothing particularly sensational but it gave me an idea of the techniques used at that time. This was before the discovery of fission, so all the radioisotopes came from the cyclotron. Lawrence made available some of the targets to me and I was able to carry out some experiments. Both he and I were enthusiastic about amateur radio and we talked about it a lot.

At that time academic jobs were very difficult to get. For every job that became available there would be a hundred applications. After I got my Ph.D. I got a job at the College of the Pacific, north of Berkeley. There I taught inorganic chemistry and physical chemistry. It worked out very fine. Teaching at a liberal arts college is a pleasurable thing. But then the War came along.

Jane and Clarence Larson with István Hargittai at the Cosmos Club in Washington, D.C., 1998 (photograph by Magdolna Hargittai).

In 1942 Lawrence asked me to come down for a very important project at Berkeley. He told me about making the atomic bomb. Of course, I had to go through all the security checks and was sworn to secrecy after I'd agreed to join the project.

My particular job was to synthesize uranium tetrachloride. We used a mass spectrometer to separate the isotopes. We ionized a volatile complex of uranium and separated uranium-235 and uranium-238. At the beginning we were separating micrograms. Lawrence told me that we needed 75 pounds to make the bomb. So we decided to build a thousand units. This was done down at Oak Ridge, Tennessee. The design and all the research came from Berkeley. In 1943 I went to Tennessee. Uranium tetrachloride is very hygroscopic, more even than phosphorus pentachloride. This causes a lot of spoiling and we had to distill it under special high vacuum. In order to get bomb grade uranium there can be only one part U-235 and 140 parts U-238. It sounds like an impossible job and it was almost impossible. It was eventually used in the Hiroshima bomb. However, in those years we were highly compartmentalized. All I knew about the broader aspects of the project was what I got from the grapevine.

Was there any point later when you were told that the secrecy was over?

President Richard M. Nixon greeting Clarence Larson, U.S. Atomic Energy Commissioner in the White House (U.S. Atomic Energy Commission, Office of Information Services, photograph by J. E. Westcott, courtesy of Jane Larson).

General Groves, the head of the whole project would visit us once in a while and he would always criticize us for working too slowly, he was always very critical. So he came for another meeting in the later part of July, 1945. That time he was all smiles and he said that we were doing a wonderful job and everybody looked around because we thought this couldn't be Groves speaking. We couldn't figure it out, but in the meantime they had done this experimental bomb in New Mexico.

However, there were lots of complications before we became successful. The chemical processes didn't seem to work. We were dealing with large volumes in stainless steel containers and because of the corrosion products we had to go through so many purifications that nothing was coming out the other end. Finally I got to *Beilstein* where they describe all the reactions and I found out that if you use hydrogen peroxide, you can separate uranium out from everything else and nothing else will precipitate. People were reluctant to change the procedure, but there was no other way out, so we used the peroxide method. That also meant that we couldn't use the iron because even a trace of iron would destroy the peroxide. The peroxide process would work but it would work in the pure solution only. We didn't quite know what to do when I suddenly remembered one of my projects from my graduate studies when I carried out a reaction under very cold conditions to slow it down. So we put refrigerating jackets around the reaction chambers and that immediately worked. The hydrogen peroxide was decomposing because the reaction with iron is an autocatalytic reaction. As soon as we cooled it down, everything was just fine and we could go on using the stainless steel containers.

Where did you find the Beilstein?

In the library at Berkeley. We had access to that library and they had taken out all the volumes of the library that had anything to do with uranium.

Wouldn't this give them away?

It was stupid, but that's what they did. Things were moving very fast. The operations of these big projects were soon turned over to big companies. This was early in the war. The University of California at Berkeley did the basic research and the desk top project and, for our part, Eastman Kodak did the practical implementation, all the engineering in Oak Ridge, Tennessee. The isotope separation by a different method, the gaseous diffusion technique, was done by Union Carbide at a different location in Tennessee. I was transferred early on to Eastman Kodak. I stayed in Tennessee to the end of the war and long after. There were three big separate organizations, the Oak Ridge National Laboratory, the Gaseous Diffusion Plant, and the Electromagnetic Plant, and all three of them were merged under Union Carbide. All three operations were successful. The gaseous diffusion process cut the cost over the electromagnetic one by a factor of ten. There was constant improvement in all these things. The upshot of it was that after a slow beginning all the problems were solved. The objective was to get the critical mass for use in the war by July 1945, and we did that. After that people thought that the atomic bombs would just roll off the assembly line. Actually there were only two available at the time.

After the war, we worked on specialized things, such as the separation of zirconium from hafnium for the submarines. Zirconium is a very good metal, non-corrosive, but it always comes mixed up with hafnium, a high cross section element, which spoils it for reactor use. It's very difficult to separate them. George Hevesy, a pioneer in the application of radiology, discovered hafnium as an element. He got the Nobel Prize in Chemistry

U.S. Atomic Energy Commission. In the middle, Glenn Seaborg, Chairman; on his left, Clarence Larson (U.S. Atomic Energy Commission, Division of Public Information, photograph by J. E. Westcott, courtesy of Jane Larson).

[in 1944 for 1943] for the use of isotopes as tracers in the study of chemical processes. We also did another separation, lithium-6 from lithium-7 for the hydrogen bomb project. A process that we worked out in this connection is now under so-called declassification. When I get it declassified, I'll send a copy to you.

Do you still have classified work from that period?

Yes, and not only I. There's some controversy about this. One of the reasons of the downfall of the Russian empire was that they were trying to match us in expenditures. Some of our processes were better by a factor of ten to a hundred, less expensive. In other words, our technology was that much superior. This is what's involved, in a small way, in these declassification processes. But it may also be just an excuse to keep them from declassification.

How long did you stay with Union Carbide?

Jane Larson with her mural in the lobby of the Chemistry Department of the University of Maryland at College Park, 1999 (photograph by Magdolna Hargittai).

From 1950 to 1955 I was Director of the Oak Ridge National Laboratory. Union Carbide operated it until the mid-1970s. In 1955 I went to New York, still working for Union Carbide. Then, in 1960 I went back to Oak Ridge as president of all the operations, not only Oak Ridge but also of the Kentucky operations. After about eight years there, I finally left Union Carbide and was appointed by the President as one of the five commissioners of the Atomic Energy Commission in Washington, D.C. That was in 1969 and we've stayed in Washington ever since.

How about now?

After I retired in 1975, for five years I was Chairman of the National Battery Association Committee. My son and I rebuilt a Karmann Ghia with an electric motor from an airplane and we took senators for rides. I helped write a bill that went through Congress giving support to organizations doing research on electric cars. The most important problem was development of high-capacity storage batteries. It's still a problem.

I also started a videotaping project called "Pioneers of Science and Technology" and went around taping all the great scientists I knew. My wife was cameraman. We have about 60 tapes now, in archives at various universities, and are still taping on occasion.

Nelson J. Leonard, 1999 (photograph by I. Hargittai).

19

NELSON J. LEONARD

Nelson J. Leonard (b. 1916 in Newark, New Jersey) is Reynold C. Fuson Professor of Chemistry Emeritus of the University of Illinois, Urbana-Champaign, and Faculty Associate at the California Institute of Technology in Pasadena. He received his B.S. degree from Lehigh University in 1937, the B.Sc. degree from the University of Oxford in 1940, following his Rhodes Scholarship there, and his Ph.D. from Columbia University in 1942. He holds a D.Sc. degree from the University of Oxford (1983). Dr. Leonard has been at the University of Illinois since 1942 and retired in 1986.

Dr. Leonard served as Scientific Consultant and Special Investigator, Field Intelligence Technical Agency, U.S. Army and U.S. Department of Commerce, European Theater, in 1945–1946. He was elected a member of the National Academy of Sciences of the U.S.A. in 1955 and has received many other honors and distinctions, including the Roger Adams Award in Organic Chemistry (1981) and the Arthur C. Cope Scholar Award (1995).

His research interests have spanned broad areas of chemistry, biochemistry, and plant physiology, including the chemistry of nitrogen-containing organic molecules, trans-annular interactions, cytokinins in connection with the growth, division, and differentiation of cells, enzyme-coenzyme interactions, and DNA/RNA interactions. He reviewed some of his research of over 50 years in a retrospective article ["The 'chemistry' of research collaboration", *Tetrahedron* 1997, *53*, 2325–2355].

The following narrative by Nelson Leonard is based on our conversation at Caltech in May 1999.*

At one time the American Chemical Society was recording interviews with people, and I was supposed to record an interview with Carl Folkers, who was 10 years older than I to the day. Going into the interview I told him that I couldn't go from cradle to grave with him. He was still so vital that I'd prefer to ask him what he's doing now, and then we went backwards. Here's where I am today.

The Division of Chemistry and Chemical Engineering of Caltech brought me here as a Sherman Fairchild Distinguished Scholar in the fall of 1991 to do whatever I wished in scholarship and to interact with the faculty. After I had been here part of the school year, the members of the Division said, "We'd like to keep you around," and they provided me with an office and the title of Faculty Associate in Chemistry. The Vice Provost renewed the appointment in 1997. I have been a member of the Freshman Admissions Committee, attended seminars, interacted with faculty and students, and have written up my research which continued at the University of Illinois through 1997. As you get older — I am now 82 — it is not possible to develop all of your new ideas. You still have ideas but you can't develop them in depth.

I have also used my time at Caltech to write review articles based upon some of our earlier discoveries, such as fluorescent derivations of ATP and related compounds. All of the fluorescent adenine derivatives, now available commercially, are being widely used. I decided that a passive way to keep up with the field which we had generated while I was at Illinois would be to write a review every ten years. That method has worked. Then, I had an invitation from *Tetrahedron* to write a "Perspectives" article. About five very nice articles of this type had appeared by the time that I was invited to submit one by the editor, Harry Wasserman. I finally decided to write about what was different about my scientific life. I enjoyed collaboration with other scientists in other places and in other disciplines, which I covered under the title, "The 'Chemistry' of Research Collaboration". Another thing that was different about my scientific career was that I felt I would get bored every decade if I continued the same work, digging deeper and

---

*In part, this has appeared in Hargittai, B.; Hargittai, I. *Chemistry International* **2003**, *25*(5), 7–8.

deeper. Thus, I usually shifted and sometimes used a sabbatical leave for the shifting of research areas.

My college career started at Lehigh University in Bethlehem, Pennsylvania, with a B.S. in Chemistry. A Rhodes scholarship took me to the University of Oxford in 1937–1939 where I was doing dipole moment measurements with Dr. Leslie Sutton, who was also experimenting in electron diffraction. I became interested in the details of chemical structure but I didn't think that I could become a good physical chemist. When war broke out in Europe and I had to return to the U.S., I ended up at Columbia University, New York, working on alkaloid chemistry for the Ph.D. degree. I determined very soon that the fun part of my life was synthetic organic chemistry and working with natural products. In 1942 I went to the University of Illinois as a postdoctoral fellow and, shortly after arrival, additionally as an instructor in chemistry. My research was concentrated on anti-malarials. As a confirmed New Yorker, I thought I would give the Midwest a year and then return to New York. However, Illinois was a fascinating place, with excellent people at the time: Roger Adams, Carl S. ("Speed") Marvel, Harold Snyder, and Charles C. Price III stayed there.

When the war ended, I got a temporary job with the U.S. Army overseas as part of an industrial intelligence unit (F.I.A.T.). Stationed in Hoechst, Germany, we were examining the research publications and research reports, for example, of the I. G. Farbenindustrie, and we started a microfilming operation. We were investigating only the scientific material and found a number of things that could be applied in American industry. The idea was, "We won the war. What do we get out of it?" Before the war had ended, the intelligence unit had started interviewing directors of research, etc., of German industry, but by September, 1945, it was obvious that details were necessary, for instance, from research reports and manufacturing procedures. One example of useful information that was obtained related to a modifier for butyl rubber. We would not have had as good synthetic rubbers in the U.S. without learning that a particular long-chain mercaptan had been used in Germany as a modifier in rubber manufacture. It had been patented, but not all patents were of the "open" type.

We entered the archives, asked for the reports, and then an Army team microfilmed them. The originals remained in place. We found it useful to hire a German librarian who believed strongly in completeness and accuracy. If a volume of reports for a particular year from a particulary branch of I. G. Farben was missing, the librarian produced it — always after a weekend. We never asked how he did it. However, we did not see any reports from

At Höchst, Germany in 1945
(courtesy of Nelson Leonard).

I. G. Farben Auschwitz, the Buna and synthetic fuel works. This was not surprising because, according to Joseph Borkin of the Antitrust Division of the Department of Justice, writing in his book, *The Crime and Punishment of I. G. Farben*, most of its records were destroyed so that the employment of concentration camp workers could be obscured. The microfilms of the research reports that we did obtain were sent back to the U.S. and were made available through the Library of Congress. However, the Cold War started very soon thereafter, so the idea of taking technical information from Germany ceased to be popular. It was considered important that German industry should be given an opportunity for revival. The decision was political and military, and that was it! But this change happened after my time there.

There was also a personal reason for my wanting to be in Europe. I had become engaged to a Dutch girl, Louise Vermey, just before I returned to the U.S. in 1939. In 1945, I met her again on a side trip from Germany to the Netherlands. She was able to come to the United States early in 1947, and we were married in May of that year. During the war we had lost contact because the Germans decided, possibly in 1944, that the Red Cross was an Allied operation, so no more brief Red Cross Letters could be exchanged. By that time I was engaged in antimalarial research at the

University of Illinois, and that work was classified. When I wrote to mutual friends in Sweden and Switzerland, my letters were intercepted by U.S. intelligence and were returned to me. My fiancée spent all of the war years in the Netherlands, but that is her stark story.

My research interests kept shifting during my career. In the first decade, my students and I worked on reductive cyclizations, electrolytic reductions, molecular rearrangements, and the stereochemistry of 1,2-dicarbonyl compounds. After my first sabbatical leave, we worked on medium-ring compounds, discovering some transannular interactions and reactions, and on small charged rings, discovering some ring-enlargement reactions of aziridinium and azetidinium salts. During a sabbatical leave in Switzerland in 1960, I started reading biochemistry, but the initiative for research came from one of my students back home, Jim Deyrup, who was working on a natural product, triacanthine, that turned out to have a 3-substituted adenine structure, 3-$(\Delta^2$-isopentenyl) adenine. Most of the adenines known up to that time were substituted at the 9-position. The 3-substitution was a nice surprise and served as a channel into biochemistry through 3-isoadenosine and its mono-, di-, tri-, and cyclic phosphates. For example, 3-iso-ATP turned out to have many coenzyme activities similar to those of natural ATP, adenosine triphosphate. An isomer of triacanthine, namely N^6-isopentenyladenine, because of its cytokinin activity (plant-cell growth, division, and differentiation) was a channel into plant physiology. I started a collaboration with Folke Skoog, Professor of Plant Physiology at the University of Wisconsin. We liked each other. He is a tough guy, and his plant physiology was really tops. We worked together for 20 years and published more than 40 papers together. During that period, Illinois made me Professor of Biochemistry in addition to being Professor of Chemistry and I served on at least one Ph.D. committee in Plant Physiology. I was happy to be able to stretch my organic chemistry into other areas.

Cooperation is also helpful in getting one's research funded. If you do mixed science yourself, someone on the granting committee is going to ask, "Is that a biochemist? Is that a chemist?" Just raising such a question may be enough to down-grade the rating; however, if you cooperate with the best biochemist or with the best plant physiologist, you may have a better chance. Granting agencies want to see a little more cooperation and a little less competition.

Another scientist with whom I had a fruitful collaboration was Professor Gregorio Weber (1916–1997). He was a great man in fluorescence. He came to the University of Illinois from Argentina by way of England, and he

Three decades of editing *Organic Syntheses*, Inc. From the left: Nelson Leonard, Henry Gilman (at his 90th birthday), and Ralph Shriner, 1983 (courtesy of Nelson Leonard).

excited all of us about fluorescence. We decided to make fluorescent derivatives of the nucleic acid bases so that they could be detected and would indicate, by fluorescence lifetime, yield, and polarization, how they were attached to an enzyme or structural protein. The time period was the early 1970s, Dr. Jorge Barrio was one of my chief co-workers, and the fluorescent derivative of ATP, namely $1,N^6$-etheno-ATP, has been the most popular in numerous applications thereafter.

We continued to do many things based upon fluorescence. We constructed a compound that was a fluorescent dimensional probe of ATP, i.e., *linear*-benzo-ATP, with the same terminal rings as in ATP but with a central benzene ring built in, thus making it 2.4 angstroms wider than the natural co-enzyme. Then, the last problem was based on a fluorescent, covalently-linked cross section of DNA consisting of five fused rings and having the same or very similar geometry to a hydrogen-bonded pair of DNA bases. It just wouldn't come apart. Dr. Balkrishen Bhat was my sustaining co-worker at Illinois in this venture. The final goal, unreached as yet, was to incorporate the covalent cross section to see, in a replicating cell system, whether you have inserted something that prevents the two DNA strands from coming apart, thus inhibiting replication, especially as in fast-growing cancer cells.

I retired in 1986, and my wife and I were intending to travel. However, the very next year she died of cancer, very fast. I decided it was better for me to get away from the empty house. I was appointed as a Fogarty Scholar in Residence at the National Institutes of Health. During a two-year period, I spent half my time in Bethesda, Maryland, and the other half in Urbana, Illinois. The next position was at the University of California, San Diego, as a Visiting Professor, and then I came to Caltech.

In San Diego, I collaborated with Dr. Leslie Orgel of the Salk Institute. Leslie had also worked for my Oxford mentor, Leslie Sutton, some years after me. We produced three papers together in which we asked questions such as, "If adenosine is made in Nature, what would happen if Nature (and we) were to make and use 3-isoadenosine instead?" I talked about this isomer earlier. "Would the appropriate derivative oligomerize on a poly-U template?" Sure enough, it does so beautifully. Leslie Orgel and I also experimented with other unnatural 3-ribosylpusine derivatives. When I came to Caltech, Jack and Edith Roberts introduced me to a friend of theirs, Peggy Phelps, and we were married in 1992. It was a blessing.

But going back to the very beginning, I was born on September 1, 1916, in Newark, New Jersey. My father was a salesman in New York and my mother was a housewife. I had two uncles who were engineers and two cousins who were also engineers. So I thought perhaps that was what you did when you grew up. I had my own chemistry set and later, in high school, I had a good chemistry teacher. At Lehigh I started in chemical engineering but I didn't like the many empirical constants they kept throwing in when they had trouble matching an equation to actual behavior. I shifted to chemistry.

As of my heroes, Roger Adams was my hero. He was still fully active when I got to Illinois although he was tied up with the war. In turn, he was to become General Clay's science advisor in Germany and General MacArthur's science advisor in Japan. At Oxford, I appreciated Robert Robinson. And at Columbia, Harold Urey was a hero.

I am very happy about my pupils and I don't like to single out special ones among my 120 Ph.D. students and 91 postdoctorates. They are well on their way, quite a number of them, and I try to keep close contact with them. There is an annual Nelson J. Leonard Lectureship at Illinois that is sponsored by my students and former colleagues. The lectureship is not in one particular subject but it is in chemistry, biochemistry, or chemical engineering, consistent with my notion of cooperation and collaboration. Now, to my loyalty and appreciation of the places I have been: Lehigh, Oxford, Columbia, and Illinois, I am happy to add Caltech.

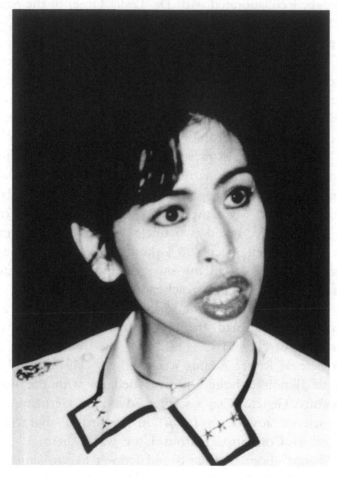

Princess Chulabhorn, 1999 (photograph by I. Hargittai).

20

PRINCESS CHULABHORN
OF THAILAND

Professor Dr. Her Royal Highness Princess Chulabhorn Mahidol (b. 1957 in Bangkok, Thailand) is President of the Chulabhorn Research Institute in Bangkok and Professor of Chemistry of Mahidol University. She is the youngest daughter of King Bhumibol Adulyadej and Queen Sirikit of Thailand. One of her interests and one of the main research lines of her Institute is bioactive natural products. To quote from the introduction of one of her papers:

"Thailand is uniquely located to represent the fauna and flora, which characterizes the biogeographic province of Indo-Burma. A number of eastern Himalaya temperate *taxa* penetrate south into the northern mountains of Thailand while the southern part is evergreen forest thus making this area one of the richest floristic regions of the world. It has been estimated that the vascular plants in Thailand include at least 10,000 species of about 1,763 genera from 245 families. The number of alkaloid-containing plants is estimated to be only about 266 species of 176 genera in 67 families based on the Thai plant names and parts of the uncompleted flora of Thailand." [Mahidol, C.; Prawat, H.; Ruchirawat, S. "Bioactive natural products from Thai medicinal plants", in *Phytochemical Diversity: A Source of New Industrial Products*, Royal Society of Chemistry, England, 1997, pp. 96–105.]

We recorded a brief conversation with Princess Chulabhorn in the Chulabhorn Research Institute in Bangkok, on July 19, 1999.* Prior to the conversation we made a tour of the Institute, which employed about 200 people, and was undergoing a large expansion program at the time.

As the time of the Princess's arrival was approaching, television crews descended upon the reception room and the evening news of all Thai channels showed the beginning of our meeting with Princess Chulabhorn.

Chemistry is not a very common profession for a princess. What turned you to chemistry?

I am probably the only princess who is a chemist. It is the idea of my parents that Thailand is a growing country, we are developing very fast, and we would need a scientist who would be far-sighted and understand every step of the industrial evolution of what we are going to develop. My parents had choices for me, they suggested physics or chemistry and I preferred chemistry. In school, at high school level, I used to hate chemistry. Then I took an entrance exam at the Kasetsart University here in Bangkok and I was accepted. At Kasetsart University I had a wonderful advisor who is a chemist and who made chemistry so fascinating for me. Her name is Professor Krisna Chutima. I did my B.S. degree at Kasetsart University, which is about ten minutes from the Chulabhorn Research Institute. Then I did my Ph.D. at Mahidol University, also in Bangkok. Mahidol University is famous for its medical faculty.

Then I had my postdoctoral training at the University of Ulm in Germany. There I was doing something completely different. As a chemist you know chemistry, of course, but we don't know what the chemicals do to our living system. Since I wanted to study our living system, in Ulm I turned to genetic engineering.

Was there any particular system you wanted to study?

It was not any particular system. From Ulm I went to study at Tokyo Medical School, which changed my life a lot. So as a result of my studies I understand chemistry and pharmacology and medicine but it took me many, many years.

It couldn't take so many years because you're still very young.

No, I'm not very young, I'm 42.

---

*The interview was conducted by I. and M. Hargittai. In part, this has appeared in Hargittai, I.; Hargittai, M. *The Chemical Intelligencer* **2000**, *6*(1), 25–28.

Princess Chulabhorn (courtesy of the
Chulabhorn Research Institute, Bangkok).

The contact between us was made by Professor Herbert C. Brown. How did you meet him?

I met him when I organized my first congress, which is called Princess Chulabhorn Scientific Congress. Professor Herbert Brown was kind enough to come and give a keynote lecture. Now we are organizing the Fourth Princess Chulabhorn Scientific Congress.

According to the mission statement of your Institute its goal is to "Improve the Quality of Life".

This is the policy of my father. He is an irrigation engineer. All his life he did everything to improve the quality of life of the poor people. At the time I was setting up my Institute I thought that I should follow in his footsteps and apply chemistry and biomedical research to the betterment of the quality of life of the less fortunate people.

What has been your special interest in chemistry?

It's been natural products chemistry. I'm always fascinated when old people are telling me about the curing and healing effects of various plants. As a chemist I couldn't just take their word for granted. I have to see it for myself what is in the plant and what is the active ingredient.

Are there plants in Thailand that can't be found anywhere else?

We have a lot of medicinal plants here in Thailand but I'm sure they can also be found around us as well as other plants although they may not have the chemists and the necessary instrumentation to investigate them.

Touring your Institute we observed that you are well equipped with NMR, ESR, IR, and other instruments.

This is from the kindness of the Federal Republic of Germany and also of Japan. They gave us a start by equipping us with the necessary instrumentation.

Is your Institute much better off than the universities in Thailand?

Yes and we accept students to come here to learn.

Do foreign scientists work here in your Institute?

Right now there isn't any but now and then we do have foreign visitors working with us here in the lab. We also send our scientists abroad to learn, all over the world, to Europe, to the United States, except to Japan because of the personal experience I had: the problem of learning the Japanese language, which makes such a stay very difficult.

Where did you learn your English?

I learned my English in Thailand but the teacher was English.

What courses do you teach?

I teach oncology, toxicology, and also biochemistry.

We've heard that you also teach chemical warfare to the Air Force personnel.

Yes.

You are wearing an Air Force uniform.

I'm employed by the Air Force.

Do you have a rank?

I'm Air Vice-Marshall.

Meeting the Princess (courtesy of the Chulabhorn Research Institute, Bangkok).

Is it very high? It sure sounds high.

It's near the top but not the top.

What do you teach in chemical warfare?

The purpose is for my students to know the chemicals, the biological agents, and to observe the circumstances around them and thus protect their lives. I'm not teaching them to kill somebody.

Chemistry has acquired a bad image all over the world.

Why? You see, now I am interviewing you.

People very often associate chemistry with pollution, with drug abuse, with chemical warfare and tend to forget that chemistry has played a fundamental role in the improvement of the quality of life. Chemistry provides more choices too but it is society that must decide what to use chemistry for. What is the image of chemistry in Thailand like?

The image of chemistry in Thailand is very good. People who study chemistry receive a lot of prestige because the subject itself is difficult and not everybody can take chemistry. I always tell people that everything around us is chemistry. We chemists do not destroy but try to protect our environment, and our work is meant to have a positive impact.

Being a Princess, when you move among scientists, when you teach students, when there are scientific discussions, even debates, can you tolerate contradiction?

Yes, I've tolerated contradiction for many years.

As member of the Royal Family you must be having a lot of duties outside your Research Institute and your university teaching. How can you cope with all your duties?

It is the kindness of my father and my mother that they see my role as a chemist more important than my role as a princess. So I don't have to go to all the ceremonial functions, only to some special ones, and my major work is at the Chulabhorn Research Institute and teaching at various universities.

Do you have children?

I have two children, two daughters, 17 and 15.

Will they become chemists?

I don't think so. The older daughter shows a lot of artistic talent. She paints beautifully and I don't think she likes chemistry.

Interviewing the Princess (courtesy of the Chulabhorn Research Institute, Bangkok).

In high school you didn't like it either.

It would be a great pleasure for me to see them become scientists, especially chemists, but we'll have to see what they would like to do.

Would you care to mention any of your heroes, role models?

I respect and have very warm feelings for Professor Herbert Brown. He is like a grandfather to me, he has the image of a perfect chemist. He taught me so much during the conference, not about chemicals but about being a chemist and about being patient.

Does your role encourage women in Thailand to go into science?

I don't think so. Nowadays teenagers prefer something easier to something that has to be a degree that is hard earned. I don't think they like that but I try to encourage them anyway.

Do you have any interest outside science?

I was a concert pianist before I became a chemist.

It's not too uncommon that interest in music and chemistry occur together. Just to mention two examples from among our previous interviewees, Jean-Marie Lehn had a dilemma whether to go into music or into chemistry. Manfred Eigen was planning to be a pianist when he was a teenager. Then he got drafted into the Wehrmacht during the last years of World War II and lost valuable years of practice. So when the war was over he decided to go into chemistry.

I think that music is very close to chemistry.

Do you have a favorite composer?

Chopin is my favorite composer.

Thinking about this interview, was there any question that you had anticipated and we didn't ask?

I didn't imagine anything. I accepted that you would come because Professor Brown had told me that it would have a positive impact. In our religion, in Buddha's teaching we learn not to anticipate anything, either good or bad, but to concentrate on the present.

Linus Pauling (courtesy of the MRC LMB Archives, Cambridge).

21

LINUS PAULING

L inus Pauling (1901–1994) was one of the greatest scientists of the twentieth century. He received two unshared Nobel Prizes. The first was in Chemistry "for his research into the nature of the chemical bond and its application to the elucidation of the structure of complex substances". The second was a Nobel Peace Prize for 1962, in 1963. There was a brief interview with Linus Pauling in the first volume of the *Candid Science* series.[1] The narrative below is based on the video recording by Clarence and Jane Larson with Linus Pauling in 1984.* We appreciate Dr. Zelek Herman's assistance in checking some facts and collecting some of the illustrations for this entry. Dr. Herman of Stanford, California, was a long-time associate of Linus Pauling.

I was born in Portland, Oregon. My father, who was a druggist, was born in Missouri, of German ancestry, and my mother was born in Oregon, of English and Scottish ancestry. I started school in a small town in Eastern Oregon, Condon, Oregon, and I was interested to notice the other day that in 1906, when I was going to school in Condon, there was another boy there, somewhat older, William P. Murphy, who received the Nobel Prize in medicine, some years later.[2] We were a small town of 500, and it is still 500 in population. My father, mother, my two sisters — a little younger than I — and I moved to Portland, where he continued to work

---

*"Larson Tapes" (see Preface). A short portion will appear in Hargittai, B.; Hargittai, I. *Chemistry International* 2005.

as a druggist. Then, my father died suddenly when I was 9 years old. Although I have some memories of him, I don't remember very much, but to some extent he influenced my life. I do know that he was very much interested in the fact that I was a good student and a vigorous reader and a thoughtful boy. Even though I was only 9 years old, he recognized that I had a special interest in learning.

When I was 10 or 11, I became interested in insects, and I got books from the library about insects. When I was 12, I got interested in minerals, and again got books from the library, and I made tables for my own use. I made some efforts to collect some minerals, not very successfully because I didn't have transportation, and our valley was not an especially good place for finding minerals. Then, when I was 13, in my second year of high school, a boy of my own age, Lloyd Jeffress said to me as we were walking home one day, "Would you like to see some chemical experiments?" I said yes, and he said, "Come on in," and I went to his home. He was an only child, and he carried out some experiments, which impressed me immensely. I became very enthusiastic about chemistry. That same day, I found a book that had belonged to my father about elementary chemistry, and I immediately repeated some experiments with materials around the house. And from there on I was a chemist.

The next year I had a year of chemistry, high school chemistry, and I followed that by half a year of physics. Also, the chemistry teacher, William V. Green, in Washington High School in Portland then gave me special supervision the following year. So I continued to carry out experiments in the high school laboratory, and I stayed after school on certain days and helped him operate the bomb calorimeter with which he determined the heat value of coal and oil used in the Portland schools. I was much impressed by Mr. Green, this teacher, and was also impressed by something that happened with the physics teacher whom I admired. He was very good to have worked out equations for Mr. Green showing how he should correct for the heat loss. It struck me as something unusual that it was possible to carry out the theoretical treatment of a problem.

I left after three and a half years without my high school degree, in the middle of the year, because I had begun in February. I went to Oregon Agricultural College although the College didn't like students to come in the middle of the year, they preferred them to come in September. But I didn't want to miss a year for that reason, and I had enough credits to be admitted even though I didn't have a high school diploma.[3] So I went to Oregon State [the Oregon Agricultural College later became the Oregon State University] to study chemical engineering. An interesting event happened about a year before. It was in connection with Lloyd Jeffress, who later got his Ph.D. in psychology in Berkeley and became head of the psychology department in Austin, Texas. At this time, when we were 15, my grandmother in Oswego said to me, "What would you like to be when you grow up?" I said, "I'm going to be a chemical engineer," but Lloyd immediately said, "No, he is going to be a professor."

I studied chemical engineering at Oregon State. First, I had to study at Oregon State because not having any money, it was the cheapest school for me to go. There was Reed College only a couple of miles from where my mother lived, but I knew you had to pay tuition there and it didn't seem that there was much chance for me to go there. Also, I didn't know that there was any profession that would involve chemistry, except chemical engineering. At this time, 65 years ago, chemical engineering was to a much greater extent taught in a practical way. The first two years the chemical engineering students were combined with the mining engineering students. I had four years of mathematics at Washington High School and time went by without my getting additional training in mathematics. So I got some instructions in mining engineering, blacksmithing, and making of drills, too. After my sophomore year, I was working as a paving plant inspector in the summer in southern Oregon.

When September came, my mother told me that we just didn't have enough money for me to return to college. She needed to continue to get

Linus Pauling (during the conversation with Clarence and Jane Larson; photograph taken from the video recording).

some support from me. So I didn't return to college, but after a month, I was offered a job as an assistant instructor, full-time, in quantitative analysis that had a very heavy load, teaching the sophomore courses in mining engineering and chemical engineering, and a more elementary course to a large number of students in agriculture. I liked the quantitative analysis course, the precision of it appealed to me.

I was 18 years old in 1919. A very interesting event occurred during this year when I was teaching. I had a desk in the chemistry library. No one else came into the chemistry library, but the journals arrived and I read. I had a little spare time despite the heavy teaching load and I read the journals. The *Journal of the American Chemical Society* came with a couple of articles by Irving Langmuir on the shared electron pair theory of the chemical bond. He referred back to 1916, to G. N. Lewis, so I got out the 1916 copy of the journal with G. N. Lewis's paper, and I gave a seminar on chemical bond theory of the shared electron pair. It was the only seminar that was given that year. A chemistry seminar was not a very common thing, and I continued to be interested in the chemical bond ever since.

A couple of years later when I was a senior, I applied for a teaching fellowship to Berkeley and some posters came from Throop College, the California Institute of Technology, it was just changing its name. I had met a couple of young fellows who had flunked out of Throop and moved to Corvallis, so I knew about Throop College that was down there in Pasadena. The head of the Chemistry Department thought that it might be a good place for me, so I applied there and to Harvard and Illinois and perhaps to one or two other universities. I received an offer of appointment from Harvard, a half-time instructorship, which would require six years for the Ph.D. That didn't appeal to me very much; moreover, I was timid about going so far away from home. The expense of travel was also significant. I received an offer from A. A. Noyes from Pasadena with a request that I decide immediately. This was not proper; in a few years the universities got together and agreed that there would be the same deadline for all universities. But I hadn't heard from Berkeley. So I thought I better take the job that was offered to me, and I wrote accepting the job at CIT (California Institute of Technology) and wrote to Berkeley and Illinois — I'd written to Harvard turning it down before — withdrawing my applications.

In January 1983, I gave the Hitchcock Lectures in Berkeley; I was Hitchcock Professor. Three younger members of the Chemistry Department spoke to me on that occasion. I'd been around Berkeley from time to time, every year almost, from 1922 on, I stayed there for a few hours on my way to Pasadena in 1922. I was visiting lecturer in chemistry and physics for five years, coming up every spring; the same time [J. Robert] Oppenheimer was coming down to Pasadena every spring. But no one had told me this story until a little over 60 years later. The story is, in the spring of 1922 G. N. Lewis was looking at the applications from applicants for a teaching scholarship of a pile of 20 or 30. He came to one of them that said, Linus Pauling, Oregon Agricultural College, and Lewis said, I've never heard of that place, and down the application went.

I was back in Berkeley, 7 years later. In 1929, I received this offer to come up to Berkeley every spring as a visiting lecturer in chemistry and physics; so it only took 7 years to reach them that stage. In fact, in 1926 or 1925, G. N. Lewis came to Pasadena, and I learned only a few years ago that he had come down to offer me a job as an Assistant Professor, but A. A. Noyes wouldn't let him. Lewis had been with Arthur Ames Noyes at MIT, where Noyes set up a research laboratory of physical chemistry. Lewis, after he got his Ph.D. with Richards at Harvard, came over to MIT as the assistant director of the research laboratory with Noyes. Noyes may have been involved with Lewis's becoming Dean of the College of Chemistry at Berkeley back in 1911. Noyes, of course, ran much of chemistry in the United States just as George Ellery Hale ran much of science in the United States. Noyes and Hale were very close together in running things during the period of 1915–1920: the National Research Council, getting the Academy building.

I've been very fortunate during my life in that several times something has happened that, in retrospect, I see, turned out to have been just the right thing to have happened. For me to have gone to Pasadena in 1922 was really most fortunate. I don't believe I could've got better training or to work under better circumstances anywhere in the world than there, in Pasadena. I came up to Berkeley, only 7 years later, 4 years after I got my Ph.D., because they needed to be brought up to date on chemical bonding.

There were remarkable teachers in Pasadena and it was a small place, a total of 300 undergraduate students and 30 or 40 graduate students and 50 faculty members back in 1922. The man, with whom I did my doctoral work, Roscoe Gilkey Dickinson was the first person to get a Ph.D. from the California Institute of Technology. He got it in 1920; then there was a couple every year until 1925 when quite a number got it in physics and chemistry.

Linus Pauling with Robert Marsh in 1960 (courtesy of Robert Marsh, Pasadena, California).

The teachers were marvelous and the classes were small. Richard Chace Tolman was one of the outstanding teachers there at CIT. I studied the two courses he gave. One was on the nature of science, a very interesting course. He may have given it only once, in the year 1922/1923. The other course was statistical mechanics; after taking it one year, I attended it the next year and the next. A number of years later I went in; I thought I would audit the course, but as soon as I came in, he turned to me and said, "Stay out," so I stayed out. He thought my presence would handicap him presenting the subject to the students who didn't know anything about it. I learned a great deal from Tolman.

My first scientific paper[4] was published in 1923 on a crystal structure. By 1925, I was publishing papers on the old quantum theory. Tolman and I published a paper in 1925 on the entropy of crystals and supercooled liquids; this was a publication in quantum mechanics.

The first book that I wrote was *The Structure of Line Spectra*. I wrote that in collaboration with Goudsmit. Goudsmit and I met in Denmark when I was there in 1927. He and I worked together for a month tackling the problem of the theory of hyperfine structure of spectral lines. I translated his thesis from Dutch to English and used that as Chapters 4, 5, and 6 of the book. I wrote three chapters on sort of an introduction to quantum theory and quantum mechanics and then three chapters on the vector model of the atom that Goudsmit had been involved in developing with the spinning electron, of course, which he and Uhlenbeck had discovered. Then, there were four chapters more that I wrote partially with material that Goudsmit had sent me from Ann Arbor, where he had in the meantime become a member of the Physics Department. That book came out in 1930.

Then, in 1935, one of my first graduate students in theoretical chemistry, Bright Wilson — who later went on to Harvard — and I wrote together an *Introduction to Quantum Mechanics*. That came out in 1935 and for 48 years it continued to be sold by McGraw Hill without any changes. Never once was it revised. Then, last year, they decided that they weren't selling enough copies to keep it in print. So it didn't quite make 50 years. For ten years it was the oldest unrevised book that McGraw Hill kept in print. Then my third book was *The Nature of the Chemical Bond*. My Ph.D. work at CIT was on the determination of the structure of crystals by the X-ray diffraction method. CIT was the first place in the United States where a crystal structure determination was made by X-ray diffraction. [C. Lalor] Burdick and Ellis, who came with Noyes to Pasadena to carry out such studies and then Dickinson's doctoral thesis was on X-ray diffraction and then he was in charge of the X-ray laboratory. This was just fine for me with my interest in the chemical bond.

In 1924–1925, I was in charge of a dozen freshmen who had been selected from a total of 125, as being probably more able ones, to an honors section. During half of that year, in their freshman year, these students carried out small researches. Noyes suggested some problems that they might attack, and I suggested some. Only one of these investigations developed into a publication, which was, of course, the first paper by the student. It was on the structure of the alloys of lead and thallium. The student was the son of our family's physician in Pasadena. He worked through the summer; after his freshman year he continued to come to the laboratory to finish his investigation, which was published in the *Journal of the American Chemical Society*. His name was Edwin McMillan. A little later he got his bachelor's degree, went to Berkeley, got his Ph.D., and became Lawrence's successor.[5]

I remember when I came to Berkeley to lecture in 1929 or 1930, Ernest Lawrence had arrived and I became well acquainted with Lawrence. Then, in 1931, I was visiting lecturer at MIT for a month, and I was asked to become chairman of the Chemistry Department — Slater had become chairman of the Physics Department, but I preferred to stay at California Institute of

Technology. When Ernest decided to marry Molly Blumer in 1931, I was an usher at the wedding in New Haven. I have a box with clippings about their wedding. Years later, Molly made some statements in an interview that rather surprised me. One of them was that Ernest felt strongly that there should be no more wars after the Second World War; I knew that he was a very patriotic man. She had been trying to get Ernest's name removed from the Lawrence Livermore Laboratory. The University turned her request down, so she got a member of the California legislature to introduce a bill to this effect. Her argument was that Ernest felt strongly that the existence of nuclear weapons required that we give up war between nations and that he himself would have not liked to see his name attached to nuclear weapons.

I was fortunate that when I went to Pasadena, I took courses of advance mathematics from people such as [Harry] Bateman and Charles G. Darwin, the original Charles Darwin's grandson. Darwin was an interesting lecturer and gave lectures, but Bateman was a great mathematician and I enjoyed his courses very much. Sometimes their mathematics was beyond me, my main interest was chemistry. I got my Ph.D. with major in chemistry in 1925; I signed up for a minor in physics, but when I got my diploma I saw they'd given me minors in physics and mathematics. In 1925, as I was approaching the Ph.D., I applied for a National Research Council fellowship, which was the thing to do. It required that one moved from the university where one got the Ph.D., so I picked Berkeley. They had no X-ray apparatus at Berkeley; G. N. Lewis had written me that they would get an X-ray when I came. Noyes said to me, "Here you have so many experimental results, structure determinations that you haven't written up for publication yet, I think it would be wise if you were to postpone going to Berkeley in order to write those papers." I said, all right. Then he said, there is something new, a Guggenheim Fellowship has come out and the head of the selection committee is coming here the next week to select a few people for the fellowship, even though there hasn't been any announcements, and you should meet Dr. Aydelotte [Frank Aydelotte of the Guggenheim Foundation], so we had lunch and dinner. Dr. Aydelotte said that they have decided that they wouldn't give me a Guggenheim Fellowship this year, but I should apply for one the first formal year applications were made. I just did what I was told, didn't really think very much for myself. That was one of the events of my life that was most fortunate.

I got married after my first year in graduate school. This was another event in my life that was most fortunate, that I got married to the right person, who was smart enough to pick me out. So we were married for 58 and one-half years. I went back for a year, for my first year as a graduate student. I was in Pasadena by myself, but at the end of the year we were married, so my wife was with me and protected me for the rest of my life, and enabled me to devote myself effectively to my scientific interests and ultimately influenced me to do more than my scientific interests. So then I applied for the Guggenheim Fellowship. I had been in Pasadena about four months and should have gone to Berkeley. Dr. Noyes said, "It isn't highly worthwhile for you to change from one laboratory to another for just a few months. The Guggenheim Fellowship would be given in April for next year. It would be better if you just stayed in Pasadena." So I said all right. I didn't have to move. He said to me, "The National Research Council requires that you leave so you can't stay in Pasadena. So why don't you resign from it and go directly to Europe and the Institute will advance you some money to take care of your expenses until the Guggenheim Fellowship comes through." I said that sounds like a good idea, we will go to Europe in 1926. I wrote to the National Research Council resigning my Fellowship. I got a very critical letter saying that here they have wasted one of their fellowships on a person who is only using half of it and it was quite improper. Well, of course, Noyes had been involved in setting up the Fellowship and I just did what he said.

He was keeping me from going to Berkeley. And other people knew this, I didn't. So my wife and I went to Europe. We left about the first of March and got there the end of March. We had the month of April in Italy. Dr. Noyes has planned out where we should go, and spend time in different parts of Italy. When we arrived in Munich, I began working with Sommerfeld. This was really fortunate, too. I stayed a year then, and then spent seven months traveling, visited Bohr's Institute, and then went to Zurich to meet with Schrödinger and Debye.

I went back in 1930 for 6 months to Europe, then I didn't get back until 1947. So the Guggenheim Fellowship didn't come through all right. With the Guggenheim Fellowship one was supposed to enclose a statement that the institution you were going to would accept you. I had written to Sommerfeld, whom I had met when he visited Pasadena and to Bohr whom I had also seen when he was in Pasadena. My memory is that I wrote on some yellow lined paper just by hand to each of these persons. I never got an answer from Bohr, but Sommerfeld answered and said yes, it would be all right, so I went to Munich. I learned much more by going to Munich than I would have learned by going to Copenhagen. Sommerfeld was a marvelous teacher, his students are outstanding, of course: Debye, Heisenberg, and Pauli. Heisenberg and Pauli took their doctorates with him. You can count how many of the theoretical physicists were their students. Various other physicists were also there. The main thing was that Sommerfeld was lecturing on wave mechanics. I arrived in Europe, just the same month, I think, when Schrödinger's paper came out, and his other papers kept coming out during the year I was there, and Sommerfeld was lecturing on this subject. It was really marvelous.

When I finally went to Copenhagen, I spent some of my time with a couple of Japanese physicists, who were working on the problem involving crystals. I was able to assist them and that worked. But most of my time was spent working on hyperfine structure problems with Goudsmit, who thought that I was a more theoretical man, better versed in mathematical physics than he was. I saw Bohr only a couple of times during the whole time I was there. There weren't any lectures being given that were at all comparable to what was presented in Munich. Of course, it was much better

Linus Pauling with his porkpie hat (photograph by and courtesy of Zelek Herman, Stanford, California).

for me to have learnt German than to have learnt Danish, too. So, this was a great time in Munich.

When I applied, in my application I said that I wanted to understand the electronic structure of atoms well enough to be able to apply this knowledge to chemical problems, to study the structure of molecules and crystals. Sommerfeld suggested a problem to me, which I worked on for a little while, which was the value of anomalous G-factor of the electron, but I didn't get anywhere with that problem.

I had a problem that I was interested in, the motion of a diatomic molecule, hydrogen chloride, in crossed electric and magnetic fields. I worked on that and published a paper in *Physical Review*. But something then happened that was really fortunate. I was reading *Zeitschrift für Physik* and I came across a paper by Gregor Wentzel. He was in Nuremberg. He had invented a way of treating atoms with many electrons, the sort of perturbation method. He evaluated the screening constants (doublets) that Sommerfeld had discovered in the course of him developing the Sommerfeld–Wilson quantum conditions in the old quantum theory. And the values that he got didn't agree with the experiment — that was the difficulty. So I read this paper with great interest, because of my wanting to do something with complicated atoms, atoms with many electrons, and I thought that since he hadn't gotten good results, I might be able to apply it — this method. I didn't just read the paper. When I came to an equation, I then developed the next equation myself. Pretty soon my equations were different from Wentzel's equations. I found at one point, when he was carrying out his expansions at the inverse powers of the atomic numbers, he just decided that there was some quantitative quantum number that he didn't need — that would be the same quantum number. This was perhaps a rather sensible assumption to make, but it was the wrong assumption. I expanded this, and my theoretical values of the screening constants agreed with Sommerfeld's empirical ones. So I took this paper for Professor Sommerfeld to see, and he said that you better show it to Wentzel. And Wentzel didn't have anything to say except that it was right. So it was published in the *Zeitschrift für Physik*. And then I went ahead, using this technique to determine ionic radii and f-values, to determine X-ray scattering powers and diamagnetic susceptibilities of atoms and ions and electric polarizabilities of atoms and ions all through the periodic table. For a year or two, I was able to exploit this treatment — approximate quantum mechanical treatment — of complicated systems very effectively.

Then I wrote papers on the principles of determining the structures of complex silicates and other complex crystals and on the theory of the chemical bond, covalent bond, in 1927. In that paper, I said that because of the resonance phenomenon, the four bonds formed by the carbon atom turned out to be equivalent, not different as suggested by the s and p orbitals. I published a note about that, a 2 or 3-page paper on that and some other results in 1928. It wasn't until 1931 that I published a detailed discussion about that, because my first treatment was so complicated that I felt that I could not convince anybody else. I found a way of simplifying it. In December of 1930 that came out as my first long paper on quantum mechanics of the chemical bond. Somebody pointed out to me the other day that times have changed since then. He had the paper by me, where it is said, "Received February 10, 1931, accepted March 27, 1931." By this time I had made such an impression on the Editor of the *Journal of the American Chemical Society* that when this paper came in he just sent it off to the printer without sending it to the referees.

Sometimes it takes longer. I wrote a paper on radioactive carbon-14 produced by bomb tests and sent it to *Science* and got it back with comments by a referee. I revised it, sent it back again and got it back again with comments by the referee. Twice I sent it back and twice again I got it back. The referee said that this estimate of 600 megatons of nuclear explosions in the atmosphere is an astronomical exaggeration. All the calculations that I made were based on guesses that I made, because the relevant information wasn't released. There was a paper by Libby about carbon-14 where there were a few numbers, then I calculated back and tried to estimate quantities. I ended up with the number 600, and I said that I refuse to make any more changes in this paper. [Philip] Abelson then printed it a year after I sent it in. But, finally it was published.

Of course, years later the information came out and my estimates were just right. In fact, in 1947 or 1952, I had forgotten which year, the government brought out its first statement about biological effects of fallout radioactivity. When this statement appeared, I was quite pleased.

Wiesner called me and said that he would like to know just how I did my calculations. On my way to Europe I sat down and wrote out just my derivations of all of the quantities. And then the government's report came out, and they had used my calculations, but the numbers were not the same since they had referred to the United States instead of the world as a whole. I had estimated that the population would continue to grow, while they assumed that the population of the United States would stay 150 million for the next thousand years. So their number looked smaller. The assumptions that I made, they accepted.

When I was in Munich, I received a letter in the spring of 1927 from Noyes saying that the Institute was offering me a position as Assistant Professor of Theoretical Chemistry and Mathematical Physics. I wrote back accepting. When I got to Pasadena in the fall, I found that I was Assistant Professor of Theoretical Chemistry. Noyes had managed to get the Mathematical Physics dropped from my title, because I think he thought the Physics Department was getting more notoriety. A little later he asked if I wanted to become Professor of Organic Chemistry — he wanted to build up organic chemistry. Well, I had one elementary course in Organic Chemistry in my junior year in Corvallis. I didn't like it. I didn't think much of organic chemistry. I had made big contributions to it, of course, with the chemical bond theory, but I still didn't like it. So I refused. I said that what I would like to be is Professor of Chemistry. He said all right, you can be. Noyes instituted most of the fundamental principles on which the California Institute of Technology was built upon and he managed to get [Robert A.] Millikan to come in as a front man to hobnob with the rich people and raise money, while he, Noyes, determined the academic principles/policies, one of which was that women shouldn't be admitted. He was a bachelor. He thought it was just a waste of energy to train women in science. It took a long time to change that policy. When I left in 1964 there were women graduate students, but no undergraduates. Now there are a number of them.

Of course I was interested in physics during all of this period and occasionally wrote a paper about some physical problem and was interested even in nuclear physics, but only starting in 1965 had I published papers in nuclear physics, nuclear structure. I was interested in inorganic chemistry and then in organic chemistry, almost entirely from the structural point of view, the question of how the properties of a substance are determined by its structure. This could be its crystal structure, molecular structure, electronic structure of the atoms which determine the other structures.

My work with nucleic acids came about through the natural outgrowth of my interests in molecular structure. First I worked on crystals — inorganic substances, simple and more complicated ones. In 1930, when I was in Germany, I learned about a new technique that Herman Mark had invented, electron diffraction by gas molecules. I asked Mark if it was

354 Hargittai & Hargittai, Candid Science V

all right if I were to build an apparatus like that. He said all right. He wasn't going to go ahead with it, he was working with I. G. Farbenindustrie — he was working mainly on practical problems. He even gave me the plans for their apparatus. So I got a graduate student to work with the shop building the apparatus, and we began determining the structures of organic compounds. During the early 1930s we got a great deal of experimental information from a couple of hundred of organic molecules, and the theory was developing rapidly. I felt pretty satisfied about the organic compounds just as I felt satisfied about the inorganic compounds.

I thought here is an interesting substance — hemoglobin. I didn't know much about biology, but I knew about hemoglobin. It had been found a few years earlier, in 1927. The molecule contains four iron atoms — about 10,000 atoms altogether, but four iron atoms in the heme groups. I have heard about this sigmoid equilibrium curve of oxygen (O_2), so I applied physical chemistry and structural chemistry to that — I worked out a theory of the oxygen equilibrium curve. That was my first paper on proteins. Then I thought, nobody knows how the oxygen molecules stick to the hemoglobin molecules. Some people say it is sort of an adsorption onto this large molecule. Other people say there is a chemical bond formed. Oxygen has two unpaired electrons, it is paramagnetic. You can pick up liquid oxygen by a magnet — liquid oxygen will hang between the poles of the magnet. I knew that. I knew that G. N. Lewis, back in the 1920s, interpreted measurements of the magnetic susceptibility of solutions of liquid oxygen and liquid nitrogen to show that there is an equilibrium between the paramagnetic O_2 and diamagnetic O_4. He had determined the equilibrium constant, the standard free energy and standard free enthalpy of the reaction. Very clever of G. N. Lewis to have done that. He discovered O_4, the dimer of O_2. So, I thought, why don't we measure the magnetic susceptibility of oxy-hemoglobin? It will be paramagnetic due to the oxygen molecules or at least there will be a paramagnetic component.

I had been getting some support from the Rockefeller Foundation for 2–3 years already. I had applied to them for some money to work on the structure of sulfide minerals. They gave me $5,000, then the next year $10,000, then $15,000 the following year. So I said that I want to study the magnetic properties of oxy-hemoglobin. They sent me $50,000. And a little suggestion that they were not really interested in the sulfide minerals, but were interested in biology. So I had a student, Charles Coryell, he had taken his Ph.D. and came to me as a postdoc fellow. He and I set up an apparatus, got some blood, and measured oxy-hemoglobin. It was diamagnetic, which showed that you had chemical bonding, but the hemoglobin without the oxygen was strongly paramagnetic, and I hadn't predicted that. This was one of those rare occasions when something has come along due to an experiment that I carried out that was a surprise to me. But the change in the magnetic properties of the iron atom permitted us to gain great insight into the arrangement of the other atoms around the iron atoms in hemoglobin. Moreover this technique of measuring magnetic susceptibility permitted us to measure equilibrium constants and rates of reactions, so over the next five years my students and I published 15 to 20 papers on hemoglobin and hemoglobin derivatives, and the method was also then used in Sweden to study heme compounds and iron proteins.

Then I thought, what about the rest of the hemoglobin molecule? [William] Astbury in England was making X-ray diffraction photographs of hair and finger nail and other people, too, starting in Japan and Germany, had made photographs of silk and wool. I took some of these photographs in 1937 and tried then to find the structure in way of coiling the polypeptide chain. Other people were trying, too, but without success. I thought, "I think I knew a lot about these atoms and how they combine with one

Robert B. Corey (courtesy of the MRC Laboratory of Molecular Biology, Cambridge).

another, but the structures that I have been predicting don't seem to be the right ones, so there must be something that I don't know about proteins." Nobody has ever determined the structure of an amino acid or a dipeptide, a simple peptide. So why don't we go ahead and do that. The Rockefeller Foundation gave us money and Robert Corey has just come that summer, in 1937, to work with me. I talked with him about this problem, which interested him. We decided to go ahead, and for 10 years at our institute with a good number of different people involved in it, we determined these structures for about ten amino acids and several simple peptides. Nobody else in the whole world had turned out a single structure for any of these fundamental substances during this whole period.

The first peptide was diketopiperazine, which is the cyclic diglycyl. The second structure was glycylglycine, then glycylalanine, then a tripeptide or two. So, ten years later when I was an Eastman professor at Oxford, I thought I better think about that problem again. I failed in 1937, here it is in 1948, eleven years later. There was nothing surprising about the amino acids or the simple peptides. They all had just the structures that I had designed to them back in 1937, but I thought I would try again and I would forget about the X-ray diffraction photographs. First, I don't have them here. But they weren't any good anyway — these fiber-diagrams [Pauling points to his hair]. Second, I'll just forget about them. Suppose, I assume the residues are equivalent to one-another. Back in 1928 I had written a paper about structural principles involving silicates and such substances. One of the principles was that the different kinds of units are to be as few as possible in number. So I'll assume that all the amino acids in the polypeptide chain are equivalent. In a course that I had from Bateman in 1927, it was shown that the most general symmetry operation that converts an asymmetric object into an identical object is rotation around some line in space coupled with translation along it. If you repeat this operation you get a helix. So I said that I haven't looked at any helical structures, I know other people have. I am not sure if I knew that then, but other people have looked at the helical structures for the polypeptide chains, but haven't found them. So I'll look at them. I took a sheet of paper, made a sketch on it, then folded the paper to get those bond angles of the α-carbon correct, and kept folding it parallel, until it came around again and I tried to form a hydrogen bond from this turn to the next turn and couldn't do it. I tried again, putting the folds in a different way, and finally got this hydrogen bond. And that was the α-helix.

The α-helix, drawing by Linus Pauling (from a posthumous publication, Pauling, L. "The discovery of the alpha helix." *The Chemical Intelligencer*, **1996**, 2(1), 32–38).

So I predicted the properties of this α-helix and the X-ray diagram. This showed the repeat in 5.4 Å. Actually that was the pitch of this helix, 5.4 Å. The X-ray diagram showed 5.1 Å and there you have about 5% error and I couldn't see how that was possible. I waited more than a year before publishing anything about it, and in 1950, a paper was published in the *Proceedings of the Royal Society* by Bragg, Kendrew and Perutz[6] on the structure of the polypeptide chain of α-keratin. They described about 20 structures, all of which were wrong. I said to Corey that we better publish about the α-helix and the γ-helix, so we sent off a short note to be printed and started writing a longer paper. But then a little later a paper was published by some others on a synthetic polypeptide that they have been interested in for artificial fibers, poly-γ-methylallyl glutamate. They spun fibers of this synthetic polypeptide and made photographs of them. They were something like the photographs you get from hair. They were different in a very interesting way. The main reflection that gave the 5.4 Å repeat didn't appear. There were two reflections off to the side, and not meridian reflections. They corresponded to a 5.1 Å pseudo-repeat rather than 5.4 Å. On the hair these reflections coalesced to form an arc. And by measuring this arc up like this, there was a 5% error, which had fooled everybody and this showed that the α-helix was right.

In the meantime there was this paper by [William Lawrence] Bragg, [John] Kendrew and [Max] Perutz with all of their structures wrong and ours right, and why. When the peptide groups attach to one another they form what is called a peptide bond in each seal. Even back in the 1930s I said that there is some double bond character to this carbon–nitrogen-part, because of the theory of resonance that I had been writing about. This double bond character requires that those six atoms lie in one plane. So we have to keep those six atoms co-planar. And then you have another group of six and you can rotate around the single bonds of the α-carbon. It makes a very simple problem — with just two parameters with my assumption of equivalence. But of course Bragg and company had a third parameter — rotation around a third bond, which made it a very difficult problem. None of the 20 structures that they described contained these planar peptide groups. So they were in error.

Lord [Alexander] Todd, head of the Chemistry Department at Cambridge was a friend of mine. When I was the chairman of the Division of Chemistry and Chemical Engineering — I became chairman in 1937 — I applied to the Rockefeller Foundation for some money to build up organic chemistry. As Noyes had said 5–7 years earlier, we ought to be doing something about organic chemistry. They gave us a million dollars on a matching grant so we were able to make some appointments. I travelled all around the United States talking to various organic chemists and offered the job to Todd who came for one term with his wife and then we offered him a permanent professorship to be Head of Organic Chemistry. On his way back to England, while he was on the ship, the British got busy and arranged for him to be offered a professorship in Manchester; then he went to Cambridge. So he told me after this α-helix affair that when Bragg read our paper, he rushed over to the Chemistry Department in Cambridge and said, "Here I came over last year to talk to you about the structure of polypeptide chains and you didn't tell me that that group is planar." Todd said, "I am pretty sure I did, I can remember quite clearly saying to you that I had always thought that the carbon–nitrogen bond had some double bond character." Of course, I am sure that is what happened, Todd telling Bragg about the double bond character, but Bragg didn't know enough chemistry to know that this meant that the six atoms lie in the same plane.

We found that the γ-helix seems not to occur in nature. It has a hole down in the middle that you cannot fill up with anything, that it is not big enough to be filled up; it decreases the van der Waals interactions stabilizing the structure. Structures in general don't have holes in them in condensed phases. It doesn't occur. But the parallel chains and anti-parallel pleated sheets also occur. Globular proteins — there have been several hundred of them studied now — all contain these units, the α-helix, the parallel chain pleated sheets and the anti-parallel chain pleated sheets in different parts of the globular molecule. So the secondary structure of proteins, that problem, was solved. This was already after the war.

During the Second World War, I was responsible investigator on 14 contracts from the Office of Scientific Research and Development on various problems. Most of them worked out pretty well, too. I had met Oppenheimer in 1926–1927, when my wife and I were visiting in Göttingen. He had gone to England after getting his bachelor's degree in chemistry at Harvard. He had done a little experimental work with [Percy W.] Bridgman on high pressure physics. He went to England for a while, but he didn't like it. I am not sure that he was a student. Then he went to Göttingen for a couple of years, worked with Born. His thesis was the Born–Oppenheimer principle relating to molecules in which Oppenheimer may have been interested again, because of his background in chemistry. So I saw him there, met him for the first time in Göttingen. Then when he came to Pasadena, my wife, Oppenheimer and I were together a great deal for a year or more and went to the desert with our youngest son, who was with us. In 1942 or 1943 he came to Pasadena and asked me to come to Los Alamos as the head of the chemistry section. The chemistry section existed, and he was having some trouble with it, I judge, and wanted me to come. I decided that I shouldn't do it, largely because of the several contracts that I had with the government.

And then, of course, I thought that I would work out the structure of DNA and started to work on it, rather desultorily, I suppose. Later on my wife said to me, "If that was such an important problem, why didn't you work harder at it?" No doubt that the sequence of nucleotides in, e.g., brewer's yeast nucleic acid overlaps with nucleotides in human beings, because several proteins have been studied. E.g., cytochrome c from brewer's yeast has polypeptide chains with about 100 amino acid residues and about 50 of them are identical with those in human cytochrome c. No doubt, the corresponding nucleic acid, the gene, has a great deal of homology with the human gene for cytochrome c.

My interest in the medical health field developed back in the 1930s when I began work with hemoglobin. In 1936, I gave a seminar talk on hemoglobin at the Rockefeller Institute for Medical Research. Karl Landsteiner, who was a member of the Rockefeller Institute — he had discovered the

KARL LANDSTEINER 1868-1943

S 3.50

REPUBLIK ÖSTERREICH

Karl Landsteiner (Nobel Prize 1930 "for his discovery of blood groups").

blood groups in 1900 and had gotten his job at the Rockefeller Institute 20–30 years earlier — was doing work in immunology and immunochemistry. He asked if I would come with him to his laboratory after my talk, and I did. He said that he hoped that I would think about the experiments he was carrying out and see if I could explain the result he was getting. I started. I got a copy of his book and thought about the problems of interactions of antibodies and antigens, and then in 1940 I developed a theory about the structure of antibodies and the nature of the interactions with homologous antigens, haptens. In 1940, I wrote a short paper with [Max] Delbrück[7] saying that the same type of interactions are responsible for the gene, that just as antibody and antigen are complementary in structure, the gene consists of two strands that are mutually complementary such that when separated, each can act as a template to form a replica of the other one. This was I think the first time that this had been stated.

The template concept went back somewhat earlier. What we had done, my students and I, in the period of 1940–1948, is to prove this without a doubt. We did this by taking chemical groups that we knew all about, such as benzoic acid group, and we put different substituents on it, e.g., a chlorine atom or a methyl group or something else in various places. We also used other groups instead of the carboxylate, e.g., to change the negatively-charged group to a positively-charged one. We were able to show by thousands of separate experiments that the antibody fits tightly around the haptenyl group. That the degree of approximation is to a fraction of the atomic diameter, a fifth perhaps. If there is a positive charge in the hapten, then there is a negative charge in the antibody. If there is a hydrogen-bond-forming group that presents a hydrogen, there is a complementary group that presents the electron pair in the antibody. All of these specific aspects of complementariness we were able to verify by experiments with the antibodies. So then I was able to reach the conclusion that the structural basis of biological specificity is a detailed complementariness in molecular structure. This applies throughout the whole of biology, explaining specificity of enzymes in catalyzing chemical reactions, specificity of antibodies and their combinations with antigens, and the specificity of genes in reproduction.

Then, when I was a member of the Committee on Medical Research that wrote the section on Medical Research in the Bush report in 1945 to President Roosevelt about what the Federal Government should do about science and medicine, I learned about the disease sickle-cell anemia. I immediately had the idea that the disease is not a disease of an organ or a cell — Virchow about 100 years ago in Germany said that there could be cellular diseases — but that it was a disease of a molecule. That the hemoglobin molecule was different from the molecules of hemoglobin in other people and something like an antigen and antibody, that it had two mutually complementary structures. The two hemoglobin molecules would attach then a third one and a fourth one, giving the long chain — long rod. These would line up side-by-side through van der Waals attractions, forming a long needle-like crystal, which grew longer and longer, exceeding the diameter of the red cell and would twist the red cell out of shape making it sticky and causing the cells to aggregate and block the capillaries and to lead to the crisis in the disease. I thought of that while Bill Castle was talking. When he got to the end of this sentence I said to him, "Do you think this could be the disease of the hemoglobin molecule?" He said no, but I asked if it was all right if I looked at some hemoglobin molecules from sickle cell patients to see? He said, "Well, what is there to stop you?" One thing to stop me was where I would get the blood. Another member of the committee was a professor at Washington University in St. Louis; I can't remember his name. He wrote to me that a student of his, a young M.D., Harvey Itano, was interning and was just given an American Chemical Society Pre-doctoral Fellowship permitting him to get a Ph.D., would I accept him? I wrote back saying that yes, I would.

I wrote to Harvey Itano saying that when he comes, I'd like him to bring some blood from a sickle-cell anemia patient and check to see if the hemoglobin is different in them than in blood from other people.

Linus Pauling (photograph by Willoughby and courtesy of Zelek Herman, Stanford, California).

He came the next fall and started by measuring the absorption spectrum that seemed to be the same as other hemoglobins. He measured the oxygen equilibrium constant — that seemed to be the same. For a couple of years he didn't get anywhere. In the meantime we were building an electrophoresis apparatus. You couldn't buy them then, it was too new, so we were building one. When it got built we carried out the electrophoresis experiment and showed that the hemoglobin was different. It turned out that the abnormality is in the β-chains. The hemoglobin molecule contains two α-chains and two β-chains. These sickle cell homozygots had in their β-chains, which have 146 amino acid residues, one residue different. The 140 amino acids in the α-chains were the same. The β-chains have the sixth amino acid residue from the free amino end different. It was different in such way as to change the electrophoretic properties. The normal adult hemoglobin has a glutamate residue there, which carries a negative charge in the side-chain. That is replaced by a valine that has a neutral side-chain, so you lose the electric charge on each of the two β-chains.

Pretty soon we discovered another abnormal human hemoglobin, Hemoglobin D, then another one, Hemoglobin E, then other people began to discover them. There are about 300 hemoglobins known today, but that was the first example showing that a human body manufactures proteins that are different in structure from those manufactured by other human beings. So we called our paper "Sickle Cell Anemia — A Molecular Disease". This was the first time the expression "molecular disease" was used. Of course, there are thousands of molecular diseases recognized now. So time went on, and when Harvey left me after eight years to go back to Bethesda, I thought I would work on something else, and I thought I got into these medical problems and discovered molecular diseases, why don't I look and see if some other disease is a molecular disease. It might as well be an important disease, I could check on cancer or I could check on mental disease. Well, everybody works on cancer, I said to myself, but nobody works on mental disease. This was in 1963 — but of course now everybody works on mental disease. OK, everybody works on cancer, but everybody also works on mental disease. So I applied to the Ford Foundation for a grant. They gave me $650,000 for a five-year project on the molecular bases of mental disease. I got some people together, and then for 10 years we worked on the molecular bases of mental disease. I formulated a molecular theory of general anesthesia during this period. We made some discoveries about schizophrenia, about mental retardation. Nothing extremely important. At the beginning of this period, I left the California Institute of Technology. I had been having troubles with them, because of my political activities, and it finally got to the point where I decided that I would leave. I resigned in November of 1963.

In 1964 or 1965, just after I left the California Institute of Technology, I ran across a work by two Canadian psychiatrists [Abram Hoffer and Humphrey Osmond] who were working in Saskatoon, Saskatchewan, who reported that they have gotten good results with schizophrenic patients by giving them vitamins. One vitamin especially — vitamin B3, nicotinic acid or nicotinamide. I wasn't especially interested since I was thinking of these vitamins as drugs. I haven't had much interest in drugs. Plenty of other people work on drugs to treat diseases. But after a while something occurred to me. This related to the amount of this vitamin that they gave the patients. This vitamin is extremely important. Back, before 1920, thousands of people in the United States and many in other parts of the world were suffering from pallegra and dying from pallegra. It was discovered at about that time that a glass of milk a day would prevent pallegra. In the early 1930s it was discovered in Wisconsin that the substance in milk that prevents pallegra is nicotinic acid or nicotinamide — vitamin B3 — usually called niacin. A little pinch — 5 mg a day — of this substance, or either one of these two substances will keep you from getting pallegra. It is a very powerful substance. Little pinch. Some of these psychiatric patients were being given ten thousand times that much. Fifty grams, fifty thousand milligrams a day — two ounces — and without any side effects.

Linus Pauling giving a lecture at Moscow State University in 1983 (photograph by and courtesy of Larissa Zasourskaya, Moscow).

Nicotinic acid is a vasodilator that causes flushing. If you take large amounts of nicotinic acid for three or four days you will flush, but from then on you may continue to take large amounts without flushing, so that the flushing reaction does not prevent people from taking large amounts. But nicotinamide is nearly as good as nicotinic acid. I thought this was really astonishing that you can have a substance that has a physiological activity over such a broad — ten thousand range fold — concentration. Doctors prescribe aspirin for people with arthritis, and sometimes the amount of aspirin they take is such that if they took five times as much they'd be dead. Many patients die from the toxicity of the drugs that are prescribed for them. Especially of course cancer patients, where the drug is given in amounts as great as possible in the hope of controlling the cancer and sometimes it is enough just by itself to kill the patient.

References and Notes

1. Hargittai, I. *Candid Science: Conversations with Famous Chemists*. Imperial College Press, London, **2000**, pp. 2–7.
2. William P. Murphy (1892–1987), Nobel Prize in Physiology or Medicine, 1934, shared with two others, for their discoveries concerning liver therapy in case of anemia.
3. Pauling was not allowed to fulfill the requirements for high school graduation of having taken two terms of American history. He wanted to take these courses

concurrently instead of sequentially, as stipulated by school authorities. However, he was awarded an honorary high school diploma from Washington High School by the Portland Board of Education in 1962. According to his official biographer, Professor Robert J. Paradowski, Pauling is the only person ever to have been awarded an honorary diploma by Washington High School. Reference: Herman, Z. S. "Some early (and lasting) contributions of Linus Pauling to quantum mechanics and statistical mechanics", in Maksic, Z. B. and Eckert-Maksic, M. eds., *Molecules in Natural Science and Medicine: An Encomium for Linus Pauling*, Ellis Horwood, New York, pp. 179–200. We are grateful to Dr. Herman for this Reference.

4. Pauling's first paper, "The manufacture of cement in Oregon", was published in *The Student Engineer (The Associated Engineers of Oregon Agricultural College, Corvallis, Oregon)* 12(1), 3–5 (June 1920). Reference: The Publications of Professor Linus Pauling by Z. S. Herman and D. B. Munro, http://charon.girinst.org/~zeke. We are grateful to Dr. Herman for this Reference.

5. McMillan shared the Nobel Prize in 1951.

6. Bragg, W. L.; Kendrew, J. C.; Perutz, M. F. "Polypeptide chain configuration in crystalline proteins", *Proc. R. Soc.* **1950**, *203A*, 321–357.

7. Pauling, L.; Delbrück, M. "The nature of intermolecular forces operative in biological processes", *Science* **1940**, *92*, 77–79.

Miklós Bodánszky, 1999 (photograph by Eszter Hargittai).

22

MIKLÓS BODÁNSZKY AND VINCENT DU VIGNEAUD

Miklós Bodánszky (b. 1915 in Budapest, Hungary) is Charles F. Mabery Professor Emeritus of Research in Chemistry of Case Western Reserve University, Cleveland, Ohio. He received his Diploma from the Budapest Technical University in 1939 and his doctorate in 1949. He chose organic chemistry during his student years under the influence of Professor Géza Zemplén, a former disciple of Emil Fischer. Zemplén was also to become the mentor of George A. Olah at the Budapest Technical University. There were several more interactions with Olah over the years. Bodánszky embarked on his first book-writing project at Olah's suggestion, he later took a professorship in Cleveland at Olah's invitation. The intervening decade between Bodánszky's diploma and doctorate was for a great part a struggle for survival for Bodánszky. As a Jew he experienced unemployment and forced labor camp, and hid from the Nazis, and he was one of those saved by the legendary Swede, Raoul Wallenberg.

Dr. Bodánszky now lives in Princeton, New Jersey. In 1989 he had lost his wife and long-time co-worker, Ágnes. His daughter, a philosopher, is also a resident of Princeton.

Fom the beginning of my (BH) studies in peptide chemistry I have used Miklós Bodánszky's books [*Principles of Peptide Synthesis* by M. Bodánszky and *The Practice of Peptide Synthesis* by M. and A. Bodánszky]. It was then exciting for me in 1999 to get into a correspondence with him asking numerous questions on my part about his life and work. Our correspondence provided the next best thing for me to a personal meeting. Later I visited Dr. Bodánszky in his home in Princeton.

Here I am quoting Dr. Bodánszky's reminiscing* about Vincent du Vigneaud (1901–1978) who received the Nobel Prize in Chemistry in 1955 "for his work on biochemically-important sulfur compounds, especially for the first synthesis of a polypeptide hormone". Du Vigneaud made it possible for Bodánszky to continue his research on peptides when he and his family became refugees in 1956. Bodánszky was 42 and an experienced scientist, who already developed methods of peptide synthesis, when he joined du Vigneaud in 1957 and a fruitful cooperation developed between them.

I was glad when you asked me about Vincent du Vigneaud because he is worth remembering not only for his work but also for his unique personality. I had had in mind to write a small book with the title *My Years with Vincent du Vigneaud*, but two university publishers to whom I sent excerpts from the planned book declined and the editor of the series of biographies published by the American Chemical Society did not even respond to my query. One of the excerpts has appeared in the *European Peptide Society Newsletters*.

Vincent du Vigneaud in the 1970s (photograph by Frank Sipos, courtesy of Miklós Bodánszky).

---

*In part, this has appeared in Hargittai, B. *The Chemical Intelligencer* **2000**, 6(4), 42–46.

I had been familiar with the name of Vincent du Vigneaud well before I first met him in Brussels in 1955. I taught medicinal chemistry at the Budapest Technical University and antibiotics formed the concluding part of the course. The discussion of penicillin included du Vigneaud's synthesis of benzylpenicillin as described in the monograph *The Chemistry of Penicillin* (Princeton University Press). Condensation of a thiazolidine with an oxazolone in order to generate the structure postulated for penicillins at that time produced, in several laboratories, materials with only minor antimicrobial activity. The Cornell group, however, led by du Vigneaud, continued the effort and secured, through a series of chromatographic procedures, countercurrent distributions and finally by crystallization of the triethylammonium salt, a sample, which was shown in a battery of tests, including X-ray diffraction, to be identical with natural penicillin-G. On reading this account I was truly impressed by Vincent du Vigneaud's insistence in eliminating any trace of doubt, in leaving no stone unturned.

Late in 1953, in a preliminary publication in the *Journal of the American Chemical Society* du Vigneaud and his associates reported the structure and synthesis of the hormone oxytocin, a nonapeptide. I thought, that this first synthesis of a biologically-active peptide should be followed by many more similar studies and decided to dedicate my work to the synthesis of naturally-occurring peptides. First, my co-workers and I reproduced the oxytocin synthesis reported by the Cornell laboratory. Some of the amino acid constituents had to be obtained by isolation; cystine from human hair, proline from gelatin. Isoleucin was synthesized in 11 steps and then had to be resolved into the pure enantiomers and purified to make it free from alloisoleucine. Because of the splendid cooperation and impatient, speedy work of my young associates, I found time for the development of a new method of activation and coupling. The nitrophenyl ester method turned out to be of practical use and was demonstrated in the synthesis of the C-terminal tetrapeptide segment of oxytocin.

With the death of Stalin in 1953, the political atmosphere mellowed in Hungary, yet it came as a surprise to me when I received a message from Bruno Straub that the Hungarian Academy of Sciences would send me to Brussels to participate in the International Congress of Biochemistry to be held in the summer of 1955; the message was difficult to believe. Nevertheless, I started to brush up on my French. In Brussels, I left a note in du Vigneaud's mailbox, informing him about my brief talk on a new synthesis of the C-terminal tetrapeptide segment of the molecule of the hormone. He came to my talk and we exchanged a few words afterward. Du Vigneaud's

Ágnes and Miklós Bodánszky (courtesy of Miklós Bodánszky).

talk on isolation, structure determination and synthesis of oxytocin was the concluding plenary lecture of the conference. This was in the summer of 1955. In December he received the Nobel Prize.

The end of 1956, following the crushed Hungarian revolution, found us, my wife, our daughter and myself, in Vienna as refugees. Inquiries to about a dozen European and U.S. scientists were soon answered, all encouraging, but, apart from an invitation to visit Glaxo in England, there was nothing tangible in the rest of the letters, except for one. Du Vigneaud invited me to join him as a Research Associate at the Department of Biochemistry of Cornell University Medical College in New York City. He named the amount of the postdoctoral stipend, the work I should do, which was research on oxytocin, and mentioned that my wife could also find work in one of the surrounding institutions, Rockefeller or Sloan Kettering (which she did indeed). Under his signature there was an additional line: "Approved", followed by the signature of the Dean of the Medical School. On March 11, 1957, I was in du Vigneaud's office. He gave me a copy of his book, *A Trail of Research*, with a dedication and I started my work in his laboratory.

When I first met du Vigneaud, in Brussels, even his exterior impressed me: a tall, well-built man in his fifties, with an imposing, aristocratic bearing. There was nothing aristocratic in his immediate family; he was brought up in a modest neighborhood of Chicago. His father was a mechanic and inventor. Du Vigneaud worked his way through college by teaching horse-back riding and later by making synthetic preparations for Professor C. S.

("Speed") Marvel. When he received the Nobel Prize, several letters came from France, claiming that they, the du Vigneauds, were descendants of an aristocratic family, marquis, who fled France during the French revolution. Alas, the Chief (as we called him behind his back) could not speak French.

His lifelong interest in peptides was awakened by a lecture given at the University of Illinois at Urbana by W. C. Rose about the recent discovery of insulin, by Banting and Best in Toronto. In his graduate studies at the University of Rochester, Professor Murlin, his thesis advisor, gave him a free hand in the choice of topic and du Vigneaud decided on insulin. Subsequently, his research involved insulin, biotin, lipoic acid, penicillin, oxytocin, all sulfur-containing materials. Sulfur became the trail he followed.

The young Ph.D. left for Germany with a two-year national stipend and joined Max Bergmann's group at the Kaiser Wilhelm Institute in Dresden. There du Vigneaud collaborated with a Greek scientist, close to him in age, Leonidas Zervas. They formed a lasting friendship. Since the ground-laying work of Emil Fischer at the turn of the century, this was the time of the first major breakthrough in peptide synthesis: the Bergmann–Zervas discovery of the "benzyloxycarbonyl" group (1932). Du Vigneaud was proud to belong to the second generation of Fischer's scientific offsprings and even the minor fact that the microanalyst of the laboratory, the lady who taught her art to the Chief, was the daughter of one of Fischer's lab assistants, was worth mentioning when he reminisced about his time in Dresden. Before returning to the U.S., du Vigneaud spent a few months with Barger (who determined the structure of thyroxin) in Scotland.

On his return to the United States Vincent du Vigneaud joined Professor J. J. Abel at Johns Hopkins University in Baltimore. Abel was the first to crystallize insulin, but unfortunately, when du Vigneaud arrived in Baltimore, there was no crystalline insulin available for studies. Abel could not reproduce his own crystallization experiment. In his own defense, he told du Vigneaud, that somebody in Syracuse, New York, did get crystalline insulin by his method and the newly-arrived young co-worker asked, "Was it not in Rochester, Professor Abel?" and Abel answered that it might have been Rochester; du Vigneaud did not tell him, that it was he, who did it in order to have homogeneous starting material for his insulin studies. Similarly embarrassed was Oscar Wintersteiner, who just received his Ph.D. under the guidance of Pregl in Graz, Austria. Pregl developed the method of microanalysis and was honored for it with the Nobel Prize. Wintersteiner came to Baltimore to analyze Abel's insulin, but there was no sample available.

The two young chemists, in collaboration with a third co-worker of Abel, Jensen, studied the insulin samples at hand, and established that in addition to cystine, already discussed by du Vigneaud in his thesis work, several more amino acids are constituents of the molecule of the hormone, and, hence, insulin is a protein. At this point Abel lost all his interest in insulin. He knew that the methods known at that time, were not sufficient for the determination of the structure of a protein. Later I met Oscar Wintersteiner at the Squibb Institute for Medical Research in New Brunswick, where he was director of biochemistry. He was one of the most productive natural products chemists of his time, with remarkable results in the field of alkaloids and steroids, including the isolation and structure determination of important glucocorticoids, penicillin, and streptomycin. Together with Josef Fried, the newly-appointed head of organic chemistry, they decided that Squibb should have a group active in peptide chemistry, and selected me to form such a group in Fried's department. Oscar Wintersteiner became a good friend, we used to meet daily. He came from Austria, I from Hungary, and we had many topics of mutual interest, chemistry, literature, music, and du Vigneaud, and I heard about their insulin period for a second time.

After his postdoctoral period with J. J. Abel, du Vigneaud remained in academia, and after brief appointments at other universities, was named Professor of Biochemistry at Cornell University Medical School in New York City where he remained until his retirement. Then he moved to the Ithaca campus of Cornell as Professor in the Chemistry Department. I stayed in contact with him and sent him one of my best students at Case Western Reserve University, Douglas Dyke, as a postdoctoral associate.

Du Vigneaud invited me to come to Ithaca to give a seminar in the Chemistry Department. It was one of the most glorious days in my life as a chemist. With a small plane we flew over the Finger Lakes at their most beautiful time of foliage. In the brilliant October sunshine du Vigneaud met me in person at the small airfield. He gave a party in my honor where I found out that all the young chemists present were working on oxytocin.

This reminded me of the day I first met him in New York City. He must have noticed my surprise when he told me that I should work on oxytocin. Frankly, I thought that in the meantime he had moved on to new exciting syntheses, perhaps that of insulin or ACTH. He said, that several times he was asked what he will study now, as he has finished with oxytocin, and that he answered these questions with "Finished? We had just started." And indeed, his judgment turned out to be sound. Already during my postdoctoral studies in his laboratory, we could show, by replacing tyrosine with phenylalanine in the synthesis, that the phenolic hydroxyl group of tyrosine was not essential for biological activity, although it somewhat increases the various hormonal potencies of oxytocin. Even more surprising was the finding by du Vigneaud, V. V. S. Murty, and Derek Hope that the omission of the N-terminal amino group, instead of destroying or at least diminishing the potency of the hormone, in fact, increased it. And through the years, many more important points of information were gathered through the synthesis and examination of oxytocin analogs. One of his associates, Maurice Manning, spent the decades that followed his time with du Vigneaud on the highly-rewarding synthesis of oxytocin analogs. Maurice is now Professor of Biochemistry at the University of Toledo Medical School.

Physiology has made considerable progress in this area. Du Vigneaud liked to call oxytocin a "baby protein" and he would have been pleased to learn that oxytocin plays an important role in motherly love. In the absence of this hormone, animals don't care about their offspring.

That neither the phenolic hydroxyl nor the N-terminal amino group is necessary for the activity of oxytocin, reminds me of an intuitive statement of du Vigneaud, in which he expressed the view, that neither one or another cystine or any other amino acid is the determining factor in the blood sugar lowering effect of insulin, but the *architecture* of the molecule itself, a well-accepted fact today, but something that was far from obvious in the thirties when he first expressed it.

I should tell you a story he told me while I was with him in New York. Some time before, he received an invitation to Chicago by Armour, the largest meat producer in the United States and also the source of medically-valuable products from animals. In order to secure their sales of corticotropin (ACTH), they wanted to explore its possible synthesis. The structure of ACTH had just become known, in part from their own research. Therefore they invited several famous scientists together, including Robert B. Woodward and Sir Robert Robinson, to solicit their opinion in making their important decision about the ACTH synthesis. The participants of this ad hoc conference agreed that ACTH should indeed be synthesized and that du Vigneaud was the best candidate to carry out the necessary research. The Chief told me that after a moment of hesitation he declined the very attractive offer. He thought that he could go back to New York and tell his associates working on oxytocin that a more important project emerged and that they should switch their attention to the synthesis of

Balazs Hargittai and Miklós Bodánszky in Princeton (photograph by Magdolna Hargittai).

ACTH. "But what am I going to tell them a few months from now if a still more important peptide comes our way? Should I tell them to abandon ACTH and turn to the new target? By this token we might as well stick to oxytocin." And how right he was. Oxytoxin still had a lot to offer while the methods for the synthesis of a much larger molecule, such as ACTH, were not yet sufficiently developed. Also, the sequence of the molecule underwent two minor revisions and last but not least ACTH slowly lost some of its glamor.

Du Vigneaud had an uncanny judgment in the selection of research projects. He knew when to start and also when to stop a project. Once he mentioned to me, that in the decision to tackle oxytocin a sentence by Otto Kamm, an investigator working at the Parke–Davis Company, had major influence on him. After being the first to separate oxytocic principle from the blood pressure rising factor (vasopressin), he remarked that both compounds appeared to have molecular weights of around 600. Later work showed that 1000 would have been a more accurate number, but the relatively small molecular weight gave du Vigneaud the necessary encouragement, promised success, just as the larger insulin molecule discouraged Abel from continuing his work on insulin. I was greatly impressed with the Chief's research acumen. Unfortunately, unlike in other matters, I was unable to emulate him. Instead of sticking to a well-going project, new peptides and new challenges tempted me time and again to start on a new route. Perhaps I could have achieved more; yet this was in my character and — in a sense — I do not regret it: it was a lot of fun.

Du Vigneaud's sense of responsibility, his standards of integrity were worth watching. His co-workers had to compose the papers reporting the experiments, but he had a sharp critical mind and made very good suggestions for changes. Editing one of our joint publications, he read the names of the researchers to whom I gave credit for their synthesis of oxytocin that followed the first synthesis at Cornell. He told me, "But Nick (standing for Nicholas) you did not mention George Anderson." I tried to defend the list of names in the references by telling him that Anderson never published a synthesis of oxytocin. The Chief rebutted, "But he told me about it." Thus I duly added the name of George W. Anderson and learned for a lifetime to be careful and to give credit whenever credit was deserved. He himself was slightly annoyed when his name was not mentioned, for instance, in connection with reduction with sodium in liquid ammonia. He said that if he would object, the answer would be that everybody knew about his contribution, but it was not true because not everybody knew it. And all this was *after* the Nobel Prize. I told this story to all my students and associates: most people are rather sensitive in this respect and do not take it kindly when their work is mentioned but their name is ignored.

Lastly, I would like to point to an important feature of du Vigneaud's research style. He always insisted that in research, no doubt should be un-clarified. All possible evidence had to be gathered to provide solid foundation for our statements. In the course of the determination of the sequence of oxytocin, Charlotte Ressler, the foremost participant in this venture, found that treatment with bromine–water cleaves the bond between the second and third residue in the chain, that is between tyrosine and isoleucine. This selective cleavage was not understood at that time. Nevertheless, after completion of the synthesis, the Chief demanded that the same experiment be carried out on the synthetic material as well.

I myself also experienced his careful attitude. The stepwise synthesis of oxytocin that I proposed and carried out with his approval (and for which he never failed to give me credit) provided us with larger amounts of oxy-tocin than ever obtained before. One day he asked me to try to crystallize the compound for the purpose of X-ray crystallography. The crystals of the available flavianate salt were silky needles, unsuited for X-ray studies.

He also gave me one of his cigar boxes (he smoked "White Owl"), containing a series of vials each with an aromatic sulfonic acid used by Bergmann and his associates, William Stein and Stanford More (both Nobel laureates), for the selective precipitation of individual amino acids. This was the basis of the solubility product method proposed for the quantitative determination of individual amino acids, a continued effort of Bergmann's group at the Rockefeller Institute. The vials were a gift from them to du Vigneaud. I tried them all, using as much as 5 mg of the synthetic hormone for each crystallization experiment. One of the acids, 4-hydroxy-azobenzene-4'-sulfonic acid, provided a salt that crystallized in lovely rectangular platelets, very colorful under the polarizing microscope. The crystalline salt was remarkably stable, it could be dried at 110°C without loss in biological potency. The Chief was pleased and I wrote up the method for a short communication in *Nature*. He appeared in my lab, the final manuscript in his hand and told me, "But Nick, in all these experiments you crystallized the synthetic material. You never did it with natural oxytocin." With this he handed over a vial containing a white lyophilized sample of oxytocin, isolated by Pierce or perhaps by Livermore, his co-workers, about eight years earlier and kept ever since on dry ice. I was uncertain whether or not the material was still intact, therefore, I went up to the second floor and asked one of our assistants who did the pharmacological testing, to have a look at the sample. Within hours I knew that it had remained fully active all those years. In the meantime I tried the crystallization, and it worked like charm. The paper was sent to *Nature*. All this, I felt, was remarkable. After innumerable comparisons, after the Nobel Prize, at a time when no one had the slightest doubt about the structure of oxytocin, I had to show once again that the natural and synthetic materials are identical in every respect. This was what it meant to leave no stone unturned. Now I could understand his caution, when he entitled the preliminary paper in 1953, "Synthesis of an octapeptide with the hormonal activities of oxytocin". Years later, I copied this title (*mutatis mutandis*) in reporting the synthesis of secretin. Similarly, after proving that the published structure of the microbial cyclopentapeptide malformin, was erroneous and having published the right structure, I continued to assemble further evidence. In such matters, I tried to follow his shining example, to remain skeptical against one's own work.

Not seldom was I told that one could not learn peptide chemistry from Vincent du Vigneaud. There was a small element of truth in such doubts; the Chief was not interested in the details of synthesis as much any more.

Yet, those with an open mind and open eyes, could learn exceptional seriousness, responsibility, a style of research rarely seen. Last, but not least, his view of the world of science was broad and full of expectation. The majority of investigators believe that almost everything is already known, only small improvements can be achieved by now. I learned from Vincent du Vigneaud, not so much from his words as from his example, that the opposite is true, that discovery is around the corner.

Melvin Calvin, 1962 (photograph by Berkeley LRL Graphic Arts, courtesy of Marilyn Taylor, Melvin Calvin's long-time secretary, and Heinz Frei, Berkeley).

23

MELVIN CALVIN

Melvin Calvin (1911 in Minneapolis, Minnesota – 1997 in Berkeley, California) received the Nobel Prize in Chemistry in 1961 "for his research on the carbon dioxide assimilation in plants". He studied at the Michigan College of Mining and Technology (B.S. degree in 1931) and the University of Minnesota (Ph.D. in 1935). Then he was a postdoctoral fellow at the University of Manchester, England. From 1937, he was at the University of California at Berkeley, rising to full professor in 1947. He became director of the Laboratory of Chemical Biodynamics at Berkeley in 1960, which was renamed Melvin Calvin Laboratory upon his retirement in 1980. Calvin remained active in research after his retirement. He received many awards and honors. Clarence and Jane Larson recorded Melvin Calvin's narrative in Dr. Calvin's office at the University of California, Berkeley, on July 16, 1984, and what follows are edited excerpts from that recording.*

I was born in Minneapolis in 1911, but very early on my father went to work for the Cadillac Motor Company in Detroit, and my whole family moved to Detroit. The first scientifically related conversation that I can recall was in grade school in the physics class. I was in the habit, as I still am, of responding to the teacher's questions almost before the question was out of his mouth. The result was that I would frequently answer questions that he didn't ask. Sometimes I couldn't answer the questions he did ask.

---

*"Larson Tapes" (see Preface). In part, this has appeared in *The Chemical Intelligencer* **2000**, *6*(1), 52–55.

I've always taught my students that, and I've done it from a different point of view, namely that it's no trick to get the right answer about some scientific question when you've got all the data. A computer can do that. A real trick is to get the right answer when you've only got half the data and half of what you have is wrong, and you don't know which half is wrong. Then when you get the right answer, you're doing something creative. It usually works and it stems from a very early time when the physics teacher said I would never be a scientist because I didn't wait for all the data to come in before I'd try to give him an answer, and I still do that. It is essential that we do that. If you don't do that, it's not a creative act. That's been basic throughout my whole life, to really try and understand all the phenomena of the world as early as possible without waiting for all the facts to be in. That philosophy can lead you also into great troubles, and it frequently does but you can make advances that way because then you won't be bothered too much by the dogma of the day.

I don't know that I had made my mind up to be a scientist about that time, I don't think I had. I was still in high school and I had not had any high school chemistry, the only course that I had that was called science was the physics course. I never had any biology course at all throughout my whole career. Later on I spent a good fraction of my time doing biology. There's been none of that formal biology in my background. That has been both an advantage and a disadvantage. It's a disadvantage in that I don't know the formal taxonomic language of biology. I have to look up everything to understand the classification systems. That's the only part that's missing. Later on, of course, I had lots of chemistry and lots of physics, mathematics, and quantum mechanics.

I chose chemistry out of high school for a very practical reason. During my high school weekends and during some of my first undergraduate week-ends I would work in a grocery store in Detroit on Saturdays, which was a big shopping day, especially in a cut-rate grocery store. The work began at five o'clock in the morning and we closed up at midnight. In the course of that day I learned all about how groceries are packaged, and I noticed that in every package of groceries, and even those that weren't packaged, some chemical role entered into the production of the product that was being sold. It was either in the canning of the food itself, in the making of the cans, in the printing of the labels, in the making of the paper, making the paper bags, everything involved chemistry. That was the time of the Depression. My father who had become a very skilled mechanic was periodically laid off. That made a big trouble for us and one of the most important factors in my decision was to find something, to learn something that people couldn't get along without so that I couldn't be fired. It didn't occur to me that I could grow food but I could certainly learn the chemistry that was involved in all this processing from the beginning to the end. Eventually, of course, I'd get into the chemistry of the growing as well.

...

I graduated from Detroit Central High School in 1927 and there was this college in Northern Michigan, which was trying to expand its clientele. In order to do so, it arranged to give a scholarship to one high school student in every high school in the state. Their only qualification was that the high school principal would certify that this was the best student in the school in that class. I got one of those in 1927. I went to what was then called the Michigan College of Mines. Later that year it became the Michigan College of Mining and Technology, and has since become the Michigan Technological University. When I was there it was a mining school and my only electives outside chemistry that I could study was geology, mining engineering, and civil engineering, and so I did all of those things and the basics, of course. I finished in 1931. I was the first chemist to graduate from there. Next I went as a graduate student to the University of Minnesota. By that time my family moved back to the Twin Cities and had a small shop there. I got a teaching fellowship.

...

I got my Ph.D. in 1935 and it was a straight physical chemistry thesis. I studied mostly electron interactions with gaseous atoms. My thesis was to measure the energy of interaction of single halogen atoms with electrons. I did that in a vacuum tube all of which we had to build. It was a time when building things was common for a graduate student. I learned the relationships between the current/voltage curves in the vacuum tube and the charge to mass ratio of the carriers. It was from that current/voltage curve that I measured charge to mass ratios and from these ratios the number of halogen atoms that was in the vacuum tube. I could calculate the equilibrium constant between the atom, the electron, and the anion. That's how we determined the electron affinity of the halogens.

...

By that time, during the course of my graduate work, I had to study the quantum mechanical theories of reaction mechanisms. The best statement about the quantum mechanical theory of reaction mechanisms that was extant at the time and was just being developed was that of Michael Polanyi.

Melvin Calvin during the conversation with Clarence and Jane Larson (photograph taken from the video recording).

Michael Polanyi had been studying reactions of sodium atoms with alkyl halides in a dilute gas. He also had undertaken a study of the reaction of the hydrogen atom with the hydrogen molecule. The way he made that measurement was to use H atoms and D_2 molecules and measured the formation of HD. He was measuring the simplest kinds of reactions, which were susceptible to first principles quantum mechanical calculations, and he succeeded in doing that and in developing what we now know as a transition state theory of reaction kinetics. His more famous pupil was Henry Eyring who preceded me in that work. By the time I got to Polanyi, he had moved to Manchester and by that time the theory of transition state had been sorted out.

Polanyi asked me to study the mechanism of activation of molecular hydrogen on platinum, starting with polarized platinum. He had the idea that you could study the reaction of hydrogen atoms attached to polarized platinum with hydrogen molecules, which were not attached to platinum. That way you'd be able to affect the activation energy of the atom/molecule reaction, and that's what he put me on. I began to study the effects of polarization on platinum electrodes carrying hydrogen atoms on the rate of exchange between the hydrogen atom and the D_2 or HD molecule. This led to a more general question, which Polanyi now posed.

Before that though you should understand who Polanyi was. He was a refugee both from Hungary and Germany. He was a surgeon in World War I for the Hungarian Army. After the war was over he realized that his interests were in basic science. He went to Berlin and that's where his physical chemistry and his ideas about reaction mechanisms were born and developed, in Berlin-Dahlem. After Hitler came to power in Germany, Polanyi left. He went to England. I went there in 1935 and spent two years with him.

Polanyi's background had some biology in it; he was aware that there were enzymes in living systems that could deal with molecular hydrogen. He thought that those enzymes — and all had metals in them — would probably be important to understand how to activate hydrogen properly. At that time he believed that the active site of hydrogenase, the enzyme, which activates molecular hydrogen and allows it to exchange with water, was an iron-porphyrin-bearing enzyme. The reason, I think, he thought that way, and I have to say, "I think" because he never did tell me, was that most of these enzymes were oxidation and reduction enzymes, enzymes that catalyzed the addition or removal of electrons from substrates. If the enzyme activated molecular hydrogen so it will exchange with the protons of water, presumably the enzyme was oxidizing H_2 to get protons and holding the electrons back somehow. When the protons exchange, they would then come back again as molecular hydrogen.

Polanyi had been studying these exchange reactions in various ways. He invented, for example, the micropicnometer to measure the density of water in order to measure the amount of deuterium in it. He would use a few tens of microliters of the water to measure its density. These micro-picnometers were little floats. The picnometer would hold a hundred or fifty microliters of water and it was put in through a microcapillary. The top of that picnometer bore a little sphere, a bulb of five millimeters in diameter. That sphere was very thin glass and flat on one side. When the picnometer was dropped in water, it would float with the water-containing part down and the bulb up. The volume of that bulb depends on the pressure. He could measure the density of a hundred microliters of water to five or six or seven places that way. That was the kind of man he was. He invented it, designed it and had it built. We didn't have mass spectrometers in those days. So we were measuring water densities that way and measuring exchange rates that way.

Polanyi had the idea that the enzymes must have some peculiar properties, which are dependent upon the porphyrins because almost all redox systems in biology that he knew about, the hemin of red blood cells, the chlorophyll of the green plants, all were porphyrin type molecules with metal centers. The hemin had an iron center, chlorophyll had a magnesium center. He put me onto that after I had been there a year and a half. He supposed that there must be something very special about this tetrapyrrolic structure which surrounds the metal and which makes it do funny things in biology. The biological tetrapyrrols are very unstable compared to the kinds of things he was used to doing.

About that time, in 1934, R. P. Linstead, Professor of Organic Chemistry at Imperial College in London, had discovered phthalocyanine. He was a consultant for ICI. ICI was making phthalonitrile, which is ortho-dicyanobenzene in glass-lined kettles. Phthalonitrile crystallizes in beautiful white crystals, but on one occasion it turned into a blue mess. Linstead determined that the glass lining in one of the iron kettles had cracked and phthalonitrile had come in contact with the iron, and this had catalyzed the cyclization of the four phthalonitriles around an iron center. He had iron phthalocyanide. That was the beginning of a new dyestuff, which turned out to be very stable, and became one of the most important organic pigments for a period of 20 or 30 years. It is known as a tetra-azaporphyrin. The bridges between the four pyrrol rings were nitrogen atoms instead of carbons that are the bridges in nature.

Polanyi told me to go down to London, find out how to make that stuff and bring it back. He gave me two weeks to do that. Polanyi then suggested to put different metals in the center and study their catalytic properties for activating hydrogen, like platinum. You could heat it up, cool it, do what you liked. I've spent a lot of time doing that and I enjoyed that very much. In so doing, I became thoroughly aware of the importance of that particular type of structure, always involving the movement of electrons and protons. Of course, the chlorophyll in the green plants, although not the same, is a very close relative of porphyrin. That also involves photochemical oxidation/reduction. That's how I got started on that business. My last experiments with Polanyi were hydrogen activation on metalphthalocyanines with copper and zinc.

I came to Berkeley in 1937. Joel Hildebrand had visited Polanyi and Polanyi knew that it was time for me to go and get a job and he recommended me to Hildebrand. Hildebrand wasn't chairman of the department; the offer came from Gilbert Lewis to come and be an instructor of the University of California at Berkeley. I came and have stayed here ever since.

The first thing I did when I came here was to learn how to teach organic chemistry because that was what I was hired for. Lewis knew a lot of organic chemistry but not in the way that the ordinary organic chemist knows it. He understood the structure of molecules in a way nobody else did. He didn't teach organic chemistry in a way an organic chemist

The Melvin Calvin Laboratory in Berkeley (photograph by I. Hargittai).

would teach it and it didn't bother him that I didn't know organic chemistry either. I had to teach organic chemistry and that's when I learned organic chemistry from 1937 on. I taught organic chemistry first for the chemists then for the biologists for about forty years. In doing so I had to learn biology because most of the time I was teaching premedical students.

My first experiment was to try to find a homogeneous catalyst metal complex that would activate molecular hydrogen. That was my personal research at that time. Indeed I did, I found a copper salt, dissolved in organic solvent, which would activate molecular hydrogen for exchange reactions and for reduction reactions, just as platinum did. It turned out to be cuprous acetate. It was the first homogeneous hydrogenation catalyst and it was my first publication out of Berkeley, in 1938. This work led to an attempt to understand, in general, homogeneous hydrogen activation. In doing this, I became even more concerned with the coordination chemistry of transition metals and how they do catalysis. That led me to thinking again about porphyrin.

In 1938 Martin Kamen and Sam Ruben, working with the cyclotron here, found that they could bombard nitrogen with neutrons and get carbon-14. They also found carbon-11 but its half-life was only 20 minutes; carbon-14 was much more useful. In the meantime they stacked huge tanks of ammonium nitrate around the cyclotron and after a while they would add CO_2 to those tanks and precipitate ammonium carbonate with regular carbon in it. I got some samples of that early on. In the meantime Ruben was busy doing some work for the Army. One of the things he was working on was phosgene, and he had an accident with the phosgene. He plunged a gas tube of phosgene into liquid nitrogen and it burst and the boiling liquid nitrogen blew the phosgene in his face and killed him. That left that whole program in limbo for the time being because Martin Kamen left also and was doing something else.

In the meantime I was involved in the Manhattan District (popularly called the Manhattan Project) as well. I had been invited by Glenn Seaborg to develop a separation method for the decontamination of irradiated uranium and also for the purification of plutonium. It turned out that the method I used was good for both. The reason I was able to do that was that I had been working on metal complexes. The method that was in use by Glenn was a precipitation method, precipitation with bismuth phosphate. It was a terrible method to work with on a large scale. So they were trying to do solvent extraction procedures and one of the solvents that was being used was tributyl phosphate. That was not very specific and my job was to develop a specific binder that would grab fission products but leave the uranium alone. I did that. I found out that uranium (VI), that is, uranyl ion doesn't complex very well. All the elements that we were worried about could be picked in a +3 or +4 state while uranium was always +6. The chelate that I built would work in one-normal nitric acid, pull them out, and leave the uranium alone. I could adjust which one to pull out by adjusting the acidity. That method brought me into contact with the whole Manhattan District and I used to go the Chicago meetings. Occasionally I'd give a little paper of my own at these meetings. It was a very exciting time, obviously. My process never reached commercial production in time to be used by the United States. The bismuth phosphate and tributyl phosphate processes got there first. By the time my technique was ready to be developed they didn't need it anymore, the pressure was off. The only people that did use it were the Canadians and the British. Eventually it came back here to be used in very special cases. That's how my connection with Ernest Lawrence also came about.

Then, one day, around V-J Day, I was coming back from lunch from the Faculty Club and met Ernest on the street and I still remember him saying, "Time to quit. Time to do something useful. Now do something with that radiocarbon." That was his attitude. He meant something like build something with radiocarbon and kill cancer with it.

So we started a radiocarbon lab within one day. We started making carbon-labeled compounds that would be useful for human metabolism, eventually to find their way to cancer cells. That's the way we started learning about the mechanism of organic reactions. Then, because of my interest in chlorophyll and how the green plants worked, it was such an obvious thing to do — here we have a tracer for the most important thing the plant handles, carbon dioxide. You feed the plant the carbon dioxide and find out where the hell it goes. I knew enough organic chemistry to know how to do it. The result was that in 1945 we began to sort out the various steps that carbon takes on its way from CO_2 to sugar. The final paper was published in 1955 — it was a 10-year period to do it. We worked in an old building and had to change a lot of things there but, fortunately, since it was such an old building, nobody cared what we were doing to it. The 60-inch cyclotron was next door to us and every time they turned it on we quit. We were counting two or three neutrons a minute and there were neutrons all over the place when they turned the cyclotron on.

We had to learn a lot of plant biology. In that 10-year period we mapped the whole route from CO_2 to sugar. There were then two more questions: one was, "What drives it?" — that's the photochemistry — and the other

Melvin Calvin on a poster in downtown Berkeley, 2004 (photograph by I. Hargittai).

question was, "What controls it? What turns it to sugar, what turns it into fat, what turns it into protein?" Those problems are still with us. I'm working on both ends of those. Some of my associates are more deeply involved in it than I.

My concerns are twofold. "Can we use plants today, plants that are not now used as crops, to solve the energy problem, to produce oil, which could be converted into gasoline, fuel?" That's one part of it. The other one is, "Can we use the knowledge about how the plant actually captures the quantum and stores it in some kind of an energy form?" It's the best photochemical converter we've got, not very good but still the best. Once we learn it from the plant we can probably run it in the lab with higher efficiency than the plant does it. We've learned now a good deal about how the plant does it and we began to simulate with totally synthetic systems, the sensitizer, not chlorophyll but something like it, the donor molecules, the acceptor molecules. The chlorophyll analog molecule hands an electron to the acceptor and then the chlorophyll analog has to get an electron from water, which it does with the help of a catalyst there. We now have a number of possible sensitizers in the middle. We have possible acceptors to make hydrogen or reduce CO_2 on one side, donors on the other side, catalysts with metals, which can take the electron away from water, and then make either oxygen, which would be a waste, or oxidize something else, which would be useful. We can do that now.

The question is, "Can we construct a unitary system that would do it all at once?" I think we now have a way to do that. That would be the ultimate answer to our energy needs, and, hopefully, we'll get it before the war for resources catches up with us. We better be able to do that fast enough to make liquid fuels, which we need on a large scale and that means hydrocarbons, to fulfill the demands of agriculture, industry, personal transport, and everything that goes with that. Even today we can use the existing plants which can make hydrocarbons, to produce them right now until we can get a totally synthetic system working. Some countries are using plants that way already, like Brazil, the sugar cane. We have to use other plants in this country but we can do it.

We can do it the synthetic way in 20 years if our population is willing to support it. That support comes in two ways. First of all the population has to believe it's necessary. Secondly, just as all other transitions into new technologies had been subsidized in their infancy, we need a little subsidy. Our farmers now are being paid to set aside land and not grow grain. Let them be paid, let them grow an energy crop on that land, and let them collect for that energy crop as well as being paid for not growing grain. In other words, don't deprive them of the subsidy that they're getting for withholding the land from grain production but allow them at the same time to grow the cash crop for oil. Once they have learned how to produce about ten percent of our needs, you can then withdraw the subsidy, and it will go on its own. After that, when we have the totally synthetic system, we won't need the plants at all. We're almost there now but it's a little further away. The plants you can do today.

These are the two kinds of activities we are involved in today. I just hope that we'll have time, and by we, I mean the scientific community of which I'm one, has time, not because I'm late in my life but because of the shortness of the peace in the community, in order to get this done before it explodes all over the world. That's what I'm really worried about.

Donald R. Huffman, 1999 (photograph by I. Hargittai).

24

DONALD R. HUFFMAN

The production of fullerene-rich soot by resistive heating made possible the development of fullerene science and technology. Wolfgang Krätschmer and Donald Huffman and their graduate students invented this simple technique and published it in 1990.[1] The first volume of the *Candid Science* series contained an interview with Wolfgang Krätschmer.[2] The present account gives Donald Huffman's perspective, based on a conversation with him at the beginning of September 1999, at the University of Arizona in Tucson. This account contains impressions from subsequent conversations with Wolfgang Krätschmer too. It is further augmented by a pictorial report of a meeting we had with Huffman and Krätschmer in which they kindly recreated the experiment in which they had produced measurable quantities of buckminsterfullerene.[*]

Donald R. Huffman was born in 1935 in Fort Worth, Texas. He did his undergraduate studies at Texas A&M University, got his master's degree at Rice University, and then served in the U.S. Army. Before resuming his graduate studies, he worked for Humble Oil Company (today Exxon). Thus, his path had taken him through Rice University and what later was to become Exxon, both of which were to play important roles in the fullerene story. At Humble, around 1959, Huffman met Peter Debye, who did consulting for Humble. Huffman's recollection of Debye's enthusiasm has been

---

[*]In part, this has appeared in Hargittai, B.; Hargittai, I. *The Chemical Intelligencer* **2000**, *6*(3), 39–43.

an inspiration ever since. It was, however, the Dutch astrophysicist, H. C. van de Hulst of the Leiden Observatory who had the most important influence on him. Van de Hulst was a pioneer of microwave astronomy and published a monograph on light scattering by small particles.[3] At one point Huffman was playing with the idea of co-authoring a similar book with van de Hulst. However, by the time Huffman and one of his former students, C. F. Bohren, put together their book,[4] he no longer thought that they needed to involve van de Hulst.

Huffman did his doctoral work, which was concerned with optical studies in solid state physics, at the University of California, Riverside. Following a postdoctoral stint at the University of Frankfurt, he joined the University of Arizona in 1967 and has been there ever since. The astrophysicists at the University of Arizona were interested in interstellar dust consisting of particles a few hundred angstroms in size. They found a strong spectral feature of the interstellar dust at 220 nanometers in the ultraviolet (UV) region and posed Huffman the following question: "What happens to the properties of solids when they get smaller and smaller. When do they lose their solid-state properties?" This question shifted Huffman's interest from the optical properties of crystals to those of small particles.

He started experimenting with evaporating metals and also carbon in an inert-gas atmosphere, generating small particles in the range of a few hundred angstroms to a tenth of a micron. He inherited the equipment he used for his experiments, and nobody seems to know who had built it originally. Already in the late 1960s, he pushed together carbon rods in the bell jar, making an arc between them. A carbon cloud was produced and he collected the soot from the walls of the bell jar and did spectroscopy on it. It was pure carbon with a distribution of particle sizes, which he determined by transmission electron microscopy. He did not have much success in trying to narrow the size distribution. The UV spectrum showed a 235-nanometer band, not far from the astronomers' observation, but it was shifted and it was also broader. This was reported in *Nature* in 1973.[5] The soot production, which began in Huffman's lab in 1968, has continued to this day.

Huffman and Krätschmer got acquainted in 1976 when Huffman was on sabbatical in Germany. He spent most of this sabbatical at the Max Planck Institute of Solid State Physics in Stuttgart, but went to Heidelberg to attend a seminar at the Max Planck Institute of Nuclear Physics where Krätschmer worked. Krätschmer first came to Arizona in 1977 to work on the interstellar dust project, but he got involved with amorphous silicates, which are important astronomically in the infrared region.

Huffman's next sabbatical came up in 1982–1983. He received a Humboldt Senior Scientist Award and spent his sabbatical mostly in Heidelberg. In order to further investigate the 220-nanometer band of the interstellar dust that the astronomers had observed, Huffman and Krätschmer set up an apparatus in Krätschmer's lab, similar to the one in Huffman's Tucson lab, and started making carbon smoke. There was an important difference between the two experiments. In Heidelberg, rather than having a gap between the two carbon rods, they brought them together and ran a current of about 200 amps through them. So whereas Huffman used the arc technique in Tucson, they were now using the resistive heating technique for evaporating carbon.

Huffman distinctly remembers when they observed, for the first time, a new feature in the UV spectrum in the form of a couple of humps at the top of the broad band. Krätschmer called this a camel feature. Although they had no idea at the time what this feature was, they later identified it as the first signal of C_{60} ever observed. They immediately started discussing what these humps could be, and this discussion lasted for eight years. Huffman was a believer and suspected from the beginning that a new form of carbon had been found whereas Krätschmer was skeptical and he was afraid that some "junk", such as traces of oil from the diffusion pump, might be the origin of the humps.

Huffman returned to Germany in the summer of 1984 to attend a meeting in Berlin on clusters. It was at this meeting that Kaldor et al.[6] reported their mass spectra of carbon clusters, which have since become famous. Huffman, like others, did not notice C_{60} sticking out there in the mass spectrum and thinks that Kaldor et al. deserve credit for their caution in interpreting the mass spectra. Huffman and Krätschmer had a poster at this meeting, but they did not report their observations of the camel features in the UV spectrum.

Huffman learned about the paper by Kroto et al.[7] in Nature in 1985 — in which they first reported the observation of buckminsterfullerene — from his new graduate student Lowell Lamb. Lamb was an avid reader of the science section of The New York Times, where the discovery was reported before the actual publication of the paper. Huffman thought at once, "That's got to be what we're making." Although Lamb immediately wanted to work on it, Huffman had no funding for it, so Lamb had to work on a different project.

Since Krätschmer maintained that what they were seeing was "junk", Huffman's role was to keep the flame alive. Nonetheless, the final and definitive breakthrough was to come from Krätschmer's lab. In 1987, Huffman filed a patent disclosure on the production of C_{60} through the University of Arizona. The disclosure carried Huffman's name alone. The recollections of Huffman and Krätschmer about this differ. Huffman remembered that he did this with Krätschmer's permission and with his name on it. However, Krätschmer told us in a separate conversation that he was taken by surprise when he heard about the patent application. It may well be that Huffman had just assumed that Krätschmer would not like to be part of it since he did not believe in it. When Huffman checked the documents in the wake of our conversation, and we talked about it again the next time, he told us that he found that the patent disclosure was indeed filed in Huffman's name alone.

Huffman did not seem to be able to produce the fullerene soot in a reliably reproducible way, and in February 1988 he was persuaded to withdraw the patent disclosure. Huffman remembers a visit by Harry Kroto at the University of Arizona around 1988 and telling Kroto about his experiments, but Kroto showed no interest in it.

We were curious why Huffman had not used other techniques for identification in addition to the UV spectra and why he did not consult with chemists. It appears that Huffman was very protective of his possible discovery. He did not want to collaborate because he thought it was something important. Huffman wanted Krätschmer and himself to crack the problem.

In the meantime, Krätschmer acquired access to a good infrared spectrometer in Heidelberg and could identify the four bands in the infrared spectrum; alas, this still did not satisfy him and he was still afraid that he was observing some junk. Huffman often reiterates that he has always been a lone wolf and felt if this was something big, he'd like to crack it. Looking back he is glad they did it this way. The interactions between Huffman and Krätschmer have always been smooth and pleasant. For years and years, it had only been Krätschmer and Huffman having a good time and doing some good work, and there was no problem because they got along well.

In the summer of 1988, there was an international conference on interstellar dust in Santa Clara, and in his oral presentation Huffman mentioned the camel feature in the UV spectrum and the possibility that it was C_{60}.

This astonished Krätschmer because he still did not think that this explanation was feasible, but because of Huffman's insistence, he began to believe in it too. By the time of the 1989 Capri conference, Krätschmer had been working actively with his graduate students on the infrared spectra, and they had isotopically enriched carbon rods and had proven to themselves that the spectral features could not be due to junk. Krätschmer presented these results at the Capri conference. He put Huffman's name on the paper, as has been his consistent habit. Then they reported the results in a paper in 1990.[8]

Kroto learned about the Capri disclosure, and Smalley must have read the *Chemical Physics Letters* communication. Kroto confirmed the production of the carbon soot having C_{60} in it to a British conference at the beginning of 1990. Kroto, being a chemist, began working on the extraction immediately. On May 15, Krätschmer called Huffman and told him, "The most amazing thing happened and you can do it yourself. You take benzene, put the soot in it, the stuff dissolves, you can filter it, and you get the C_{60} solution." Then it was possible to crystallize it. Within one hour of Krätschmer's call, Huffman reproduced the experiment of the extraction. Then they knew that they had it.

Two days later, Huffman was scheduled to go to Paris and he was musing to his wife, "Here we are sitting on the biggest discovery of our life, we've got something that nobody in the world has ever had, and anything we do is brand new, and I've got to leave town." Although Lamb was still working in Huffman's lab, he was working on something else. Huffman was still hoping that he and Krätschmer could crack the problem, and he did not want to see it spread to too many people. He still thinks that he was correct in that because the more people there are involved in a big thing, the more fights there will be. Nevertheless, he put Lamb on the project and he went off to Paris. He left a whole list of things for Lamb to do. It included measuring the density and the X-ray diffraction spectrum. After the Paris meeting Huffman went on to visit Krätschmer in Heidelberg, and they began writing the paper. Their habit was that Krätschmer would type the text and Huffman would find it agonizing because "Wolfgang would argue over every little English thing." This time, Huffman decided to do the typing at Krätschmer's computer.

Huffman considers the high point in his scientific life the moment when he went over to the geology department in Heidelberg to run the X-ray diffraction spectrum and the plotter started producing the spectrum, it was so beautiful. As a solid-state physicist, he was most interested in the crystal structure. At that point he realized that they had a third crystalline form of carbon, in addition to graphite and diamond. He found it mind-boggling. He did not identify correctly, though, which of the two closed-packed structure it was, and it is not correct in their historic paper.[1] Everything else is.

At that point, Huffman brought up patenting again. This time the patent was in both their names, Huffman and Krätschmer, through the University of Arizona and the Max Planck Society, and the patent had to be filed before the paper was sent in. It was a hectic summer for Huffman, especially when Krätschmer left for vacation. Huffman thinks it is a cultural thing with Europeans that they go on vacation when vacation time comes, no matter what. Huffman had to fight the thing through, the patent and the early stages of the paper. Although they were afraid that Kroto would scoop them, it did not happen. The paper sailed smoothly through the review process.

As for the patent, it is still being fought, and this may be going on for a long time to come. Huffman is not optimistic about their getting the American patent rights any time soon but he is optimistic that when they get them, they will be retroactive. He draws his optimism from the example for the patent for the laser, which is now held by Gordon Gould, rather than by the two Nobel laureates. Gould got a patent based on his 1957 notebooks, long after the initial patent had been given to Schawlow and Townes. Gould did a very good job in documenting his findings. The royalties were retroactive, and Gould now has a corporation whose main business is to collect the royalties on every laser that has ever been made.

The success and potentials of fullerene science and technology were greatly determined by Krätschmer and Huffman's invention of a method for the laboratory preparation of fullerenes. Huffman distinguishes between the periods before and after the Nobel Prize was awarded to Kroto, Smalley, and Curl when he describes their recognition of the importance of the invention. After 1990, there was a lot of competition for recognition. There were clearly two very important groups. The 1994 Award by the European Physical Society recognized the contributions of both groups by including Huffman, Krätschmer, Kroto, and Smalley. Although there were various speculations, it was realized that only three people could share the Nobel Prize. So there was a lot of competition. Huffman feels that Krätschmer and himself, by their nature, could not be the winners in public relations.

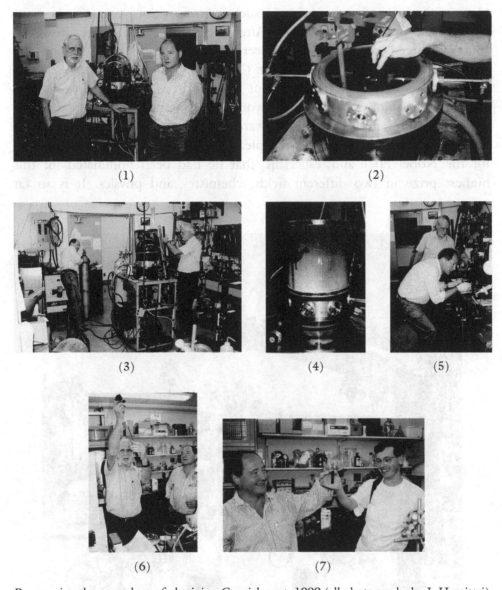

Re-enacting the procedure of obtaining C_{60}-rich soot, 1999 (all photographs by I. Hargittai).

(1) Huffman and Krätschmer before the experiment with the clean bell jar between them.
(2) The graphite rods — somewhat displaced position — in the apparatus.
(3) Huffman is ready to start the experiment while Krätschmer is adjusting the helium pressure.
(4) The bell jar is rapidly getting fogged as the resistive heating experiment is producing the fullerene-rich soot.
(5) Krätschmer is collecting the soot with Huffman looking on.
(6) Huffman is dissolving the soot in carbon disulfide with Krätschmer looking on.
(7) Balazs Hargittai helps Krätschmer hold up the test-tube with the C_{60} solution.

He understands, of course, that in such a competition, you do not pour great praise on your competitors. After the 1996 Nobel Prize, however, things changed, and the winners were gracious.

As for whether there is life after C_{60} for Donald Huffman, he thinks that C_{60} was only a perturbation but not the dominant thing in his life, although he thinks that his wife would give a more honest answer to such a question. He is looking forward to doing more research when he retires from teaching. He finds real pleasure in just having been mentioned for the Nobel Prize and, especially, that he had been nominated for this highest prize in two different fields, chemistry and physics. It is so far

Donald Huffman and Wolfgang Krätschmer under the archway at the University of Arizona in Tucson, 1999 (photograph by I. Hargittai).

beyond everything that a country boy from Texas would have ever expected to have happen to him. "It has been a great ride," he adds.

* * * * * * * * * *

There is an intriguing archway on the campus of the University of Arizona in Tuscon. It is called *25 Scientists* and was inaugurated in 1993. It is made of welded steel and is painted in bright colors. The figures on the archway represent some important branches of science and some discoveries. One of the central units shows a buckyball, somewhat compressed flat with the carbon atoms and the bonds between them painted on a blue background. Two figures are holding this buckyball and one of them resembles Donald Huffman. This is no accident because the artist, George Greenamyer modeled it after Huffman, who was pleased to serve as the model for the figure. When we (Balazs and István) arrived in Tucson at the end of August 1999, Wolfgang Krätschmer was just visiting Huffman and we asked them to pose for us beneath the archway. They graciously complied. Besides, they recreated for us the experiment that produced the C_{60}-rich soot.

References

1. Krätschmer, W.; Lamb, L. D.; Fostiropoulos, K.; Huffman, D. R. "Solid C_{60}: a new form of carbon", *Nature* **1990**, *347*, 354–358.
2. Hargittai, I. *Candid Science: Conversations with Famous Chemists.* Imperial College Press, London, 2000, pp. 388–403.
3. Van de Hulst, H. C. *Light Scattering by Small Particles.* Wiley, New York, 1957. Dover edition, New York, 1981.
4. Bohren, C. F.; Huffman, D. R. *Absorption and Scattering of Light by Small Particles.* Wiley-Interscience, 1983.
5. Day, K. L.; Huffman, D. R. "Measured extinction efficiency of graphite smoke in the region 1200–6000 Å", *Nature* **1973**, *243*, 50–51.
6. Rohlfing, E. A.; Cox, D. M.; Kaldor, A. "Production and characterization of supersonic carbon cluster beams", *J. Chem. Phys.* **1984**, *81*, 3322–3330.
7. Kroto, H. W.; Heath, J. R.; O'Brien, S. C.; Curl, R. F.; Smalley, R. E. "C_{60}: Buckminsterfullerene", *Nature* **1985**, *318*, 162–163.
8. Krätschmer, W.; Fostiropoulos, K.; Huffman, D. R. "The infrared and ultraviolet absorption spectra of laboratory-produced carbon dust: evidence for the presence of the C_{60} molecule", *Chem. Phys. Lett.* **1990**, *170*, 167.

Alan G. MacDiarmid, 2002 (photograph by Magdolna Hargittai).

25

ALAN G. MACDIARMID

Alan G. MacDiarmid (b. 1927 in Masterton, New Zealand) is Blanchard Professor of Chemistry at the University of Pennsylvania and Scholar in Residence and Chairman of the Advisory Board of the Nanotechnology Institute of the University of Texas at Dallas. Alan MacDiarmid shared the Nobel Prize in Chemistry for 2000 with Alan J. Heeger of the University of California at Santa Barbara and Hideki Shirakawa of the University of Tsukuba "for the discovery and development of conductive polymers".

Alan MacDiarmid studied at Victoria University College (University of New Zealand) where he obtained his B.Sc. and M.Sc. degrees. He earned his Ph.D. at the University of Wisconsin (1953), followed by further graduate studies with H. J. Eméleus at Cambridge University in England where he earned a second Ph.D. (1955). He has been at the University of Pennsylvania since 1955. He has received many honors and honorary appointments, including the Order of New Zealand (2002) and he has been a member of the National Academy of Sciences of the U.S.A. (2002) and the National Academy of Engineering (2002).

We recorded our conversation in Dr. MacDiarmid's office at the University of Pennsylvania on March 20, 2002.*

Let's start at the beginning.

When I was about 10 years old, I found some of my father's old chemistry books from the late 1800s when he was studying to be an engineer. I

---

*In part, this has appeared in *Chemical Heritage* 2003, *21*(1), 8–11.

found these books very intriguing, but I couldn't really understand much of the chemistry in them. I used to cycle into the center of Lower Hutt, in New Zealand, in whose suburbs we lived. I went into the public library where one of the new books was *The Boy Chemist*. I took out this book constantly for about a year and did just about every experiment in it. They recently reprinted 30 copies of this book.

Do you think children would be interested in doing those experiments today?

First of all, the title of the book would create some concern. Today it would be called something like "Chemistry for the Young Person". The book would still arouse interest amongst certain people because it did some fun things. It gave a description, for example, of how to make invisible ink, using lemon juice, ammonia, and other common things. There were more sophisticated experiments in it as well. Many of the chemicals in *The Boy Chemist* I could get locally and for those unobtainable, my father would go to a chemical supply house in Wellington. I delivered milk in the mornings and the money I earned from my milk round was used to buy the chemicals.

At about the same time I read this book, I became interested in photography. I bought cheap black-and-white printing paper and going to the bathroom at night, I would expose some old negatives, immerse the printing paper in the developer and watch the image appear. Then I washed the paper with the fixer to remove the non-exposed silver halides. I found the mystery of chemistry intriguing.

Chemistry has not been popular lately.

Dupont Company used to have a logo, "Better living through chemistry". Then some time ago they dropped chemistry from it because chemistry has such a bad connotation.

Unfortunately, all too often, chemistry is not being taught properly. Often, the task of teaching introductory chemistry is given to the youngest, most inexperienced faculty. It is, unfortunately, considered by some to be degrading and not intellectually stimulating. I feel that such a philosophy, where it does exist, must be changed. We have to stimulate the interest of younger people. Last year after receiving the Nobel Prize, I specifically requested to teach another freshman chemistry course which I greatly enjoyed.

I always say that when you stop learning you start dying. I was 50 years old when I got involved with the research that led to the Nobel Prize. In order to understand the problems I was dealing with, I had to teach myself a whole area of solid state chemistry, electronic structure, and bonding and processes in the solid state about which I knew absolutely nothing. My previous interest was silicon work, inorganic chemistry. My interest in conducting polymers though goes back to the time of my Master's thesis work in New Zealand, where I was working on tetra(sulfurnitride), S_4N_4 which forms orange-colored crystals. That's when I fell in love with colors. It was about 30 years later that Alan Heeger, my physicist colleague here at the University of Pennsylvania at the time told me about a sulfur nitrogen polymer, poly(sulfurnitride), he had read about that formed a golden film with metallic properties and asked me if I could prepare some of it. This we did and the IBM people eventually showed it to be superconducting at very low temperatures.

Then, when I was a visiting professor at Kyoto University about 27 years ago, I gave a seminar at the Tokyo Institute of Technology on the sulfur-nitrogen polymer. After my talk, as we were having tea, Hideki Shirakawa showed me a silvery film, which was polyacetylene. I at once invited him for a year to my lab in Philadelphia. Upon my return home, I talked with my contact at the Office of Naval Research, who had been supporting my research for a long time, and asked him for approximately 23,000 dollars to support Shirakawa as a postdoctoral fellow. He was a little reluctant because I was asking for money to do polymer organic chemistry and I had no experience in organic chemistry and I was not a polymer chemist either. My proposal was merely based on my curiosity in an organic film that was silvery. Nonetheless, I got the money I had asked for and Hideki could come. When he came, we measured the conductivity, which was low and as we purified our samples, the purer they became, their conductivity got smaller and smaller, but when we made them impure with bromine on purpose, their conductivity increased enormously!

Did not Shirakawa test conductivity back home in Japan?

He was not particularly interested in conductivity, he treated his polyacetylene with chlorine and he was interested mainly in the mechanism of chlorination.

He was the one who discovered the material.

That is right. He was the one who first synthesized this silvery polymer through a misunderstanding between the Korean and Japanese language. Shirakawa had been polymerizing acetylene gas using the standard Ziegler–Natta catalyst, producing black-brown, rather uninteresting polyacetylene. Then he had a new graduate student from Korea and Shirakawa asked this student to make the catalyst, tetrabutoxy-titanium-triethylaluminum, of so many millimolar concentration in toluene. The graduate student came back after a couple of days and showed that the stirring bar would not go around in the reaction flask due to the presence of some globs of silvery-pink jelly. He asked the student what he did exactly. The student told him that he did exactly what Shirakawa had told him, he used the catalyst, tetrabutoxy-titanium-triethylaluminum, which he had prepared in so many molar concentration in toluene. Thus he made the catalyst a thousand times more concentrated than Shirakawa had directed him.

So the Korean graduate student made the discovery.

The student did the experiment but Shirakawa interpreted the results.

Was the student invited to the Nobel ceremonies?

I think Shirakawa had lost contact with him. Of course, your remark touches an interesting aspect of this discovery and, generally speaking, of scientific discoveries. Who is the discoverer? Is it the person who does something mechanically or is it the person who realizes its significance? Quite often the person who does the mechanical operation in the lab is also the person that realizes its significance. Sometimes it is not the same person. We can even raise a broader issue, what does one mean by the term "scientific creativity", or is it a contradiction in terms? In my own research, we found that we could make certain types of polymers and convert them to metallic conductors. Is that scientific creativity? In our case, somebody within the next 10 years would've found out exactly the same things that we'd found out. But take Beethoven composing a symphony. How many years would you have to wait for another Beethoven to appear and write another Beethoven symphony? In my opinion, "scientific creativity" could be a very controversial term.

Suppose you had not visited Shirakawa, you would've gone into scientific oblivion soon.

And probably also Shirakawa and Heeger. As a matter of fact, Shirakawa had given up work on polyacetylene and by the time of my visit he was concentrating on the degradation of polymers in the environment. He returned to polyacetylene because of my visit and then my invitation that brought him to the University of Pennsylvania. We both benefited from and exploited each other's interactions.

Hideki Shirakawa during the Nobel
Prize Centennial in Stockholm, 2001
(photograph by I. Hargittai).

How did Alan Heeger get involved in this project?

Alan and I were working on poly(sulfurnitride). I was doing the chemistry
and Alan was doing the physics. Then Shirakawa came over and we were
getting these amazing results. I didn't know enough about the basic physics
techniques and asked Alan to join in. Actually, before asking him, I'd asked
another member of the Physics Department whose name I will not mention.
I told this person what we'd done, that we found these big changes occurring,
and he said, "Alan, this is just a junk effect." He thought it was a mess,
it was not crystalline, and he advised me, "Don't touch it." He declined
to collaborate with me. Alan Heeger was more adventurous and he was
willing to give it a shot. I think you have to take calculated risks.

Did you have any difficulty in publishing your results?

Our very first paper on poly(sulfurnitride) I submitted to the *Journal of
the American Chemical Society*. The manuscript had been in for quite a
while when I called the editor. It turned out that he had sent it out to
a number of referees, who were rather skeptical. Finally, the editor decided
to act as a referee himself and accepted our paper. He later became president
of the American Chemical Society.

Of course, we need not be very hard on the referees or on our first
physicist colleague who had declined cooperation with us. There was a time

when polymer science was considered to be a contradiction in terms. Solid state chemistry also had a difficult path. In the good old days, if you didn't get a stoichiometric composition, it was considered bad chemistry, and you better go back to your bench to re-crystallize, and re-purify your substance until you got a stoichiometric composition. It was not considered to be a nice and clean type of chemistry, rather, it was considered to be a good place where poor-quality work could get done and could get easily hidden. Things have changed considerably.

When I was a visiting professor in New Zealand in 1999, I learned that the kiwi is not only a fruit, it is also a bird, and Kiwi is the unofficial name of a New Zealander. During my visit I heard a lot of complaints about the insufficient support for science in New Zealand. Do you think that your Nobel Prize has had a beneficial effect?

This is a tremendously interesting question. The answer is, I believe, yes. My Nobel Prize has stimulated interest in New Zealand science. Incidentally, the population of New Zealand is 3.8 million, it is smaller than the population of greater Philadelphia. I believe the government has probably exploited my Nobel Prize as much as possible to promote interest in and support for science. For example, a MacDiarmid Chair in physical sciences was instituted about a year ago at Victoria University in Wellington, my alma mater. Just a few weeks ago, the MacDiarmid Institute of Materials Science and Nano-technology was officially dedicated. It coincided with the award by the New Zealand government of several thousand New Zealand dollars for a five-year period, which is being used to get the Institute going.

Your Nobel Prize has made an impact beyond Philadelphia and New Zealand.

The Chinese have built a beautiful research institute for me. It is the first research institute in China named after a foreigner. I am the director of this institute. Back in 1999, they made me Honorary Lifetime Professor at Jilin University in Changchun, China.

What do they expect of you?

This will be a show-case institute for China, an institute for ideas and the exchange of people. The emphasis is on people. A beautiful building alone would not suffice. I had had personal experience in this. About 50 years ago I just finished my Ph.D. at the University of Wisconsin. There was this beautiful, shining, state-of-the-art building at Wisconsin. Then I went to Cambridge, U.K., for my second Ph.D. When I walked into the lab in Cambridge, my heart fell. It was old and cold, the windows were dirty, you couldn't see out of them, and there were droplets of mercury in between the parquet blocks on the floor. But the Faculty was outstanding and they attracted outstanding students and postdocs from throughout the world. In those cold, dirty labs, they had state-of-the-art equipment and they also had an excellent library. This experience has molded my thinking ever since. Science is people more than anything.

The buildings may have been worn out, but the state-of-the-art equipment and the excellent library are more than just people.

If you get the right people, they will agitate, they will move heaven and earth to bring in the money to do the type of studies that they need to do. I also say, vision without money is a hallucination.

Speaking about Wisconsin, it has been one of the leading schools in the United States considering the number of Nobel laureates in the sciences, connected with the University of Wisconsin one way or another. However, they tend to associate with Wisconsin for relatively brief periods of their careers and then they move on to somewhere else. It does not seem to be able to hold them on the long run.

I don't know enough about the historical aspects of Wisconsin, but there are universities, and I will not go into specifics, that spend an enormous amount of money, sweat, blood, and tears to get good budding faculty members. Once they have joined, the same universities will unfortunately not always spend sweat, blood, and tears to keep them.

You have mentioned the new opportunities in New Zealand and China. Anything closer to home?

I have a very nice new opportunity at the University of Texas at Dallas where at the present time I am a science and technology advisor to the President of the University of Texas at Dallas. I am also a chair at the Nanotechnology Institute there and a visiting scholar in residence. I have my office and administrative assistant in Dallas and fly down there quite frequently. I will be 75 shortly and I feel that I have at least a good 10 years more. My health is excellent and my research is going well working on nanofibers of conducting polymers and throw-away plastic-paper electronic circuits, using conducting polymers. This is the way of the future. I have had about 25 issued patents and I have a company based on some of these patents. That raises another point. Basic scientific research and involvement in its technological applications are not necessarily incompatible. I have been involved with basic, fundamental research that has the promise of possible technological application. During the 25 years of my involvement in basic research on conducting polymers, we have come up with about one issued patent a year (I had had three patents before embarking on this work). It is fundamental research, yet it has an application for technology.

You now have three MacDiarmid institutes, one each in New Zealand, China, and Texas. Is there any interaction between them?

We have worked out a method of interaction between these institutes on a bilateral basis that is being signed during these weeks. The Chinese institute coming into being is especially intriguing. China has been behind the Western powers in materials science and technology. With this institute, China would like to show that it can be a world leader in this area, and if it can start on an equal basis, it can even surpass the Western powers in the future. This is seen as a real opportunity for China. Whether this will happen or not we will learn in the future.

Are they equal in nanoscience and nanotechnology at the start?

No, China is three or four years behind. However, nanoscience and technology is poorly defined. There was a special committee associated with the National Science Foundation, which defined a nanomaterial about two years ago. According to its definition, a nanomaterial has at least one dimension of 100 nanometers or less. I like this definition very much. One concern I have is that some people might survey their past work and might select the portion of it that deals with small dimensions and will re-label their projects and ask for more money under this umbrella. Using and misusing this as a buzzword is a real danger.

You have been a Nobel laureate for a year and a half. How has your life changed?

I've had less sleep in the last 18 months than ever before. The adrenaline content of my blood has been higher than ever before, and I've never worked so hard before as I have during these past 18 months. My main concern is that the Nobel Prize is harming my research. I don't have enough time for discussion with my students. At the time the prize was announced, I was working on a manuscript with a German postdoctoral fellow, Kurt. Following the announcement, for days and days I kept postponing our next discussion with him. At one point Kurt became real angry and told me, "Dr. MacDiarmid, this Nobel Prize came at a very inconvenient time." I don't have a large group, never had. When I was in Cambridge, in Harry Emeléus's group, the rule at Cambridge was that no faculty member should have more than 6 Ph.D. students at any given time. I have had a small group but kept very close contact with my students. I have encouraged them to telephone me any time of any day of the week if there is an urgent development that needs to be discussed immediately.

Do they?

On occasions. The worst was once at 2 a.m. on a Sunday morning. Now I have an answering machine.

May I ask you about religion?

I was born into a strongly Presbyterian family in New Zealand. Later my family changed to Christian Scientist and most of my life until my college years I was a Christian Scientist. By the time I went to the university, I no longer believed in the basic tenets of Christian Science. However, the concepts I have learned have really directed my whole life. Since then my religion has been whatever is the nearest church where my children could go to Sunday school most easily. I believe in God although I don't know in what sense. Every night before I go to sleep, I still say a little prayer like children do. I strongly believe in the effects that emotions have on the body chemistry, on the body's natural immune system. If one can achieve the appropriate state of mind, it will assist the body's natural immunochemistry. I am religious also from the point of view of my research. I do my homework, but I find thoughts and research ideas flowing into my mind without me having anything to do with it. My mind merely acts as a conduit, as a pipeline for research ideas. Whether it is something supernatural, I don't know, but I like the old Chinese proverb that says, "I am a very lucky person and the harder I work the luckier I seem to be." This sums up for me the relationship between religion and creativity.

Alan J. Heeger, 2004 (photograph by I. Hargittai).

26

ALAN J. HEEGER

Alan J. Heeger (b. 1936 in Sioux City, Iowa) holds the Presidential Chair and serves as Professor of Physics and Professor of Materials at the University of California at Santa Barbara. He heads a research group at the university's Center for Polymers and Organic Solids. Alan Heeger shared the 2000 Nobel Prize in Chemistry with Alan MacDiarmid of the University of Pennsylvania and Hideki Shirakawa of the University of Tsukuba "for the discovery and development of conductive polymers".

Alan Heeger grew up in Omaha, Nebraska. He was an undergraduate student at the University of Nebraska. He went to graduate school at the University of California at Berkeley. After completing his Ph.D. in Physics in 1961, Dr. Heeger taught and did research at the Department of Physics of the University of Pennsylvania for two decades. He was made Professor in 1967 and served as laboratory Director and Vice Provost for Research. He moved to Santa Barbara in 1982 where he has been ever since.

Among his many awards and distinctions, he is a member of the National Academy of Sciences of the U.S.A. and a member of the National Academy of Engineering of the U.S.A. He holds about 50 patents and is on the board of several high-tech companies.

We recorded our conversation in his office on February 8, 2004.

There are three Nobel laureates at the University of California at Santa Barbara, and this is unusual. How did it happen?

The first prize was to Walter Kohn, which was in 1998. Walter is a theoretical physicist; his work had been widely recognized, and he received the prize in chemistry. Then in the year 2000, I received a prize also in chemistry, having also a physics background. In that very same year Herb Kroemer, who is a physicist and an electrical engineer and whose appointment is in electrical engineering, was awarded the prize in physics. This is an interesting university, a place where interdisciplinary science is the culture, and not just the way we operate.

You did your prize-winning research in Philadelphia.

I came here in 1982. The original work started at the University of Pennsylvania a few years before that, in 1976–1977 with Alan MacDiarmid and Hideki Shirakawa. But it was particularly pleasing for me that the citation on the Nobel Prize was for the discovery and development of conducting polymers. The discovery is a moment or at least a singular time, but often, and certainly in this case, the development of this class of materials, the development of the science, the beginning of the technology took 25 years. Much of the work was done here after I was here. We started our company UNIAX in 1990 and brought that to flourish, which was the first commercialization of this class of materials.

Why did you move here?

It looked like an opportunity, which I would regret not to take.

You mean California?

California is nice but that was not the point. The offer came at a point in time when the Institute of Theoretical Physics had just been formed. When I was approached, I had the sense that something special might happen here and that I might be able to play a critical role in that. It was somewhat of a risk because the University of Pennsylvania is a well-established research university and the University of California at Santa Barbara at that time was certainly not. It was a decision that I am very pleased that I did make. We have seen this institution coming from UCSB, "University of California Sunny Beach", to a really world-class institution of international impact. Physics, engineering, and the science department are world-class now.

My observation has been that there are schools of great tradition that keep people around, like Princeton University, whereas some other great school, like Wisconsin, have been connected with a considerable number of famous scientists who appear there only in passing. Do you think that Santa Barbara will be able to keep its best people? It will also be a hard act to live up to have three Nobel laureates in the period of three years.

If I look at some of my colleagues here, I wouldn't be surprised to see several more Nobel Prizes in the coming years. We have been both wise and very fortunate in the people we brought in. A university is its faculty and there is a tradition here having to do with interdisciplinary science and that approach to science will be a continuing effort over the next decades.

I apologize if my questions don't always sound too tactful, but I have the impression, also from having talked with Alan MacDiarmid, that UPenn may not have quite appreciated you and him. My impression was that the other Alan received more opportunities from outside of his own school then in his home base, from places like China and Texas.

They tried very hard to keep me. UPenn is a great institution, but a lot of great scientists have gone through UPenn also. Walter Kohn was there too. You never know how these things go. I certainly look back upon my years at UPenn as very special. It was that time when I formulated this concept of deep interdisciplinary science. I was in the Laboratory for Research of the Structure of Matter at UPenn from the moment I came there; it had just been formed and in the early 1970s I became director of that Lab. That laboratory was set up to be an interdisciplinary place for science and engineering, primarily physics, chemistry, and materials. As a scientist working in that Lab and even more so as its director for 8 years, I became deeply involved with interdisciplinary science, with crossing over into chemistry. When people ask me, what do you do, my answer is, Scientist. I enjoy that, I think that way, and I feel very comfortable talking about chemistry and I am learning some biology these days.

How far into biology?

I am working now on DNA, biosensors, proteins, new things. I just found that the process of moving to something new is exciting. Basically, I am a risk taker. I enjoy this process of reaching out. I like to go skiing and come to the edge, look down, and do it. It is the same kind of thing when you go into a new field.

Is there still any risk for you today to take?

It's a bigger risk. If I made a fool of myself in science 30 years ago, it was just me. If I make a fool of myself now, everybody will know it, and it will be bad. Science by its very nature is risky. When you write a paper, you say, "I am going to tell you that I understand this and this is right." You are doing your best, but you don't know. I pride myself with doing forefront leading edge research and especially since it is inter-disciplinary, that is, you are reaching out into something where you really don't have all the knowledge, it's dangerous.

Did you ever have your papers rejected?

I had many papers rejected, sometimes because they misunderstood it, sometimes because it is too far out. One of my favorite referee's reports, on a paper that I actually liked a lot, said, "This paper is spherical nonsense."

Meaning …

That it makes no sense from any point of view. We all have these experiences. I have had plenty of grant proposals rejected also. If you propose to do something really new, the referee can always find many different reasons to say that this wouldn't work.

Would you please give a summary of the importance of the conducting polymers from the point of view of basic science and from the point of view of applications?

From the basic science point of view, there are a number of issues. First of all, it really changed the whole world of materials in the context that there were no such things as metallic polymers. This was a term that did not and could not exist. It goes back to the work in the 1970s of Nevill Mott and Phil Anderson working on the metal-insulator transition. I was interested in that as were many people. I was also aware of Peierls's work on the fact that one-dimensional conductors are stable to phase transition to become insulators. This idea then of trying to make real systems which were sufficiently anisotropic to be viewed as one-dimensional conductors, was to me a very interesting possibility. I got started in this whole field by a rather subtle mathematical issue having to do with one-dimensional physics. We worked on two levels. First we were playing with molecular structures, flat planar molecules that have the ability to stack. Then I remember very clearly all of a sudden realizing that if we could do this in polymers, then we would have not only that beauty of very high anisotropy of quasi one-dimensionality, but we would have the potential of interesting materials for applications.

It was a new kind of matter.

That's right. We saw very early in our initial discoveries with MacDiarmid and Shirakawa, for example, electrical conductivity, which was the fundamental discovery, and one could see the potential for applications. Immediately afterwards we discovered the electrochemistry of these materials and we could see all kinds of further applications. We saw that the optical properties changed as we doped these materials, so we went from semiconductor, which was opaque in the visible to a metal which was opaque in the infrared, but transparent in the visible. At that time the materials were still unstable, they were not processable, so there was a long way to go. Many people contributed to each step along that way, putting functional groups on the side chains, for example, and so on. It was the decade of the 1980s when these materials were becoming more mature. Then the device physics started in the 1990s, LEDs and diodes were made, and one could see that this had many possibilities. Today we have conducting polymers that have all the characteristic properties of a metal, conducting polymers that can be textbook examples of metals. But these conducting polymers can be kept in solution, for example, polyaniline can be kept in a bottle as dissolved in toluene, it is a typical solution. Then you let toluene evaporate and you have a metal. Which real metal can do that? There are lots of good metals, but the conducting polymers have some additional properties that the real metals do not. Their solutions are inks and you can print them, just to mention an example of application. We can print displays, we can print solar cells, large areas, low cost.

At which point did you realize that applications would be forthcoming and at which point did you start filing patents?

We began to see that early on. Some things we even anticipated too early that never came to fruition. Early on we made some batteries using these materials but that never became an industrial application. On the other hand, in 1987 I remember, we made here the first diode by casting a film from solution. That seemed like such a simple thing that I didn't even bother to write a patent, and of course that was foolish because the diode is a fundamental electronic component. Now I know that I could've written a patent that could have been very broad. We had early patents on some materials that were discovered by someone else at Cambridge, but we were right there with a number of important contributions and some patents. We continue to do that but there was no one point where you see everything. We made contributions to the discovery of soluble semi-conducting and metallic polymers. Many other people contributed to that as well. You see these things happening and all of a sudden the field has emerged to a new point.

When did it first occur to you that you might receive the Nobel Prize for this work?

Shortly after the initial discovery of doping. I knew it was important.

It was 25 years ago.

It was 25 years but it was not, as I said earlier, a completed discovery; there were discoveries along the way, all the time. It was certainly not a boring period.

When did you start paying attention to the October announcements of the Nobel Prize?

People started saying to me that one of these days I would get the Nobel Prize in the 1990s.

Did you wonder who would be the maximum three people involved?

I did not grasp that question early on. I had no doubt in my mind that Hideki Shirakawa made important contributions early on and in some sense he set the foundation which enabled us to go forward. There was also no doubt in my mind that Alan MacDiarmid and I worked very close together, so it was the right group. We were the pioneers. This is not to say that many other people didn't make important contributions.

Originally, these materials were just a curiosity.

You could do science on them but they were not yet ready for processing. The most exciting thing was to watch the field taking off. We made the initial discovery in the late 1970s and by the early 1980s there were international conferences with hundreds of people and in the 1990s with thousands of people, and it goes on today. I started a company here in

Santa Barbara in 1990, it's UNIAX. Initially I started it with a colleague here, Paul Smith who is now at ETH in Zurich in the polymer department. He had a strong background in materials but had not had a good sense of the importance of processability, etc. Our initial idea was making metallic polymers processable and we succeeded in doing that. Then we moved into the area of polymer LEDs, light-emitting devices, and that became the focus of the company, which was quite successful. In the year 2000 that company was acquired by Du Pont. The year 2000 was a very good year for me. We sold the company and I received the Nobel Prize. The only thing that might have been better if it happened in the opposite order; the company might have brought in more value. I am no longer involved with that company but I am involved with several other commercial ideas. Once you've done this once, it's a Californian entrepreneurial disease. I am thinking of and working on starting another company.

Who is your hero?

There are many. Charles Townes, discovered the laser, became an astrophysicist, he is now 88 years old and going strong. We met last year in St. Petersburg. Watson and Crick's discovery was incredible. This one

Alan Heeger with John Bardeen (1908–1991, Nobel Prize in Physics 1956 and 1972) and Robert Schrieffer (b. 1931, Nobel Prize in Physics 1972) (courtesy of Alan Heeger).

beautiful understanding, look at what it did. That discovery had many interesting aspects and one of them is that people think that in order to make progress, you have to know everything. We scientists know that this is not true. You don't have to know everything, you don't even have to know much; you may need to know a little and you can start thinking, trying to understand something, and Watson–Crick is an interesting example, because neither of them was an expert. When I was at the University of Pennsylvania as a young faculty, I had the good fortune to have as a colleague Bob Schrieffer who came in the same year. He had been at the University of Illinois at Urbana and even then everybody knew that one day there would be a Nobel Prize for BCS [Bardeen–Cooper–Schrieffer theory of superconductivity], so even as a young man he was a famous scientist. The point I am making is that if I have to single out one person from whom I learned good taste in science, it is Bob Schrieffer. Having good taste in science is important. Almost everything you do in science is typical, so one of the most important things is the nature of the problem you choose. Is it a problem that will have impact, will it be important? You can't learn that; the only way to learn that is working with people whom you can mimic. I published some very important papers with Bob Schrieffer but more than that, I learned a great deal about good taste in science from him. He is now in Florida but we still see each other. We have now a tradition that a group of us, five or six couples get together for New Year's Eve every year. The group originated from Philadelphia.

Could we talk a little about the history of the discovery of conducting polymers? I sent you my interview with Alan MacDiarmid. Did you have a chance to read it?

I'm sorry I did not. So I can give you a completely independent view of it.

Just to be fair, I had also a little correspondence about it with Hideki Shirakawa.

I had been working in this area of quasi one-dimensional conductors since the early 1970s. These were initially materials that I mentioned before, stacked organic molecules with good pi–pi overlap along the stack. Then in about 1975, all of a sudden, this material poly(sulfurnitride) emerged. It was very exciting because it was along the lines I was working even though it was also very different. Although it was interesting to me, I had no way to deal with that, but I wanted to have some of that. I had heard MacDiarmid give some talks in the Materials Research Lab at UPenn. He was mostly involved with silicon chemistry, but I also remembered his name some way connected with sulfur–nitrogen chemistry. So I called him up and told him that I wanted to get together and discuss some science with him. I still clearly remember when we met; it was a nice November day and we did not have many in Philadelphia, but it was a sunny, beautiful day. It was late October or early November because the days were becoming short.

Which year?

Probably 1975. I went to his office and I was telling him about "SNX" and how this was a wonderful material. It was a metal and wouldn't it be great to study this and that I knew that he had some interest in such materials and I went on and on and on, for a long time. We talked, but he was not interested. The reason he was not interested was because I was saying $(SN)_x$ and he was hearing Sn_x. He was not very impressed that tin is a metal. It was such a funny beginning.

But he let you talk.

He is a polite man. He thought I was crazy, but he is a polite man. However, once we got passed that, we were on. He had a beautiful background for this problem. He had worked on S_4N_4 and we got into that. We wrote a number of nice papers and developed a relationship. That was the important thing, not so much that the poly(sulfurnitride) was an interesting project but that the two of us got together. It was interesting; we wanted to learn, we each of us wanted to reach out across this deep valley of interdisciplinarity. We would get together on Saturday mornings, it was not group meetings, it was just the two of us, to talk about this kind of science. I remember clearly that I tried to teach him about the metal-insulator transition. In order to do that I wanted to choose a simple system, so I suggested to consider a chain of hydrogen atoms. Alan said, no. It doesn't work. It doesn't exist. I remember coming back another week, suggesting CH as the basic unit and that he accepted and we talked about it without my knowing about polyacetylene. Very shortly afterward he went to Japan on a month-long trip and he was giving lectures about poly(sulfurnitride). Alan is a very visual man, he loves color, and he showed photographs of his poly(sulfurnitride) golden films, and he carried little vials with the crystals of the substance.

After one of those lectures, Shirakawa came up to him and told him that he had some silvery films. When Alan came back and told me that it was polyacetylene, we talked about that. We were able to get Shirakawa into Philadelphia within a few months. I remember this as if it was yesterday.

Are you familiar with the Shirakawa story? How did that happen?

I know all those stories. It's a wonderful story that he had been working on polyacetylene and he, as all others working on polyacetylene, had a black powder that was not easy to deal with. And it was not very interesting either because of that. Then he had a Korean visitor who misunderstood what he said in Japanese and instead of making the catalyst in the millimolar concentration, he made it in molar concentration and out came something very different. It was a gel, which was difficult to deal with. It's an amazing story but what is really wonderful about it is that maybe 99 out of 100 would have said, "Stupid." Shirakawa didn't do that; he said, "This is interesting." When I look back at this, here we were, a physicist with no credentials in chemistry, an inorganic chemist who did not have any serious knowledge in organic chemistry as a professional. Yet we were trying to start this new field only because Shirakawa had created a solid foundation for it. They had done all conceivable measurements, infrared spectroscopy, X-rays, and characterized it in every way they could. For Shirakawa, that was the work of his life. It was his goal trying to make polyacetylene better.

Do you know the name of that Korean co-worker?

I should know but I don't remember.

I think we all should know his name. It is Dr. Hyung Chick Pyon as I learned it from Shirakawa. Don't you think he might have merited to get invited to the Nobel ceremony?

I don't know. I don't know anything about their relationship.

Of course, we should not exaggerate the importance of his contribution because if it had been up to him, the experiment might have disappeared into oblivion. But for the story of the discovery, it was essential.

I never thought about it and you may be right. I don't know. In any case, shortly after Shirakawa arrived in Philadelphia, he walked into my office. We talked about polyacetylene and it was obvious that this might be an important system. It had an interesting electronic system. It was not our discovery, it had had extensive literature. We talked about doping as a means of inducing the formation of a metallic state. He had already had some vague idea about that. He had also seen some anomalous infrared absorption and he didn't know why. It occurred to me that it was because of doping. In fact, this is how we came to the idea of doping. We talked about this and had this idea of charge-transfer doping, right on the spot there. We laid out the concept and then turned it over to a postdoc of mine, C. K. Chiang. We set out a plan for doing also infrared experiments; a student of mine was doing far infrared measurements, and it worked.

Did you invite Dr. Chiang to Stockholm?

I did.

Did you ever discuss the history of your interactions with Alan MacDiarmid?

I told you about how we got together.

I have published an interview with Alan MacDiarmid in which he mentioned your jointly working on poly(sulfurnitride), he doing the chemistry and you doing the physics. Then he told me about Shirakawa and his coming to Philadelphia and they getting amazing results. Again, he said, he did not know enough about the basic physics, so he asked you to join in. But, he added, before asking you, he had suggested cooperation to another physics professor in Philadelphia, but this colleague turned him down. This physicist even warned Alan not to touch these messy substances. Alan said that you were more adventurous and you were willing to give it a shot.

We all have our own memories of those times.

MacDiarmid's story is very dramatic in that his physicist colleague tried to protect his reputation that working on these messy substances might ruin.

This goes back to what I said earlier: anything you do in science is dangerous in the sense that you put yourself on the line.

MacDiarmid told me that this colleague is still at UPenn.

I know the story; I know it all. But I want to make a point: in that situation the rigor of the work that Shirakawa had done made it possible for us to go forward. Otherwise we would not have been able to go forward.

If Alan had not gone to Japan, had not given his talk at Shirakawa's department, Shirakawa had not attended his presentation, nothing would have come out of Shirakawa's discovery.

That's absolutely true. I remember going to Japan not long after that, Alan and I went together — and I think what I am going to say is still true today — people viewed us as coming there, seeing this gem, pulling it out and bringing it to America. And in some sense this is true. That couldn't have happened without this unique combination of history and people and interactions, the poly(sulfurnitride) story, and so on.

There is a widespread feeling among many Europeans too that American scientists are so much more powerful that it is better to develop a discovery without the Americans knowing about it too early. This may not be a valid fear because there are so many small groups in American universities that are not very different from their European counterparts. The story of conducting polymers though may look like an example of the powerful Americans developing a Japanese finding. However, my impression is that Shirakawa's discovery might have disappeared into oblivion without the Philadelphia connection.

We were ready. It's amazing. We were thinking about this kind of problem and we were ready. At least in my memory we had talked about the CH polymer. So when Alan came back and showed me Shirakawa's sample, we were ready. It all happened very quickly. Alan arrived in September of 1976 and the famous experiments where we did the doping and got the ten orders of magnitude increase in conductivity were done in November of the same year.

Was Shirakawa there by then?

Yeah.

So you didn't just bring over the discovery and the sample but the discoverer as well.

We brought him in. That was wise, it was the right thing to do historically too. There was a great deal of courage involved in that act shown by

Shirakawa. He was a young man and to reach out the way he did was very uncharacteristic in the sense of Japanese style. So he was very courageous as well. He may have had a tenured job but he was certainly not the senior professor at his place and he may have been criticized for what he had done. He was with us only for one year and made a big impact. Our collaboration, the three-way collaboration continued when he went back to Japan. But then — as such things happen — it died off over some short time. He sent us samples and the cooperation continued even after I had moved here. He sent over his colleague — who is now in his position in Tsukuba — to work with us. He spent a couple of months with me, helped us to get started, so we continued to interact for quite a few years. Shirakawa retired not long before the Nobel Prize and stopped doing science.

And you and Alan MacDiarmid?

It's difficult to have a close interaction at long distance. We always have remained close both as friends and as interested in science that each of us continued to do, but we didn't collaborate a lot after I left. He has remained very active in science.

Coming back to conducting polymers, you mentioned biological applications. Would you tell us more about it?

I had expected for many years that conducting polymers should have a role in biology and medicine. There were a couple of things that looked interesting. They might be important in nerve repair and there is a growing literature on the subject and I am looking into it right now. Another idea is to use conducting polymers as light harvesters. We did some experiments in the early 1990s by putting a C_{60} molecule near a luminescent polymer. C_{60} is a good acceptor of electrons. If you photo-excite the polymer, kicking an electron from the pi band up to the pi-star band, without the C_{60} being present, in many cases you get strong photo-luminescence. We found that the energy level of the LUMO of C_{60} was just right so that if C_{60} is present when you make this photo-excitation, the electron should transfer from the polymer into C_{60}. Indeed, it happens and it happens so fast that the luminescence is severely quenched because the electron is now separated from the hole. We worked to time-resolve that and in our early experiments we could see that the electron transfer happens in less than a picosecond. We detected that with a pulse laser. Subsequently it time-resolved and it is fifty femtoseconds.

Then, one of my former students who was at Los Alamos at the time, discovered that you could use this as a biosensor. They used water-soluble semiconducting polymers. They put a quencher, something that would quench the luminescence by electron transfer. Then they hooked it on to a biological ligand. The quencher was charged and went near the charged polymer and quenched the polymer. When the antigen came, or whatever was on the ligand, it pulled that quencher away, and you got luminescence. The thing that was really interesting and still somewhat mysterious is that the transport of energy along the polymer chain is very efficient; a single quencher will quench the whole chain. This capability of doing light harvesting plus the specificity that you can build into the biological realm, antigen/antibody, etc., lead to the whole concept of biosensors.

For the last two years we have been doing here DNA sequence detection. We are not trying to sequence the genome but I want to know the sequence of a genetic disease, whether one has this genetic disease or not, is it genetically-engineered food, is it my product or your product, and so on. I want to know if it is anthrax of bio-terrorism or it is just the flu. It may be just a 20-base sequence and you can tell but you need to read the bases.

How do you know which 20-base section to check?

That's for the biologist to know. There are now libraries which tell you. So we have been working on this luminescence scheme, a very elegant approach, involving DNA. Another project is related to electrochemistry. I had learned a little electrochemistry in connection with conducting polymers and electrochemical doping. We use a DNA which has on one end all Gs and on the other end it has all Cs. It forms a loop on itself. Then we attach that loop to a gold electrode and we put a redox label on the other end. When it's in the loop, it's held down close to the gold and you get strong electrochemical redox; you can see it. When you bring in the complementary DNA, it competes with the loop, it opens up the loop, takes away the ferrocene molecule or whatever else may be there to about a hundred angstrom distance, and turns it off. You can read the sequence using single molecule conformational change. We are very excited about these schemes. Most of my research group — and I have a 20-member group — is still working on conducting and semiconducting polymers, but I am most excited about the biosensor staff. In addition to my group in which the postdocs and students work directly with me, I have close collaboration with colleagues.

Can you give me a ballpark figure of your annual budget?

More than half a million and less than a million.

Do you spend a lot of time on writing grant applications?

Less time than I used to spend before the Nobel Prize.

What are your funding agencies?

I have support from NSF, NIH, the Army Research Office, Air Force, several industrial sources. Then we have here in Santa Barbara wonderful block grants. The first we got from the National Science Foundation, a large grant for materials science. A block grant brings in, say, 5 million dollars a year, it is locally administered, we manage the money here, we set up our own groups, this is very nice. Then, a few years ago, a Japanese chemical company, Mitsubishi came here and wanted to set up an institute. They have put in two and a half million dollars per year for 5 years. It wasn't just for me, it's broadly materials, not just conducting polymers. Just this year the Army Research Office is putting here an Institute for Collaborative Biotechnology and I am working in it with the biosensor work. The block grants give you flexibility; when you need something you can quickly respond and also, they tend to solidify collaborations. We work together not only because we want to but also because there is a funding source that helps make it happen. It puts a lot of confidence into this group of people.

I would like to ask you about your family background.

On either side of my family I was the first with an advanced degree and one of few who even had a college degree. My father was a business man in small town Iowa and died young. My mother did not have much schooling but was very intelligent. She assumed and never gave it a second thought that my brother and I would go to university. She had a strong feeling for education that is often the case in Jewish families that came from Eastern Europe. But there was no one in our family as far as I know who was a great rabbi or scholar.

Your present family?

Ruth and I were married when I started graduate school; I was 22 and she was 20. We went off to Cornell at the beginning, but I completed it at Berkeley. When I finished we already had two children. This was

Ruth and Alan Heeger in Stockholm, 2001 during the Nobel Prize centennial celebrations (photograph by I. Hargittai).

in the early 1960s. Postdoc-ing was not so widespread as today and I was able to move from the Ph.D. into the Assistant Professor position at the University of Pennsylvania, and got started right away. I was quickly promoted there.

How about religion?

It's not a big part of my life; I am not religious. I am a Jew, I am part of that culture, but I never go to services. We raised our sons as Jewish, not as religious.

Did you have hurdles in your career?

Because my father died young — I was 9 and my brother was 2 years old — we didn't have much money, I didn't go to Harvard, I went to the University of Nebraska. It gave me a good start and I went off to world-class places. Everything has gone smoothly since then.

Did the Nobel Prize change your life?

Yes.

In what way?

Every way. In opportunities and expectations. Just the other day there was a site visit here from one of the major institutes that I had mentioned in our conversation. I was asked to give a speech at the end of the day; they asked me to give an inspirational speech, and I did, and I felt good about it. When a Nobel laureate goes to Asia, mothers bring up their children for almost like a blessing. People treat me differently than they used to. There is also responsibility associated with that because the Nobel Prize and the whole Nobel tradition should be maintained. In Santa Barbara, it was not only important for me as an individual, but was important for the University. This community somewhat under-appreciated what was happening out here. The relationship between the University and the community was not a tight one. But after these Nobel Prizes, people still stop me on the street today and say, congratulations. It changed my life in subtle and direct ways.

What do your sons do?

Both of them are involved in science. The older one, Peter, is a medical doctor who no longer is practicing medicine but has become an immunologist; he does transplant immunology at the Cleveland Clinic. The younger one, David is a neuroscientist and does functional MRI. I have written papers with both of them. My oldest grandson is entering university this fall.

Any message?

In science — as in other areas of human endeavor — you need to take risks. Only then when you go out into something you don't know, are you going to find something interesting.

Jens Christian Skou, 2003 (photograph by I. Hargittai).

27

JENS CHRISTIAN SKOU

Jens Christian Skou (b. 1918 in Lemvig, Denmark) is Professor Emeritus at the Department of Biophysics of the University of Aarhus, Denmark. He received half of the Nobel Prize in Chemistry in 1997 "for the first discovery of an ion-transporting enzyme, Na$^+$,K$^+$ATPase". The other half of that Nobel Prize was shared by Paul D. Boyer and John E. Walker "for their elucidation of the enzymatic mechanism underlying the synthesis of adenosine triphosphate (ATP)".

Jens Christian Skou received his M.D. degree from the University of Copenhagen in 1944 and his Doctor of Medical Sciences degree also from the University of Copenhagen in 1954. He received clinical training at the Hospital at Hjørring and at the Orthopedic Clinic in Aarhus. He held appointments as Assistant Professor (1947), Associate Professor (1954), and Professor (1963) at the Institute of Physiology of the University of Aarhus. In 1978–1988 he was Professor at the Institute of Biophysics at the same university. His distinctions — in addition to the Nobel Prize — include various Scandinavian and European awards, honorary doctorates and memberships in learned societies. Among them, he has been a member of the Danish Royal Academy of Sciences (1965) and the Academia Europaea (1993), and a foreign associate of the National Academy of Sciences of the U.S.A. (1988).

We recorded our conversation in Professor Skou's office at Aarhus University on September 20, 2003.*

---

*In part, this has appeared in Hargittai, B.; Hargittai, I. *Chemistry International* **2004** *26*(2), 14–17.

First, I would like to ask you about the importance of sodium and potassium pumps.

In the 1870s it was shown that there is a difference in the concentration of sodium and potassium inside the cell and outside the cell. The potassium concentration is higher in the cell than outside and the reverse is true for the sodium concentration. The question was how to explain it. Part of the explanation was given by Donnan in 1913. He showed that if you have a cell which contains proteins which cannot pass the cell membrane and there is an ion pair, for example, potassium chloride, which can pass the cell membrane, then at equilibrium the product of potassium and chloride in the cell will be equal to the product of potassium and chloride on the outside. But as there must be electro-neutrality on the inside, the concentration of potassium must be higher than the chloride concentration, because part of the potassium concentration is used to neutralize the protein negative charges. On the outside, the product consists of equal components. If you have two components in a product with unequal size, which is equal to a product of two components of equal size, then the sum of the two unequal components is higher than the sum of the two equal components. This means that there is a higher concentration of potassium + chloride inside than outside, which means a higher osmotic pressure. Furthermore, there are the proteins in the cell, which also adds to the osmotic pressure. Equilibrium can only be obtained if water can be prevented from flowing in. As the cell membrane is permeable to water, the only way to establish the equilibrium is by adding some ions on the outside to compensate for the high osmotic pressure inside the cell, and that ion is sodium. That is why the sodium concentration must be higher outside than inside. But the problem was then to explain why sodium is not distributed like potassium. The answer to this was, until 1939, that this is because the membrane is impermeable to sodium.

But sodium is smaller than potassium.

Yes, but hydrated sodium is bigger than potassium. Even so, it was a puzzle to understand how the membrane, which consists of fatty molecules, could be permeable to potassium chloride. In any case, it was observed and it was measured. But it was difficult to get reliable values for intracellular sodium partly because of the high content of sodium in the extracellular tissue. So it was accepted until 1939 that the membrane was impermeable for sodium.

Then it was shown by some American scientists, Fenn, Steinbach, and Heppel independently that the frog muscle membrane is permeable to sodium. From this, another American scientist, Dean wrote a theoretical paper on electrolyte distribution in mammalian cells. He suggested that the only way to explain the difference in the sodium concentration on the two sides of the membrane was that there were pumps in the membrane, which could expel sodium from the inside. This was very difficult for many scientists to accept at that time, especially for chemists. It was difficult to accept that the membrane, which according to the present view consisted of a bimolecular layer of lipids, could have an active component, which could convert chemical energy into work. Much discussion was going on. The foremost defender of the view that the membrane is impermeable to sodium was Conway from Ireland. He did a number of experiments together with Boyle. They varied the extracellular concentration of potassium of frog muscles, measured the distribution across the membrane and compared this with the calculated distribution assuming that the membrane was impermeable to sodium. The measured and calculated values agreed for values of extracellular potassium from 28 moles and up, but not at the lower physiological values. In spite of the inconsistencies they took the result as a support for the view that the membrane is impermeable to sodium. It is a typical example of a research where you are preoccupied by a certain view and try to prove that it is correct instead of asking, "Could there be other possibilities?" The other possibility was that the distribution of sodium is not as Conway and Boyle assumed, an equilibrium distribution, but as suggested by Dean, a steady state distribution with the membrane functionally impermeable to sodium. The paper was published in 1941 in the same year Dean published his paper suggesting the pumps. Experiments in the following years showed that Dean's view was correct, there were pumps in the membrane. Conway reluctantly gave way. When I met him in 1961 he admitted that there may be pumps in the membranes, but the low permeability of the membrane to sodium is the most important. He was right in the sense that from the point of view of the pump's energy consumption it is important that the membrane has a low permeability to sodium. 20–25% of the basic metabolism is used for pumping sodium out of the cells. What we did not know in 1961 or, more correctly, what we started to know in 1961 is that the energy used for keeping sodium out of the cells is not wasted. The gradient for sodium from outside the cell to the inside which is created and sustained by the pumps in the cell membrane is an energy source, which — by carrier molecules in the membranes — is used for transport of other substances in and out of the cells against their gradients driven by the gradient for sodium.

The pump is not only important for the distribution of the ions between the single cell and its surroundings, which is necessary in order to solve the cells osmotic problem due to the presence of proteins inside the cells, but also for a number of other functions. In the intestine the gradient for sodium created by the pumps is used for absorption of sugar and amino aids. 170 litres of water is filtered from the blood to the kidney. Of this 1 litre is excreted as urine while 169 litres after being cleaned for waste product are reabsorbed to the blood. The driving force for this is osmotic gradients created by the sodium pumps in the kidney. The potential across the cell membranes which is about 70 millivolts positive to the outside is due to the sodium and potassium gradients across the membrane created and sustained by the pumps. The membrane is more permeable to potassium than to sodium. This means that potassium leaks out faster than sodium leaks into the cell and this leads to a diffusion potential across the membrane, which increases to a size of about 70 millivolts positive to the outside. At this potential, the rate of the leak of the two ions across the membrane is equal. The membrane potential in nerves and muscles is the basis for the nerve impulses. The nerve impulse is due to a short-lasting localized increase in permeability for sodium and — as sodium is in a higher concentration outside than inside the cell — to an influx of sodium and — as sodium carries a positive charge — to a depolarization of the membrane potential. This is followed by a decrease in permeability for sodium and an increase in permeability for potassium, which flows out of the cell, and by this repolarizes the membrane potential. Following this the pumps will pump the sodium which had flown into the cell out of and the potassium into the cell. In the brain more than half of the metabolism is used for pumping ions.

What is the role of these ions in the cell?

A number of enzymatic processes in the cell requires the presence of potassium. Sodium has a different role, it just compensates for the osmotic pressure.

How does this relate to our intake of sodium and potassium?

Normally we get much more potassium than sodium because all our food stuffs contain a high concentration of potassium. We usually add extra sodium

Young Jens Christian Skou
(courtesy of Jens Christian Skou).

artificially to our foodstuff. It gives it taste and it is also necessary since we do not get enough sodium in our food.

At which point did you join this field of research?

I joined this field in the beginning of the 1950s. By then the existence of the ion pump had been accepted. But it was not known what the nature of the pump was. I got my M.D. in 1944 and started my internship at a hospital in Hjørring, in the northern part of the country. I wanted to become a surgeon. While I was in the surgical ward, I became interested in the action mechanism of local anesthetics. At that time we did not have anesthetists; it was just in its beginning as a speciality in Denmark. Nurses did the narcosis and used ether or chloroform. It is very unpleasant, not only for the patient but also for the person who administers it. This is why, whenever it was possible, we used spinal anesthesia. For spinal anesthesia, anesthetic narcotics are used, which are neutral substances like ether and chloroform, whereas local anesthetics are weak bases, which in a water solution dissociates into an ionized part which is water soluble and a non-ionized part which is lipid-soluble. From the pharmacology curriculum I knew that Meyer and Overton independently around the turn of the previous century had shown that there is a correlation between the lipid-solubility of narcotics and the narcotic effect. I wondered whether something similar existed for local anesthetics. I also wondered whether it was the ionized or the non-ionized part in the local anesthetic that was the active ingredient. I decided to use this problem as the subject for a thesis later on.

After two years of clinical training, there was a position vacant at the Orthopaedic Hospital in Aarhus and I got it. After one more year there, I applied and received a position at the Physiological Department of Aarhus University. In Aarhus, I started to study the effect of local anesthetics. I measured the minimum concentration of 5 local anesthetics and of butyl alcohol as a representative of a narcotic which could block nerve conduction in the ischiatic nerve from a frog, and how the blocking potency depended on the pH. I then compared this to their solubility in lipid. There was a certain correlation, but not nearly as good as for the narcotics. So I went on and looked for a possible correlation between the capillary activity of these substances and their narcotic effect. Again, there was some correlation but not good enough. Then I read about Langmuir's work on mono-molecular layers of lipids on a water phase in a book *The Physics and Chemistry of Surfaces* by N. K. Adam.

Schulman had described how drugs added to the water phase beneath a monomolecular layer of lipids penetrated up into the monolayer and at a given area of the monolayer increased the pressure in the monolayer. I realized that a phospholipid monolayer is similar to one half of the cell membrane, and that is why it could be used as a model of the water-lipid interface of the cell membrane; the cell membrane is a double layer of phospholipids. I extracted lipids from the ischiatic nerves from frogs, added a drop of the lipids dissolved in an organic solvent on the water surface of a Langmuir trough, and observed that after the evaporation of the solvent the lipids formed a monomolecular layer if the area of the surface was big enough. In the Langmuir trough the pressure the monolayer exerts on a floating barrier can be measured as a function of the area of the monolayer. I adjusted the area of the monolayer to a pressure of 10 dynes per centimeter, and observed that the pressure in the monolayer increased by adding local anesthetics to the water phase beneath the monolayer. The pressure increased by increasing the concentrations of the local anesthetics, and the more potent the local anesthetic the less was the concentration necessary to give a certain increase in pressure. The relative blocking potency of the 5 local anesthetics varied by a factor of 1 to 920 from the least to the most potent, which means that the most potent blocked the nerve conduction in a concentration which was 1/920 of the concentration necessary for the least potent. The minimum blocking concentration of the 5 local anesthetics gave an increase in pressure in the monolayer which varied from 3 to 10 dynes per centimeter, which means by a factor of about 3, compared to a variation in blocking potency of 920. This suggested a correlation between the ability to block nerve conduction and the ability to penetrate into and increase the pressure in the monolayer of lipids extracted from the nerves.

I used these experiments and results for my thesis, which was published as a book in Danish. Afterwards I wrote it up in 6 papers published in English. But I wanted to continue this work to see what could be the connection between the penetration and pressure increase in the monolayer and the blocking of the nerve impulse. This was in the late 1940s and by that time Hodgkin, Huxley, and Katz had published their papers on the mechanism of the nerve impulse. As mentioned previously the nerve impulse is due to a localized short-lasting increase in permeability to sodium which leads to an influx of sodium and depolarization of the membrane, followed by an efflux of potassium which leads to repolarization. As the local anesthetic in blocking concentrations has no effect on the membrane potential a possible connection between the pressure increase and nerve block could be that the local anesthetic by the pressure increase blocked the opening of the membrane for sodium, and thereby for the influx of sodium and, consequently, the initiation of the nerve impulse. The structural basis of the opening of the membrane for sodium was unknown. In my view, it was most likely that it was a protein which by a change in conformation could open for sodium and that the local anesthetic by penetrating into the membrane blocked the conformational change. In order to show this I should incorporate the protein into a monolayer of lipids then add a local anesthetic to the water phase and see if the penetration into the monolayer had an influence on the conformation. There were, however, two problems. One was as mentioned that it was unknown whether such a protein existed (it was later shown that it does). The other was that there were no methods available to measure a conformational chance of a protein in a monolayer. That is why my idea was to incorporate a protein which had enzymatic activity in the lipid monolayer and then see if penetration of a local anesthetic into the monolayer had an effect on the enzymatic activity, and if so to take this as an indication that there was an effect on the conformation of the protein.

What do you mean by the conformation of the protein?

Three-dimensional structure.

When was this?

At the beginning of the 1950s.

At the time when Linus Pauling was discovering the α-helix.

Yes. But I didn't know. My problem was to find an enzyme with a high turnover number as the amount of protein I could incorporate into a mono-layer was very limited, and if possible an enzyme which was membrane bound. One of my candidates was acetylcholinesterase. It is a membrane bound protein, which is prepared from the electric organ of the electric eel. I had no access to electric eels, but I knew that David Nachmansohn at Columbia University prepared the enzyme from electric eels he got from South America. In his view the increase in permeability of the nerve membrane for sodium was due to an enzymatic reaction in which acetylcholine was involved. Acetylcholinesterase is the enzyme which hydrolyzes acetylcholine after its action. This was a hypothesis, which we came to know was not true; and even at that time people were very skeptical about it. My choice

Jens Christian Skou at sea in the 1970s (courtesy of Jens Christian Skou).

of this protein had nothing to do with this theory; my choice was because it had a high turnover number, was membrane bound and I had a method for measuring the hydrolysis of acetylcholine. I wrote to Nachmansohn asking if I could come in August and September 1953 and prepare some enzyme. He answered that I was welcome but he would not be in New York in August because he used to spend his summers in Woods Hole at the Marine Biological Station. So he suggested to me to come to Woods Hole first and then return to New York with him in September. I accepted although I had no idea of what to do in Woods Hole. Woods Hole turned out to be a great experience to me. In Aarhus, a 19-year-old university then, the scientific milieu was very poor. At our department, we were three young doctors with no scientific background who tried to do research for a thesis. Our professor was a very kind man who was very concerned about our social situation, if we could exist on our salary, had a reasonable place to live, and so on. But I cannot remember that I ever discussed my scientific work with him. Woods Hole was a shock to me; for the first time in my life I realized that science was a serious affair and not just a hobby for young doctors who wanted to make a clinical career.

Was there anybody who made an especially strong impact on you?

There were many. There was Szent-Györgyi, Ochoa, Wald, Cole, Curtiss, Grundfest, and others. Huxley came by and visited Grundfest's laboratory where I spent my time. Nachmansohn who shared a laboratory with Grundfest was not doing experiments. He was a very social person and he arranged parties, especially beach parties, and he introduced me to many of the scientists at these parties. Scientists interested in neurophysiology came from all over the world to work in Woods Hole in the summer time because there was access to squids, and squids have a nerve fibre, the so-called giant axon which has a diameter of 0.5 to 1 mm. It can therefore be used for experiments which cannot be done on the normal very thin nerve fibers. I did not take part in the experiments, but looked, listened, and learned. When the fishermen could not catch the squids, there was no activity in the lab. I then spent the time in the library reading and from a book written by Nachmansohn I learned that Libet in 1948 had shown that there is a magnesium-activated ATP hydrolyzing enzyme in the sheath part of the giant axon from squid. I knew that ATP is the energy source in cells and I wondered what could be the function of an ATPase in the membrane. I decided to take a look when I came home, not only because I was curious, but also because an ATPase, being in the membrane, had to be a lipoprotein, and I was looking for such a protein for my monolayer experiments. I spent September at Columbia in New York with Nachmansohn preparing acetylcholinesterase. In the beginning of October I returned to Aarhus and started experiments on monolayers of acetylcholinesterase. I prepared monolayers of the protein and was able to measure the enzymatic activity in the monolayer, and observed that the enzymatic activity of the monolayer was pressure dependent. The next step was to prepare a phospholipid monolayer containing the protein and see if addition of local anesthetics to the water phase could influence the enzymatic activity.

Before that, however, I wanted to examine the enzymatic activity of the magnesium ATPase I had learned was in the nerve membrane. I had no access to giant axons, but decided to use membranes from nerves from crab legs instead. The crab nerves are not single fibres as the giant axon, but consist of many single, very thin nerve fibres. They have a similarity with the giant axon in that they have no myelin sheath. I contacted a fisherman south of Aarhus who sent me some crabs. The nerves were isolated from the legs, broken to pieces by homogenization, and the membranes were isolated by differential centrifugation. The membranes were then added to a test-tube containing ATP and magnesium and the hydrolysis of ATP was measured.

How much crab did you need?

I got a box with about 200 crabs once a week, so before I had finished the experiments I had used several thousand crabs.

Who paid for them?

It cost very little. When the fishermen collect the nets, they have beside the fish also a lot of shore crabs, which are a nuisance for them. They usually throw them away. I also tried the big, deep-sea crabs, but the amount of nerves I got from them was much less in proportion to their weight than what I got from the small shore crabs.

It must have been a lot of work.

It was. I had two laboratory assistants to pull the nerves out of the legs. I needed nerves from many legs before I had enough material for an experiment. A problem was how to kill the crabs. We started with a hammer but if you hit the crab with a hammer, there are crab pieces all over the room. Finally, we cut the legs with a sharp pair of scissors above a big pot with boiling water and dumped the crabs into the pot, which killed them immediately. The procedure gave us another problem — the smell of boiled crabs. The Institute could be localized on the campus from the smell.

Why not kill the crab first and cut the legs after?

Because it would have denatured the protein in the nerve.

Today this might not be possible to do.

Probably not, but at that time we did not consider it as a problem. The process was very fast in any case.

Why were only the legs used?

Because that's where the enzymatic activity had been localized by Libet. But from my later experience, it must also be in the cell membranes, but at that time this was unknown. It was easy to isolate the membrane pieces from the nerves in the legs by differential centrifugation. Anyway, I found that the membrane pieces in the presence of magnesium hydrolyzed ATP to ADP and inorganic phosphate. As my interest was the blocking of the nerve conduction, and as this, as mentioned above, involves effects of sodium and potassium, I tested the effect of sodium and potassium on the enzyme activity. Sodium gave a slight increase in activity in the presence of magnesium, while potassium had no effect.

I also discovered that I could not reproduce my findings in these experiments. In some experiments I got a higher activity when I added sodium or potassium. But I could not reproduce it and did not understand why. I did experiment after experiment, varied sodium, varied potassium, varied magnesium, and tested the effect of calcium, but without getting an answer to why I sometimes got a higher activity and sometimes a lower one from addition of the cations. It took me about a year and a half to get an answer to the problem. I got ATP as a barium salt. In the beginning I prepared it from minced rabbit muscles, precipitated it as the insoluble barium salt, later on I bought it as the barium salt. For the experiments it was converted to a water-soluble salt on an ion exchange column. As I had found no effect of potassium and very little effect of sodium on the enzyme activity I did not bother if the ATP was converted into a sodium or a potassium salt. After I had tried to find a solution to my experimental problem for a year and a half, I decided to compare the effect of a sodium salt and of a potassium salt on the activity and found to my surprise that addition of the potassium salt gave an increase in activity but the sodium salt only had the usual low effect. I analyzed what was in the medium and realized that there was sodium in the medium. In other words, I discovered that the enzyme required a combined presence of sodium and potassium for activity. It had not occurred to me, and was a surprise. I did not know of any other enzyme, which required a combined effect of the two cations.

What was the explanation?

The interpretation of the experiments is that there are two sites on the enzyme for the cations. On each site the two ions compete with each other. On one site, the affinity for potassium is high compared to the affinity for sodium while on the other site, the affinity for potassium compared to sodium is low. Potassium on both sites does not increase the activity, while with sodium on both sites there is a slight increase of activity. If there is sodium plus potassium in the medium potassium in low concentrations replaces sodium at the site with the high affinity for potassium but not at the site with the low affinity for potassium, which means that there is potassium on one site and sodium on the other, the situation which gives the high activity. It is because of the different affinity of the two sites for potassium relative to sodium that it is possible in the test-tube with no sidedness of the preparation to see the combined effect of the two cations.

I did not know what the function of the enzyme was. Had I known that it was the sodium–potassium pump, and had I known the literature about the pump, I would have known that the pump pumps sodium out and potassium into the cell and for its activity requires an effect of sodium on the inside of the membrane and of potassium on the outside, a combined effect of the two cations. This had been shown on intact cells, but was unknown to me. I was primarily interested in the effect of local anesthetics on the impulse mechanism, the influx of sodium and the elimination of potassium. So my first reaction to the observation of the combined effect of the cation was: "Could it be that this is the sodium–potassium channel in the membrane?" But I knew from Huxley's and Hodgkin's experiments that the opening and closing of the membrane to sodium and potassium was voltage dependent, yet here I had an enzymatic activity, which was ATP dependent. My conclusion was then that my observation had something to do with the active transport of sodium and potassium.

I knew very little about sodium and potassium transport. It was not my field. I was a medical doctor interested in returning to complete my medical training to become a surgeon. So I had to look into the literature to see what I could learn about active transport. I wanted to see what was known about the substrate for active transport. I found a paper by Hodgkin in which he and Keynes showed that poisoning of a giant axon with dinitrophenol, cyanide or azide, stops the active transport of sodium out of the nerve. It was known that these poisons stop formation of energy-rich phosphate esters in the nerves. As ATP is an energy-rich phosphate ester, this was to me besides the effect of the cations a support for the view that the enzyme was involved in the transport of the cations across the cell membrane. I summarized my findings in a paper with the title "The Influence of Some Cations on an Adenosine Triphosphatase from Peripheral Nerves" [Skou, J. C. *Biochimica et Biophysica* **1957**, *23*, 394]. I considered including the words "active transport" in the title but found it too provocative, instead, I ended the paper by saying that the crab nerve ATPase seems to fulfil a number of the conditions that must be imposed on an enzyme which is thought to be involved in the active extrusion of sodium from the nerve fibre. Due to lack of reference to active transport in the title very few people interested in active transport read the paper.

Were you the sole author of the paper?

Yes, I was. I had no co-workers, and I had not discussed my work with others, not even with my professor. I wonder whether he knew what I was doing. His main concern was the smell of the boiled crabs, and his concern was only whether I couldn't have chosen another test object. I was, as mentioned, a medical doctor and not trained as a scientist. Had I been, I would probably have started to read literature about membrane phenomena, like active transport, when I started looking for the membrane ATPase, and would then have known that for the active transport a combined effect of sodium and potassium is necessary. That perhaps would've saved a year and a half of experimental problems.

My school English was not good enough to write a paper in English. I had published papers in English before but had professional help. This time I decided to do it myself with help from a fellow scientist in the department who had spend a year in the U.S.A.

Didn't you think that you should go somewhere for a year?

No. I was interested in finishing my scientific work and go back to the clinic and do clinical work. That was my interest.

Was this the kind of work that would be done in order to become a surgeon?

In order to have a career in the clinic, you had to have a thesis. Of course, this work was already going beyond my thesis work, which was, as mentioned earlier, the effect of local anesthetics on the membrane. After this I could have stopped and returned to clinical work. In fact I had applied for a clinical position and had got it. When this happened, I told my wife that I would like to spend my spare time on continuing my research because I had become so interested in solving some of the problems I was working on. My wife told me that it would be more appropriate to stay in research since that is where my real interest seemed to be. So I withdrew from the clinical position, in fact, in subsequent years I did this several times, and I just kept going on with my research. I found the monolayer experiments especially interesting. In some ways the work on the ATPase was a kind of digression. I could hardly wait to return to my monolayer experiments.

In 1958, I took part in an international biochemistry conference in Vienna. I gave a paper on my monolayer experiments because that was what I was really interested in rather than the ATPase project. At the meeting I met again Robert Post of Vanderbilt University whom I had known from our time together in Woods Hole. We told each other about our work. He had been working on active transport of sodium and potassium in red blood cells. I also told him about my work and that I seemed to have identified the pump. I could see in his reaction that my work on the pump was more interesting than the work on monolayers I reported to the meeting. His first question was whether the enzyme was inhibited by ouabain, to this I could only answer, "What is ouabain?" He told me that Schatzmann in Switzerland in 1953 had shown that cardiac glycosides, of which ouabain is the most water soluble, are specific inhibitors of the active transport in red blood cells. I immediately telephoned my lab assistant in Aarhus to get some ouabain from the pharmacy and test it on the enzyme. A few days later Post visited me in Aarhus, and I could tell him that ouabain inhibited the enzyme. That convinced him that it was a pump. That gave me also further evidence that it was the sodium pump.

By reading the literature, I learned that the red blood cells are the classical test objects for active transport. The advantage is that they are single cells and it is therefore easier to measure transport of the ions across the membrane by using isotopes. I had prepared to look for the enzyme in red blood cells.

Post asked me if he could go on with these experiments when he returned home and I agreed. He had experience with this test object and I had not. In 1960 Post published his paper [Post, R., *et al. J. Biol. Chem.* **1960**, *235*, 1796] on the red blood cells in which he showed that there is a correlation between the effect of sodium and potassium on the activity of the Na,K-activated ATPase isolated from the red blood cells and their effect on the fluxes of the ions across the red blood cell membrane. It was important support for the view that the enzyme was responsible for the active transport of the cations, that it was the pump. The activation by potassium in the presence of sodium is very low in membranes from red blood cell, not nearly as pronounced as it is with membranes from crab nerves. But Post managed to show the effect of potassium, perhaps helped by the knowledge from my experiments that it should be there. I was lucky by choosing the crab nerve membranes as test object because the potassium activation in the presence of sodium is very high. With most other tissues the membrane pieces form closed vesicles, which means that in the test-tube there is no access for the cations to both sides of the membrane. As sodium is necessary on the inside of the membrane and potassium on the outside for the combined effect of cations the effect cannot be seen when the membrane pieces form vesicles. Membrane pieces from crab nerves do not form vesicles, or do so to a much lower extent than membrane pieces from other tissues. This is so probably because they do not have a myelin sheath, the extra lipids around the nerve. Had I started on other tissues I would probably not have seen the combined effect of the cations. I learned later to open the vesicles from membranes from other tissues by treating them with detergents.

When I visited the United States in 1963, the Editor of *Physiological Review* asked me to prepare a review article on the ATPase. The review was published in 1965 [Skou, J. C. *Physiol. Rev.* **1965**, *45*, 596]. It gave all the evidence showing that this enzyme had all the characteristics needed for the enzyme to be the sodium pump. It was named the Na,K-ATPase.

Was it this paper that was the principal basis for the Nobel Prize?

It was the 1957 paper as far as I know.

What did you work on after these seminal papers?

On the pump. But after 1965 I got co-workers, and there were people all over the world who started to work on the enzyme.

All the time?

All the time, yes. There was enough to do. To establish that the Na, K-ATPase was the pump, was only the beginning. The problem was to understand how the enzyme could couple a reaction with ATP to the transport of the sodium out of the cell and potassium into the cell against electrochemical gradients. For this it was necessary to know the kinetics of the reaction. And to know the structure of the system and the structural changes related to the reaction. To get information about the structure it was necessary to have a pure enzyme. It took 7 years to find a method to purify the enzyme in the membrane. With the pure enzyme structural information has been obtained. The amino acid sequence of the protein has been determined, and the folding of the protein chains in the membrane. From the structural studies a picture has begun to emerge of the structural changes related to the transport process. From the kinetic information it has been possible to set up a model for the transport reaction. It is, as mentioned, not my work, but results of work done by co-workers and by scientists from all over the world. In 1973 the first international meeting on the Na,K-ATPase was held in New York, and from 1978 the international meetings have been held every third year in different places in Europe, Japan, North and South America. The proceedings from these meetings give good information about the development of the field.

You waited 40 years from 1957 until the Nobel Prize materialized.

Yes, it was 40 years. Luckily.

When did you think first that it might bring you the Nobel Prize?

During the years I had taken very little notice of who got the Nobel Prizes and for what, and had not speculated in getting it myself. I had got so much recognition from the scientific community that I was satisfied. I did not work to get prizes. When I got it my concern was if I really deserved it. It is better to deserve and not receive than receive and not deserve, the best is of course to deserve and receive. I am glad that I got it that late because afterwards it took me two years while I could not do anything else than being occupied by what followed the Prize. It would've spoiled my research possibilities had I received it at an earlier time.

You received half of the 1997 chemistry prize. The other half went to Boyer and Walker. Was there any connection between you and Boyer and Walker?

No connection. I was reluctant when I got it because I felt that there were others in the active transport field who should have shared it with me. This was I. M. Glynn in Cambridge who has done some fundamental work on establishing the concept of active transport and Robert Post in Nashville whose work has been of the utmost importance for the development of the Na,K-ATPase field.

Did you tell them what you thought?

Yes.

What may have been the reason?

Glynn wrote back to me saying that the reason was probably that the Nobel Committee connected the work of the formation and the use of ATP. Boyer and Walker showed how ATP is formed and I showed how at least some of the ATP is used. Apparently, that was the connection. Of course, they deserved the prize for their work. My problem was that I felt that I got too much of the honour by being the only one in the transport field who was honored.

Do you think that the Nobel Prize is a good thing for science?

The most important in my view is that it attracts attention to science. You sometimes wonder why it is that the Nobel Prize is so prestigious. There are prizes that are as big from the point of view of money, but the Nobel Prize is The Prize. It may have to do with the history behind the prize, the age, 100 years, that the Nobel Committees are very meticulous in their choice, with the festivities around the presentation of the prize, and the way it is announced world-wide. My daughter read about my prize in the local newspaper in Kathmandu.

There are also problems. First that it is a limited number of fields that can obtain a Nobel Prize. Next that the number of worthy candidates seems to be bigger than the possibilities. It is so attractive to get this Prize that this leads to disappointments and bad feelings. I once was with a group of scientists in a plane on the way to a meeting when one of the scientists said that he deserved a Nobel Prize, and complained that the Committee in Stockholm was not positive to him. I was flabbergasted. There are scientists who cannot sleep from excitement in the beginning of October waiting for the call from Stockholm, which may never come. I doubt that the Prize has an effect on the way science is conducted or

on subjects chosen. It is not possible to foresee if your research will lead
to a breakthrough. The Prize may strengthen competition about priority.
You might assume that it would lead to envy. Personally I have only had
positive responses. A problem can be that you as a Prize winner suddenly
become a known person whose opinion becomes important on any matter
and whom many draw on for many different reasons. Examples include
giving lectures for scientists and in high schools, which I enjoyed and found
important in order to stimulate the interest for science, and in many other
connections; becoming honorary members of societies and academies, which
all emphasize that it has no economic consequences, but then a year later
you get a letter offering help from a lawyer to change your will in their
favor; writing letters and signing declarations for political, cultural or scientific
support, and for support of charitable organizations; sending pictures with
autograph to people all over the world, etc., etc. The first two years after
the Prize I was fully occupied by these activities, and now 6 years later
I still daily receive e-mails asking for such favors. I had been emeritus for
13 years when I got the Prize, and could therefore use the necessary time
for these activities but if it had been when I was scientifically active it
would have been a problem.

Nobel's intention was to give it to younger people in order to help them.

I know. He didn't mean to give it to 79-year-old people. But I was glad
I got it that late. After all, could you end your scientific career in a better
way? It's a great thing to get it.

It was also a great thing for Denmark.

I had tremendous press coverage nation-wide, and the university has made
good use of it too.

Did the Danish Queen invite you for an audience?

No, as I many years ago had refused to receive a decoration, this was
not expected. I was invited to a royal party, but I said no to the invitation.
Monarchy is in my view an anachronism in a modern democratic society.

*I have read your autobiography and you write in some detail about
the Second World War. For me it was especially interesting that you
wrote about what the Danes did to save the Danish Jews in 1943. Why
did you write about this story in your autobiography?*

Jens Christian Skou standing next to the street sign that bears his name in Aarhus, Denmark, 2003 (photograph by I. Hargittai).

The occupation had a tremendous effect on our whole life situation. The rescue of the Jews was a light in a dark time. It made such an impression that it was possible in spite of the situation with the Germans hunting the Jews to bring most of them to security in Sweden. I did not personally take part in the rescue operations. I had no contact with the people involved, nor had I any contact with the Jewish community. Some of the students in our group disappeared and from this I realized that they were Jews. It had never occurred to me. Whether people were Jewish or not was not a question I ever thought about. Most of the Danish population cheered the people who at that time were unknown for their rescue of the Jews.

How was your life impacted by the German occupation?

During the war we lived in the atmosphere of the occupation and it had an impact on all of us. In my medical student group we knew that one of the students was an informer for the Germans. He was liquidated. Because of fear of revenge from the Gestapo against the group, teaching was cancelled. Many of our teachers who took part in the resistance against the Germans disappeared, went underground or escaped to Sweden. We managed to get our final exam in the summer of 1944. But we did not dare to assemble to sign the Hippocratic oath, but came one by one to a secret place. I started my internship at a hospital in the northern part of the country. In the surgical ward the head of the department had escaped from the Gestapo to Sweden. The next in line in the department was anxious to teach me how to operate for appendicitis and the like, which was unusual for someone who had just started on his internship but pleased me because I intended to become a surgeon. Eventually I realized that the reason for this was that he was involved in receiving weapons from England delivered by plane at night. When we were on night duty together and he had to leave to receive weapons he wanted to be sure that I could take over. He was later arrested by the Gestapo, but he survived.

You have said that you did not have mentors. Do you have heroes?

No. My knowledge about science outside the field I was interested in was poor when I was young, and so was my knowledge about other scientists.

And today? Do you have heroes?

No, I don't think so. My knowledge about people in science is still sparse. I've been interested in my own field and not much about other fields. However, I have many other interests. Science has not been my life in the sense that I did work in science mornings and evenings. I started in the Physiological Department. Once one of the deans of the medical school asked me why he never saw lights in our department in the evenings. I told him that we didn't work in the evening. We worked during the day, 7 or 8 hours, then it was important to get home and relax.

Wasn't there any time a question that wouldn't let you sleep?

No. Never. I went home at 4 or 5 o'clock to take care of my children. Maybe my wife would say that I worked in the evenings but it happened

Jens Christian Skou and his daughter in a winter resort (courtesy of Jens Christian Skou).

very seldom. If you are disciplined and work concentrated during the day
that should suffice. I worked 7 or 8 hours concentrated.

How about your co-workers?

Same. I accepted that, but I required that everybody be there in the morning.

And no work during the weekends?

No, no.

And your students?

The same. But I have had very few co-workers or students. If you look at
my publication list, you will see that most of my papers carry my name

alone. After 1965, after the publication of my review, the situation changed somewhat because a number of people became interested and came to the department. During the 1960s the University got more money, we were expanding, there was a lot of construction on the campus, and we received new positions. When I started in 1947 at the University we did not apply to funds for money. The Medical School got a certain amount of money from the University, which got it from the state. The money was divided between the departments in the Medical School, and the amount we got at the Physiological Department was sufficient to cover the expenses for our research. I could do my experiments without asking for money; true, they were inexpensive experiments.

I became chairman of the department in 1962. We were 4, the professor and three young doctors. During the 1960s we became 25 scientists in the department. I had the dilemma whether to make it a membrane department or spread the people among different fields. I chose to cover as many areas as was practicable. Only a small fraction of the department did membrane work. When a new scientist came to the department, and wanted to work on the Na,K-ATPase they either came with their own problem, or if not I suggested a problem. But then I expected them to go on on their own. They could of course ask for advice or help but not work with me. The engagement is much higher when it is your work and your publication.

In 1972, we got new rules for the University. The professor is no longer the born chairman, the chairman was elected and could be anyone in the department, even a non-scientist. Luckily, that took the administrative burden from me, although not right from the beginning because they elected me chairman.

So at the beginning, when I was a young scientist, we had a very secure financial situation for our work. There was no pressure to publish and it took me years before I published my first paper. I could take my time, do my work, didn't have to ask anyone for money. Of course, I had to teach. It was a heavy burden because we were four people in the department and we had an intake of 140 medical students a year. Our scientific work was mostly outside the semester, that is, outside two times four months. The situation changed in the 1960s when we got our first national fund, for which we had to apply for money for bigger equipment; we still received money for the department from the University for the daily expenses. It was not until the late 1970s that we had to apply for money for everything because the University was no longer in the position to give us the money.

So when a young person joins the department today...

I shall return to your question. In general I am very skeptical about funding systems where you have to find all the money for your research from funds. To write application for a grant which often is every second or third year is very time consuming. It is valuable in between to consider what you plan to do, but as it is impossible to foresee how your research will develop, you know that half a year or a year later you will probably not be doing what you wrote in your application. It is not only time consuming to write the applications, but also to evaluate them, meaning that highly qualified scientists spend a lot of time on evaluation, time which could be used better on research.

Second, you never know if next time you will receive support. That is why it is important to be able to present results at the next application, which may tempt you to select problems you know can give results. It can be a hindrance for new thinking and for testing new ideas which may or may not give useful results, but which is so important for the development of basic research. Renewal of the grant also increases the pressure for publication, which may be too early.

Third, the funding system favors successful research. This, of course, deserves support. But with 50 to 55% of the applications worth supporting but with money only for 15 to 20% as it is in our country, this leaves little room for new thinking. It is often the young who get the new ideas that move borders. It is very difficult for them with few or no previous publications to obtain money to work on their own ideas. Instead they must join teams who have money and work on the ideas of the head of the team. It is a hindrance for free research. Good research requires quality but also originality and engagement. And it stimulates the engagement to work on your own problem and be the author of the paper rather than to work on the problems of the head of the group and be one of many authors on the publication. With the present funding system it is important that heads of groups let the ideas of the single members develop and let them work independently on them, and not just focus on their own ideas.

Fourth, there is a tendency that the politicians use the funding system to direct the research. The money is put in subject-earmarked boxes. In basic research it must not be the earmarked subjects which determines the research, but the ideas of the scientists. To let the money determine the research subjects leads to lost possibilities or in the worst case to mediocrity.

This does not mean that we should not have funds. It is necessary for covering the need for expensive chemicals or equipment. But the funds ought to be a supplement to basic amounts of money to the departments sufficient for the daily expenses, so the single scientist in the department independently can choose their research subject and work on it without having first to find a grant.

If you started your career today ...

I would become a surgeon. Absolutely. I would not work under conditions where you need to apply for grants to buy a pencil. And I doubt that the coming generations will accept such conditions.

You lost your father when you were 12 years old. What impact did this have on your life?

I'm not sure I can describe it. Of course, it was a great loss. We were four children. Fortunately, it had no effect on our financial situation. It probably made me more independent, I knew what I wanted to do and it made me make decisions.

May I ask you about religion?

Ellen-Margrethe and Jens Christian Skou in their home, 2003 (photograph by I. Hargittai).

I was born into a family, which was religious, Danish Protestant. It meant a lot for me in my youth. Although I still like to go to church, I'm not sure whether I am a believer anymore. Christianity is important to me because much of our values in society are based on it.

You have had a privileged life. Did you ever have any big challenge?

We lost our first child. Shortly after she was born, we were told that she had no connection between her gallbladder and her intestine and that she would live less than a month. She survived for one and a half years. We knew that we would lose her and we lost her.

What has been your main concern recently?

I have been very vocal about the need of a better funding system to support research in this country, especially to give the possibility for young people to work on their ideas. Other scientists have joined in and we have had some results but not yet enough.

Have you become a politician?

To a small extent, certainly less than my wife who has been very active in politics for 12 years on the County Council, and is well known for her activities. She is a Social Democrat. She may be better known at least in Aarhus than I am.

Paul C. Lauterbur, 2004 (photograph by I. Hargittai).

28

PAUL C. LAUTERBUR

Paul C. Lauterbur (b. 1929 in Sidney, Ohio) is Professor and was for many years Director of the Biomedical Magnetic Resonance Laboratory, University of Illinois in Urbana. He and Peter Mansfield of the University of Nottingham shared the 2003 Nobel Prize in Physiology or Medicine "for their discoveries concerning magnetic resonance imaging". Paul Lauterbur received his B.S. degree in Chemistry from the Case Institute of Technology, Cleveland, in 1951 and his Ph.D. also in Chemistry from the University of Pittsburgh in 1962. Between 1969 and 1985 he was Professor of Chemistry at the State University of New York at Stony Brook. He has been at the University of Illinois since 1985. He is a member of the National Academy of Sciences of the U.S.A. (1985) and has received numerous awards and other recognitions. He has the Gold Medal of the Society of Magnetic Resonance in Medicine (1982); the Albert Lasker Clinical Research Award (1984); the European Magnetic Resonance Award (1986); the National Medal of Science (1987); the Kyoto Prize for Advanced Technology (1994); the National Academy of Sciences Award for Chemistry in Service to Society (2001); and others.

We recorded our conversation on February 1, 2004, in the Lauterburs' home in Urbana, Illinois.

You have not been a Nobel laureate for a long time yet. Would you care to comment on the changes you have already experienced or are anticipating in your life?

Have you read Michael Bishop's book *How to Win the Nobel Prize*?

I reviewed it for Nature.

He expresses some of the sentiments I feel about it. Some Nobel laureates stay deliberately in science and continue their research as any other scientist. Others become professional Nobel laureates. Michael Bishop has gone into administration.

Though not immediately, only after a while.

Jim Watson changed his working style completely after the award and Francis Crick has tried very hard to stay away from doing anything that depends upon having won the Nobel Prize. He has stayed very active in doing science.

Did you find it strange that you received the prize in physiology or medicine rather than in chemistry?

The impact of my work has been primarily in the area of medical diagnosis. If I had to guess I would have guessed that it would be in that area.

What was the first thought that pushed you into this direction of research? You must have answered some similar questions before.

I try to find new words every time and give you a perspective. There was something readily understandable and that was my concern for sacrificing animals in operating on them. There was also something else that captured my interest even more and that was that this was a form of imaging that was different from anything that had been done before. That's also why the first version of my paper to *Nature* was rejected. It was because I left out cancer and medical diagnoses and concentrated only on physics, on science, basically. That was also why I gave it a strange name, NMR Zeugmatography.

Did Nature *give you an explanation for the rejection?*

I have the letter somewhere but I have not looked at it for some time. It was just the standard text, something like to the effect that we don't see why you make such a big fuss about it.

But it was published nevertheless.

It was [Lauterbur, P. C. "Image formation by induced local interactions: Examples employing nuclear magnetic resonance", *Nature* **1973**, *242*, 190–191].

How did you convince them to publish it?

I wrote them a long, impassionate letter and offered to put in more speculation about possible applications to enforce the idea that there may be some interest in it. Someone else later told me that my manuscript was then sent out to a new referee whose comments were not terribly encouraging but then he added something to the effect that this seems to be crazy but I had never done anything crazy before.

That means that by then you had established your reputation.

I had been very active in nuclear magnetic resonance for years, ever since I was a graduate student. I was 43 years old when I submitted the manuscript to *Nature*.

I made an interview with Richard Ernst in 1995 and he mentioned that you did the first real imaging experiment in 1972. He did not give me the impression that he found it extremely important although when I asked him for some pictures to illustrate his interview, one of the photos he sent me was the NMR image of his own head. Then, in 2002 he wrote a Foreword to a collection of papers in NMR spectroscopy, [Current Developments in Solid State NMR Spectroscopy, Springer-Verlag, Wien, New York, 2003] and he sounded upset that no Nobel Prize has been given yet for NMR imaging. He mentioned two fields where there should be yet a Nobel Prize, magnetic resonance imaging and solid state NMR. I remember what he wrote verbatim: "The disrespect for MRI in Stockholm is particularly difficult to understand."

I have known Richard Ernst for many years and I remember very clearly when he was first proposing two-dimensional NMR. We were at a meeting in Switzerland and I came up to him afterward and I told him how interesting I thought that was and promising, but he said, "No, no, no, this is nothing, anything you can do in 2D you can do in 1D." I was very appreciative of his innovation in that area, as I had been of his innovation of Fourier transform NMR. He has always been a generous man with everybody. His first instinct was to give all possible credit to anyone else.

It was somewhat strange though that he was selected alone for the Nobel Prize.

I heard him give a talk shortly afterward and he gave me the impression that he was surprised too. He discussed in great detail the inspiration he had from West Anderson when he was at Varian, from the Varian brothers, ...

He did not tell me about the Varian brothers, but he told me that Varian did not follow up his Fourier transform innovation.

In some respects, it was a very conservative company and for many years it was operating with no competition. So they were not hungry to jump at new ways of doing things. Bruker was. It was a great shock and surprise, particularly to the technical people at Varian when Bruker suddenly came out with practical Fourier transform NMR.

How did you originally come across NMR?

I'm not sure of the exact sequence, but it was in the early 1950s. I was at the Mellon Institute in Pittsburgh. There were two things that happened in the early 1950s. One was that a group from Varian came out to talk to a research group at Mellon Institute and I had the chance to listen to some of these discussions. They were talking about the possibilities of NMR. The other was Herb Gutowsky from Illinois came by to give a seminar. He was studying what you could tell from NMR possibly about the charge distribution in substituted methanes.

In the liquid?

Yes. It was only in the liquid that people could do high-resolution NMR in those days. He was talking about making chemical shift measurements in methyl chloride, methyl bromide, methyl iodide, and all the various combinations. I had been interested for many years in the relationship between silicon and carbon compounds. I knew that I could find a way to obtain or synthesize most of the corresponding silicon compounds, like tetrachlorosilane, trichlorosilane, dichlorosilane, even silane itself. I went up to Gutowsky after he gave his seminar and proposed that I could provide some of these compounds — they were commercially not available — if he could make this kind of measurement that he was describing. Perhaps we might learn something about the relationship between carbon and silicon.

Were you a graduate student at that time?

Paul Lauterbur as a high school
student (courtesy of Paul Lauterbur).

I was a full time employee in a research group, which was funded by
Dow Corning Company at Mellon Institute and I was just taking some
graduate courses on the side in my spare time. I had a bachelor's degree.

What was it that intrigued you about carbon and silicon?

That was something I was wondering about since I was in high school.
I had my home laboratory. I don't remember the age when I first got
a chemistry set. In my laboratory I had lots of chemicals that came from
chemical companies and apparatus I bought for a little bit of money I
made. That was from the time I was a freshman in high school.

Was there a family inspiration in your interest?

My interest just developed over time and I don't remember any special
incident.

What did your parents do?

My father was a mechanical engineer and my mother was a homemaker.
She had a couple of years in college, but nothing special.

No friend, book or teacher steering you in this direction?

There were books around and encyclopedias. I had many scientific interests
that developed over the years. When I went to college, I wasn't even sure

that I wanted to major in chemistry. I thought I might want to major in physics instead, but I stayed with chemistry. It fits with the Nobel Prize that I was never concerned with what they called what I did. I did what I was interested in, regardless whether it was chemistry or physics or biology as long as it was interesting.

You said that you had been upset how animals were treated. There are many others feeling like that but then people go onto the street to protest or occupy a building where animals are used for experimentation, or find other ways to protest. Very few people would develop a new physical technique to spare them.

My approach in general is to not get involved in things when I cannot make any difference by getting involved personally. But if there is something that I can do personally, then I take action. What I told you was a psychological trigger, but not a real motivating force that kept me going afterwards.

What was the continuation of your meeting with Gutowsky?

He agreed that we could try a collaboration. I then spoke to my superiors at Mellon Institute, and it was probably the wrong way around, but they gave me permission to do what I had already promised to do. I imagine I was probably sometimes a bit intolerable in that way.

So how did it go?

Before we could actually implement our plan, I was drafted into the Army. It was the time when people who had had deferments for educational purposes or something like that but not really, really critical, got drafted. The idea was that these people with deferments were getting older and it was more and more difficult to draft them.

This was at the time of the Korean war.

Yes. So many draft boards said, there should be no more deferment and they set the standards higher. Fortunately, I had a bachelor's degree and two years experience of research. It was just enough qualification for me to be put in a special program called the SPP, Scientific and Professional Personnel. This program included misfits of various kinds, like Ph.D.s, semi-literate people, some not terribly well-balanced individuals, some who were too old, in general those ended up there who would not make good infantry. So I had a chance to go through various tests and I was assigned — as a private — to an Army chemical laboratory in Maryland. It was nominally run by civil servants; they were regular, long-term employees. In practice, for example, one man who came to work in the laboratory to which I was eventually assigned, had just gotten his Ph.D. from E. Bright Wilson, Jr., at Harvard. There were also two organic chemists with Ph.D.s from Harvard. The real scientific expertise lay with the soldiers who were assigned to the staff. The civil servants who were competent to understand the situation, made full use of these people in order to do the jobs that they were hired to do themselves.

Did they work on meaningful problems?

Yeah. They were hoping to make compounds of fluorine that would be even more nasty than nerve gases. It turned out though that everything they worked with was pretty much inert. So they worked with nerve gases, phosphorus compounds, there were irritants, mustard gases, a panoply of such materials that people worked on in those days. They also worked on various other things, very amusing, for example on body armor, hanging it on goats and shooting at them. That was called biophysics by the Army. It was while I was working in a medical laboratory on some nerve gas research that a nuclear magnetic resonance spectrometer was acquired in another laboratory. Because I knew a little bit of the subject — although I had never done any experiment — I became an instant expert. I was able to get transferred along with a friend into that laboratory. Eventually I got four publications out of my service in the Army. We went ahead and did things that would result in publications and we also went ahead and did things that would simply contribute to the research programs of other people in the laboratory.

Would not the Army object to your publishing papers in the chemical warfare program?

We published some general things, surveys of chemical shifts, analyses of complicated spectra, things that were not directly related — as I saw it — to killing people. Some of the things may have been a little more relevant than they thought. For example, the people who were doing synthesis were sure that they knew what the structures were of some compounds that contained sulfur and oxygen in some phosphorus-containing materials. One of their products, about which they were convinced that they knew the structure, contained two widely-separated phosphorus resonances of equal quantity. It was immediately obvious from the little that was known at that time that what they had was a mixture. There was a paper about phosphorus resonances by Gutowsky. But they didn't believe it, of course, they were the experts. I was just a soldier with a technique that they had never heard of. Gradually though we managed to help the credibility of the laboratory a good deal. We could also solve problems that they could see no other way they could solve. I had also given a seminar in Pittsburgh about the use of nuclear magnetic resonance to characterize polymers, especially rubbers. Some of the work I was doing at Mellon Institute was on silicon polymers and I had a chance to do some experiments and do a lot of thinking on these matters, and I decided that there was something very interesting to follow up there.

What happened when you got out of the Army?

I had two choices. One was to become a full time graduate student in Gutowsky's group at Illinois and the other was to go back to Mellon Institute and to persuade them to buy an NMR machine. Mellon Institute at that time was essentially an independent organization except that it was connected to the University of Pittsburgh. That was before it joined the Carnegie Institute in the other direction, across the street. Because of the connection to the University of Pittsburgh, I was able to take graduate courses for free when I was an employee of Mellon Institute. I talked to the management of the Institute, which had a program independent of those that were funded by individual companies, and tried to convince them that they needed a nuclear magnetic facility in the building. They did not have terribly good judgment when it came to this matter and they decided that because they had not needed NMR in the earlier years, they would not need it now. But, fortunately, the management of Dow Corning had a director of research who had been in part successful because he had been a great champion several years before of infrared spectroscopy. He had also been told that NMR would not be needed because people had not used it before. This director was very proud of his insight into analytical techniques. They decided that I could join their group and they would buy an NMR machine. This was an attractive offer because it gave me independence and I also preferred doing some work rather than sitting in classrooms.

From calculations, I got that it should be possible to see silicon-29 resonance, which was a great interest to a silicone company. That tipped the scales. What I ordered from Varian was a machine specifically to have the capabilities of doing silicon-29. There was a group supported by Pittsburgh Plate Glass at the Mellon Institute with which I collaborated and they were doing solid state NMR. I had to fly out to California for the machine, I did test experiments there and showed that it would indeed work. I also realized that if we could see silicon with this machine then we could also see carbon in a similar way. There was a lot more scope in carbon NMR than was in silicon NMR; there was a wider range of electronic features; different kinds of compounds; double bonding; triple bonding; and other peculiarities. I decided to take a look at carbon spectroscopy and to spend a good deal of time on that as well as doing a certain amount of silicon NMR.

This work led to the publication — more or less simultaneously with some other guy in an industrial laboratory — of the first paper on carbon-13 NMR. I had the opportunity to give a lecture for the Research Council of Canada, which was in Ottawa; the most prominent people of the field were present, including the authors of the standard book in the field, Pople, Bernstein and Schneider. Pople later won the Nobel Prize for his theoretical chemistry and Schneider later became the head of the Research Council of Canada. I had given my talk up there, so they knew about my work. I don't know who refereed my paper and whether it had any influence on its publication, but I have my suspicion. In any case, it appeared in the *Journal of Chemical Physics* and started a great explosion in the work on carbon-13 NMR although many people said it was pointless.

Why?

Because the samples had to be big and had to be pure compounds, the peaks were not very sharp, and the sensitivity was low. It was not practical as a trace technique, by any means, but it was practical to sort out some of the effects that occurred in series of compounds with different substitutions and the like. It could even be used with enriched carbon to solve structural problems. Anyway, it was the first big increment of some kind of recognition.

So you did not come to Illinois at that time.

When the Dow Corning group at Mellon Institute decided to buy this machine, I abandoned any idea of going to Illinois. As it happened, I'd also run into Gutowsky in the Army. He was a consultant to the same laboratory that I worked in. As far as I could tell, he had offered a very dim opinion of my abilities because I asked a stupid question about NMR when he came by for a visit one day. He lost no time in telling me, not in detail but rather dismissively, that he thought it was a stupid question. He died recently, a few years ago. He was one of the major figures in the NMR field as a whole.

He was a Wolf Prize laureate.

He was widely regarded as a leading candidate for a Nobel Prize.

Not for solid state?

No, for liquids. I have my own idea who deserves one for solid-state work.

???

I don't want to advertize it. There is more than one person.

I have interviewed John Waugh.

He is certainly someone who would be considered by any knowledgeable person in the field. When I gave my first carbon-13 lecture — John doesn't like this story — in Canada, he stood up — he was then a young Assistant Professor at MIT — and he said that he knew about some similar work from Harvard and my numbers were all wrong. Of course, I was just nobody, a part-time graduate student, and the Professor from MIT tells you that Harvard says you are all wrong, you must be concerned. The first thing I did when I came home, I went into my lab to check my numbers. I was right. It was a common kind of error that caused his misconception. Perversely, it helped me build my self-confidence when a famous expert says you are wrong and I found it was not so.

It was doubly unkind, first to tell you that others had already done what you were reporting on and adding that you were wrong and they had been right.

What they had done was measure the carbon–hydrogen coupling constant, which, of course, produces a doublet — on using some enriched carbon — on either side of the residence of the carbon-12 bonded to hydrogen, in the hydrogen spectrum. What he remembered was one half of the splitting, from the center to the left or to the right. I reported the true coupling constant, which was the full splitting from both sides. What he was saying was that I had the coupling constant all wrong by a factor of two.

Did he specify that it was twice what it should be?

He just said that it was all wrong. But this is just an anecdote; John has done some very good work; and we have been friends for many years. Last spring one of his former students invited me for a talk, Warren Warren at Princeton. He has been working on very innovative experiments in solid-state NMR, even in implementing quantum computers using NMR. You should be familiar with him.

It's interesting to recall those days. John is a rather sardonic individual and he has often been quoted saying that NMR was dead. I think, at least partly, that indicates that someone feels that he is not going to make any new major contributions. When he was awarded the Bruker Prize some years ago, someone challenged him during the acceptance address after the eulogies: "John, how do you justify getting a prize like this when you always say that NMR is dead?" His answer was, "It is dead, it just occasionally twitches a little."

I was once describing to him some work I was trying to do. We discussed NMR over some drinks one evening and he said, "It won't work." I asked him to explain to me why it wouldn't work. He repeated, "It won't work," and he added, "Anyone who solves that problem, will win a Nobel Prize, and you don't have one in you." I think that it's good to have an outspoken relationship with people.

I met him in Ahmed Zewail's laboratory at Caltech. This was before Ahmed's Nobel Prize. I found it a little embarrassing to be present when he curtly told Ahmed that he was wrong in something. Not everybody takes criticism as graciously as you apparently did.

John is a very bright guy and he is also very outspoken. When he thinks of something, he does not censor his own speech. He just tells people when they are wrong.

Or even when they are not, as your example showed.

But I appreciate his outspokenness.

In your Nobel lecture, you said something that the whole idea about NMR imaging came to you one evening and the next morning you had your notes witnessed.

Yes.

I would like to ask you, how did this happen and why did you think of the need to have the notes witnessed?

While I was once having dinner with Don Vickers — he was one of those whom I invited to Stockholm — he was the one who had actually been doing the experiments on wet tissues provided from Don Hollis's laboratory. As we discussed that, I could be more specific than I would have been, had I been there all by myself, I convinced myself that it seemed like a really good idea.

A good idea, what?

Imaging. That one could use magnetic field gradients in some way to encode special positions and that would open up a vast cornucopia of applications. So I went out to a store on that evening and bought a notebook. Either later that night or next morning — I forget which — I wrote down in very general terms the idea. Since I had worked in an industrial lab where the habit was to have notes witnessed for potential patent purposes, I thought to do that.

Was the lab of Dow Corning at Mellon Institute an industrial lab?

Yes, it was and such practices were commonplace there.

You were no longer working for them at the time when the idea hit you.

No, it was a decade before. So I asked Vickers to put there a date and his signature.

Did you take out a patent?

One of our other colleagues was a patent attorney. So, of course, I discussed with him the possibilities of patenting things. He decided that he would work with me for a percentage of any patent fees that would actually come out of this interaction, but we then had a falling out in some business matter in the company. He was a director and I was a director and we had a disagreement.

Did you work for a company at that time?

At the time I was spending the summer with a small company that I was a consultant to and a member of the Board of Directors. It was called NMR Specialties, Incorporated.

How did you become a board member?

Unfortunately, the founder of this company had engaged in some practices that the company's banker was enraged about, clearly not the sort of things that any respectable company should be doing. We had two choices for the company, close it down that instant at a specially convened board meeting or find someone who would take over the direction of the company. Since I was an academic and free for the summer from my regular academic duties, I was elected by acclimation to that position. That's why I was there at the time that these experiments were being done. But another member of the board was a patent attorney. We discussed these things and came to an arrangement, but as president of the company, I had to make business decisions of various kinds and one of them was not very favorable for this gentleman. So we disagreed about one of my decisions. This was after I had returned in September to my job at Stony Brook, and this patent attorney returned all the papers that we had prepared for patent claims and refused to continue with these matters.

Did you then go to another person?

At that point I went to my university, the State University of New York at Stony Brook; the percentage of whatever came out of it was better than nothing. But the organization that made patent disclosures for the university ruled at the very last minute that it would cost more to apply for a patent than they could ever conceivably be making from it. Of course, this was wrong by several orders of magnitude. I thought it was too late to start it all over again, and decided to just get on with my work and make a virtue out of necessity, and made my laboratory entirely open to anyone who wanted to come and see what we were doing, discuss these matters, inspect my apparatus, and do whatever they wanted to do.

In retrospect, do you regret this decision?

Having a patent would have — as I found paradoxically later — accelerated industrial development.

Why is that so?

Some of the companies I have talked to had no interest in doing anything without being covered by an enforceable patent.

Why?

Because they wanted the advantage of keeping competition away. So having a patent would've helped accelerate industrial development.

When was all this happening?

Mid-1970s.

Was patenting then as widespread at universities as today?

No, but they still had policies at my university that you could apply for patents and share the proceeds or they could be licensed, for example. However, these policies were mostly untested because, as you said, it was yet at the time of early days in that kind of thing. So, as I said, I made a virtue out of necessity and the result was that the development was faster in some academic laboratories because I was proselytizing everywhere, all over the world, inviting people to my laboratory, and so on. That kind of openness probably contributed positively to the later decision in Stockholm. Another person that I invited to Stockholm was the head of my department at Stony Brook at the time when I did this work. He was on a commission that later changed the rules and eliminated this evaluation company from the State University of New York regarding patent applications, because of their performance in this case as well as in other cases. His name is Francis Bonner, he is the founding chairman of the Stony Brook Chemistry Department. So my whole commercial connection was through that company, NMR Specialties, and nothing worked out with the medical industry, but you can't have everything.

Somebody warned me that you may not want to talk about Dr. Damadian. [Here reference is made to an unprecedented vocal protest by Dr. Raymond Damadian following the announcement of the 2003 Nobel Prize in Physiology or Medicine from which he was left out, according to him, unjustly.]

They were precisely, absolutely correct. I really don't want to. It would just be an example of "he said" and "I said" and "you said", and so on. It would go on forever.

Do you know him in person?

Yeah but have not had personal encounters with him over the years. I know there will be no end to this controversy.

I have also heard that you said that you would not accept the Nobel Prize if he would be included.

Some people have said that and I would rather not add fuel to the fire.

Could you have possibly said that?

I would neither confirm nor deny it. I had a telephone call the other day, a reporter from a campus paper called who had been talking to a Fonar representative [Fonar is Dr. Damadian's company], who made a certain claim, and he wanted to know my reaction to it, and I said, no comment.

I think that even if you had made such a statement concerning any conditions for accepting or not accepting the Nobel Prize, the Nobel Committee would have disregarded it. This is also why I found it hard to believe that you might have made such a statement.

I know from my personal conversations in Stockholm that those people are very jealous of their autonomy.

Of course, one may say things in a company of friends and it would be different, for example, to make such a statement in a talk or have it printed.

Certainly, I have never encountered anyone claiming that I had made such a statement in writing or in a public place. Concerning the whole story, there is a book about it by Don Hollis called *Abusing Cancer Science*. Would you like to have a copy?

Yes.

We'll give you a copy. Hollis self-published it and it is now out of print. Commercial publishers were unlikely to be willing to take a chance to have lawsuits. The disputes go back for many years and I can refer you to this book.

Did you expect to receive the Nobel Prize?

I was not surprised that it happened, but I didn't expect it to happen in 2003.

Did you expect it to happen before or later?

People used to say to me, "Don't you have your Nobel Prize yet?" Naturally, I could not say that I never thought of it. Then people used to tell me that they had nominated me for the physics prize, for the chemistry prize, or for the physiology or medicine prize. But you never know whether people are telling you the truth. Some people are clearly just trying to express friendly thoughts.

Of course, there is a secrecy warning in every invitation for nomination.

But some people like to disregard it; they don't think secrecy is scientifically justified in such matters.

I have looked up your website on which you list the chemical origin of life as one of your interests. Is this a recent interest?

Yes and no. It dates back to my high school years as a kind of general interest. But as a serious working occupation, it started several years ago. I think I have a potentially useful new idea that Mother Nature might approve of.

Would you like to talk about it, or you would rather not?

It would be too soon. I don't like to talk about things that may or may not be done. I am basically an experimentalist and until some of my experiments confirm my ideas, I would prefer not to talk about it.

Do you have a student working on it?

I have several undergraduate students working on it.

Do you interact with others who are investigating the chemical origin of life?

I've talked to a few people who have overlapping interests. One of them is Carl Woese who is Professor of Microbiology on our campus. He received the 2003 Crafoord Prize [*awarded by the Royal Swedish Academy of Sciences in fields where there is no Nobel Prize; it is also handed out by the King of Sweden; its monetary value is half of that of the Nobel Prize*]. I have also talked to other prominent people, like Manfred Eigen, who picked up a Nobel as a young man, and to some others. My wife thinks that I should be more public about my ideas.

The University of Illinois at Urbana has just got two Nobel Prizes, one is yours in physiology or medicine and the other is Anthony Leggett's prize shared in physics. But Urbana is rather far from the Northeast or California and it takes quite a drive even from Chicago to get here.

We have our own airport. One of our chancellors — who later on was the president of Caltech — once said, "We suffer from the country's disease of bicoastalism." The chemistry department here is of high quality but is not regarded as being in the same league as Harvard or Stanford. Our school is very strong in engineering and it provides an ideal environment for dedicated studies; there are few distractions around.

May I ask you about religion?

I don't have much to say. I was raised in a Catholic family but decided that I didn't believe in that stuff anymore at about the time I was in college.

Do you talk about it freely?

It is not usual in a university setting, where most of my friends and acquaintances are. I know there are a great many people who do not make any clear distinction between being an evil person and not going to church. I am not a "crusading atheist".

Do you have heroes?

No. Something came up in Stockholm and I thought about it and I do regard the Wright brothers as heroes; they were people who went about their work in a straightforward way, in a focused way, and just did the job. My feeling was accentuated by the fact that I grew up near Dayton, Ohio, where the Wright brothers did their work, so there is a sense of connection there. Another connection is that their sister went to Oberlin College as did our daughter.

How about mentors?

Unlike most people whose autobiographies I read, I don't have a long series of mentors whom I regard as people who directed and shaped my life. I did not continue my studies right after my bachelor's degree because I did not like the idea of sitting in a lot more classes. I saw the chance of getting into a laboratory and doing some work.

So it was not for financial reasons.

No. Only then did I decide to take some courses at the University of Pittsburgh. As a result, I've given in to one invitation that I would've never otherwise; I'm giving a commencement address at the University of Pittsburgh this spring. Anyway, I started to work with a research advisor in the Physics Department while I was registered in the Chemistry Department. We did not get along very well; our scientific styles were much too different. He lost interest in his own work basically and soon left for another career, and I was left without advising. The head of the Chemistry Department told me not to worry but continue my work, which I was doing physically at Mellon Institute, and he just signed my papers, and so I could fulfill the formal requirements for my degree. He was not involved in nuclear magnetic resonance, but he thought that I was scientifically doing the right things.

Originally I was not thinking of getting a Ph.D. Once after I had given what I thought was a very good seminar talk, it turned out it was actually a recruiting talk to a major university and they were surprised that I did not have a Ph.D., and they could not make me an offer. That is what made me think that it might be a good idea to spend some time and effort to acquire this magic document. I did, and this was by putting together some of the papers I had already published. Then, when I had some difficulties with things I wanted to do that turned out were not permitted where I was working, I decided that I needed a better place to work. Among the alternatives that presented themselves was the very new State University of New York at Stony Brook. I became interested and they made me an offer. As it turned out, they made an offer to me for Associate Professor. So I was never a full-time graduate student, never had a research advisor for the work I was doing, never had a postdoc position, and never was an Assistant Professor. I never was in a situation of working in a field in which you have an inspiring mentor to work with whom you admire.

Paul Lauterbur in the Stony Brook days
(courtesy of Paul Lauterbur).

Did you ever get letters of gratitude from patients?

I did.

Before the Nobel Prize?

Yes.

So people knew about you and about your contribution.

Oh, yes. I don't know exactly where all this stuff comes from, but people come up to me in a shop and tell me about their daughter and thank me for the fact that she is alive today. It is a good feeling.

And after the Nobel Prize?

I even got a free lunch in a local restaurant. That was after the Nobel announcement. In exchange I had to sign four menus for the staff. Also, *Nature* earlier had published a full page ad congratulating me, referring to the paper I published in *Nature* in 1973. Of course, 30 years ago they had rejected the first version and published only the second version. They recently came out with a book containing some 20 most influential papers for the past one hundred years published in *Nature*, and mine is among them.

It seems to be easier to publish mediocre papers than groundbreaking papers.

It is very natural because for a truly original paper there are no real peers to judge it. The experts are very proud of their expertise and disappointed when something proves to be original that they had not thought of it themselves and they like to look for flaws and tend to think that it may not be true; there may be all sorts of psychological reasons. When I submitted a proposal to NIH, they examined it at a study session — the usual way they review proposals. Although people are not supposed to talk about it, someone told me that at first the reviewers were negative because they found my proposal crazy. Then someone said that maybe because it is crazy they should take a second look, and, fortunately, they did. The consensus was that they still found it crazy but they could not find anything wrong with it, so they decided to fund the proposal.

> At this point we involved in the conversation Dr. Lauterbur's wife who was present at the recording from the start. She is Dr. M. Joan Dawson (b. 1945), Professor of Molecular and Integrative Physiology, Biophysics, and Neuroscience at the Center for Biophysics and Computational Biology of the University of Illinois at Urbana. She received her Ph.D. degree in Pharmacology from the University of Pennsylvania in 1972.

How did you meet?

We met at a meeting at Oxford University. I was at the time at University College in London and I was collaborating with people in Oxford. I am American, but lived a dozen years in the U.K. I was postdoc at the time we met in 1978. Paul gave a talk about imaging and he showed pictures of green peppers and red peppers and he diagnosed a disease in green pepper. The image showed a tumor in the green pepper and I was very impressed with that. I had never heard of NMR imaging before.

What was your background?

I did my undergraduate studies in the School of General Studies at Columbia University and received my bachelor's degree there, I have a

Paul and Joan in their home, 2004 (photograph by I. Hargittai).

Ph.D. in Pharmacology from the University of Pennsylvania. After my doctoral studies, I went back to Columbia University as a postdoc before going to London for another postdoctoral study there. I gradually migrated to physiology from pharmacology as I was working with Douglas Wilkie, a muscle physiologist in the Department of Physiology of University College in London. I was also doing NMR physiology *in vivo* cooperating with people in Oxford.

I have interviewed George Radda of Oxford University at the time when he was the head of the Medical Research Council in London.

I know him and we had a paper together. I was interested in making chemical and biochemical measurements of metabolites non-invasively, which was a very attractive possibility.

So you attended Paul's lecture ...

I was single and Paul was in the last stage of a divorce, but I don't remember exactly how we met during that meeting.

It was easy for you to notice him, he was the lecturer, but did he notice you?

He certainly did. We had some conversation. Then, we just kept meeting in different places around the world for a while. He was very much in demand as a speaker and people were also excited about the work I was doing. Also, Paul developed the habit of always stopping by at University College in London when he was in Europe. Then when I went back to the U.S. for a visit, I sent him a letter telling Paul that I was coming, and I misspelled his name. Then we started going on vacations together and I was happy with that. I was a bit surprised when Paul suggested that we get married. He told me that he was on the train in Italy when it suddenly struck him that what he really wanted to do was to marry me and have children with me. I still have the letter. We got married in 1984, so I was 40 years old by then and I wasn't eager to get married. I had never been married before. We have one daughter who was born in 1985.

Please, tell me about your science.

Thank you. People hardly ever ask me about it. At the time I met Paul, I was doing studies of metabolism using frog muscle as a model and I recently went back to these studies full-time. In the meantime I had done a lot of collaborative work. Once I became competent in *in vivo* NMR spectroscopy, I was in a lot of demand as a collaborator. In order to tell you what I want to do with the rest of my career, first I have to tell you that in biology metabolism is thought as being controlled by enzymes that have on and off switches depending on signals from the environment. However, less attention is being paid to thermodynamic limitations in fundamental metabolism, such as glycolysis and oxidative phosphorylation. Unfortunately, biologists don't know much about thermodynamics and the textbooks sometimes add to the confusion. I hope to make clear the role of thermodynamics in metabolism. For example, in such a simple process as glycolysis, the rate and extent of glycolysis is dependent on thermodynamic limitations. If I can get such things made clear to people and have the textbooks changed, I'll consider my work useful.

Did you have tenure when you came here?

Yes and I already had tenure at University College in London.

I asked because it is very difficult when a woman is tenure-track while having a child.

It was still tough going while Elise was growing up.

Paul Lauterbur and his wife, Joan Dawson, and their daughter, Elise in 1987 (courtesy of Paul Lauterbur).

How did you solve it?

Nobody ever solves it. I did something that I keep telling my daughter not to do, which was that I took up all the household responsibilities and financial responsibilities. Elise was seven-month-old when we came here and I was working full-time at that point. There was a nursery and we also had a babysitter. That was one thing getting married and having a child late, I had money enough to get care for her when I was not available.

How do you cope with the limelight on your family although you may be in its shadow?

Thank you for noticing that.

I did not notice it, I just supposed.

Of course, I am in the shadow and I have been throughout our marriage. I surprised myself by being quite happy to have Paul have all this recognition and not being jealous and upset by it. I thought that I would be jealous and upset by it, but I'm not. The negative thing is that it does involve me in a lot of kinds of work that is not something that I would like to be doing. For example, Paul does not have a secretary and during that period from the day of the announcement of the Nobel Prize, from the minute we received that phone call from Stockholm until after the Nobel ceremony, my time was totally taken up with basically being his administrative assistant.

But that was only two months.

But that is always like that. There are always additional things that come up because of Paul's prestige.

Is there anything with which he is helping you?

He is a darn good scientist and he is one of my best physiologist colleagues. I can talk to him about my finds and get thoughtful responses.

How about your colleagues at work? Is there sympathy, jealousy, joy? You must be traveling a lot.

Certainly, no one tells you if they are feeling jealous, although I think there is a little bit of that, but there are also people who are genuinely pleased and happy for me.

Do you think that your life is changing on the long run?

I think — and Paul is going to be mad at me for this — but I think that the fact that Paul has a lot of potential traveling is going to change my life quite a bit. Not that we did not travel before but he does have health issues now. I prefer that when he travels he has someone with him.

Not necessarily you?

Me or one of his kids, just make sure that somebody is there if something happens.

Does Elise participate in any of this?

When Elise was old enough to talk or actually not yet old enough to talk, and Paul was doing a lot of traveling, I was staying home most of the time. She cried when one more time her dad left for the airport. I told her that your daddy really loves you and that he really wants to be with you. And she said, "Daddy loves airplanes." When she became a little older, 4, 6, 8, she was proud, because she realized that her father was special. But she was also a little bit jealous with the people who were taking him away from her. I never asked her about how she felt growing up with a father who was a superstar scientist. I'm not sure if she would tell me. I know though what she tells about the problem with growing up with two professors in the family: you cannot ask even a simple question without getting a lecture in response.

Gunther S. Stent, 2003 (photograph by I. Hargittai).

29

GUNTHER S. STENT

G unther S. Stent (b. 1924 in Berlin, Germany) is Professor Emeritus of Neurobiology at the University of California at Berkeley. He arrived in Chicago in 1940 as a refugee from Nazi Germany. He received a Ph.D. in physical chemistry from the University of Illinois at Urbana-Champaign in 1948. He then studied as a postdoctoral research fellow of the U.S. National Research Council at the California Institute of Technology (Caltech), the University of Copenhagen, and at the Pasteur Institute in Paris. He participated in the early development of molecular biology and has written several influential books.

His textbook *Molecular Genetics* is a classic. He is also the co-author of *Phage and the Origins of Molecular Biology* (with James D. Watson and John Cairns). He edited the critical edition of James Watson's *The Double Helix*. He has made contributions to the history and philosophy of science. His latest book *Paradoxes of Free Will* won the 2002 John Frederick Lewis Award of the American Philosophical Society.

Among his distinctions, Dr. Stent is a member of the National Academy of Sciences of the U.S.A. (1982), the American Academy of Arts and Sciences (1968), the American Philosophical Society (1984), the Akademie der Wissenschaften und der Literatur (Mainz, Germany, 1989), and the European Academy of Arts and Sciences (1993).

We had several sessions of recording at the Budapest University of Technology and Economics on January 13–15, 2003.*

---

*In part, this has appeared in *Chemical Heritage* 2004, 22(1), 2–6.

Your autobiographical memoir Women, Nazis, and Molecular Biology *reads like a novel, but it describes, as I take it, your family background and life up to your postdoctoral studies in molecular biology. We cannot presume, of course, that all of our readers have read your book. So could you please provide a summary of your childhood, youth, and early professional training? And, please augment it with your graduate school years.*

I was born in 1924 in Treptow, an unfashionable suburb of Greater Berlin. Few Jews lived in Treptow, whose anti-Semitic, petty-bourgeois milieu was light years distant from the avant-garde action — the Albert Einsteins, the Max Reinhardts, the Marlene Dietrichs, the Bertolt Brechts — to which the Berlin of the Weimar Republic owed its cultural glamor. My father, Georg, was a native Berlin Jew, who owned one of the largest bronze statuary and lighting fitting factories in Germany. My mother, Elli, came from a family of well-to-do, assimilated Jews in the Silesian city of Breslau. She suffered from an inherited manic-depressive disorder and took her own life when I was eleven.

My religious upbringing, in so far as I had any, took place in the context of anti-Zionist, Germano-Christianized Reform Judaism. There were no observances of Hebrew ritual in my home: no Sabbath candles, no Passover seder, no matzoth. Judaism became a major factor in my life when I was nine, when Hitler came to power, and I began to fear for my life, as I watched his storm troopers marching through the Berlin streets bawling: "When Jew-blood spurts from our knives, we'll all have twice-better lives."

I attended Berlin public schools until the fall of 1935, when the Prussian minister of education decreed that the Jews should take care of the education of their children. So my elder sister enrolled me in a private Jewish school, whose main educational goal was to prepare its students for their exodus from Nazi Germany. Thus we received intensive training in the three languages that we were most likely to encounter abroad as emigrés: English, French and Modern Hebrew. I stubbornly resisted learning Hebrew, but I had acquired a fluent command of both English and French (alas, with a life-long, indelible German accent) by the time I escaped (illegally, by stealing across the "Green Frontier" between Germany and Belgium) on New Year's Day of 1939.

My plan was to join my sister, who had emigrated to the States with her husband in 1936. But I had to cool my heels for fifteen months — first in Belgium and then in England — until I received my U.S. immigration papers. Finally, in March 1940, I sailed for New York from Liverpool in the third-class bowels of an ancient Cunard Line steamer.

I moved in with my sister, who lived in Chicago's Hyde Park district, which had turned into one of the "Fourth Reich" settlements of German and Austrian Jewish refugees, similar to Washington Heights in Manhattan and Swiss Cottage in London. On my enrollment in Hyde Park High School, the Vice-Principal gave me a few tests and discovered that I had no academic skills. I couldn't do fractions, let alone geometry or science. Although I was almost sixteen, he made me start as a freshman, graduating class of 1944. Yet, it turned out that I did have one very useful academic skill. I managed to work out a scheme to beat the Chicago system of awarding academic credit units. By accumulating them at a furious rate, I made it through the four-year curriculum of Hyde Park High in less than two years, still appallingly ignorant of most academic subjects, except for German and French.

In the fall of 1942, I went down to Champaign, to enroll in the University of Illinois as a freshman in Chemical Engineering. I had never heard of this calling, but its name suggested to me something brand new, something futuristic. "Gunther Stent, Chemical Engineer" had a nice ring to it, and I figured it would set me apart from your run-of-the-mill college graduates. As it turned out, I didn't like chemistry, until, in my junior year, I took my first course in physical chemistry. Professor Frederick T. Wall's lectures were dynamic, lucid and well prepared, and I found his presentation of chemical thermodynamics captivating. In contrast to inchoate inorganic or organic chemistry — not to speak of the boring engineering courses — physical chemistry appealed to me as a logically coherent discipline, whose theories are expressible as mathematical relations. So I switched my major to physical chemistry.

Upon graduating from Illinois in January 1945, I applied to work for a Ph.D. in physical chemistry at Caltech under Linus Pauling. He was my scientific hero because his *The Nature of the Chemical Bond* was the first textbook that I actually enjoyed reading in all my thirteen years of doing time in German and American schools. But Caltech turned me down; so I resigned myself to staying on at Illinois and accepted Professor Wall's offer to do a Ph.D. under his direction, while working as a Research Assistant on the Synthetic Rubber Research Program of the U.S. War Production Board.

The mission of the rubber program was to develop procedures that would make synthetic rubber tires as good as, or even better than, tires made from natural rubber. Natural rubber consists of a homopolymer, in which hundreds of isoprene monomers are linked end-to-end, whereas synthetic rubber (Buna-S) is a co-polymer of equal proportions of butadiene and styrene monomers. One of the flaws of Buna-S that impaired its elastic properties was that the synthetic butadiene–styrene co-polymer molecules varied greatly in their length, whereas natural isoprene homopolymer molecules are of uniform length. Professor Wall assigned me to work out a method by which Buna-S could be resolved into a series of fractions, each of which would contain co-polymer molecules of uniform length. I decided to try a modification of the thermal diffusion column invented by K. Clusius and G. Dickel in 1938 for resolving gaseous mixtures of atomic isotopes, such as $^{12}CO_2$ and $^{13}CO_2$.

I built a column 2 m tall, consisting of two concentric steel tubes separated by a narrow, 1 mm gap filled with a toluene solution of the synthetic co-polymer, the outer tube being cooled by cold water and the inner tube being heated by hot water. It worked as I had hoped. After letting the column run for two days, downwards from the top, the gap contained Buna-S molecules of ever-greater chain length. Mathematical analysis of these data then allowed me to provide an entirely novel way of determining the molecular length distribution among the original synthetic co-polymer molecules.

I was jubilant: it was my first success as an experimental scientist. I had finessed Mother Nature and made her do my bidding! Perhaps my device would become known as the "Stent column". When people would congratulate me, I would feign modesty and point out that the idea was actually pretty obvious.

I proudly presented my findings at a national meeting of all War Production Board Rubber Research groups in the spring of 1946. There was no need for me to feign modesty. My talk aroused only mild interest; my colleagues thought that my method would never provide a practical way of producing Buna-S molecules of uniform chain length on an industrial scale. My results were never published, and my liquid phase thermal diffusion method for resolving polymer molecules of different lengths vanished without trace from the corpus of science. This was a pity, because, unknown to me, biochemists studying proteins and nucleic acids in the 1940s were in great need of techniques for separating differently-sized molecules present in extracts from living cells. My thermal diffusion column could have served beautifully. In retrospect, I have no doubt that it would have made a stir among biochemists had I gone ahead and adapted it for their purposes. But by the 1950s, when I finally became aware of the opportunity I had missed, much better molecular separation methods had become available.

This episode proved to be only the first of several instances in my career when I hit on a good thing that could have gained me substantial fame — but didn't. Many scientists try to sell sour grapes by claiming that someone who claimed credit for it stole one of their brilliant original ideas. Such thefts do occur, of course, but not all that often. The more banal cause for the failure to get due credit for one's discovery is, as in my case, the lack of personal qualities needed to have it make an impact. Originality and inventiveness, though necessary, are not sufficient for making a mark in science: one has to have also the intuition, stamina, and, above all, the self-confidence necessary to exploit one's discoveries and present them as a salable package.

How did you, trained as a physical chemist, become interested in biology?

In my second year of graduate studies, my friend and mentor Martha Baylor, a postdoctoral research biologist in charge of the Illinois Chemistry Department's Early American electron microscope, suggested that I read *What Is Life?* a then recently-published brief tract by the famous Austrian physicist Erwin Schrödinger. She thought that I would be interested in what the co-discoverer of quantum mechanics had to say about the connection between thermodynamics and biology. At Hyde Park High I had found botany and zoology terminally boring, but I took her advice.

Schrödinger announced that a new era was dawning for the study of heredity, thanks to some novel ideas put forward by Max Delbrück, whom he identified only as "a young German physicist". How, Schrödinger asked, do genes manage to preserve their hereditary information over the generations? Following Delbrück's then ten-year-old proposal that this stability derives from the atoms of the gene molecule staying put in "energy wells", Schrödinger proposed that the gene-molecule is an *aperiodic crystal*, comprised of a long sequence of a few different, over-and-over-repeated basic elements. The exact sequence pattern of these elements would represent a "code", by means of which the hereditary information is encrypted. Thus Schrödinger was the first to put forward the concept of the genetic code, one of the most important ideas of the 20th century life sciences. Schrödinger had no idea of the molecular nature of this code. He speculated, however, that "from Delbrück's general picture of the hereditary substance it emerges that living matter, while not eluding the laws of physics as established up to date, is likely to involve hitherto unknown 'other laws of physics'. Once they have been found, they will form just as integral a part of this science as the former."

Max Delbrück presenting a theory of Stent and Élie Wollman's (courtesy of Gunther Stent).

What is Life? had a tremendous influence. There were many young physicists and chemists who, after the Manhattan Project and after Hiroshima, were looking for new challenges. Schrödinger painted a picture of the future for them and called their attention to the big mystery of the gene. I have called *What is Life?* the "Uncle Tom's Cabin of Molecular Biology" because it changed history by arousing the affects of its readers. [When President Lincoln met Harriet Beecher Stowe, the author of *Uncle Tom's Cabin*, he said to her, "So you're the little lady who started the Civil War!"] Although professional biologists were contemptuous of *What is Life?* when it appeared, it started molecular biology.

As a mere Ph.D. candidate in physical chemistry in his twenties, I was too green to be suffering from the post-war professional malaise of my elder colleagues. Yet I was so captivated by the fabulous prospect that by studying genes I might turn up "other laws of physics" that I resolved to join the search for the aperiodic crystal of heredity. Delbrück, the young German physicist, had probably been drafted into the *Wehrmacht* and been killed during the war. But perhaps there were people in America working along these lines.

Good news reached me in the summer of 1947. Delbrück was not only still alive, but he had just been appointed Professor of Biophysics at Caltech — my academic dream place, home of my hero Linus Pauling. I wrote to Delbrück to ask whether there was any possibility of my working under his direction in Pasadena. I was thinking of applying for a new type of National Research Council (NRC) Postdoctoral Fellowship sponsored by the Merck Chemical Company. According to an announcement in *Science* magazine, the fellowship's purpose was "to provide special training and experience to young men and women trained in chemical or biological science who wish to broaden their fields of investigational activity." What sort of biophysical problem could I be working on in his laboratory if I were awarded a Merck Fellowship? Delbrück replied that he was not in a position to state in any detail the type of problems he was going to work on next year. But he was thinking of doing some experiments on phototaxis in purple bacteria, which might be a good opening for the study of excitatory processes.

I didn't know the meaning of "phototaxis" or of "excitatory processes", had never heard of "purple bacteria", and was totally in the dark about what all this might have to do with genes and leading me to the discovery of other laws of physics. But I thought that there would be plenty of time to find out what Delbrück's proposed project was all about, in the unlikely event that my fellowship request would be granted. In my application I declared that I hoped to apply my knowledge of physical chemistry to the study of biophysical problems, with special emphasis on the investigation of life processes from the point of view of thermodynamics and reaction kinetics. To that end, I intended to study the general nature of excitory [sic] processes under the direction of Professor Max Delbrück at the California Institute of Technology. I had the good sense not to let on in my application that I had a hidden agenda, namely looking for other laws of physics.

Many months later, I received a telegram asking me to come to New York — all expenses paid by the NRC — for an interview with the Merck Fellowship Board. My sky-high exultation over this marvelous news subsided as soon as I began to think about the interview. It wouldn't take more than one or two incisive questions by the Board to establish that I knew nothing about the "excitory" processes on which I was proposing to carry out advanced postdoctoral research by studying the phototaxis of purple bacteria. Obviously, I had no idea about how all this was going to lead me to novel insights about life processes from the point of view of thermodynamics and reaction kinetics.

My fears were not unfounded. The Merck Fellowship Board consisted of six formidably distinguished, awe-inspiring senior scientists. It included the Chairman A. N. Richards, President of the National Academy of Sciences and ex officio High Priest of American science; the geneticist and future Nobel laureate, George W. Beadle, Chairman of the Caltech Biology Division; Detlev W. Bronk, Professor of Biophysics in the University of Pennsylvania and President of the NRC; Hans T. Clarke, Professor and Chairman of Biological Chemistry at the College of Physicians and Surgeons of Columbia University; George O. Curme, Director of Research of the Carbide and Carbon Chemicals Corporation; and René Dubos, the famous bacteriologist at the Rockefeller Institute for Medical Research.

I was in deep trouble as soon as Chairman Richards asked his first question.

"So you want to go into biology; what do you plan to do?"

"I want to study excitory processes to test whether the Second Law of Thermodynamics applies to living systems."

"You mean excitatory processes, don't you?"

"Yes, Sir. I think so. Yes, I do."

"How are you going to do it?"

"I'm going to study the phototaxis of purple bacteria."

"How? And how's this going to tell you something about the applicability of the Second Law?"

"I'm not exactly sure. Professor Delbrück suggested that this would be a good experimental material for my project."

Upon this answer, the Board members frowned and shook their heads in disbelief. After I proffered a few more obviously unsatisfactory responses to the questions of other Board members, René Dubos finally asked me sarcastically:

"Then, if I understood you correctly, your proposed postdoctoral studies in biology at Caltech would have to be at the — (pause and emphasis) — *undergraduate* level. Isn't that so?"

"Yes, Sir. I guess so."

After this response, I was dismissed summarily and shuffled out of the room totally humiliated. Three days after I got back to Champaign, a telegram came that said that they had given me the NRC Merck Fellowship, as one of only seven awardees among a total of forty-six applicants. I could only conclude that the unsuccessful thirty-nine were even more appalling frauds than I.

A few weeks after I was awarded my fellowship, Delbrück asked me to meet him in Chicago. He enchanted me: lightening-quick on the uptake, funny, and amazingly well informed on a wide range of subjects. He seemed to know everybody, especially the all-time greats of quantum physics, onwards from Max Planck through Niels Bohr to Werner Heisenberg, Wolfgang Pauli, and Paul Dirac.

When the time finally came to discuss my future projects, Delbrück didn't mention phototaxis of purple bacteria or sensory excitation at all.

Instead, he asked, "I take it that you want to work on phage?"

"Yes Sir, that's exactly what I want to work on. But could you refresh my memory as to what 'phage' is actually all about?"

"No need for that now. You'll find out what it's all about soon enough at Cold Spring Harbor Laboratory on Long Island. You're going to spend the summer there and take the Phage Course. In early September, we'll all head out West, to Pasadena."

On my arrival at the Cold Spring Harbor Lab in July 1948, Max (which is what everybody there called Delbrück) introduced me to James Watson, a twenty-year-old graduate student, who had also been fascinated by Schrödinger's *What is Life?* Watson was working for his Ph.D. at Indiana University, doing research on the effects of X-rays on phage with Salvador Luria, whom he had chosen as his mentor because Luria was a collaborator of Max's.

I didn't like Jim at first, because — my junior by four years and a mere graduate student — he treated me as an equal, acting as if his opinions were just as good as mine. But before long, I came to terms with the depressing fact that Jim's opinions were almost always right and mine almost always wrong, whenever we disagreed about some scientific matter. And so we became life-long friends.

Would you please paint an impressionistic picture of Delbrück's background?

Of the many remarkable people I met during my career, none seemed more secure in his person than Max. He was born in Berlin in 1906 into the Prussian intellectual and scientific aristocracy. Hans Delbrück, the foremost German historian of war, was his father. Adolf von Harnack, the friend of Wilhelm II and founder of the Kaiser Wilhelm Society for the Promotion of the Sciences, was his uncle; and Justus Liebig, the great 19th century chemist, was his maternal grandfather.

Max studied physics in Göttingen in the late 1920s, but flunked his first Ph.D. examination. He passed on a second try in the following year and was awarded a postdoctoral fellowship to go to Copenhagen to become a disciple of Niels Bohr. It was Bohr who influenced Max to take

Niels Bohr (standing at the left), Gunther Stent (standing fifth from the right), James Watson (standing third from the right), Herman Kalckar (standing between Stent and Watson), and Élie Wollman (squatting on the right) (courtesy of Gunther Stent).

an interest in biology, by persuading him that the presently known laws of physics might not be adequate to account for the phenomena manifested by living creatures, especially their self-reproduction. So while studying genetics one might make a contribution not only to biology but also to physics, by discovering some hitherto unknown physical laws that make it possible for like-to-beget-like.

After his stay in Copenhagen, Max went back to Berlin and got a job with Otto Hahn and Lise Meitner as a physicist at the Kaiser Wilhelm Institute for Chemistry. To prepare himself for the discovery of other laws, Max joined a discussion group led by the Russian geneticist, Nicolai Timoféeff-Ressovsky, which met at the Kaiser Wilhelm Institute for Brain Research in Berlin-Buch. As a result of these discussions, Max published a paper authored jointly with Timoféeff and K. G. Zimmer in 1935, in which they drew attention to the mysterious stability of our genes, maintained at the body temperature of 37°C over the millennia. They proposed that genes must owe their chemical stability to being molecules. This paper remained virtually unknown, until wide attention was drawn to it a decade later by Erwin Schrödinger in his *What is Life?*

In 1937, Warren Weaver, director of European operations of the Rockefeller Foundation, offered Max a Rockefeller Fellowship to spend a year or two with Thomas Morgan's group at Caltech, to gain professional competence in fruit fly genetics. So Max left his Berlin employers Hahn and Meitner just before they discovered nuclear fission. According to Max, they might have discovered it sooner, and maybe Germany might have developed an atom bomb before the end of the War, if he hadn't given Hahn and Meitner bad advice.

Not long after he arrived at Caltech, Max decided that fruit fly genetics was too complicated for him. And when he ran into another postdoc there working on bacterial viruses, or bacteriophages (nicknamed "phages"), he decided that it would be easier to find those other laws by studying their self-reproduction than that of fruit flies.

Did Delbrück have a good knowledge of chemistry?

No. On the whole, he was contemptuous of chemistry, and especially of biochemistry (and of biochemists). He acknowledged that metabolism was important for biology, but he didn't believe that biochemists could be of much help in solving the mechanism of self-reproduction and hence the mystery of heredity.

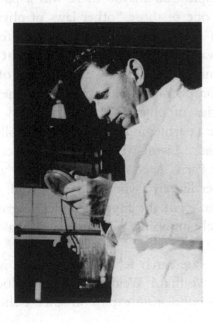

André Lwoff (Nobel Prize 1965)
(courtesy of Gunther Stent).

By the time the Phage Course was over, I felt I had become an expert bacteriophagologist. I had imbibed the conceit of Max's Phage Group that there was no point in paying any attention to the work of our predecessors or of contemporaries external to the "Church", as the French microbiologist, André Lwoff, referred to the coterie of Max's disciples. Reading publications lacking the Church's imprimatur was worse than a waste of time: the unsubstantiated claims based on poorly designed experiments presented by such confused heathen outsiders would just put wrong ideas in your head.

How did you fare at Caltech?

Caltech lived up to my fantasy of a palm-tree studded Academic Nirvana: a double tier of adobe-colored California-Mission-style laboratory buildings facing a subtropically landscaped central mall, stretching for half a mile between two Pasadena streets, set off against the sunlit San Gabriel Mountains peaked by 10,000 foot Mt. Baldy and populated by brilliant minds, like Linus Pauling, George Beadle, and Max.

My research project was one of the few Max could have picked for which my training as a physical chemist happened to have eminently qualified me. One of the phage strains studied by the Phage Group fails to attach to its bacterial host cell unless it has been previously "activated" by contact with the amino acid tryptophan. Max suspected (or maybe hoped) that the hitherto known facts about this activation process were not compatible with ordinary physico-chemical principles. So maybe there was a paradox hidden here which might lead us to one of those "other laws of physics".

As much as I was hoping to run into a biological system manifesting an "other law", I feared that this was not one of them. I thought that I wouldn't have much trouble coming up with an explanation of the seemingly bizarre tryptophan activation phenomenon within the framework of ordinary house-and-garden theories of physical chemistry.

Max informed me that I would have a partner in my project, Élie Wollman, a young French bacteriologist from André Lwoff's Department of Microbial Physiology at the Institut Pasteur in Paris. According to Max, Élie and I were going to complement each other like liverwurst and rye bread. He's got the bacteriology, of which I haven't got the foggiest and I've got the math and physics, of which he is largely innocent. Together, we'll make the perfect phagology sandwich.

By the fall of 1949, there were six research fellows working in Max's lab — Élie Wollman, Jean Weigle, Wolfhard Weidel, Renato Dulbecco, Seymour Benzer, and me — a virtual population explosion since I showed up as Max's first Caltech postdoc a year before. We formed a close-knit circle, with Max as our *spiritus rector*. Dulbecco would presently succeed in extending the method of plaque assay we used in phage work to animal viruses, which would set the stage for quantitative studies on animal viruses to fathom their intracellular reproductive cycle. He also opened the era of animal virus genetics by isolating virus mutants and developing techniques for mixed infection of single animal cells with two or more genetically different mutant viruses. For these contributions Dulbecco would be awarded the 1975 Nobel Prize in Physiology or Medicine.

Benzer, who had received his Ph.D. in solid state physics from Purdue University, was my classmate in the 1948 Cold Spring Harbor Phage Course. He, too, had been seduced by Schrödinger's *What is Life?* and hoped to get started on finding the aperiodic crystal of heredity in the Caltech lab of the protagonist of Schrödinger's book. Within a few years, Benzer would convert the fuzzy concept of the Mendelian gene of classical genetics into its precisely defined, latter-day molecular-genetic version. I will always believe it a shame that Benzer was not included in the set of Nobel laureates honored for laying the foundations of molecular biology.

For us members of Max's research group, there was no clear separation between our professional and our private lives, because Max's benevolent (or, in New Age California-Speak, "caring") interest in his disciples was all-inclusive. He not only guided our scientific work in the lab, but also supervised, not to say intruded in, what would normally be considered one's private, after-hours affairs, such as choice of girlfriends, partying, going out for concerts, plays, movies, or dinner, and camping. As our *pater familias* Max considered it his business — if not actually his duty — to inform himself about all facets of our lives. He was a stranger to the concept of privacy.

Exchanging my anxiety-ridden sovereignty for an insouciant thralldom under which I could leave decisions about my professional and private activities in Max's hands appealed to me. To give up all that freedom and personal responsibility for making choices with which I had been saddled since my mother's death was a relief. It was like being in the Army, where every soldier, downwards from the Chairman of the Joint Chiefs of Staff, takes orders from a superior authority figure, who is held accountable for the commands one obeys.

I would like to ask you about that exceptional group of young people at Caltech, when you were Delbrück's postdoctoral student there.

In addition to Élie Wollman, Jean Weigle, Wolfhard Weidel, Renato Dulbecco, and Seymour Benzer, whom I have already mentioned, there were Carleton Gajdusek, Benoit Mandelbrot, and Jack Dunitz, and, collectively, all these men had an even greater influence on my intellectual and characterological development than had Max.

Benoit, the mathematician-inventor of the "Mandelbrot Set", was the one of our set who would become the most famous. Hidden from the Nazis in his childhood as a Jewish–Polish refugee in a monastery in wartime France, Benoit had no manners when he showed up at Caltech as a graduate student. Élie Wollman and his wife, Odile, befriended and civilized him. Although I like him and we are good friends, Benoit's brilliant, albeit self-centered, persona is not universally admired. As I have heard it said about another famous scientist: "You can rely on him: he is always there when he needs you."

It was an extraordinary group. What makes such a group come together?

I think it was the renown of Caltech's faculty and its unique, paradisiacal environment.

It took Wollman and me most of our second year at Caltech to write three papers presenting the results of our experiments on tryptophan activation and to hone our theory to account for them. I would never again devote as much effort and care to any of the couple of hundred other papers and essays I eventually published, and I consider these three papers with Wollman as my best. Max predicted that, some day, they would become famous classics. Alas, only a few people read them when they came out — we did get one fan letter from an immunologist in Australia — and they have long since been forgotten.

As I had feared, there was no need to invoke "other laws of physics" to explain the seemingly bizarre dynamics of tryptophan activation. We managed to devise a model based on conventional physico-chemical principles that accounted for all the data. It was a forerunner of the "cooperative" models of the complex interactions of small molecules with enzymes and other protein molecules put forward a few years after our papers, which have formed the basis for understanding the regulation of protein function ever since. As far as I know, no contributor to the vast literature of cooperative protein interactions ever cited our tryptophan activation model.

Where did you go after completing two years of postdoctoral work at Caltech?

My first choice would have been to stay on at Caltech after my Merck fellowship expired, for the rest of my life, if possible, even if it had to be as one of those perennial non-faculty hangers-on with whom Caltech was crawling. But Max did not ask me to stay on.

My second choice would have been to go to Paris, to continue my collaboration with Élie Wollman at the Pasteur Institute. My motivation for that second choice was more affective than scientific, because I had become so dependent on Élie and his wife, Odile, that I could not see myself facing life on my own without their emotional support. Alas, Élie's *grand patron* at the Pasteur Institute, André Lwoff, refused to accept me. He told Élie that in the coming year there would be no room for me in his department, since he had just accepted a new young *assistant* by the name of François Jacob with whom Élie would have to share his *labo*. Perhaps there might be room for me a year later.

Before long I came to believe that Lwoff's real reason for not accepting me for the coming year was that he didn't want any static from wise guys from Max's Phage Group in his lab at that time. Lwoff was well along the way towards establishing the reality of *lysogeny*, the phenomenon of the permanent propagation of some types of phage by their living host bacteria. Max had decreed *ex cathedra* that lysogenic bacteria do not exist and that the perception of lysogeny is a self-delusion attributable to the sloppy bacteriological techniques practiced by its *aficionados*.

It so happened that in the 1920s and 1930s Élie's late bacteriologist parents, Emanuel and Elizabeth Wollman, were pioneers in the study of lysogeny. They showed that lysogenic bacteria propagate their phages in a non-infective form. The elder Wollmans had been on the staff of the Pasteur Institute before the War. Despite being Jewish, they did not bother to hide during the German occupation of Paris and continued their work in their lab at the Institute. They were arrested at work and disappeared in one of the German extermination camps. Élie had gone underground, fought in the Resistance and joined Lwoff's team at the Pasteur Institute after the War.

Lwoff followed up the elder Wollmans' findings, and by the early 1950s had solved the enigma of lysogeny. He confirmed that lysogenic bacteria do perpetuate phage genomes in a non-infective form, which he called *prophage*. Activation, or "induction", of the prophage generates a crop of infective progeny phages, which are released upon disintegration of the lysogenic host bacterium. It took a long time for Max to accept Lwoff's clarification of lysogeny (for which Lwoff was awarded the Nobel Prize in 1965). And — at least so I am convinced — that was why Lwoff was not keen on having any of Max's disciples like me in his lab until Max officially granted Lwoff his *nihil obstat*.

With Paris, my second choice for the next working place, out of the picture as well, Max suggested that I move to Copenhagen to work in the lab of the Danish biochemist Herman Kalckar. At first, I was dismayed that Max would recommend that I work with a biochemist, because I knew that he had very little use for biochemists. So I figured that Max couldn't have a very high opinion of my promise as a creative scientist. But I felt much better after I found out that he had made the same proposal to Jim Watson, of whom, as I knew, Max thought very highly. Max told us that it might do us some good to learn DNA chemistry from Kalckar, because, maybe, DNA *does* have something to do with genetics. He didn't realize that Kalckar actually knew very little about DNA. His specialty was adenosine triphosphate (ATP) and its provision of free energy for driving biochemical reactions. I suspect that Max thought at the time that being composed of ATP-like nucleotides, DNA provides the free energy for driving self-reproduction of proteinaceous genes, in chromosomes as well as in phage.

So following Max's suggestion, Jim and I applied for (and were awarded) NRC Merck and American Cancer Society postdoctoral fellowships, respectively, for going to Copenhagen.

Did Jim have much of a background in biochemistry?

No. And neither had I.

You were at least a physical chemist.

Yes, I was. But nobody at Illinois was interested in physical biochemistry during my student years there.

What happened to you and Jim Watson when you got to Copenhagen and joined Kalckar's lab in the late summer of 1950?

Kalckar was very distracted because he was about to leave his wife and head for Naples with a young American woman postdoc in tow, who had arrived in his lab at the same time as Jim and myself. Moreover, it turned out that Kalckar neither knew much about DNA, nor was he very interested in it. So Jim and I decided that we would leave Kalckar's lab and move in with Ole Maaløe, the Head of the Danish State Serum

Institute's Standardization Department. I knew Ole because he had spent a few weeks earlier that year in Max's laboratory at Caltech.

Mindful of Max's advice that maybe there *is* something useful that can be learned by studying DNA, Jim and I set out to examine the metabolism of phage DNA in Ole's well-equipped laboratory. To that end, we carried out radioactive tracer studies on the fate of the parental phage DNA and on the synthesis of the progeny phage DNA in the infected bacterial host cell. Our results were not exactly world-shaking, but they helped bring into focus the intracellular transactions of phage DNA that were in want of understanding.

Did the two of you share the same lab?

Not only did Jim and I share the same small lab at the Serum Institute, but we shared it also with the immunologist, Niels Jerne. He held a junior position in Ole's department and was in the early stages of the experiments with immunized rabbits that would eventually earn him the Nobel Prize in 1984 for his development of the selective theory of antibody formation.

You told me that in his younger years Jim was very much given to imitating people he admired. Whom did he imitate?

When I first met Jim at Cold Spring Harbor in 1948, I was surprised when he told me that he was from Chicago's South Side (where I too was from), because he didn't talk like a Southsider. He spoke with a weird accent. I thought that, maybe, he was a foreigner who had learned English in a Berlitz school. It turned out that Jim was imitating Roger Stanier, the brilliant Canadian microbiologist whom he had met at Indiana University. Stanier was born on Vancouver Island, off the Coast of British Columbia, where they speak a peculiar, old-fashioned kind of English.

Did Jim imitate Delbrück at some point?

Not in his speech, but in Max's Berlin-style short haircut and some mannerisms. By the time of his discovery of the DNA double helix, Jim had changed his hair styling to the long, Cambridge-style locks.

When did Jim's interest turn to the three-dimensional structure of DNA?

Halfway through our year's stay in Copenhagen, in the fall of 1950, Sir Lawrence Bragg had been invited by the Danish Royal Society for the

Promotion of Science to give a public lecture. Jim and I went to hear Bragg, who began by saying that he had intended to speak about the work going on in the Cavendish Laboratory in Cambridge, especially about the work by Max Perutz on hemoglobin and by John Kendrew on myoglobin. But just the week before he left for Copenhagen, Bragg received a letter from Linus Pauling. In that letter Pauling described how he had found the secondary structure of proteins, the alpha-helix.

So Bragg thought it would be more interesting for the Danish colleagues to hear about Pauling's breakthrough rather than about the progress that his own people were making.

How did Jim react to Bragg's lecture?

As Bragg was describing Pauling's alpha-helix, I noticed that Jim was getting more and more agitated. As soon as Bragg had finished his presentation, Jim turned to me and said that *this* is what we should be working on. Doing the Pauling number on DNA and finding its three-dimensional structure. He thought that we were wasting our time working on DNA metabolism, which wasn't going to get us anywhere.

I didn't agree with him. I thought that the study of phage DNA metabolism was much more promising than working out the three-dimensional structure of DNA, for which, in any case, I believed that Jim was even less qualified than myself. What was that structure going to tell us that would be of any help in solving the problem of genetic self-reproduction? I didn't see the connection.

How could Jim see the connection?

Because he is a genius, that's how. For me, a genius is somebody who sees connections that I don't see.

Have you met any other geniuses?

Niels Jerne was one, and maybe an even higher grade of genius than Jim was.

Anyone else?

Linus Pauling, Francis Crick, Seymour Benzer, François Jacob. Of the three great French biologists that I came to know well — André Lwoff, Jacques Monod, and François Jacob — I liked and respected Jacob the most. He

Jacques Monod (Nobel Prize 1965)
(courtesy of Gunther Stent).

is a real hero for me, not only for his scientific and literary achievements (such as his books *The Possible and the Actual* and *La Statue Interieure*) but also for his personal history. On my way to Budapest for this interview I took my son, Stefan, to the Museum of Military History in Paris. There we found on exhibit memorabilia about Jacob's activities as an officer in the Free French Forces during World War II, including the document issued after the War that bestowed on him one of the highest honors, namely membership in the illustrious Order of the *Compagnons de la Libération*.

In the mid-1960s — when the French government was reducing the level of support of basic scientific research — Jacob attended a reception given by President Charles de Gaulle at the Elysee Palace for the surviving *Compagnons*. De Gaulle spotted François and said, "Ah, there you are, Jacob! What are you up to these days?" "I work at the Pasteur Institute, *mon Général*." "And do you have everything there you need?" "No, *mon Général*." "Well, I'm happy to hear this. Carry on with the good work!"

When I first met Monod I liked and admired him very much as well. But we fell out over his book *Chance and Necessity* when I reviewed it in the *Atlantic Monthly* and wrote that I found it phony. Monod was very upset by my review. He told me that my main purpose in life seems be to tear him down.

Was Delbrück a genius?

No. I think he lacked the hallmark of a true genius, namely having many brilliant, original ideas.

What were Delbrück's main merits?

Leadership, absolute integrity, and inspiring people by clearly explaining complicated matters. He could instantly recognize an original idea when he saw one. His rationality was supreme. Whenever there was evidence for something and he had to draw logical inferences, he rarely made a mistake. But if not all the answers were in, or if they were ambiguous and he had to use intuition for making a judgment, then he was often wrong. Yet his beneficial social influence in science was very great, and his being awarded the Nobel Prize was undoubtedly due to this role.

Not long after we heard Bragg's lecture, Jim told me that he was actually going to try to work out the 3-D structure of DNA.

What did you think of Jim's plan?

I thought he had gone off his rocker. What did he think a mere biologist like him could do about working out the 3-D structure of DNA, when a physical chemist like me wouldn't dare to wrestle with that problem?

What happened then?

At the end of our Copenhagen year, Jim persisted in what I considered at that time his megalomaniac lunacy. He decided to spend his second Merck Fellowship year at Bragg's Cavendish Laboratory in Cambridge, to acquire the necessary skills in X-ray crystallography for working out the 3-D structure of DNA.

But the University of Chicago embryologist, Paul Weiss, who had become chairman of the Merck Fellowship Board in the meanwhile, refused to approve Jim's plan of moving to Cambridge. Weiss instructed Jim that if he were really serious about working out the structure of DNA he must go the Stockholm, where there were people working on the biochemistry of DNA, rather than wasting his time on its X-ray crystallography. Jim defied Weiss and went to Cambridge anyway. So Weiss cut off Jim's fellowship stipend.

Max and Salvador Luria, who, like most of Jim's friends (including myself) didn't expect that he would actually succeed in his ambitious project, managed to persuade the March of Dimes Foundation to provide financial support for Jim's stay in Cambridge. In later years Weiss had the nerve to claim that *he* was one of the discoverers of Jim Watson and provided the funds that made possible what is now widely regarded as the twentieth century's greatest biological discovery.

Let's get back to Bragg's lecture. Did Bragg mention that he, Perutz, and Kendrew had reported 20 models for the protein structure that all turned out to be wrong? [Proc. Roy. Soc. 1950, 203A, 321–357]?

Since at the time of the Bragg lecture my knowledge of X-ray crystallography was even more superficial than it is now, I don't remember what it was exactly that Bragg said. But I doubt that I would have forgotten his mention of 20 incorrect structural reports by his minions, Perutz and Kendrew.

Perutz described their errors in his book *I Wish I'd Made You Angry Earlier*. This title quotes what Bragg actually said to Perutz when Perutz informed Bragg of his failure to find the right protein structure and told Bragg how angry he was when he saw Pauling's paper. On the basis of this experience, Perutz vowed never to resort to modeling again. He told me that even though Pauling had been so successful with modeling in working out the alpha-helix, no modeling had gone into Perutz's eventually working out the structure of hemoglobin [*Candid Science II*, p. 288].

Was there a difference in their methodologies in that Pauling used methodologies of which the British did not approve?

I do remember that, according to my informants, the British did not regard Pauling as a very serious X-ray crystallographer. They considered his procedure as depending too much on guesswork, as well as on model building. Pauling was seen as insufficiently analytic, in contrast to Perutz and Kendrew, who believed only in real facts, as revealed by the direct interpretation of X-ray diffraction patterns. Model building and inspired guesswork were simply not in.

Not quite. There was a major difference in that Pauling utilized the fact that the peptide bond is planar and that restricted the possibilities for his model. That came from the knowledge of the length of the peptide bond and his resonance theory. Perutz did not have this information although Alexander Todd had told them that the peptide bond was a partial double bond. For Perutz this did not translate into planar bond configuration, whereas Pauling knew that the double-bond character corresponded to planarity. Later, Watson and Crick followed Pauling's model-building approach rather than that of Perutz, even though they did their work in the Cavendish.

Could it have been that Perutz was a little overrated?

You are the first person who ever asked me this question. Since, as I mentioned before, his main work on molecular structure lies outside the realm of my professional competence, I am not able to provide you with an informed judgment. Why are you asking?

When I discussed with Perutz his development of the heavy-atom substitution method for the solution of the phase problem in protein crystallography, it was my impression that he did not give adequate credit to the real originators of that idea. He did not mention those people who had first suggested this method for solving the structure of small molecules, nor did he mention those other people who had suggested to him to try out this approach for solving the structure of proteins. Of course, Perutz was undoubtedly the first to apply the method to the solution of the structure of proteins, and he was immensely successful in doing so.

Horace Judson reports in his book, The Eighth Day of Creation, *that Perutz was the only person who ever asked him for money for granting an interview, a fifty-guinea donation to the laboratory recreation fund. Judson paid it.*

My own, slightly negative feelings regarding Perutz, are based on his writings on social, political and scientific topics, which I found generally shallow and poorly developed. For instance, a few years ago he trashed Schrödinger's *What Is Life?* in the *New York Review of Books*. His article seemed to lack philosophical depth and suggested to me that, despite his general acclamation as the British panjandrum of Molecular Biology, Perutz had little feeling for the ideological context in which that discipline arose.

I had always taken it for granted that Perutz was a Jew, until I found out in a strange way that *he* didn't consider himself to be one. While I was spending a sabbatical leave at the Cavendish Laboratory in 1961 he was invited by the Weizmann Institute to give some lectures in Rehovoth. He told me that he was taking his family along, and to save money on the fare to Israel, he worked out a complex itinerary. They were going to fly from London to Cyprus (which was still under British rule) via British European Airways on a cheap, intra-Empire fare. Then they would take a boat to Lebanon, and travel by bus from Beirut via Jordan to Israel, which they would enter via the Allenby Bridge across the River Jordan.

I admired his ingenuity and asked him how he was going to get the certificate required by the Jordanians of all travelers crossing the Allenby Bridge, attesting that they are not Jews. "Nothing simpler," Perutz answered, "our clergyman will certify that we are Anglicans."

When we visited Wittgenstein's grave in Cambridge, it was in an Anglican cemetery and we found there the graves of Perutz's parents. I told Perutz about this the next day. And he confirmed that his parents were Christian converts.

How much did you know about DNA when you moved to Copenhagen?

Almost nothing. As far as I remember, the first time I ever heard of DNA was during my stay at Cold Spring Harbor in the summer of 1948, at a seminar given by Rollin Hotchkiss from the Rockefeller Institute in Manhattan. Hotchkiss spoke about the DNA-mediated transformation of bacteria, by means of which Hotchkiss's teacher, Oswald Avery, had shown in the early 1940s that DNA is the genetic material. There had been hardly any mention of DNA-mediated bacterial transformation in the Phage Course, or in any of the other seminars presented by visiting scientists that summer. In retrospect, this seems strange because the physical and chemical nature of the gene was of keen interest not only to the members of Max's Phage Group but also to most other people then summering at the Cold Spring Harbor Lab.

Evidently, Avery's discovery, though known to almost everyone at the Cold Spring Harbor Lab (but not to me) had made little impact on the very people interested in the nature of the gene. There are at least two reasons for this lack of impact. One reason was the doubt that Avery's "transforming principle" was really pure DNA. People remembered the embarrassing error that had been made by the famous German chemist, Richard Willstätter, in the early part of the 20th century, when he claimed that enzymes are not proteins. He had based that false claim on the demonstration of catalytic activity in what he believed were protein-free solutions, unaware that they did contain proteins at concentrations so low that they were undetectable by the analytical methods available to him at the time. So Alfred Mirsky — a colleague of Avery's at the Rockefeller Institute — asserted that Avery's allegedly protein-free DNA preparations were probably contaminated by genetically-active protein.

The other reason for doubting Avery's interpretation of his transformation experiment was that at that time DNA was believed to be a monotonous macromolecule like starch, whose chemical structure and composition is the same, regardless of its biological source. This belief made it impossible to imagine how DNA could be the carrier of genetic information.

A dramatic demonstration of the delayed acceptance of Avery's experiment occurred as late as the spring of 1952, when the British Society for General Microbiology held its annual meeting in Oxford, which Jim and I both attended. Luria had been invited to give one of the plenary addresses, whose text he had submitted to the organizers a week or two before the meeting. When Luria was denied a U.S. passport for the trip to England because of his leftwing politics, the British organizers asked Jim to stand in for his absent teacher and read Luria's paper.

One of the main points of Luria's paper was that his and Cyrus Levinthal's electron-micrographic images of phage-infected bacteria strongly suggested that its *protein* is the genetic material of the phage. So they proposed that the phage DNA is some kind of "glue" that holds the protein together when the mature, infective phage particle is "baked" at the end of the eclipse period of the intracellular reproductive cycle.

Shortly before the meeting, some exciting news from Cold Spring Harbor had reached Jim in Cambridge. Alfred Hershey and his young assistant, Martha Chase, had shown that when the phage infects its bacterial host, only its DNA enters the cell. The phage protein remains outside, devoid of any further role in the reproductive drama about to ensue within. Thus the phage DNA rather than the phage protein is obviously the carrier of the phage genes.

So what was poor Jim to do? He decided to present Luria's text in its full original version and merely to mention at the very end that a minor revision of Luria and Levinthal's main conclusion was actually called for. The genes of the infecting phage that are responsible for directing the synthesis of progeny phages happen to reside in its DNA and not in its protein.

How did it come about that we members of Max's Phage Group immediately accepted Alfred Hershey's and Martha Chase's claim in 1952 that the phage DNA, rather than the phage protein, is the carrier of the viral genes? Why had we not accepted for all those years Avery's interpretation that DNA is the carrier of bacterial genes and yet immediately accepted Hershey and Chase's inference that DNA is the carrier of phage genes, even though the margin of possible experimental error was much larger in Hershey and Chase's experiment than in Avery's?

The explanation of this puzzling psycho-sociological fact is that in the meanwhile Erwin Chargaff had shown that the composition of DNA is not monotonous and does vary according to its biological source. Now one could easily visualize how genetic information is inscribed in DNA as a specific sequence of the four kinds of iterated nucleotide building blocks, whose long string makes up the giant DNA molecule. In other words, DNA turned out to be Schrödinger's "aperiodic crystal" composed of a succession of a small number of different elements, the exact nature of their succession representing a "hereditary code".

When did you become aware of Chargaff's findings?

As most of my colleagues, including Jim, almost as soon as he reported them in the early 1950s. And that was the reason why, according to Jim's memoir, *The Double Helix*, it was his learning of the Hershey–Chase experiment in the spring of 1952 that drove him and Francis Crick to intensify their efforts to work out the structure of DNA. The Hershey–Chase experiment was a benchmark in the history of molecular biology.

Why was the Hershey–Chase experiment a benchmark and not the Avery experiment?

Because we accepted the Hershey–Chase experiment, but not Avery's experiment, as proof that DNA is the genetic material.

Why didn't you accept Avery's experiment as proof?

Not because we paid heed to Mirsky's claim that Avery's DNA preparations might be contaminated with genetically-active proteins, or by any other possible technical errors. In fact, from a technical point of view Avery's transformation experiment was much cleaner than Hershey and Chase's. But we didn't accept it because we heeded the epistemological warning of the British astronomer Arthur Eddington that one should not put too much faith in experimental results until they are confirmed by theory. In Avery's days, there was no way in which one could conceive how DNA *could* be the substance of heredity, since DNA was considered to be a monotonous polymer like starch. Who could have believed that starch is the genetic material, even if a dozen different experiments had implied that it is? As for proteins, however, it was well known by that time that they are long chain molecules made up from 20 different kinds of amino acids, which can follow each other in almost any order.

Chargaff's chemical analyses changed all this. Chargaff had shown two things. For one, he found that it is not the case that the basic building block of DNA is a "tetranucleotide" which contains all four bases in equal proportions, which is what people believed prior to Chargaff's analyses. For the other, he found that the actual base composition, i.e., the percentages of adenine (A), guanine (G), cytosine (C) and thymine (T) in a sample of DNA, varies with its biological source. This finding opened up the possibility that DNA encodes genetic information and made Avery's transformation experiment credible.

That means that Chargaff's analyses of the base composition of DNA made a decisive difference.

Absolutely. Strangely enough, though, Chargaff did not take credit for making the Avery experiment credible. Instead, he claimed that he discovered nucleotide base-pairing in DNA and that Watson and Crick merely "popularized" it. What Chargaff *did* discover was the *compositional equivalence* of [G] = [C] and [A] = [T]. In the paper in which he published this finding there was no mention of "base-pairing", or any other remotely equivalent structural concept. What he *did* say was that "whether this is more than accidental cannot yet be said."

But there were some sophisticated people who did believe that Avery's discovery was valid; they believed it right away.

Who were they?

The British. The Royal Society presented its highest award to Avery, the Copley Medal, in 1945, and its President said in his speech that Avery had shown that the genetic substance was "a nucleic acid of the desoxyribose type".

I can believe that they gave him the Copley medal, but I doubt that they gave it to him for showing that DNA is the genetic substance.

But that is exactly what they gave the Copley Medal to him for.

I'm amazed. Do you have any proof? Did they use these very words?

Yes. You can find them in Sir Henry Dale's Anniversary Address to the Royal Society in 1945.[1]

This is not well known.

It certainly doesn't seem to be well known to the people with whom I usually argue about this case.

It may be that it is not well known because it happened outside of America, and, besides, who reads laudations?

For me the strongest proof of the lack of general appreciation of Avery's experiment has always been the dog that didn't bark is the *Festschrift* that the Genetics Society of America brought out in 1950 to commemorate the 50th anniversary of the rediscovery of Mendel's laws. To that *Festschrift*, the then paramount philosopher–theoretician of the gene, Hermann Muller, contributed an article in which he discussed at length what he considered to be all the possible material incarnations of the gene, without mentioning DNA at all.

Do you then agree that the Royal Society was very foresighted in 1945?

Yes, but it all hangs on one of Sir Henry's phrases.

But isn't that more than what anybody else said?

Yes, it certainly is. But how did Sir Henry Dale know? He must have had some advisors. It would be interesting to know who his advisors were. This is very surprising. None of my adversaries have ever cited this. You are the first one.

I'm not your adversary.

No, no, of course, not. When Watson presented Luria's lecture in Oxford as Luria's stand-in, Luria and Levinthal still believed that proteins are the genetic material. It was 1952, long after Avery's paper and past Chargaff's discovery of the variability of DNA composition. What was still missing was the Hershey–Chase experiment.

Let's get back to your story.

After we spent a year in Copenhagen, Jim moved to Cambridge and I moved to Paris with my fiancee, Inga Loftsdottir, an Icelandic pianist to whom Jim had introduced me at his boarding house in Copenhagen and who was going to study piano under Reine Gianoli at the *Ecole Normale de Musique*. We were married in Paris in the spring of 1952.

Was Jim also looking for a wife?

He was looking for a girlfriend. He didn't find one in Copenhagen, even though there were many attractive young women working as technicians at the Serum Institute. But there was little social contact between them and us members of the senior staff, who (much to our regret) had their meals in a separate dining room.

You ended up at Berkeley. How did that happen?

In the summer of 1951, towards the end of my stay in Copenhagen, just before my fiancée and I moved to Paris for the second year of my post-doctoral experience, there was an international Poliomyelitis Congress in Copenhagen. Niels Bohr gave a dinner for a select few of the participants, including his scientific son Max and myself as Bohr's scientific grandson via Max. I happened to be seated next to the Nobel laureate Wendell Stanley, who had just moved from the Rockefeller Institute to Berkeley, where a big Virus Laboratory was being built for him. Since I would have to support a wife before too long, I asked Stanley whether there might be a job for me in his new Virus Laboratory when I return Stateside a year hence. He said, yes there might be, and why don't I write to him when he gets home. So I wrote, and a few weeks later he responded, offering me the lowest level non-faculty research position, which I gladly accepted.

When, many months later I finally arrived in Berkeley, I learned from some of my future colleagues that Stanley had hired me without knowing much about my attainments. He reckoned that anybody he had met at Niels Bohr's house must be a pretty good scientist, an adornment for his new Virus Laboratory.

How did you become a faculty member?

Roger Stanier, the Canadian microbiologist whose weird accent Jim had imitated when I first met him at Cold Spring Harbor, had joined the Berkeley faculty as Professor of Bacteriology. He arranged for my appointment as a tenured Associate Professor of Bacteriology, to teach Berkeley's first course in Bacterial Genetics, with Stanley's Virus Lab still paying my salary.

Would you care to single out something from your science at Berkeley to tell me about?

I think that my best work — or, at least that which I enjoyed most — did not deal with the molecular biology of phages and bacteria, the research on which I devoted the first twenty or so years of my career in Berkeley. Instead, it dealt with the neurobiology and developmental biology of leeches, which I took up only in mid-career. In that switch of research interests I had been influenced by Max, as I had been influenced by him when, as a nascent Ph.D. I switched from physical chemistry to molecular biology before that discipline even existed.

In the fall of 1949 — during my second and last year at Caltech and three years before Watson and Crick's discovery of the DNA double helix — Max announced to his disciples that the search for the mechanism of biological self-reproduction was now in "good hands". What he meant by this locution was that the quest for its solution would soon be over and that our present line of work was about to turn boring. He revealed to us that the future of vanguard biology now lay in understanding the brain, which he considered to be the last frontier. So to prepare us for that future, he assigned to each of us a set of neurobiological papers that we were to present to our colleagues.

I drew three papers on a theory of human hearing by the Hungarian biophysicist G. von Bekesy. They were full of complex equations relating to resonance and fluid dynamics and completely beyond my understanding. So I decided that neurobiology was not for me and that I would stick with research on bacteria and phages, even if, according to Max, that subject had only a limited future as an *avant garde* activity in the life sciences.

By the late 1970s — twenty years after Jim's and Francis' discovery of the DNA double helix — Max's prediction had finally come true. Molecular biology had grown from an avocation of a small clique into a boring, ecumenical mass movement. Max himself had turned to the study of sensory perception (with an unfortunate choice of the fungus *Phycomyces* as his research material, which, in the end, led him nowhere). And Seymour Benzer took up research on the neurogenetic control of insect behavior. So I decided to switch to neurobiology as well, and, on Seymour's recommendation, was accepted by Stephen Kuffler (the head of the Neurobiology Department at Harvard Medical School, who tended to regard molecular biologists as conceited know-it-alls) during a sabbatical leave from Berkeley.

Can you tell me something about Stephen Kuffler?

Steve, as everybody called him, was born in Hungary in 1913 and studied medicine in Vienna. He emigrated to Australia in 1937, where he began his work as a neurobiologist in the Sydney laboratory of the future Nobel laureate, John C. Eccles. In 1945, Steve left Australia for the United States, where he served successively on the faculties of the University of Chicago (1945–1947), and Johns Hopkins University (1947–1949), until he reached the apex of his career with his move to Harvard Medical School in 1949.

Steve was a superb experimentalist, teacher, and leader, who played a critical role in the eventual development of neurobiology into a megascience. In that regard, Steve resembled Max, who played a similar founder's role in bringing about the eventual hegemony of molecular biology.

Max, as well as Steve, came to be perceived — at least, by their hapless rivals — as a *capo di tutti capi* in either of the two Mafias formed by their disciples, who would eventually run many, if not most departments of molecular biology and of neurobiology in American government laboratories and institutions of higher learning.

These similarities notwithstanding, there were important differences between Max and Steve. Max had the mindset of a theoretician, who is interested in data only insofar as they provided hardcore proof or disproof of theories. Steve, however, had the mindset of a pragmatist, who is interested mainly in phenomena *per se* and regards theories as mere mnemonic crutches that help one remember complicated empirical data.

This divergence in epistemological attitudes between Max and Steve is captured by the distinction between the hedgehog and the fox, to which the Greek poet, Archilocus of Paros, drew attention in the eighth century BCE. More famously, it was the subject (and title) of a book-length essay by the British philosopher, Isaiah Berlin, that appeared in 1953. According to this distinction, the fox knows many little things, whereas the hedgehog knows only one big thing.

Steve, the fox, studied the multifarious aspects of diverse metazoan nervous system. To that end, he experimented with a broad range of animals — invertebrates as well as vertebrates, from worms and insects to primates. He chose each of his diverse working materials for providing him with an especially favorable preparation for the investigation of some particular aspect of nervous function, such as vision, muscular contraction, nerve regeneration and impulse generation, in which he happened to be interested at the time.

But Max, the hedgehog, was mainly interested in one thing, namely the central principle that differentiates living matter from dead matter, of which the capacity for self-reprodction had long been the most mysterious aspect. Since all creatures share that capacity, Max restricted his experimental studies to a single (very convenient) biological system, namely the bacterial species *Escherichia coli* and its bacteriophages. It was only toward the end of his career as an experimental scientist — when Max had become interested in Steve's bailiwick of neurobiology — that he switched to the study of the phototropic response of the fungus *Phycomyces*.

On joining Steve's research group, I was lucky once more. My first preference would have been to work with David Hubel and Torsten Wiesel, the Watson and Crick of Neurobiology, future Nobel laureates and star members of Kuffler's department. But since I lacked the skills in neuro-anatomy and neurophysiology needed to work on the brains of Hubel and Wiesel's cats, Steve suggested that I join John Nicholls, who was working on the neurobiology of leeches, the blood-sucking annelid worms. Their nervous system is vastly simpler than that of cats, and — thanks to Nicholls's tutelage — after a few weeks I had managed to acquire the minimal skills needed to work with it.

What did you achieve during your sabbatical stay in Kuffler's Department at Harvard Medical School?

I produced only one paper as a direct result of my stay at Harvard. In that paper I published an attempt to provide a neurophysiological explanation of Donald Hebb's hypothesis of learning by modulation of synaptic strengths. I presented my theory to the assembled members of Steve's department, who received my talk with obvious hostility. They reproached me for having wasted a whole hour of their precious time on speculative moonshine.

My colleagues' unfavorable reaction to my debut as a neurobiologist notwithstanding, I went ahead and published my theory in the *Proceedings of the U.S. National Academy of Sciences*. In terms of reader response, this theoretical article was the most successful paper of my entire career. It made the honor list of most frequently cited papers compiled by the citation index of *Current Contents*.

On my post-sabbatical return to Berkeley, I set up a neurobiological laboratory and decided to try to work out how the nervous system of the leech generates its swimming rhythm. When I began this project I was not aware that I was about to continue a study that had been started in the *quattrocento* by Leonardo da Vinci. Leonardo was interested in the locomotion of animals, including the eel-like swimming movement of leeches. As indicated by a set of Leonardo's drawings in one of his notebooks (whose artistic quality does not quite match that of the *Mona Lisa*), Leonardo understood that alternating convex and concave contractile deformations of the dorsal and ventral body wall generate the leech's rearward-traveling body wave. The rearward travel of that body wave drives the leech forward, by pushing against the water.

Within three or four years, my co-workers — foremost among them, William Kristan and Otto Friesen — and I had worked out the basic mechanism of leech swimming in terms of identified nerve cells and their connections. We showed that a chain of segmentally-distributed circuits of nerve cells generates the alternating convex and concave contractile deformations of the dorsal and ventral segmental body wall. These circuits turned out to owe their oscillatory activity rhythm to the principle of *recurrent cyclic inhibition* first proposed by the Hungarian neuroanatomist, G. Székely. In the early 1980s, after Kristan had established his own laboratory on the San Diego Campus of the University of California, he began a highly successful study of the leech's overall control of the swimming rhythm with his own postdocs and graduate students.

Meanwhile, I had set out to investigate the embryonic development of the leech nervous system, for which purpose David Weisblat and I devised a novel method for tracing the origin, or precise line of descent of identified nerve cells of the mature animal from the fertilized egg. This method consists of injecting an identified cell, or blastomere, of the early leech embryo with a labeled tracer substance (such as a fluorescent dye) and then examining the distribution of the injected tracer over the cells or tissues of the later embryo or of the nearly mature juvenile animal. These experiments showed that the fate of the cells of the leech embryo is highly *determinate*, in contrast to the high degree of *indeterminacy* of cell fate in the embryos of complex vertebrate creatures, such as mammals.

Your book entitled Women, Nazis, and Molecular Biology; The Memoirs of a Lucky Self-Hater, *which you published in 1998, shows that you thought about yourself quite a lot.*

Although I am not given to much self-evaluative introspection about my motivations, I do feel guilty about the way I treated the women who figure in my memoirs. This is especially the case for its main character, Hildegard,

Gunther and Molly Stent in their home in Berkeley, California, 2004 (photograph by I. Hargittai).

the actress with whom I open my story on the first day of my postwar return to Berlin in 1946 and close my story on the day before my departure from Berlin for Copenhagen in 1950. All my life I had feelings of guilt about the abominable way I treated her.

Were your feelings toward her superficial?

Yes and no. She was a lovely woman and I was very fond of her, but not in love. The facts that her parents had been Nazis and that she once had a Waffen SS officer as her lover had nothing to do with my inability to love her. Hildegard just didn't match my ideal type of woman, which Lore, the other woman from Berlin whom I later met on the Kitzbühel ski slopes, did match. So I did fall in love with Lore, but in the end married Lore no more than I married Hildegard (who, on mature reflection, would probably have made a better wife than Lore).

Is Hildegard still alive?

Over the years, I tried a few times to find her but never succeeded. Since I didn't know the family name of the French Officer whom she presumably married in 1950, there was no way I could trace her. Her maiden name was Thews, which is not a very common German name, and she had a kid brother whose first name was Gerhard. So I searched the Internet for a Gerhard Thews. I found one listed in Wiesbaden and called him a few years ago, asking for his sister's address. "So sorry," Mr. Thews replied, "I don't have a sister."

Your book was a self-published book, wasn't it?

Yes, despite my reasonably good connections with German, French and American publishers, I was unable to find one willing to bring it out. I couldn't even find an agent willing to *try* to place my manuscript. I suppose the main reason why no publisher wanted to take it on was the mention of Jewish self-hatred in its title. Although Jewish self-hatred was widely discussed before the War, it became taboo subject in the wake of the Holocaust — especially in Germany, where people of goodwill fear that the concept can be misused to justify anti-Semitism. The reasoning goes that if the Jews hate themselves then it's no wonder that everyone else hates them as well. So I had to bring out my book under my own imprimatur.

But your environment in Germany conditioned your self-hatred when you were growing up in the 1930s.

True enough. But this argument is too sophisticated for the anti-Semitic street. There is a substantial literature on Jewish self-hatred, and there are those who say that understanding Jewish self-hatred in Germany could help understand the psychological stresses of blacks in America. If you are being told as a child, all the time, that you are bad, you are bound to believe it.

Maybe, once the Germans are able to face all facets of their past, it will be possible for a German publisher to republish your book. Did you ever talk with Delbrück about the Nazi times in Germany?

Not very often. But he did tell me how the National Socialist German Union of University Teachers blacklisted him in 1936 for his failure to satisfy the criteria that had to be met by a National Socialist university teacher. Max had signed up for the *Dozentenlager*, which was a political indoctrination camp that the Nazis required aspiring *Dozenten*, or assistant professors, to attend before being granted the license to lecture at a university. One day, after his class had solemnly intoned the sacred Nazi anthem, the "Horst Wessel Lied", Max pointed out to the instructor that the last two lines of the first stanza, *Kameraden die Rot Front und Reaktion erschossen, Marschiern im Geist in unseren Reihen mit*, are semantically ambiguous. Since the relative German pronoun "*die*" can mean either "whom" or "who", two alternative readings are possible:

(1) Comrades *whom* Reds and Reactionaries shot, or
(2) Comrades *who shot* Reds and Reactionaries.

"So," Max asked, "were our dear, defunct SA comrades victims, or were they killers of Reds and Reactionaries?"

The flabbergasted camp director reported Max to the National Socialist German Union of Docents, which in turn, blacklisted Max in an edict found after the war in the Berlin University archives.

Did Delbrück ever ask you why you left Germany?

There was no need for him to ask, since he knew without my having to tell him that I am a Jew. In my first few years in the States I did try to keep secret my shameful status as a German-Jewish refugee, especially in the anti-Semitic ambiance of the University of Illinois, where living accommodations for Jews and Gentiles were totally segregated. In some respects, the social and sexual isolation of the Jewish students at Illinois was even more complete than in pre-Nazi Berlin. Moreover, I was told that there never were and there never would be any Jewish chemistry professors at Illinois.

In any case, it would have been impossible to conceal my non-Aryan origins from Max. He was very inquisitive and nosey about personal matters and asked blunt questions that a discrete person would have never asked. Yet, we never talked about my family background.

You write in your memoirs that you are no longer a self-hater. How did you get rid of your Jewish self-hatred?

My liberation began at Caltech. It was the first place I had ever been in my life where it didn't make a difference, or wasn't even of any interest, whether you were a Jew, no more than whether you liked to ski or to listen to music. It just didn't make any difference. But it was only when I visited Israel in 1967 — just after the Six-Day War — that, for the first time in my life, I was actually *proud* to be a Jew. Unfortunately, the reasons for my sudden pride may not have been all that salubrious. I felt at home in

Israel because I experienced it as a blend of two regions to which — in Freudian psychoanalytic jargon — I am still cathected. To California, because of its geography and to Prussia, because of its civilization.

My feelings of Jewish pride began as soon as our plane from Rome crossed the Israel border. The ground beneath us was brown on the Arab side and green on the Israeli side. My fellow Jews had managed, as the Arabs had not, to turn the desert into a garden! Once on the ground, I discovered that I remained a Prussian militarist, since I admired the stalwart Israeli Defense Force. I saw it as the kind of army that the Germans would have liked to have had but never did because, contrary to received anti-Semitic opinion, the Jews make better soldiers than even the Prussians. Frederick the Great would have been proud to have the IDF in his service.

I also appreciated the social structure at the Weizmann Institute of Science in Rehovoth, which I experienced as half Kaiser Wilhelm Institute and half Caltech. After each of the lectures I gave, the members of the audience asked questions in a strict order of academic rank: eminent professors first and lowly graduate students last. During our stay in Rehovoth, the Institute threw a costume party for the children of its faculty and staff to celebrate the Purim Holiday. Most of the little boys came dressed as soldiers and most of the little girls as army nurses. My Berkeley colleagues would have been shocked by this scene, but I liked it.

Moreover, on visiting a kibbutz I realized that this Utopian institution is rooted in German romantic notions about the nobility of tilling the soil, as opposed to such allegedly ignoble, socially parasitic occupations as the practice of law and banking, which, according to anti-Semitic belief, is favored by exploitative Jewry.

You edited the Norton critical edition of Jim Watson's Double Helix *and it has been highly successful. How did this book come about?*

Norton has a highly respected and successful set of critical editions, which provide an inexpensive source of the works of such all-time greats as Darwin, Machiavelli, Thomas More, St. Paul and Thoreau. Jim is the only living author whose work was ever included in this awesome series. Since I had previously published in *The Quarterly Journal of Biology* a review of the reviews of *The Double Helix*, I was a natural choice for the editorship of its critical edition, i.e., for the post of literary stooge for my still living friend, Jim. He needed a stooge, because literary etiquette frowns on your editing a critical edition of your own work, especially if you don't happen to be dead.

Does Jim take criticism well?

Yes, if it is justified criticism. By now, he is very secure in his person. He can't be easily insulted, just as Francis Crick can't.

Are you secure in your person?

I don't know. But certainly much more so than I used to be, especially ever since I was elected to membership in the American Philosophical Society twenty years ago. Come to think of it, I *did* manage to insult Francis.

How did you do that?

Thirty years ago, I contributed an essay to a special issue of *Nature* dedicated to the *twentieth* anniversary of the double helix. In that essay I claimed that Francis believes in God. I started my essay by quoting Salvador Dali: "And now the announcement of Watson and Crick about DNA. This is for me the real proof of the existence of God." To lend support for Dali's insightful inference, I presented a linguistic analysis of some of Francis' writings, in which I substituted the word "God" whenever Francis had used the word "nature". As it turned out, this one-for-one substitution did not change the essential meaning of the text.

Was this a joke on your part?

A semi-hemi-demi joke, not a total joke. In any case, as Sigmund Freud's analyses of jokes have shown, really good jokes can be told only about subjects that people take seriously, such as sex and religion. By that criterion, it was actually a good joke.

Did Crick respond to your allegation?

Not publicly, but he let me know that he didn't like it.

Why was he upset if he is so secure?

Because, as he has written somewhere, science attracted him as a vocation because it shows that religion is bunk.

So, is Francis then an atheist?

He says that he is, but I doubt it. In 1947, when I was a twenty-three year old graduate student in physical chemistry and served with the U.S. Military

Government in Occupied Germany, I attended Max Planck's last public lecture (at which he collapsed and died a few days later) in the Westphalian town of Elberfeld. The title of Planck's lecture was "Science and Religion", and its main thesis was that you couldn't *be* a scientist and an atheist at the same time. On first hearing Planck's thesis I thought it was ridiculous. But interested all the same by what Planck had said, I started reading accounts of the Babylonian origins of science. Then Planck's thesis began to make sense to me.

Eugene Wigner says that physics does not endeavor to explain nature, it only endeavors to explain the regularities in the behavior of objects. Further, he says that the regularities in the phenomena which physical science endeavors to uncover are called the laws of nature. Thus he calls for a great degree of modesty in our aims for our science.

The metaphorical designation of the regularities in the phenomena that science endeavors to uncover as "laws of nature" has its roots in Babylon in the eighteenth century BCE. The Babylonians, whose king, Hammurabi legislated the first explicit laws governing the social behavior of *persons*, developed the idea that the author of the regularities in the behavior of *objects*, or "laws of nature", is a divine legislator, whom they called "Marduk", a.k.a. "God".

Do you believe in God?

I don't know whether I do, even though I believe that I as a scientist ought to. But, by way of a paradox, I do believe, as Planck did, that all scientists *have* to believe in Him.

Even though they say they don't?

Yes, because it's easy to say so. But if they would think about it more deeply and consider the metaphysical origins and intellectual history of their vocation, they will come to a different conclusion. Einstein's response to the question whether he believes in God (posed to him on his first visit to America by New York's Chief Rabbi) was that he believes in the God of Spinoza. By that he meant that he believes (as did Spinoza) in a Marduk-like entity that created the world but no longer intervenes in worldly events. Therefore there is no sense in praying for His help or forgiveness.

   Niels Bohr was one of the few five-star scientists who really *was* an atheist — and not merely paying lip service to atheism. Bohr's friendly, yet unrelenting arguments with Einstein regarding the validity of quantum mechanics arose not from differences in their *physical* beliefs but from differences in their *metaphysical* beliefs. Einstein insisted that Werner Heisenberg's quantum mechanics and its Uncertainty Principle could not be a complete description of reality, because, as he put it, God does not play at dice. But Bohr had no problem with the Uncertainty Principle, because he believed that there is no God and that he *does* play at dice.

You were present at the cradle of molecular biology in the 1940s. Now, at the dawn of the twenty-first century we have entered the post-molecular biology era. The label molecular is no longer needed because all of biology has become molecular biology. What is next?

In prognosticating the future of the biological sciences in the post-molecular biology era, when molecular biology has disappeared as an identifiable specialized discipline, one should bear in mind a deep general principle of the history of science. According to that deep principle, the easier scientific problems are generally solved before the more difficult ones. Thus four difficult, long unsolved problems that had been of central concern for biologists ever since Aristotle founded their discipline in the fourth century BCE were finally worked out during the recently-ended twentieth century CE. They are *metabolism*, *heredity*, *embryonic development*, and *organic evolution*.

All four of these ancient problems owe their definitive resolution to the pioneering application of molecular biological ideas and techniques. But in line with the deep historical principle, we can expect that the leftover biological problems that still await their solution in the coming twenty-first century CE are likely to be more difficult than any that have been solved thus far.

One very difficult unsolved problem is the *origin* of living matter, which still lacks a credible, coherent proposal for its solution. Perhaps, being a historically unique event that left no traces, the origin of life may be *intrinsically* insoluble. Despite its obvious importance (and the certainty of the award of a Nobel Prize for its solution), few biologists seem to be working on the origin of life, most likely because of its apparently hopeless intractability. In any case, if it *is* ever solved, molecular biological principles are bound to have played a key role in its solution.

Probably the deepest, as yet unsolved, biological problem, and hence likely to be one of the very last to be resolved, is *consciousness*. In fact, until recently, the problem of consciousness appeared so deep that it seemed to be a philosophical rather than biological problem.

The first difficulty that any would-be student of consciousness encounters is the explication of its *meaning* in sufficiently concrete terms so that it can be investigated scientifically. Thus my Berkeley colleague, the philosopher John Searle, has defined consciousness as consisting of "inner, qualitative subjective states and processes of sentience or awareness". Consciousness, so defined, begins when we wake up in the morning from a dreamless sleep — and continues until we fall asleep again, die, or go into a coma, or otherwise become "unconscious". According to Searle, consciousness comprises all of the enormous variety of awareness that we think of as characteristic of our waking life. It includes everything from feeling pain to perceiving objects visually, to states of anxiety and depression, to working out crossword puzzles, playing chess, trying to remember your aunt's telephone number, arguing about politics, or to just wishing you were somewhere else.

The main reason why consciousness has tended to be considered out-side the realm of biological problems is that it differs in three essential aspects from other phenomena of the natural world. They are the *qualitative character*, the *subjectivity*, and the *unity* of conscious experience.

The qualitative character of conscious experience is reflected in the dif-ferences in the "*feel*" of diverse conscious experiences, such as the redness of the setting sun as compared to the smell of the ocean at the beach. And the subjectivity of conscious experience arises from the necessity of the presence of a conscious living observer for the existence of such a thing as a qualitative "feel" in the first place. This is to be contrasted with the objectivity of such natural phenomena as the spectral properties of the light of the setting sun or the chemical components of the air on the ocean shore. For their existence no conscious living creature *is* required. Finally, the unity of a conscious experience consists of the wholeness of the perception of the various features that constitute a particular experience. For instance, the experience of sitting on the beach at sunset is composed: the color of the sunlight, the smell of the ocean, and the pressure of the sand on your buttocks. Its qualitative character and subjectivity makes conscious-ness an even greater challenge than the origin of life, which is an objective phenomenon that does not involve any subjective "feel" and which, far from requiring the presence of any conscious living creature, even excludes it by definition.

These aspects that make consciousness different from all other natural phenomena do not exclude it from their realm, however. Since consciousness is the product of processes that occur in our brain, understanding it is obviously a biological problem, albeit an especially difficult, fascinating, and troublesome one. For that very reason, the study of consciousness has recently become very *a la mode* among the romantics in science, the Faustian types who constantly measure themselves against the infinite. They include Francis Crick, who has been working hard on the consciousness problem for the last ten or so years. Crick proposed that the search for what he called "*the neural correlate of consciousness*", or NCC, is, — or ought to be — the main agenda for people in their quest to understanding consciousness. Unfortunately, however, neither Crick nor anyone else has yet been able to put forward a proposal for the nature of the NCC that promises to lead to a neurobiological understanding of consciousness, as defined by Searle.

To get started on fathoming the scientific basis of some difficult phenomenon, it is often helpful to consider the antonym of its name, to clarify the distinction between it and another phenomenon that is different from, but related to it. So the antonym of "conscious" is "unconscious". Unfortunately, there exist two different meanings of "unconscious", which is at least partly responsible for the conceptual confusion that has brought on the philosophers. The more commonly understood meaning, to which I will refer as "*comatose*", denotes the global absence in a person of awareness, sensation, and cognition. This is the meaning of "unconscious" that is of interest to anesthesiologists, who have some understanding already of the phenomenon in terms of the neurobiology of the brain stem gateway to the cerebrum. But the study of the comatose state is unlikely to lead to an understanding of the individual neurobiological processes that are responsible for the diverse mental phenomena of consciousness for which Searle provided some examples.

Psychoneurologists are therefore interested in the less commonly understood meaning of "unconscious", to which I will refer as "*subliminal*". This term denotes the *selective absence* of awareness of some particular sensation, memory, or emotion in an otherwise fully conscious person. The reason for avoiding the term "unconscious" is that one of the phenomena of great interest to psychoneurologists is that designated as "*performance without awareness*". This term refers to the condition of a class of brain-damaged subjects who can carry out certain mental tasks whose performance requires their use of recently acquired knowledge of which they deny being aware, even though these subjects are otherwise fully conscious.

Sigmund Freud used the ambiguous term "unconscious" in his description of the selective absence of awareness (or "repression") of remembered experiences, whose role in personality formation was one of his main psychoanalytic propositions. Freud's case would have been better served by an antonym of "conscious" other than "unconscious", such as "*subliminal*".

There is also a phenomenon designated as the *selective presence of awareness*. This term refers to the occurrence of several alternative percepts of the same sensory input in a fully conscious subject. This phenomenon is generated by *ambiguous figures*, such as simple drawings that may evoke in the viewer an alternation in the awareness of two different percepts. The best known examples of such ambiguous figures are the *Necker Cube* and Rubin's *Vase*.

In the case of the Necker Cube, the viewer is aware of two different three-dimensional percepts of a two-dimensional drawing, which flip back and forth without settling on either one. And in the case of Rubin's Vase, the viewer is aware of the percept of the drawing that flips back and forth between a vase and the profiles of two girls facing each other.

The neurobiological mechanism of the flip-flop of ambiguous figures has not yet been worked out, but it is apparent that it would provide an opportunity, at least in principle, for locating the NCC. For the visual sensory input into the brain remains exactly the same throughout the viewer's gaze at the ambiguous figure — and hence also the activity pattern of the nerve cells responsible for its preconscious processing of the visual input. But the neuronal activity pattern of the subject's NCC ought to manifest a periodic change whose frequency is equal to that of the flip-flopping of the alternating percepts.

One of the leading contemporary investigators of consciousness, the neuropsychologist Antonio Damasio, has formulated a subtler explication of consciousness in terms of the psychologically central concept of the *self*. On the basis of his studies of the cognitive deficits of brain-damaged subjects, Damasio has divided the phenomenon of consciousness into three distinct sub-categories. He designates them as *proto-self*, *core self*, and *autobiographical self*.

The proto-self is a gathering place at which the electrophysiological input from the sensory organs first enters the brain, whence it is conducted to successively higher brain centers for the abstraction of sensory percepts. The mental processes of the proto-self are subliminal and, strictly speaking, not a part of consciousness. But they are a necessary prelude to it.

At the second, or core self stage, which is still subliminal, the sensory percepts generated by the proto-self are integrated to generate a dynamic map of the physical structure of the person in her many dimensions. The core self is a transitory entity whose content changes from moment to moment and has no memory of its past contents.

It is only at the third, or autobiographical self stage, that sensory percepts reach consciousness and give rise to self-awareness. The autobiographical self does have memory, so that persons remember what happened to them in their past and construct out of these memories what François Jacob referred to poetically as their "interior statue". In other words, it is thanks to their autobiographical selves that persons are aware of whom they are.

As formulated by Damasio, the concept of the autobiographical self seems to imply that one cannot be anybody but oneself. This proposition cannot be of absolutely general validity, however, because on stage great actors seem to be able to become someone other than themselves, i.e., manage to escape from their autobiographical selves. The inability of Clark Gable, or John Wayne, or Woody Allen to do so probably explains why they were not great actors. Maybe they weren't even real actors at all, since they seemed to be unable to dissociate themselves from their autobiographical selves. No matter in what role they were cast, they were always the self-same Clark, John, or Woody.

Damasio derived his three categories of consciousness from the study of patients with diverse brain lesions. He found that subjects who have no proto-self are virtually vegetables, lacking most mental traits characteristic of humans. Subjects who do have intact proto and core selves but lack an autobiographical self do have some human mental traits but are still not able to function as autonomous persons. Only those subjects who are in full possession of an autobiographical self meet a necessary, albeit not always a sufficient, condition for a neuropsychologically intact interior statue.

Damasio's work follows in the tradition established towards the end of the nineteenth century by the first neuropsychologists of language, Carl Wernicke and Paul Broca. They were the eponymous discoverers of the two areas of our cerebral cortex that are dedicated to speech. One, Wernicke's Area, is critical for the *comprehension* of speech, while the other, Broca's Area, is critical for the *production* of speech. Both areas require consciousness for their function. Wernicke and Broca did their work by locating "natural experiments", i.e., unfortunate subjects who suffered brain lesions, due either to an injury or to a cerebral stroke, and were unable to perform specific linguistic tasks.

Studies of a rather rare category of persons with slight damage in some very restricted area of their brains dedicated to the processing of sensory information have provided one of the most promising neuropsychological approaches to the problem of consciousness. Damasio would concede an autobiographical self to such persons who suffer from what appears to be a paradoxical impairment of their conscious perception of some particular kind of sensory stimulus. The psychoneurologist Lawrence Weiskrantz, who studied such persons intensively, has designated their condition as "*performance without awareness*".

Awareness is a phenomenon that is more narrowly circumscribed than the kind of global personality phenomena related to the concept of the self formulated by Damasio. Studies of this thematic reduction of the consciousness problem represented by performance without awareness have shown that under certain conditions, an otherwise fully conscious person can perform a certain task requiring recently-acquired knowledge without that knowledge having reached the consciousness of the subject's autobiographical self. Thus this abnormal specific exclusion of a particular piece of knowledge from the autobiographical self, i.e., of its *subliminal* presence, holds out the promise of dissecting the normal neural pathways leading to specific conscious experiences.

One example of performance without awareness is the condition bearing the oxymoronic name "*amnesiac memory*". Subjects afflicted with amnesiac memory cannot remember, let alone provide a verbal account of, their recent experiences. Yet, they are able to demonstrate recall of such experiences when examined by means that do not require their verbal response. In other words, they are subliminally aware of some past experiences without being consciously aware of them. For instance, a subject suffering from amnesiac memory who has been shown the letter *A* cannot say at a later time which letter it was that he saw. But that subject *is* able to identify an *A* as the letter he was shown by pointing to it when it is presented along with a set of other letters.

Another striking instance of performance without awareness has been given the oxymoronic name "*blindsight*". People who are afflicted with blindsight have suffered a brain lesion — typically from a head injury or cerebral stroke. When tested for their visual capacity they report that they are unable to see anything at all, i.e., that they are blind. Or, in some less severe cases, they say that they are unable to see a substantial part of what would represent the normal visual field. Yet precise psychophysical tests carried in a laboratory setting show otherwise. In these tests blindsighted subjects are asked to identify some features of a visual display by responding to questions about it by body movements (such as pressing a button) rather than by spoken words. Such tests reveal that these "blind" persons are, in fact, able to locate accurately the spatial positions of "unseen" visual stimuli. In other words, although the blindsighted subjects do perceive the visual stimuli after all, they are not consciously aware of having seen them.

A neurological explanation of the blindsight phenomenon is now available in terms of three different neural routes along which the visual information is transmitted from the eyes to higher brain centers for processing and interpretation of the visual input.

The first and second of these routes are designated as the *seeing pathway* because they are presumed to pass through, or send collateral nerve connections to parts of the brain dedicated to the production of conscious awareness of sensory percepts. Thus the seeing pathway is presumed to provide visual information to Crick's postulated neural correlate of consciousness, or NCC.

The third route, designated as the *on-line pathway*, bypasses the cortical areas dedicated to the production of conscious awareness of the visual information. The on-line pathway heads directly for the somatosensory cortex, where the visual information it carries is integrated as well and passed on to the motor cortex for the command of body movements. Since the on-line pathway by-passes several intermediate cortical processing areas, it takes much less time for the visual information it carries to reach the somatosensory cortex than the visual information carried by the seeing pathway.

As it turned out, the blindsight phenomenon was long known to expert tennis and baseball players, who can return a fast serve or pitch before they have actually seen the ball they are hitting. They are able to do so because passing the visual input information from the eye to the somatosensory cortex takes less time along the subliminal on-line pathway than along the seeing pathway.

The existence of the two separate visual pathways — one indirect, aware, and slow, and the other direct, subliminal, and fast — bids fair to provide an opportunity for finding the neural correlate of consciousness. It ought to be possible to identify the NCC by comparing the neural activity patterns of various cortical areas during similar visual experiences of normal and blind-sighted subjects. Thus far, this obvious approach has not provided the sought-after information, largely because of the difficulties entailed in making highly-localized recordings of nerve cell activity from the living human brain.

Several novel methods for imaging cerebral nerve cell activity became available towards the end of the twentieth century. They hold out promise for gaining a better understanding of some long mysterious cognitive functions of the brain. Prior to the introduction of these recently-developed imaging techniques, there were only two very limited procedures available for the neurological study of cognitive brain functions. The older of them consisted of examining the performance of subjects with particular cognitive impairments while they were alive and then, after their death, performing brain autopsies to identify their neuroanatomical deficits. This was the procedure by which Carl Wernicke and Paul Broca identified the cortical areas involved in the comprehension and production of speech in the latter part of the nineteenth century. The other procedure, electroencephalography, or EEG, was developed in the 1930s. It consists of placing electrodes on the scalp and recording the medley of electrical signals emanating from the underlying brain tissues. The EEG method can reveal whether an unconscious person is or is not brain-dead, but it can only provide very crude information about the localization of cerebral activity.

One of those novel imaging methods is *positron emission tomography*, or PET scanning. The principle of this PET scan procedure is to label the brain of a subject with the short-lived fluorine radioisotope, ^{18}F, which, as it undergoes radioactive decay, emits a positively charged electron, or *positron*. Upon the positron's collision in the brain tissue with an ordinary, negatively charged electron, the positron and the electron mutually annihilate each other and thereby generate two gamma ray photons that travel in diametrically opposite directions. The two paths taken by the gamma ray photon sister pair as it travels out of the brain tissue are recorded by a helmet-like spherical array of radiation detectors surrounding the subject's head. A computer program then converts the gamma ray photon emission pattern recorded by the spherical detector array into series of two dimensional images (or optical sections) of the cerebral pattern of positron release by the decaying ^{18}F radioisotopes. This method of three-dimensional recording and computerized rendering of the data as a series of two-dimensional sections is called *tomography*.

To perform a PET scan of a subject's brain, the subject is injected with deoxyglucose labeled with the ^{18}F radioisotope. Deoxyglucose is a structural analog of glucose, the main source of energy fueling the operation of the brain. Deoxyglucose, which lacks one of the five-hydroxyl groups of ordinary glucose, is taken up but not metabolized by the cerebral nerve cells. Consequently, the higher the rate of activity of a cerebral nerve cell, the higher is the rate of its accumulation of ^{18}F-labeled deoxyglucose, as revealed by the tomographic radiation detector. Thus the PET scan method can provide not only fairly precise images of brain anatomy but also of the rate of nerve cell activity in different anatomical structures, and hence of their function.

The development of the tomographic brain scanning method represents a benchmark in the history of neuropsychology — just as the Hershey–Chase experiment represented a benchmark in the history of molecular biology — since it can be performed on awake human subjects without requiring any surgical interventions. Its advent was one of the main reasons why President George-Bush-the-Elder declared the years 1990–2000 as the "Decade of the Brain". For analysts of brain function were enabled to localize brain activity with much higher accuracy than ever before and without invasive procedures that interfere with brain function or endanger the subject's life. Moreover, investigators of mental processes can now observe the functioning of the living brain while their subjects think, perceive, and initiate voluntary actions. So it does not seem out of the question that one of these days cleverly designed comparative applications of these novel brain-imaging methods to blindsighted and normally-sighted subjects might reveal the NCC at last.

But perhaps mankind would be better off if it should turn out that the NCC *cannot* be found and that the challenge for the twenty-first century of explaining consciousness cannot be met. For the subjectivity of conscious human experiences and their opacity for objective observation from the outside by other persons is an essential component of our concept of what it means to be an autonomous human being. If — God forbid! — it should ever come to pass that people carry hand-held brain scanners that allow them to view and interpret other people's NCC, and thus have direct objective access to their consciousness, mankind would surely experience the greatest change in its interpersonal relations since *Homo sapiens* appeared on the terrestrial scene.

Note

1. Here is what Henry Dale, President of the Royal Society (London) said on the occasion of announcing Oswald Avery's award of the Copley Medal in Dale's Anniversary Address to the Royal Society, 1945 (pp. 1–17): "Here is a change to which, if we were dealing with higher organisms, we should accord the status of a genetic variation; and the substance inducing it — the gene in solution, one is tempted to call it — appears to be a nucleic acid of the desoxyribose type. Whatever it be, it is something which should be capable of complete description in terms of structural chemistry."

John E. Sulston, 2003 (photograph by I. Hargittai).

30

JOHN E. SULSTON

John E. Sulston (b. 1942 in Bucks, England) is a Staff Scientist of the Medical Research Council (MRC) Laboratory of Molecular Biology (LMB) and the Wellcome Trust Sanger Institute, Cambridge, United Kingdom. He was co-recipient of the Nobel Prize in Physiology and Medicine for 2002 together with Sydney Brenner and H. Robert Horvitz "for their discoveries concerning genetic regulation of organ development and programmed cell death".

John Sulston earned a B.A. degree (1963) in organic chemistry and a Ph.D. (1966) in oligonucleotide synthesis from the University of Cambridge, England. He was postdoctoral fellow at the Salk Institute for Biological Studies in San Diego, California, 1966–1969 in the field of prebiotic chemistry. He was first appointed Staff Scientist at the MRC LMB in 1969, and the appointment became permanent in 1974. In the period of 1992–2000 he was Director of the Sanger Centre (now the Wellcome Trust Sanger Institute). He has been a member of the Human Genetics Commission since 2001.

John Sulston was elected to the Royal Society (London) in 1986 and was knighted in 2001. He has received many other distinctions and awards.

Our conversation was recorded in the home of the Sulstons in Cambridge, on April 26, 2003, during the 50th anniversary celebrations of the discovery of the double helix structure of DNA. It was in the evening of a busy day. The recording was sandwiched in between a whole day of talks at the MRC LMB and the evening dinner party for all the LMB alumni and current members. On our way to the party, we stopped at the Sulstons' home together with Bob Waterston who was staying there. We had one hour for the recording and we both felt that the questions and the answers had to be tailored to stay within this one hour. As it turned out, at the end of one hour, when the sixty-minute tape ended, we were done.

John acted as Bob Waterston's host and John's wife Daphne was away, in New York. The Sulstons just had a grandchild and John was not there only because of his commitments in Europe. Daphne Sulston just retired; she'd been for many years a university librarian in the Applied Math Department of Cambridge University. The Sulstons have two children, Ingrid (born in 1967) and Adrian (1969). Ingrid is a biologist with a Ph.D. from Berkeley, but her professional interest is in interactive museums; taking now a break for having her kids. Adrian is not married; he has a position with a software company in Edinburgh.

You gave one of the talks yesterday at the meeting celebrating the 50th anniversary of the discovery of the double helix. You sounded passionate about the importance of making the human genome information freely accessible to researchers.

Did I? I think I am pragmatic about it. I have become, shall I say, definite about it, although you could call it passionate if you like, but I would call it more of a definite position than a passionate position, because it seems to me so simple. It is essential that human genome information is in the public domain. The reason for speaking, if you like, passionate about it is so people understand what the issues are. They think that it's just another batch of data, but we have to consider the principle of how scientific data are handled. This is also why I allowed myself briefly to criticize *Science* magazine in passing because they actually aided the idea of withholding information from a private company, while at the same time giving them the right to publish their paper in *Science* magazine. To me it is the antithesis of scientific publication because in the case of sequence papers the sequence is the data; it's not something extra, which you can take or leave. The extent of this disagreement indicated to me that I must talk about it.

To me the line seems to be blurred between private research by drug companies and university research.

So I have to explain more about the connection to the human genome. If it would've been the case that a private company was simply sequencing the human genome, or any other genome, as they do, for their own purposes, keeping it private and that's the end of it, that's fine. Then the public domain continues to do those important genomes, which are universally valuable, in its own way, and this has happened many times. I just give you one example. *Staphylococcus aureus*, I gather, has been sequenced ten or maybe twenty times by different companies, because it is important especially for overcoming resistance of that organism. That's fine and I have nothing against it and I would never say that their sequencing should be released. The difference with Celera [Craig Venter's company] was that it deliberately positioned itself as being instead of public domain and issued misleading statements about the extent to which they would release their data. They claimed to Congress [of the U.S.] that they would release their data every quarter, for example. In fact, many American commentators believed that that's actually what happened. You can go back to *Congressional Records* where promises were made to release the data every three months, and it never happened. I'm obviously summarizing all sorts of things that are described in *The Common Thread* [by John Sulston and Georgina Ferry]. They also went and lobbied Congress to try to have public sequencing of the human genome shut down. They told Congress that NIH funds should not be used in such a project. So this is different because they are not doing something on the side that does not effect us all. They are saying that it's appropriate for the human genome to have its sole source in the hands of an American corporation and everybody who wants access to it have to pay and those who can't pay, of course, can't use it. As I've pointed out, there is an additional disadvantage, had that all happened, even the people who do pay couldn't communicate with each other and publish freely.

Does this happen with any other medical research?

You are quite right about the blurred line. It happened again with rice, for example. There was an American company that was the only one sequencing rice, but, fortunately, their semi-released publication was duplicated by the Chinese. They did put out their data at the same time and made them available freely. Concerning medical information, I gave examples during the talk, quoting Myriad [Myriad Genetics], for example, but I didn't quote the fee Myriad charged for the sequence of the breast cancer gene, which was USD2,500. That's an example in medical research, in clinical practice where a company is using a monopoly, which was gained by establishing a patent portfolio in this case, using a lot of public data because the discovery of the breast cancer genes was very largely done in the public domain. Myriad just managed to establish a few little bits of it. They established a U.S. portfolio and they managed to shut down all other commercial sequencing or any sort of testing those genes in America. So American physicians who want to have that test have to pay USD2,500 to Myriad, which is probably about ten times what's necessary. They are also trying to extend that all over the world. They've managed to get the Europeans, the EU, unfortunately, to issue one patent, which is now in litigation, and they are also trying to do this in Toronto, to extend their patent to Canada.

Was your getting into the Human Genome Project a consequence of your having been involved with the C. elegans *work?*

Oh, yes, directly. And that's where Bob [Waterston] comes in too; the worm was equally divided between the two labs and so was the funding. For twenty years this has been going on with Bob. We began with the mapping and then we continued with the sequencing. I began working with Sydney [Brenner] in 1969; Bob joined Sydney's group a little bit later. At that time we did both genetics and cell biology, and our particular cooperation on the genome started in the early 1980s.

Getting back to 1969, what was your position?

My position was initially a one-year staff member, very much like the stories we were hearing today that people were just coming for a year to the Laboratory of Molecular Biology and they might have stayed a little longer if there was money. It was just the same for me. Before, for two and a half years, I was in Leslie Orgel's lab at the Salk Institute in California. There was the possibility of trying to get back there and there were other possibilities of jobs, but in the end we decided, my wife liked it here too, we decided to continue here in Cambridge.

Sydney was the group leader.

Absolutely. He was also joint head of the division of cell biology with Francis [Crick] and as far as the research group went, he was the head.

How long did you stay as staff member under his supervision?

Forever, but it also depends on what you mean by his supervision because we did not define these things clearly.

This is why I'm asking.

I got an extension first of all; first I had one year as a second postdoctoral stint at LMB. Then I had an extension of that for three years, and then somehow I happened to be in a slightly different position from most of my colleagues in that I had no desire whatsoever to run a group. I was very much doing things on my own; in collaboration, of course, but not wanting to be head of a group. So I remained in a pretty solitary sort of way, working, for example, on the cell lineage project. I collaborated with others and after a while I got a postdoc, Judith Kimble, but I can't really say that I was head of anything.

Were you still under Sydney?

I was under Sydney because he was now head of division, but we never worried much about the titles.

When did you feel that you became an independent investigator?

At some point, and I've forgotten when it was, I got tenure, and I suppose that was the moment.

When was it?

I don't know. We could look it up.

You were still working on the worm.

There were a series of things. That's my attitude; I was very happy pottering about the lab; in fact, I did feel that I didn't get a lot done. My first publication was with Gerry Rubin, curiously. We did a little thing on the arrangement of genes in the yeast genome; I just sort of assisted him really. Then Sydney got me to measure the size of the DNA and that was something I did on my own. The thing that I actually had title for was getting the determination of the cell lineage under way. It began about the mid-1970s with establishing what the larvae lineage has to do with the eventual nervous system of the worm. Working together, John White and I were able to establish that particular cell types came from particular branches of the cell lineage. Looking back, it was a great break for me although it did not occur to me at the time because I was still running around doing all sorts of things that I thought were interesting, particularly about the worm. But I'd found something that people considered to be mine, something that I was good at. It went on gradually, with several of us working on different aspects, until I made the decision to tackle the late embryonic cell lineage that nobody had managed to solve before although many had tried. Once I'd done that, I had a substantial piece of work, and that brought me the Fellowship of the Royal Society, the embryonic cell lineage. That came out in the early 1980s. At that point I personally was done with the cell work, but I was not quite sure how I wanted to proceed. As it happened, the notion of the genome fell into the vacuum in my mind and I thought it would be the thing for me to do and I switched to that.

Did you change work place?

I did because the Lab was always rearranging. I remember as I got displaced from the place where I was doing the cell lineage. I was writing up my findings and occasionally went back to do some small experiments with Nichol Thomson.

What followed your cell biological experiments and your work on cell lineage?

In the 1980s, I worked on the genome map, initially with Alan Coulson and then very quickly joined by Bob Waterston and that's when our collaboration started. In the 1990s came sequencing.

John Sulston and Bob Waterston in the Sulstons' home, Cambridge, 2003 (photograph by I. Hargittai).

That's when you left LMB.

In a way I never left LMB, but that's right, in 1992 I changed for the Sanger Institute. We started sequencing in 1990 on a small scale and in 1993 we moved into human sequencing. At that time the big lab out in Hinxton was being built and we moved there. In terms of physical moves at the LMB itself, I spent my initial years in a big lab, having just a little bench space. Then I moved to a very small room for doing the cell lineage work, then I had a little bit of office space for a while. Then with the physical map, and this is an interesting question about independence that you were pressing me on, when we began the physical map, Alan and I, in 1982, I said it to Sydney who was head of division, that this would be a good thing to do. This is when he gave me a little bit of extra space. The thing is about the Lab, the distinction between the group leader and the division leader is not so great as you would imagine in a typical university setting especially in an American setting, where the group leader is absolutely king. It's not like that; it's all done by people pushing and persuading, and when you are doing a piece of work you have your own space, but you'd better not be gone on holiday too long because somebody else will come and take over your space.

Sydney could have received his Nobel Prize for a long time and it was apparently a difficult question which of his achievements to choose to award the prize to him for. Finally, the worm studies were selected. In that case it was natural to select you and Horvitz as co-winners. On the other hand, the Human Genome Project could also be selected for the Nobel Prize in which you were very much involved and such a consideration might have given a push for Sydney's Nobel Prize.

You just cited two scenarios out of a possible thousand or so. The thinking process is in the Karolinska Institute and I am too involved to view it rationally. I am not willing to speculate about this and I have somewhat mixed feelings on it anyway. I have quite a lot of pleasure but also quite a lot of doubt in connection with this Nobel Prize.

What would that be?

Whether I ought to have one. I don't regard myself as being that sort of exalted and I feel that I'm being lucky, if you like. Also, I feel that I'm being raised above my position, but that's something I've always felt throughout my life. I felt that when I got tenure and when I was elected to the Royal Society; I always felt that way. This is not a story, just the facts. In the light of that it's especially difficult for me to comment on the Nobel Prize. I absolutely welcome Sydney because he is one of these people who has contributed a lot in all sorts of ways and it's interesting to see how almost everybody really does come through to say yes.

It was very interesting to listen to some people today who had for decades been fighting Sydney, especially during the 1970s about policy in the Lab. In the end, Sydney is an extremely clever guy, extremely charismatic, so even people who fought him for a while come around and say this is a great guy. What I'm saying is that there was extremely strong support for Sydney having the Nobel Prize. But why was it delayed so long? It's this business about perhaps being involved in a number of things, absolutely the reverse of Sanger who was doing his own thing, always in one thing in a very clear way whereas Sydney was always involved in many different things. He may have missed the prize for being associated with other people. For the moment I speak about Sydney and will come back to myself in a moment. I think the worm is probably the one thing of Sydney's where this is appropriate because he really started something with the worm. He himself had many doubts during the 1970s, I know. He became very interested in computing and he actually regarded the worm like the sorcerer's apprentice, like the magic spell, something that he'd started and thought he should stop it at some point because it was just burgeoning out without doing anything particular.

Then, there is someone like Bob Horvitz who absolutely knew what he was doing when he came to Sydney; he knew he was going to do something important and he did. Sydney just couldn't call back the spell. Bob did a huge amount of work in America, he did more than anybody else and propagandized the worm into America, giving many, many seminars about what he was doing, he quickly got many postdocs and they studied all sorts of cell division mutants and found all sorts of things.

As we then change into the 1980s, Bob began this rather new line with the programmed cell death. There is a man called Ed Hedgecock who was instrumental in getting this thing going. So there were many people who could've been included in the prize like the early cell death workers, Wyllie, for example, why isn't he there? Anyway, Bob Horvitz has done a huge amount of work. So then I ask, why am I there? and I think it is for the cell lineage more than anything else. I must say that everybody has been very nice about it. One saving thing for me about it is that I've had one hundred percent nice e-mail about it from so many people. There are sometimes controversies, but I have not heard about any in connection with our prize. They even commented on it in Stockholm. The Karolinska people said when I arrived that they often get dissenting e-mails, but none has arrived in our case. That was very nice.

Then you were asking about the human genome in this connection and all I can say is that what I know is from gossip and also from deliberate statements that people have made to me, which are repeatable though not attributable. There have been a number of nominations for people to get the Nobel Prize in all sorts of combinations for genomes, including the human genome and including the worm genome, probably including the *Drosophila*, although that one I did not hear myself, and there were various combinations of people involved with these projects. Before I actually got the Nobel Prize, when anybody asked me, did I think there was a Nobel Prize in the human genome, I said, no, it's a technology, not a breakthrough, it's not a discovery. I would still say that, but now I really didn't want to be quoted on that simply because it's obvious, it looks like I got in, nobody else can. So it's not a good remark for me to make, but, nevertheless, in an objective way, that's what I feel. It's a very complex thing, it goes back all the way to Fred [Sanger], it comes through the people who built the sequencing machines, those who made the high throughput possible, then those who managed the labs, who created the computation, and so on. The real bottom line is that it would be too soon to think about a genome prize because the dust has not settled yet.

So the bottom line is that there should be a prize for it, but not right away.

Maybe, maybe not, but anyway not right away. In Stockholm, they drew the line carefully. Look at the citation; there is no mention whatever of the genome. My feeling is that they would push it off as long as possible to see where the dust settles. I also know that there is a lot of pressure, a lot of lobbying going on. OK? That's what I know.

When you got involved in the C. elegans *work, what did you think, what was the goal?*

To learn how an organism is constructed from its genes. Sydney said himself today that as with any experimental system, you don't necessarily expect to answer all the questions, but the idea was to use the worm to tackle that in as many ways as we possibly could.

Now I would like to make a big jump and start at the beginning. As we are quite limited in time, I wanted to have some most important questions out of the way first.

Quite.

You are Sir John. What did you receive it for?

It was for services to genomics.

Was Sydney offered it too?

I don't know but if he had been he would have probably turned it down. Sydney, like the other great ones, has much higher awards. Sydney is a C. H. [Companion of Honor, restricted to 48].

Aaron Klug is Sir Aaron and he is also O.M. [Order of Merit, restricted to 24].

Max [Perutz] turned it down and Fred [Sanger] turned it down too. They turned it down for various reasons. I did speak with Max about it, shortly after I accepted mine. He said he'd turned it down because he didn't want to make a distinction between him and the staff members of the Lab. Then several of us have accepted it, John Walker has accepted it and so has Paul Nurse. Whereas there are only 24 O.M.s and 48 C.H.s, there are hundreds of knights. It used to be for people in industry who had made some money and supported the government. It's not altogether a savory bunch, the knights.

Do you ever introduce yourself as Sir John, say, when ordering a hard-to-come-by theater ticket?

I would never do such a thing. I discussed the knighthood extensively with my wife who is more left wing than I. We really went around and around; it was a joint decision. There were two good reasons for taking it. One is that it is becoming good as things are becoming more democratized and it's not bad that scientists join in this general thing. There is a certain amount of approval for knights in the population, it's a people thing, it's pretty good. I find that around our village: knights are an advertisement for science. That's one thing, so they would know that scientists are being knighted. The other thing is when you go to some places, it does open doors; not getting into the theater but it is regarded especially in government circles so it helps when you want to speak with somebody, for example. So not knowing exactly what I would do, I thought, well, I may later regret it if I turn it down. Now I sometimes have the feeling that it has created a separation for people who don't really know how to address me. There is a little bit of embarrassment. I always say to people, don't worry about it, just call me John.

You mentioned being a left-wing person even in your talk.

Did I?

You said something like, I don't want to sound too left-wing. So what does it mean for you?

One shouldn't really use these terms. So what is this left-wing thing? What does it mean to me?

Where do you position yourself in the political spectrum? I wouldn't have asked you this, but you had brought it up.

What I mean is that, for an illustration, one likes to use his title or not, just what we discussed a moment ago. Social things if you like. It may not be very sensible to call it left-wing.

Are you to the left from the Labor Party?

Yes. The most serious point is this, and may be one should be more specific every time one uses such a term. There is a question about the extent to which we should commodify everything, privatize, settle everything with money in a completely free-market economy. The other extreme is where there should be extreme regulation or at least everything should be common goods and not very much private goods. Of course, neither extreme makes sense; all our society lives somewhere in the middle and we are near where we should be. But still we are swinging too far at the moment towards private ownership, and forgetting that we need the commons also.

Our conversation is taking an interesting turn, but were you pro-Soviet, for example?

I was not pro-Soviet and was too young for that anyway.

Not by your age. You were born in 1942.

That's true. You know why I said that because I was much less political in a sense of having any interest in it in the 1970s and the 1980s than I've been in the 1990s. The reason is that I got into this public versus private issue due to this specific problem. But I've always voted either Labor Party or Liberal [Liberal Democrats], depending on particular issues. For example, I have voted LibDem for quite a while now because I always thought that proportional representation was a good idea.

Would you tell us something about your childhood?

I was born in Bucks, very near London. Then we lived in Hertfordshire and I went to public [meaning private] school. My father was a priest [Anglican]; he did not have a parish; he spent his working life as one of the organizers of a Missionary Society in London. He was also a people's priest and very well loved. In the weekend he always participated in the services, preaching to people and so forth. So he did a lot of parish work, but not as the head of the parish as a vicar would. My mother was a teacher; she taught at grammar school for many years, both before and after we were born, my sister and I. When we were growing up, she took time out.

The young John Sulston
(courtesy of the MRC LMB Archives).

My father tried very hard to instill his religion into me and I started out to be religious. It was a slow and rather painful process in the second half of my teens and it was very sad for him that I gradually abandoned religion. I didn't feel that I really got away from it until I was at the university. It was difficult because one has to search after all for morality and the justification for one's life and way of living. The single strongest thing, although the whole thing was very complex, was when I came to realize that there were many great religions in the world. It seemed to me that none of them should be really depended upon; I saw no particular reason to have faith in the Christian rather than Hindu or Islam, let alone to choose between the various Christian faiths. Why would I be Anglican, Catholic, Methodist, or whatever. At the same time, and it began at puberty, there was the growing independence, I felt very strongly, very passionately, I suppose, in the power of science. I was not so stupid as to imagine that science would be able to solve everything; I knew perfectly well that neither at that time nor in the immediately foreseeable future would science address what's going on in our head although I have confidence that it will one day. Neither does it provide morality. What it does do is to provide a picture of the Universe, which is self-consistent, it's growing all the time and it's absolutely and wonderfully solid compared with believing things written in an old book. And so out of that comes for me a sense that we should depend on ourselves, think as rationally as we can about what we are and what we can become. This, I think, is humanism.

That was my switch, but some principles, some morality, have stayed with me from my father's teachings. Not all however: there was a spectrum of views. For example, my father believed quite strongly in the hierarchy of society. He believed that it was appropriate that some people were wealthy and would have servants, not that he was wealthy and we didn't have a servant at home, but he felt that that kind of situation was perfectly fine. It was the establishment view. I used to argue with him about it; I was always trying to be more democratic. It's not an absolute position, just a comment that he seemed to be happier with the stratification of society.

When was it when you became interested in science?

It has always been with me; it started before I could even talk. I can see it in my grandson; he is a baby and he is already very manipulative. He loves building things. When I say I always was a scientist that's what I mean.

Manipulative? Isn't that a bad word?

Doing things with his fingers, doing things. You have to be careful about the context with this word. We use this word all the time. Genetic manipulation, for example, is a neutral term although it may be used politically. Only when you say somebody is manipulating another person is it negative.

Could you pinpoint a book, a teacher, or a chemistry set that turned you to science?

Clearly, many people contributed, but I always wanted to have these sorts of things; when they asked me what I wanted for Xmas, or birthday, or where I wanted to go, it was always something that had something to do with science or engineering. My parents treated me exactly as we treated our children, which is to look and see what the child wants to do and encourage them in doing that, unless it's something destructive, obviously.

You studied at Cambridge. Was it difficult to get in?

I had good grades, but I had to work hard to get them and I tell you why. I never had any trouble with schoolwork, it was quite easy for me, and I always had fairly high marks. But then, for some reason and I still don't know why I did it, I insisted on giving up math and doing biology. We couldn't do all sciences; we had to choose. We did chemistry and physics and we either did math or we did biology. I insisted on doing biology and it was a very odd choice because at that time biology was not taught very strongly at school and it wasn't thought of as being a strong academic discipline. I had some trouble in getting to grips with it whereas I was fine with math and that was the reason why I gave it up. I'm sure it was that I had a growing interest in biology. So I didn't get fantastically good grades the first time when I was in the A level, but I had a year in hand so I was able to retake biology, and at the end I got to Cambridge with a scholarship. Then it was all easy until I got to my third-year exam, when it again became difficult. I got through them with a lot of work.

As you became a researcher, was there any single determining factor in it, a venue or a mentor?

There was a whole series of mentors and they all contributed quite a bit. The people I in particular remember in science is one of my schoolteachers,

John Sulston, Alan Coulson, and Sydney Brenner (courtesy of the MRC LMB Archives).

the guy who taught me physics. He had a sense of a mind reaching out
and got me excited. You know how exciting it is to be taught that you
can weigh a planet, you can calculate how faraway a star is, even if you
can't do it actually, you can feel how the mind can reach out. Isn't that
incredible? Our minds can reach out through the light years. It is fantastic.
Once you learn that, you realize that there is no real limit to the power
of our thinking. We have to learn; it's not that we can figure out everything, but we can learn to think about anything. Other people have described how they realized the boundaries of their ignorance, but when
you realize these boundaries, you also realize that there is so much to do.
There's nothing artificial about it. With all the religions, they put a fence
around you; they say, this is the answer, we got all the answers and you
look it up in the book. It's not like that with science; you realize the
boundaries of your ignorance and that there is an infinite domain beyond
those boundaries. That happened to me in school.

My university years as an undergraduate were not so great, though I
had a nice chemistry supervisor. Then I became a graduate student, and
my supervisor, Colin Reese was great because he just gave me things to
make in chemistry. I enjoyed manipulating the molecules. Then I went
to Leslie Orgel and that was very intellectually stimulating because Leslie
was working on prebiotic chemistry, the origin of life and we all talked
about that, about big ideas, about how this would've happened.

Had you stayed with Leslie Orgel, wouldn't it have been a dead end for you?

Yes, it would've been, I know, but the point is that I gained from that experience. The whole point about mentoring is the inspiration you receive, and I thought a lot about this question. I know many people would simply answer this question with a name of so-and-so. I didn't think that for me it was true at all. It was not that I had no mentors, I had lots, one after the other.

Leslie Orgel is a brilliant person. Why, do you think, he stayed with that line of research about the origin of life?

It's a great question for a polymath because you have to think about everything. It's difficult to measure the success with such a project. But I believe that at some point we shall, not possibly know exactly how life originated because that would be a historical matter, but at least have some plausible scenarios. That's what Leslie is working for and I find it a wonderful topic. It's not suited to me because I'm a nuts and bolts man.

When I visited him, he had already spent decades on this problem, but I felt as if he was on the run, catching up with things.

On the run?

As if he has not yet found his peace.

You can't tell with Leslie. He has this intellectual, slightly cynical attitude in everyday things, he likes arguments of an intellectual kind. He has this approach that all is a kind of game of cricket.

I have no right to say this but he appeared to me as if he hasn't fulfilled the promise he had carried and as if that made him nervous.

I don't know. It's a very negative thing to say about a person because it implies that everybody has to reach some kind of star quality in the eyes of the world.

Can't one feel that one didn't live up to one's potentials? You seem to have lived up to yours.

I'm an over-fulfilled promise. You have to look back and when you do, you'll see that Leslie has contributed hugely to the discussions of his topic.

I never doubted that, I merely told you about my impressions of him, what he projected to me, and not about his actual achievements. He just did not impress me as being at ease.

I am reasonably at ease in my skin and you are saying that Leslie is not. It's an interesting observation of yours, I must say. I don't think I would like to pursue this and you said I could stop at any point.

And I'm not trying to press you at all.

I wouldn't like to pursue this because he is one of my mentors, one of my many fathers, if you like, and I see him as a wonderful guy. I'm sure you can take anybody, you can take Sydney, for example, have him focus on one thing and not so much on others. He has contributed so much in many ways.

Of Sydney I had the impression that he was perfectly at ease, he is not on the run, he is still full of what he wants to say and appears to have a schedule that allows him to elaborate it without rushing anywhere.

I don't think either that Sydney is on the run, but you could argue that he should've done things in a different way. I want to say that I respect all these different ways people are running their lives; they have all contributed hugely and Leslie contributed hugely to science (and to me) so I feel that's fine. It would be nice to have everybody happy, but, inherently, people achieve things when they're dissatisfied.

Sydney was also one of your mentors and helped you ...

... and many others. But you are raising a very interesting point.

I have to ask you my next question and I apologize, but we are on the run in this interview. Did you ever have any adversity that you had to overcome in your career?

Any adversity. [Long silence.] Yes, in the sense that life goes up and down, are you asking about when I was particularly unhappy or challenged?

It's good for you if you don't quite know what adversity may be. Like Sydney grew up in an unfriendly environment in South Africa, just an example.

It's a good example. I certainly felt at some point of not having achieved much. I was actively looking for a route out of this Lab some time in the mid-1970s, because I felt that I had not done enough to justify my position. I was very impressed by the Lab, I wasn't overwrought by it, it didn't have the effect of making me going into a corner, but it did have the effect of making me feel uncomfortable. I had a great friend, Michael Wilcox, who died many years ago, of cancer, and we used to talk about this and I remember how we discussed this, how we could never live up to the achievements of the past. We saw the Lab as arrogant, but I remember saying that I thought its arrogance was justified. This was shortly before I started working on the embryonic cell lineage and it is ironic because I had done things before too with which people have credited me, but to me it just hadn't amounted to much. My self-estimation at that point was not very high and I was in some despair. It was not due to personal circumstances, my conditions at the Lab were always great, but I felt of getting out, just getting a job. Then somehow things started moving again. I've noted in myself that I'm most productive when I am in some measure of despair. You can see the connection. For some people despairing is difficult to handle, for others they just start doing something and the distraction works to their advantage. But it was not a happy thought. I was a little bit like that and it did help me to get going because it increased my level of determination. Once it started to run, after only a few months, and it was a year and a half job to do the embryonic lineage, so after only a few months, I reached the plateau, because I knew how to do it. It was terrific, and it was just coming out. It was one of the nicest, most productive phases of my career.

The other adversity was not despair, but it was very definitely a challenge. It was the announcement of Celera and the gradual dawning that these people were really going to try and take the human genome into private domain. That was absolutely a piece of adversity and this is why you said I felt passionate but I just felt definite about it. I'm just quibbling with words, but I knew we had to stop them. It was a recognized conflict and there certainly was a lot of despair and many of the collaborators at that Cold Spring Harbor meeting felt that they were not going to make it.

I'm sure you discussed this with Jim Watson.

Sure we did and especially Georgina and she had more quotes from Jim in the book. I know very well what Jim thinks about it; he goes around

István Hargittai and John Sulston in Cambridge, 2003 (photograph by Magdolna Hargittai).

now saying this is shit. This is a good Jim quote. I don't say it, Jim does.

I just bought your book but haven't read it yet.

I'm sorry to refer to it so much, but it's useful because the story is there with all the quotes that Georgina collected.

You are now a little over sixty with a tremendous career behind you. What could be a challenge for you today?

I am still very much haunted on the short term; I'm still trying to clear space to think. Daphne and I are definitely trying to clear some space because there has been a lot of pressure; it has just been one thing after the other; we can't think, or I can't anyway. We are going away in May and we are going to stay away and not be in touch for a while; that will be the first among a few more things we need to do. Maybe something will come out. I feel right now the same as I felt at the end of the cell lineage project.

Do you feel emptiness?

I want to feel emptiness; I want to feel empty; I want to have space; I want the noise to stop and go away. We finished the worm; we cleared away the human genome; other people are finishing it now. If you ask me what mostly disturbs me or mostly stimulates me, both, it's the larger aspects of this phenomenon of privatization. Once you are sensitized to the existence of a problem which you believe in, then, of course, you see it everywhere. For example, I opened the paper two days ago and I read that the World Health Organization is kind of being held at ransom by the American Sugar Corporation. This is because the World Health Organization issued a report saying that no more than ten per cent of a healthy diet should consist of sugar. It seems pretty sensible to me; it even sounds like a lot of sugar. But the American Sugar Corporation is coming in, and I don't remember the details, but its message is very clear that they're trying to get the American Administration to block subscriptions to WHO unless its reports are withdrawn and the recommended sugar content was raised to twenty five per cent! That makes me mad.

Would it have made you mad one year ago?

One year ago yes, five years ago no and that's what I mean by getting sensitized.

But one year ago you were not a Nobel laureate. Today, you can do more about it.

Precisely. So I have an obligation to really examine my position and see what I can do, if anything.

You are in that process.

I am in that process, exactly. There are lots of new things and that's why I need space to think and read. It's like the knighthood, the Nobel Prize I considered very briefly rejecting it, but that was not a serious consideration, whereas for the knighthood it was a serious discussion. In both cases, the reason for accepting is because one hopes that one can do something useful with it. An honor is worth nothing unless you use it.

We've run out of time and I appreciate your candor, I feel it's genuine.

Sure, it's genuine; I always say what I think.

It was your choice to step down as director of the Sanger Center.

I judged, and many people think it was not a bad judgment although other people think it was a terrible judgment depending on how they see things work out. I stepped down at a good point in time. We had got either done, in the case of the worm, or completely funded, in the case of the human genome; thus the two big projects that I'd set out to do were built up. The place had a good reputation so that was a good idea to depart as director; I still have an office there, and to let new blood come into the Lab, so that we don't become endlessly mono-cultural.

The Center is named after Fred Sanger who is still around. He had retired from LMB but does not go there.

He comes a little bit to the Sanger Institute and he was quite pleased to have it named after him. It used to be called Center but now it is called Institute. Fred feels some pride in that. When I had to ring up to ask him, can we name this Institute after you, Fred said, yes, but it better be good.

That was an honest answer.

That was an honest answer and it was also, for me as the new director, a terrifying answer. We just had fifteen staff at that point. For some reason, of all the scientists that I heard about during my school days, and this was the *annus mirabilis*, 1953, of course, we heard about it in school, people said, wow, British scientists doing all these wonderful things. For some reason, I do remember that it was Fred who made the impression on me. He was always very mild, unassuming, always doing very solid things. I do find it very remarkable that I'd fallen for this connection with him.

Is there anything you would like to add?

No, I think you've covered it pretty much.

Renato Dulbecco, 1975 (courtesy of and © The Nobel Foundation).

31

RENATO DULBECCO

Renato Dulbecco (b. 1914 in Catanzaro, Italy) is Distinguished Research Professor and President Emeritus of The Salk Institute of Biological Sciences. He studied at the University of Torino and received his M.D. degree in 1936. After war service in the Italian Army, he worked as an Assistant in the Institute of Anatomy of the University of Torino and studied physics at the same university in 1945–1947. In 1947–1949, he was a postdoctoral fellow under Salvador Luria at the Department of Bacteriology of Indiana University in Bloomington. From 1949 to 1963 he was at the California Institute of Technology (Caltech), first working with Max Delbrück and eventually as Professor. From 1963 he has been associated with the Salk Institute, but also held positions at the University of Glasgow (U.K.), the Imperial Cancer Research Fund Laboratories in London, and the University of California at San Diego.

Renato Dulbecco shared the Nobel Prize in Physiology or Medicine in 1975 with David Baltimore and Howard M. Temin "for their discoveries concerning the interaction between tumor viruses and genetic material of the cell". His many other distinctions include the Lasker Basic Medical Research Award (1964), membership of the National Academy of Sciences of the U.S.A., and foreign membership of the Royal Society (London).

The video recording was made by Clarence and Jane Larson in La Jolla, California, on March 25, 1986.* Dr. Dulbecco checked and corrected the text in July 2004.

---

*"Larson Tapes" (see Preface).

Family background, childhood

I was born in 1914 in Italy. There is something interesting in my background because I was born from a mother from the south of Italy, from Calabria and my father was from the north, from Liguria. These are two civilizations, which are very, very different in cultures and everything and I was born as a hybrid. These two cultures have different characteristics; for instance the southern people are more imaginative and tend to be extroverts and the northern people are solid, they work hard. This is not to say that the southern people do not work hard. Looking back at my life, these two qualities, that is, to have imagination some time and work hard most of the time were very efficient for me.

I was born in the south; my father was a civil engineer; in Italian it was called "Genio Civile". He worked for a government organization, which builds bridges and roads and so on. He was there because years before there had been a very severe earthquake in the south of Italy and several villages and towns had been erased. So the government had undertaken the reconstruction of these villages and towns with the modern building methods. My father was a specialist of something that at the time was very new, the reinforced concrete technology. They built everything this way and apparently that worked very well. We did not live very long in Calabria where I was born because my father had to go into the army; there was war; and my father worked in a factory where they were making guns and bullets for the war effort. We went to the north, to Torino and we stayed there through the war and I don't remember anything from that time.

At the end of the war we went where my father was born; the town is called Imperia, near San Remo, at the French border. I went to elementary school there and to high school, called Ginnasio-Liceo. At that time I became interested in science for the following reason. In the town where I lived, there was a meteorological station. It belonged to a government network and it was also a seismological station for detecting earthquakes, which are quite frequent in Italy. This station was run by an amateur, a pharmacist. He did this in his spare time. My father knew him quite well. When I was in the Liceo, in the equivalent of the tenth grade here, my father introduced me to him, just at an accidental meeting. The pharmacist was interested in what I was doing and told me to come to the Observatory. So I went there. He loved to tinker; he loved to do things with his hands;

he had all kinds of instruments to measure the winds and barometric pressure and humidity, and he sent the data daily to Rome. Then he had a big pendulum, I don't know, maybe 50 feet high, with a big mass at the bottom with a mechanical registration on a cylinder covered with soot. Every day he had to renew the thing and check the tracing, and send back the information.

At that time — it was 1928 — radio was just beginning. I got very interested about radio and wanted to build one for myself. I bought a book and studied the principles of the radio and the circuitry and the tubes, and I built a radio. First I built one with a crystal and I remember when I listened to it for the first time, it was absolutely incredible, with the earphone, and there was sound and suddenly, music came through. My mother was a lover of opera and I gave her the earphone. Then I moved on to make a real radio with tubes.

Next, I thought that the seismograph was too antiquated. There could be a more modern seismograph using electronics, so I built one using variational capacity, and it worked. In about a week, there was a distant earthquake, which the big machine did not feel at all and mine did. Then I met someone who belonged to the central office of the seismological survey in Rome; he was so excited that he had it published. I was about 15 years old. Something was pushing me in that direction.

University studies

When I finished high school, I had to go to the university and we had to choose, which university. Most people in our region went to Genoa because it was close. But my father had studied in Torino so he thought that I should go to Torino too; he loved Torino; it is a nice city. I went to Torino and started studying medicine. That was also a difficult decision because my inclination was to study something like physics. I liked physics very much. On the other hand, my mother said that her uncle was a great surgeon and that it is a fantastic career and she convinced me to become a physician, so I went to medical school. One good luck I had at the time was the following. In the second year of medical school at that university the students could go as intern; an intern would go to a lab and work there even in the second year. I had the great fortune to go to the laboratory of a great man whose name was Giuseppe Levi. He was a Professor of Anatomy, but he was a very modern man, he knew biology, he was interested in growing cells in culture, which, at that time, nobody knew. He was a modern biologist at the time. He got me interested in the science of biology rather than in medicine and that was a decisive influence.

In the course of the medical school I changed; I went from anatomy to pathological anatomy because that was more important for medicine. It is still true; you make the diagnosis, then you go to the anatomy people, the body is there, you start opening it up, and see whether the diagnosis is right or wrong. I remember some fantastic cases when the doctors made mistakes. This was at the university hospital; the people sent the bodies from the Department of Medicine. Everyone had to be autopsied. To be admitted to that hospital, it was a condition. I remember a case when they brought in a cadaver with a very impressive diagnosis, which was a disease of the pituitary gland, which is at the base of the brain. As the cadaver was being wheeled in, a man who was working there just helping the autopsy, whispered to me, "cancer of the stomach." It was, of course, a completely different thing from what the doctors had diagnosed, and he had no medical qualifications, just a long experience. In the autopsy, first the skull was opened and the brain was taken out and there was the bottom of the brain and everything looked perfectly normal. The physician was there and he was very nervous already. Then the autopsy proceeds and as soon as we get to the stomach, there is a big cancer of the stomach. At that time the means of diagnosis were very limited; lots of biochemistry and the hormone studies, which are now routine, didn't exist. Things were very crude.

War times

I finished medical school in 1936 and I had to serve in the Italian Army for two years, as a physician. By the time I left, it was 1938 and the war broke out in 1939, and soon I was back in the army. It's a different story and I won't talk about that. I was a physician with the Italian Army in Russia; I barely salvaged my skin. I got back from Russia in 1944 and the war ended in 1945; there was about a year and a half in between, and I became a partisan. I didn't want to go back to the army; by that time the Germans had taken over, and I was hiding in the hills near Torino and worked as a physician for the partisans. I set up a dentist's studio to help them with that too because they needed that as well. It worked out very well; I became a good dentist in a matter of months. I ended up on the first City Council of Torino. During the war there were these clandestine organizations consisting of various groups, representing various tendencies. I was in one of them and when the Allied troops came in and the Fascists and the Germans left, we automatically took over the city. I stayed there only for two months because I am not a politician.

It was yet before the changeover, but when it became clear that it would occur, we organized a first aid system throughout the city. We knew that there would be fighting in the city. I went around to the houses and the ladies were fantastic; they were willing to let their houses become first aid stations. They were fantastic because had they been caught, they probably would've been killed; there were no middle ways at that time. However, the Germans left without doing any harm fortunately, they didn't blow up the bridges. It all happened overnight and by the morning it was a different world. The following evening, the lights went up in the city after over two years having been absent. I was enthusiastic and everybody was enthusiastic at that time.

Start in research

I went back to work with Levi, but I still didn't know what to do. I was interested in science, yet I was a physician and I had this experience during the war, so I felt a little inclination in that direction as well. It was touch and go; I knew I would do one or the other. In Levi's laboratory, there was also Rita Levi-Montalcini. She became a very well-known scientist. She used to do experimental embryology. She took chicken embryos, took one piece and moved it to another place and followed how this change affected the development. Her studies helped a great deal to understand the development of embryos. She convinced me to do something similar rather than abandoning the more medical aspects. She was very good with her hands and I didn't want to challenge her because I knew that she was excellent. I thought to do it in a different way, using irradiation to interfere with the development.

A friend of mine was a radiologist and he gave me one of these radium needles that they use to treat cancer. I applied this needle to the early embryo and followed its development. I noticed two things. One was that some cells that were characterized in the chicken embryo as the primordial or the germ cells, and which come from the outside of the embryo, migrate into the embryo, then they localize in two regions at the side of the spinal cord from which the gonads develop. These were then to become the ovary or the testicles, depending on the sex of the embryo. However, these cells did not reach there, and this was the first observation. The second was that all these embryos developed as males; there were no females. This was very interesting and it has not yet been worked out fully. The presence of the germ cells within these structures is required for the development. As in all vertebrates, the two sexes differ because of the two sex chromosomes, X and Y. In us, the female has two Xs, and the male has one X and one Y. In the chicken, it is the reverse. The male has two X and the female has X and Y. If you look through the development of sex, the homo zygotic sex, the one that has the two Xs always stands to predominate. In a human, if there is interference, the one most likely to develop is the female rather than the male. In the chickens the reverse is the case. So without the help of germ cells, only that particular type of sex could develop that corresponds to double X.

Giuseppe Levi was a great professor, very critical, very imaginative, and had tremendous influence. Everybody immensely respected him because he was such a clean person. During the Fascism, he was an anti-Fascist — one of only a few people who dared to say it — he was an admirable figure. He was also respected as a scientist because he was so straightforward and his suggestions were so nice, and he was productive and enthusiastic. He sent this manuscript to the Accademia dei Lincei, which would be the equivalent to the National Academy of Sciences of the U.S.A. They published their proceedings. This gave me a push and from then on there was no other possibility for me than to become a scientist, a biologist. I also learned tissue cultures, which proved to be very useful for me later on. So there were many wonderful opportunities in my early career.

Then I did another thing, which turned out to be really excellent. I did not know exactly how to get into science and which kind of science to do. Thinking about what I knew about science, I realized that with my medical background, it was very little. The medical background does not give you any preparation to work in science, it makes you into a physician but nothing else, especially the one I had. So I wondered about how I should enlarge my background. As I had done these experiments with radiation, I thought that radiation would be a very good tool to dissect biological functions. I talked to people, especially to Rita Levi-Montalcini and they all were saying that I ought to go maybe study chemistry, maybe study physics. Finally, I decided to study physics because I liked physics from the beginning. So I enrolled in physics at the University of Torino.

I went back to school; during the day I worked in the Institute of Anatomy as a Teaching Assistant and taught histology to students and microscopic anatomy. Then I went to classes and studied at night. This went on for two years, I took the exams, but then a decisive event happened. This was decisive because all these other events had happened before; otherwise it wouldn't have been decisive.

Indiana University, Salvador Luria, phage research

The decisive event was this: Salvador Luria came to Torino. Luria had studied at the same university in Torino; he was a year ahead of me, so I knew him a little bit. Then, at the beginning of the war, there was persecution of the Jews and being a Jew, he left, and finally came to this country [the United States]. In time he became well established as a biologist; he worked with phages, bacteriophages, bacterial viruses. In 1946, he came to Torino, for a visit, and he went to the Institute to see the people who were there. Thus there was an occasion to meet him. He asked me about what I was doing and I explained it to him, how I was using radiation and that I was studying physics. He was enthusiastic because he had done the same; he had started the same way. At the time he was using radiation just to study the genes of the phage. He had also studied physics, had spent a year in Fermi's laboratory in Rome. There was an affinity there between us and he suggested to me to come to his laboratory and he said, "I would give you a fellowship." That was a fantastic thing because in Italy, what could I do; I could have done nothing in spite of my being enterprising and having lots of good will, and I was working day and night, but the opportunities weren't there. So I said yes, and I came here in 1947 and joined his laboratory in Bloomington at Indiana University.

That was a wonderful place for biology at that time. H. J. Muller was there; he already had the Nobel Prize on the genetics of flies, and there were also other outstanding people like Tracy Sonneborn and Ralph Cleland. It was a real center for genetics; probably the strongest center in the United States, even in the world at the time. I took a course by Muller on genetics; I remember that my English was so poor that I could not really follow his lectures at the beginning. There was a German woman who took her notes in German and she would let me use her notes; I could read German. So for some time I studied from her notes in German. By the end of the term I had no problem to follow the course. It was also the time when

Jim Watson came. We were in a small lab, Luria had a little cubicle to prepare his lectures, there was Jim Watson and I and an assistant, a girl. So Luria, Watson and I were together in this little room and we could not escape talking and discussing things. It was in the years 1947–1949.

Then I made the first discovery, which was very good and it made me known already in the first year I was there. The discovery was the following. Luria worked with bacteriophages and he had discovered a phenomenon, which he thought was recombination; and in the end, in fact, it turned out to be recombination. If you take a bacteriophage and irradiate it with ultraviolet light, then you infect the bacteria with this irradiated phage, the number of bacteria producing phage depends on how many phages infect the bacteria. The number was disproportionately higher for those who had more than one. If you had one, you may get 10^{-4} survivals, but if you had 10, you may get one-tenth survivals instead of getting 10^{-3}. Obviously, there was some cooperation within these crippled phages. I studied this phenomenon; I made a curve, and made predictions on the basis of certain theories, all this together with Luria.

Then I remembered something from what I had learned in Cold Spring Harbor. I had spent a summer in Cold Spring Harbor. Many biologists went to that laboratory and learned a lot there. I went there in the summer of 1948. We worked there in the laboratories and I was doing an experiment together with Luria. I noticed some irregularity as we worked with bacteria plated in cultures in round Petri dishes. There was a layer of solid agar, a nutrient, and then we put the bacteria on top. The bacteria grew faster, making a kind of lawn, but if there was an infection, that would produce phage and the phage would infect other bacteria and that would appear as holes in the lawn. This means that a lot of phage particles were there and they would multiply. It was a way to assay the sample. I irradiated the bacteria and prepared the cultures and I always made two parallel dishes. I prepared the two cultures in an identical way but the two cultures had different numbers of plaques although they should have been identical. I could not solve the problem by the time we had to leave Cold Spring Harbor.

We went back to Bloomington and then an idea hit me. When we prepared the two dishes, we left them on the same bench for a short time and then we placed them in the incubator. First we thought that the temperature in the incubator could make the difference as one plate was above and the other below. I now put the two plates into two different

Renato Dulbecco, talking with Clarence Larson in 1986 (photograph taken from the video recording).

incubators at different temperatures, but the results did not change. So the non-uniformity could have only originated from the time the dishes were left on the bench. Above the bench there was a big fluorescent light and I thought, maybe it was the light that made the difference. I made an experiment in which I placed one plate under the light and covered the other with a piece of black paper. That made a big difference and proved that the phenomenon was reactivation by light.

Caltech, Max Delbrück, phage research continues

This was an intriguing phenomenon. It was already known that bacteria could do that, but nobody thought that this would happen to the phages as well. Then I had some fun in determining which part of the spectrum of the light caused the effect, which I did with making filters and so on. I first thought that it was the nucleic acid that was affected by absorbing the light, however, it turned out to be something, which was a porphyrin. Obviously something was happening in the bacteria and not in the virus itself. This was a good thing in any case because it attracted the attention of people to my work. As a result, I was asked by Max Delbrück to Pasadena, to Caltech. He knew me because we had met in Cold Spring Harbor. He developed theories and I was also trying to create mathematical theories and to test them on my experiments. We had lots of fun together. He asked me whether I wanted to go to Pasadena. It was a very difficult decision because Luria was so good with me, not only as a teacher but also he was like a father although he was only a year older than I. When I came I did not know English and I didn't understand the culture of the society and he guided me in my initial steps. But I had now this opportunity. I didn't know what to do and I talked with Jim Watson and he said, "You must go to Caltech; Caltech is the best school of biology in the country." I thought I had no choice, but to do it. I told Luria that I was sorry but I had to go, and Luria was very sad.

So I came to Max and I continued my work with phages, but what really interested me at the time was the relationship with this phenomenon between light and reactivation. There was then the other one that Luria had discovered, the cooperation of particles, and I was interested in how these two things fit together, how they interfered. After I had done a lot of work, I came to the conclusion that everybody accepted that they were two independent phenomena. They could add up, they were probably additive, but were independent. This was about two years after I had come to Caltech, and it was also essentially the end of my work with bacteriophages. There was another accidental circumstance, which was the following.

Animal virus research begins

A gentleman from San Marino, the rich community near Pasadena, Colonel Boswell had herpes zoster shingles, a terrible disease, very painful. He was a good friend of the President of Caltech, Lee DuBridge and talking together the Colonel asked DuBridge what he knew about this and Lee said that we didn't know much about these viruses. He told him about our work with bacteriophages that we understood more and more but nobody did any good work on viruses like herpes zoster and so on. The Colonel said, "Why not? I want to give you, Caltech, a fund to start work in these viruses." He hoped that out of this something might come out which would alleviate his pains. DuBridge handed the thing over to Max Delbrück. One day Max called me and Seymour Benzer who at the time was also working with phages too in Max's group. He explained the situation and asked whether either of us was interested in getting involved. Seymour said that he was not interested; he preferred to stay with phages and it was a wise decision for him.

I, however, always liked new things, so I said, "Yes. I am going to do that." This put together my medical knowledge and my knowledge of tissue culture that I learned from Levi. So I accepted this. We decided that I should go around the country to visit laboratories where people knew whatever was being done with animal viruses in order to see what we could do at Caltech. I spent three months going from one laboratory to another and visiting especially those that used tissue cultures. There were not many because most of the work with animal virus work was done with actual animals or embryos. I learned quite a lot about the techniques we could use and to choose which virus I wanted to do anything with; and I learned the characteristics of these viruses. When I went back to Caltech, Max asked me to write a summary of my findings and make a proposal.

My conclusion was that the reason the work in this field did not progress was that the field lacked a good assay system. I went through the statistics and showed that the assay people used was so abominable in terms of precision; it lacked completely precision; unless there were enormous differences in the parameters, it was impossible to see the effects. I suggested that we should try to develop a system similar to the one that we used for bacteriophages, with animal cells. I thought that this could be done using cultures. This was accepted and I started working in that particular area. I first tried to use the cultures available at the time, but, fortunately, there was a development that came from a laboratory at NIH where Wilton R. Earle, a pioneer in modern tissue culture techniques, showed that you could take a piece of tissue, and you could disrupt this by mechanical means, and make a suspension. You could put the suspension on a Petri dish, the cells would stick and would make a nice, uniform layer.

I thought that that was what I needed and that I would start with that. Then I had to get a virus that was really pathogenic, something that really killed the cell. People weren't very enthusiastic, but I convinced them that it wasn't so bad after all, and they gave me a laboratory corner of a sub-basement to isolate my work on the viruses. For the virus, I chose equine encephalitis. I chose that because it was highly pathogenic for the cells and killed them very effectively. But we worked carefully. In a few months I succeeded in building up the system where everything worked. When I started making real experiments, I remember that the first day there was nothing going. On the following day, there was again nothing, but when I held the plate against the light, under certain light conditions, there was a beautiful plaque. I went to call Max and said, "Come and see it." He realized what was going on and came and I showed him the culture and he said, "What date is it?" and I didn't even know what date it was. Anyway, that was a great beginning.

Work on poliomyelitis

This opened every sort of things. Once we had this method, we could really go and study every parameter of viral duplication, the action of antibodies, and so on, and this went on for several years. We did lots and lots of good work, and extended this to poliomyelitis because at the time that was the malady de siécle and there was support by the National Foundation for Infantile Paralysis, as it was called at the time. It had funds ready to support promising research and they actually asked me to see whether the system would be applicable to poliovirus. We very rapidly succeeded in working on polio; we isolated mutants of the polio-myelitis virus. At this time Albert Sabin was working on his vaccine, and this gave him means by which he could purify his viruses, to obtain pure strains by making these plaques because we had shown that there were beautiful plaques. He could just pick a plaque and there was no problem whatsoever; in a simple way he got what he wanted. He could then analyze any number of different strains for their properties. Those were the mutants, which, again, helped him characterize these mutants. This brought us into a fruitful collaboration with him for a little while. He was in Cincinnati and I was here, but there was contact by corres-pondence. That was at the concluding steps of his efforts and he came out with his vaccine very soon afterwards. That was the critical factor for him; he had these variants that he accumulated but they were obtained by mass transfer populations of viruses. He tried to statistically separate things by means he had, by dilution. But he could never be sure because for viruses especially like polio the efficiency of infectivity versus the number of particles was so low because most particles don't participate in infection. Under these conditions he really needed something like that. Without that, he probably would've not been able to do that with polio. He realized that our technique was a useful approach and he used it. So this is the work we were doing in the field of viruses.

Tumor viruses, Howard Temin and Harry Rubin

Then there was another shift, a major shift, when I moved from these viruses which kill cells to another class of viruses which induce tumors in animals. This started because at the time I had a large lab and lots of people in my lab, and there was a veterinarian in the lab, his name was Harry Rubin. He was interested in some chicken viruses, which caused leukemia in chicken. He came to my lab to work; he was following up the work that he had done before. We talked a lot and had a lot of exchange with him, and I got increasingly interested in this type of viruses. Also at the same time, a grad student came, Howard Temin, and he also became interested in these viruses. Harry and Howard joined efforts and they developed a method for assaying, which was called the focus method by which the viruses — instead of making a plaque — form a little proliferating focus in the culture. A group of cells becomes different in characteristics and start growing. It's a little tumor really in the culture, and they did this.

Howard Temin was very intelligent and he had the ability to see beyond the experiments, which lots of people don't have. I remember when he came to discuss his thesis and I, of course, was his major advisor, and Max was there too on the committee. Howard explained what he had done and he came to the conclusion that probably this virus established some kind of permanent relationship with the cells, with the genes of the cells. He was thinking of a phenomenon of a lysogeny, which was known at the time with phage. The gene in phage establishes a permanent relationship with the DNA of bacteria, the genome of bacteria. So he suggested that that would be the case with the virus as well. I remember that by that time Max's eyes looked very hard because he didn't like that; he felt that Howard did not have any evidence; that that was speculation; Max was very much solidly attached to the data and didn't want people to go beyond the data. I defended Howard; his intuition was good and interesting and he should not be discouraged from speculating this way.

Research philosophies

The discussion did reveal a difference in philosophy. Max's philosophy was formed 20 years before; he was a physicist and his approach was formal; according to him you can make a theory, make a mathematical model, make a prediction and then you make an experiment and if your prediction follows the curve from the experiment then there is reason to believe that your prediction was OK. With this approach, there is no other way to do an experiment whereas in the time that we are now talking about [late 1950s], there was already intuition that there are other ways to do experiments. First of all, there are molecular approaches that would be coming soon for which you didn't have to make mathematical curves, but you could look at the molecules and you could look at the phenomena, maybe in not so quantitative ways, yet they would give you just as good results. It was not even called yet molecular biology. I remember that just at that time Jim Watson came to Max's lab and in talking with him, he and I were discussing this very point that the molecular era was coming and that we really should start thinking in terms of molecular biology. But when we discussed this with Max, Max did not like that because he thought it was premature; he thought we did not yet have the necessary things in our hands. There were two philosophies. Max was very solid, attached with the data whereas the new generation was looking forward, not backward.

Research on tumor viruses continues, work on polyoma

We remained interested in these tumor viruses even after Harry and Howard had left. But I did not want to stay with the old system as I always had the inclination to take on a new system, something new that I can do from the beginning and do everything; that's what I like to do. So in around 1958, I was looking for another tumor virus more suitable for what I wanted to do. Right at that time someone at the NIH–NCI discovered a virus, which could produce tumors in mice. The virus was called polyoma because it caused many different kinds of tumor. I thought maybe I should look at this virus; I asked them and they sent me the virus. The question was whether I would be able to assay the virus because only then could I do quantitative studies on it. I succeeded in preparing suitable plaques using mouse kidney cultures. Although it took a longer time than for the other viruses before, it worked and it was a very reproducible assay, suitable for quantitative work. By that time Marguerite Vogt had joined me. She was a former associate of Max Delbrück, who was more interested in the virus work than in the genetics that Max was interested in at that time. She was very good in tissue culture; she learned it in my lab and became the best person to do it.

We thought that we should see whether we could obtain changes with this virus, which would be similar to the changes that Harry and Howard had obtained with their virus. We shifted to another animal, to the hamster because we knew that polyoma could induce cancer in hamster but would not kill the cells. That seemed to be the right system. So we made the cultures for the hamster embryos, infected them with the polyoma virus and wanted to see what happens. For a long time nothing happened and we kept religiously transferring the cultures week after week. Then one day, Marguerite came to me and said that "This culture was very yellow," meaning the color of the indicator of the medium, indicating high acid production, and tumor cells usually tend to do this. Thus it was an indication that something was going on and there was a lot of replication of the cells much more than in the control cultures. We decided that this was a transformation; we had the tumor cells in vitro. For the tests we took newborn hamsters to see whether they would develop a tumor and they did. Thus we had viruses with tumor producing ability and tumor transforming ability in addition and we could quantitatively analyze it with the plaque assay. That started a long series of work, which lasted for a decade.

The next question I was interested in was which kind of nucleic acid was there in the polyoma virus, whether it was RNA or DNA. The reason is the following. The virus that Temin and Harry worked with was an RNA virus and it was difficult to see for Temin's idea according to which the virus would establish some permanent association; the RNA and DNA did not make associations. We did not know of any association between RNA and DNA, the two making joint molecules. We thought that we would be more interested in having a DNA virus, which would be more likely to form an association with the hamster DNA and it would also be easier to prove it. At that time I had an Englishman in my lab, John Smith, who was a very good biochemist. I asked him, "Why don't you look at the nucleic acid of the virus, you could purify it and you could make a determination." It turned out to be a DNA virus.

I then decided that this was the virus I wanted to work with. The first work we did was to look at the properties of this DNA. I collaborated briefly with Jerry Vinograd, a physical chemist, who was in the Chemistry Department. He was interested in nucleic acids in general. One thing that he noticed is that if you centrifuge the DNA, extracted from the virus, instead of giving just one batch, as you would imagine, it gave two batches. It was not clear why. It was as if there was a double molecule, two molecules. At about the same time, Jim Watson had done some work with the papyloma virus, which is related to the polyoma virus. He found the same thing, that is, that centrifugation produced two batches. That seemed very strange. I couldn't understand how there could be two molecules and I supposed that there was a different explanation for it.

What I did was I looked for another method to determine the size of the DNA molecule. That was the time when Morton Mandel developed a column method using a different principle to determine the size of DNA. We set up the system and looked at the two forms and they did not look different, they looked the same. Thus according to one criterion the molecules looked different, according to another they looked the same. Obviously, there was some structural reason, not length. I noticed very soon that if I used a small amount of DNAase, the one that seemed to be just one molecule before, generated the one that seemed to be two molecules. That seemed impossible, you can't make two out of one. I did a double experiment and my proposal was that this was actually a ring, a circular molecule. I thought that DNAase would cut through the molecule and would make it linear and the linear molecule would go slower and the circular molecule — since it is more compact — would go faster. The slower molecule would appear to be twice as big. The size of these molecules was about five million. At the time the ultracentrifuge could be run at a hundred thousand gs, not only the analytical but also the preparative ultracentrifuges. So I made this proposal for the circular molecule, but, unfortunately, I had bad luck because I couldn't go one step further. I should have looked at these two molecules under the electron microscope; they were two pure molecules separated from each other. I talked with the people at Caltech who had an electron microscope at the time asking them to help me, but for some reason I could never convince them to do it. However, Vinograd succeeded in doing it with electron microscopy; he could show that the two forms are both circular; one was circular and super-coiled and the other was relaxed. He made a very beautiful piece of work. But by then I had left the field of nucleic acids, because I moved to La Jolla in 1963.

The Salk Institute

The reason I came here was interesting. I had many conversations with people like Jacques Monod and Leo Szilard. Jonas Salk had been talking to them about his idea of making this institute; he always had wonderful ideas because he is unconventional and he would go after ideas nobody would go after, and he has the energy to follow them up. So we talked about this institute, we thought that there would be a small group, we would choose each other, so it would be a congenial group, and the idea seemed very attractive and very interesting, a new adventure, and I was always adventurous.

The first group was Szilard, myself; for a while Seymour Benzer seemed that he might join us, but he did not get along too well with Jonas, so he did not come. We had Mel Cohn and Lennox. Jacob Bronowski was an addition who came slightly after our group had come together. The whole idea behind this institute was that it would not be solely hard biology, but it would be broader; there would be a philosopher, someone interested in the philosophy of science, and that's why we got Bronowski. He did very well and he was especially very good in explaining things. He did a marvelous job in his book *The Ascent of Man*. It's a masterpiece in this respect.

The first year we did not have any lab here, so I went to Glasgow for a sabbatical. I spent a year there during which I continued to be interested in the polyoma virus. I had a graduate student of mine there, Mike Freed, who came there with me. We tried to see what we could do; we followed several avenues. One thing that I tried to find out, why certain cells, which are affected by the polyoma virus, are stimulated to grow. We wanted to see whether enzymes were involved. I chose an enzyme to investigate and there were competent biochemists whom we could ask for advice. I started making some assays and we had some evidence that something did happen. When I came here, I started serious work in that direction and I had a very good postdoctoral fellow, his name was Lee Hartwell. He became a very well known yeast geneticist at the University of Washington. Together we started chasing some of these enzymes.

Viral and cellular DNAs

I also started looking at the DNAs made in the cells; I knew that lots of DNAs are made in the cells, and I wanted to find out whether these were viral DNA or cellular DNA. We expected the viral DNA to multiply and thought that there might be lots of viral DNA made in the cell. We set up a system to distinguish the sizes of the DNA molecules. The viral DNA had a defined size, whereas the cellular DNAs broke and had a broad distribution in size. Thus we could distinguish one DNA from the other without applying any specific marker, because uniformity characterized one and heterogeneity the other. Soon I made experiments in that direction and it became clear that the DNAs made in the cell were not viral. The viral DNA did not replicate at all. So the virus caused replication of the cellular DNA. We also looked at the enzyme involvement in DNA synthesis and the virus was stimulating the formation of these enzymes. We worked on these problems the two of us, I and Lee Hartwell.

It became clear that we had a new phenomenon in front of us, namely, this virus, which produced these important changes in the economy of the cell. The next question was, how does the virus do this? What happens to the viral DNA when the viral DNA enters the cell? People had tried to answer questions like this by looking at whether you can extract infectious DNA, which would make a plaque? However, you could not extract infectious DNA; it seemed to disappear. However, my experience with bacteriophage made me weary of such interpretation because I knew one case in bacteriophages when something similar happened, but the phage DNA was not lost because it could be gotten back later.

There was a new kind of biochemistry at the time, concerned with DNA sequences. That was the time when Paul Dotty at Harvard University had shown that you can melt the DNA double helix, separate the two strands at high temperature, and then you could anneal it, lower the temperature and the two strands would get together and re-form the double helix. If you took two DNAs that had sequences in common, the sequences could anneal each other. Accordingly, you could ask the question, whether the viral DNA could persist in the cell? So we purified the viral DNA, we made it radioactive, we knew how to do that, and then we made the annealing experiment with DNA extracted from the cells. We found that the viral DNA was there because the annealing experiment was positive. Then we could show that it was there all the time, no matter how long time has elapsed, the viral DNA was there. There was thus a permanent association that Temin had predicted long before.

The next was to see how the viral DNA was incorporated in the cell to become the same molecule, how the genes of the virus were incorporated into the genes of the cell. We had to work hard for that. The idea was that we should take the DNA of the cell and keep it as big as possible. Then we centrifuged it, hoping to separate the cellular DNA, which would be heavy and would go down to the bottom of the tube, from the viral DNA, which was smaller and would stay in the middle of the tube. There were complications of many kinds; the main issue was how to get the DNA of the cell intact. The extraction as was done at the time, we knew, would break the cellular DNA to pieces and it would be difficult to distinguish them from the viral DNA and arrive at conclusive results especially because there was so much more cellular DNA than viral DNA. Therefore we decided to use a different approach. I got a good idea talking to John Lett, who was a radiation biologist. He was interested in measuring breaks in cellular DNA. He made a sucrose gradient with alkaline, at pH 12; then he would put the cells directly at the top of the gradient. The cells would disintegrate in the high pH and he would centrifuge it down, and the cellular DNA remained intact, denatured but intact. It collected at the bottom of the tube and it was quite uniform. There were not many breaks under these conditions because there were no mechanical shears. We adopted this system and did some experiments, but still we had problems. The separation was not very good for another reason. The polyoma DNA in completely intact form, when it is put in an alkaline gradient, denatures, the strands separate and they form a ball and the ball goes down very fast. So I again had to talk to other people. I learned the usefulness of talking to other people very early in my career. I was talking to Bob Sinsheimer who was at Caltech at the time, now he is at Santa Cruz. I explained to him my problem and he suggested putting more EDTA [ethylenediaminetetraacetic acid], a chelating agent with heavy metal to bridge. I went back to my experiment and we added lots of EDTA, and everything worked beautifully. The separation was accomplished, it was perfect, and we had the cellular DNA at the bottom. Then we took a fraction of it and tested it by hybridization with a probe of the polyoma DNA and we gained conclusive evidence that the two DNAs were integrated.

It was always my forte that I always looked for the latest technological advance. That's what really determines everything. You have a better technology and you get better results than by using worse technology. I was always very keen on that and it always paid off.

We then continued our work trying to achieve the ultimate result, which was to show that the viral DNA could be recovered and could give rise to a virus again. We did succeed in doing that. That came from a different experiment for which I had the opportunity by having in my lab someone who came from Harris's laboratory in Oxford. They had shown that you could fuse cells. They used a fusing agent, a virus, to produce hybrid cells.

We thought to take hamster cells in which the virus could not replicate because there was something missing that was needed for the replication of the virus. We also used a mouse cell, in which we knew the virus could replicate, and we fused them together to see whether this hybrid would produce the virus. We did this experiment and it worked. We now had the evidence not only that the DNA went in and became integrated, but that it was still there and could be taken out and give rise again to an infectious virus. This completed the cycle of our experiments. It was a tremendously exciting time. I remember that the conclusive experiment to show integration was done in my absence. I had to go to Cold Spring Harbor for a meeting and Joe Sambrook, who was the main person to run this experiment, was left there to finish the last experiment. He was supposed to send me a cable to say what the result was because I was going to talk about this result. The meeting was going on for a few days and I still did not hear from him and I thought that maybe it didn't work. On the day I was supposed to give my talk, just after I started talking, a telegram arrived saying that it worked. We had lots of fun. This was essentially the end of the work we did in this field.

Human cancer research

In 1972, I went to London. It seemed to me that I'd done so many interesting and good things in this field with the polyoma virus that if I would go beyond I would have to apply new technologies, but the problems were not as well defined as before. It would be a tremendous effort to see clearly where we should go. As I had worked for so many years in cancer research, but avoiding cancer, I thought maybe I should take a look at some real cancer. I went to London to the Imperial Cancer Research Fund in the center of London in Lincoln's Inn Field. One of the reasons I went there was that they supported a unit at Guy's Hospital, which was dedicated solely to breast cancer. It was a little emotional because Seymour Benzer's wife developed breast cancer and everything was so difficult — the diagnosis, prognosis, treatment, everything was so nebulous. I felt that this was really a field that needed attention.

I did not give up entirely work on the viruses and did a few other interesting contributions in London. It was in the direction of how the genes were expressed and which parts of the genes were involved with the viral DNA. We were also interested in which kind of proteins was made by the virus. We knew that there are new proteins because they could be attacked by using antibodies. We isolated the proteins and showed that there were two main proteins rather than one as was thought before. There were proteins called the large, the middle, and the little one, and we made a proposal that the main protein was the middle one. First of all, we could trace it to the cell membrane and I always felt that the cell membrane was important in these events. Second, there were mutants that did not transform and did not make this middle protein. So there was a strong correlation. Later on our proposal was confirmed by other means.

Personally, I was more involved in studying breast cancer. At the beginning I had to learn a lot, which I did; it was a new field for me. Then, after five years, I came back here. When I came back to Salk, I continued in that direction and took also a new direction that came out of a meeting. Armand Hammer, the great financier, is interested in cancer, has been for a long time. Some years ago he decided to support some workshops here at Salk on specific subjects. I've been involved in them and in the beginning we concentrated on monoclonal antibodies. The antibodies are made by the cells and each lymphocyte makes one antibody, but there are a hell lot of different antibodies because there are so many different lymphocytes making them. The idea of Köhler and Milstein was to isolate one lymphocyte, fuse it with a cell of a myeloma tumor cell and the resulting hybrid cell becomes immortal. This cell will continue to produce the antibody made by the original lymphocyte. By itself, the lymphocyte would never yield a population of the antibody, but this way large amounts of the specific antibody can be produced. This is the monoclonal antibody. I adapted this technology to study breast cancer in a more experimental way in rats. There was a chemical induction and we wanted to define the cells in which the tumor developed. We now understand the rat system, both the normal development of the cells and the development of the tumor cells, but it was a lot of work.

Now I am working mostly with human breast cancer, using the same approach. We make monoclonal antibodies and we have interesting results but still lots of work have to be done. They are interesting in two ways.

We have reagents, which can recognize breast cancer; they are important for diagnosis and they are widely used by pathologists. The other thing is that there are molecules, there is at least one molecule that is present in the early development of breast cancer and then it disappears. This molecule connects cancer to an early stage of development. It explains why it makes a difference whether a woman has a pregnancy at a young age or a late age for whether she would have breast cancer or not. These are events that are divided by 30 years. We try to understand what it means. Another thing that opened up in this work is the following. In the breast there are lots of benign breast diseases by various names, like nodes, most of them with no consequence. We have noticed that some of these have markers of malignancy. We don't know yet, but maybe we have a tool by which we might be able to tell from analyzing these nodes that the woman who does not have breast cancer is going to have breast cancer. Such studies take time; a whole lot of people and following it up is not so easy. It has to be done by pathologists in hospitals and we don't have hospitals here. But there are other people who are also interested in these problems and we hope that we will find out. That's all I have to tell you.

On the Human Genome Project

My opinion comes from my perspective from cancer research. In cancer research, there have been tremendous achievements in recent years. One of them is the discovery of oncogenes, which are genes in normal cells, but something goes wrong and they become cancer genes. When oncogenes were discovered, it was thought that they would answer the problem of cancer because they would explain the mechanism. But now that we have followed these things for a number of years, we are increasingly convinced that this is not the case. They give us an important clue, but there are lots of mysterious things happening in addition to that. The thing that seems to be most mysterious is the sequence of events that happen between the beginning of cancer and the final expression of the malignant cancer. The evidence seems to tell us that it is probably oncogenes that are not what is involved; there are other genes involved. My question is how do we get to these other genes. There are many different approaches, but what we hit against is the ignorance of the human genome all the time.

In a way, everybody does the experiments in the following way. We start from a phenomenon, an effect, for example, there is an enormous cancer cell and we find that the cancer cell has a protein that the normal cell does not have. Then we can go back to the gene that makes this protein. It's lots of work but usually people succeed in doing it. Then we have to find out what this gene is doing, how it is involved. So usually we proceed from the peripheral phenomenon to the gene.

However, in cancer research, this approach, in effect, is not possible for the following reason. First of all, we would like to study real cancer rather than what happens in the tissue culture. The real cancer cannot be translated into tissue culture because the real cancer is enormously heterogeneous. It is made up of cells of such a different array of properties. If you make a culture, which may come from one such cell, OK, you have one cell, but what about the other 50 different types? Real cancer is a mixture of a hundred different things and you can't get a hundred different answers because that is no answer. That's why I have developed the idea that the work should proceed in the opposite direction. You should start from the genes and once you know all the genes you can ask all the questions whether gene A or gene B or gene C is present in this particular cell? This you can do microscopically. You can hybridize them and use the same principle that we have described before, except that you can do it on the cell. Then you can see whether the radioactivity is present in that cell or not. In this way you could trace any gene that you could think of. In other words, you could express any gene that you had.

In effect, the problem of cancer is intimately related to the problem of development, which is the same thing. An organism is heterogeneous; it is not made up of one cell type. If you want to understand about genes important in development, this approach to start from the phenomenon and go back to the genes is very difficult. It would be much better to start from the gene and go back to the effect. I have made a proposal and my proposal is made without consideration for time or cost. We should first know all the genes, the human genome, and I specified that it should be the human genome and not the mouse genome because human cancer is very different from the mouse cancer. We are interested in humans; we are not interested in mice; we are interested in mice only if they can help us to understand man. So why don't we go directly to the human genome? It would be just as difficult to do the mouse genome.

That was my proposal that we should try to sequence the whole human genome and once we have all the genes that exist in the human genome, we have all the probes, we will have the understanding of how the genes are organized, we might be able to see genes which are related and probably controlled in the same way, that is, expressed in the same cell. Using these guesses, trying to fish out genes, we could then go by the cytological method and see whether they are actually expressed in the cell. Then we can make a kind of catalog telling us which genes are expressed in which cell. If we have that, we can really go to cancer.

In breast cancer, for example, we would know that cell A has these genes according to the catalog and cell B has these other genes according to the catalog. Looking at the breast cancer we could see which genes are expressed. If two cells side by side express different genes, it doesn't bother me because we will know that different cells may express different genes. In addition, once we have the genes, I can do the real crucial experiment: I can take cells from a human cancer which will grow in mice and I can see whether I can use a reagent, which will neutralize a certain gene. If that gene is important for this final stage of the cancer, I should be able to make the cancer cell into a normal cell. There is a so-called negative strand RNA by which you synthesize an anti-gene, so to say, and this anti-gene RNA goes to the cell and prevents the actual gene. This has been done in a number of cases and has been shown to work.

You can still ask the question whether it is actually the gene that transforms the normal cell into a malignant cell? So you can do the other experiment, you can put the gene into a normal cell and see whether this will make a normal cell malignant? And see how many genes it takes to achieve this result? You can do everything. Having this opportunity, the question of cure and therapy becomes a question of many different possibilities. Once you know which gene it is, you can find out what it does, you can find out what proteins it makes, and test for their functions, which give you the antibody. You can trace those proteins; find out what they do — a whole new world opens up in cancer research.

I have talked about this with many people and everybody would like to have this information. But everybody is skeptical about whether we should do it because it costs too much and who is going to pay for it? It should be done on an international basis and it should also be done in an industrial way. In effect, it is a boring thing to do that. The interesting thing is not to do it; the interesting thing is to have the sequences, and then to use them. However, lots of people do lots of things because it's their job, not because they like to do it. The technologies are improving, and I have seen the technologies improving throughout my life, so I have absolute faith that the technologies can be developed by the people whose business it is to develop the technologies. They will come up with the answers because it's their money, they'll make money on that. So the two things go hand in hand. If there were a commitment and somebody came up with the money to do it, I believe that the human genome could be done in less than 10 years. I am absolutely convinced that the time has come.

* * * * * * * * * *

Paul Berg

A Reminiscence of my Association with Renato Dulbecco

Paul Berg (b. 1926) received the Nobel Prize in Chemistry in 1980 "for his fundamental studies of the biochemistry of nucleic acids, with particular regard to recombinant-DNA". *Candid Science II: Conversations with Famous Biomedical Scientists* contained a conversation with Dr. Berg (pp. 154–181) in which he mentioned (p. 164) the influence of the year (1967–1968) he spent with Renato Dulbecco on his research career. This prompted us to ask Dr. Berg in July 2004 to expand on Dr. Dulbecco's impact on his work.

Not having been one of the *cognoscenti* of bacteriophage or animal virus biology, I was unaware of Renato Dulbecco's accomplishments in those fields. My own interests for years had been in exploring the enzymatic mechanisms of transcription, the assembly of amino acids into proteins and more particularly the enzymatic and structural specificity of aminoacyl-tRNA formation. However, in 1965 or thereabouts, I made a decision to change my research field. I was intrigued by the question of whether the genetic regulatory processes discovered in bacteria during the 1950s and early 1960s would also explain regulation in mammalian cells. It seemed to me that studying that question with mammalian cell cultures might provide some answers.

Because of the relatively small size of the viral genomes, studying the expression of their genes following infection seemed to provide a reasonable approach. About that time, Renato came to speak at Stanford about his work with the recently-discovered polyoma virus, notable for its quite small genome; about 5.3 kbp. Besides being able to propagate in cultured mouse cells, polyoma virus also produces tumors when it infects mice. Most importantly, Renato had shown that polyoma could recapitulate its oncogenic property by transforming cultured hamster, rat and human cells into tumor-like cells. To me, learning the molecular details of polyoma's replication and its tumorigenic capabilities seemed like a worthwhile endeavor. But as I had no hands-on experience in that field, I contacted Renato about the possibility of spending a sabbatical year in his laboratory at the Salk Institute, a request he welcomed. Perhaps he was surprised, even chagrined, when I arrived in September of 1967 with my long-time research assistant, Marianne Dieckmann and a post-doctoral fellow. Francois Cuzin, who had earlier arranged to join my lab after completing his thesis at the Pasteur Institute in Paris, also decided that work on DNA tumor viruses would be interesting.

That year was filled with the excitement that inevitably comes from learning about a new biological system. Under the close attention and supervision of Renato's long-time associate, Margarete Volk we learned the technical skills of culturing mammalian cells and of following the course of viral infections pretty quickly. She imbued us with the cardinal principle of cell culture work, namely that the care and feeding of the cells had priority whether day or night. Getting used to the longer interval between conceiving an experiment and getting a result took some doing but soon we learned to have several different experiments going simultaneously. The work went well and by the end of the year we published some interesting results about the nature of the integrated viral genome in a mouse transformed cell line. But the real excitement emerged from the continuous discussions about the data that we and other visitors in the lab were generating. Renato's lab office was alongside the Stanford contingent's benches and conversations flowed easily back and forth between the two places. His vast experience and insights were frequently manifested by his tantalizing "what if" questions. During our stay at the Salk Institute, I developed a warm relationship with Michael Stoker, a distinguished animal virologist from England, who was soon to become the Director of the Imperial Cancer Research Fund in London's Lincoln's Inn Field. Through that association, I spent several months during several summers of the early 1970s working at the ICRF. André Lwoff also spent half a year in Renato's lab and I found him to be more affable than his reputation had led me to expect.

At the end of a year we returned to Stanford and I chose to continue that line of research except with SV40, a human analogue of polyoma. Within another 2 years that work became the dominant activity in my lab. The most important outcome of my year in Renato's lab was that it set me on the path to construct recombinant-DNAs. Based on discoveries and conversations that were ongoing in Renato's lab and being well aware of the contribution that transducing viruses had had to molecular biology of prokaryotes, I began to think about how the genetic constitution of cultured mammalian cells could be altered. One thought was to explore the possibility that SV40 could be used to transport foreign genes into mammalian cells. Considering that the amount of DNA that could be carried by SV40 virions is limited and that cells could be transformed with SV40 DNA itself, I decided to use SV40 DNA as the vector for inserting new DNA into mammalian chromosomes. David Jackson, Robert Symons and I succeeded in constructing the first recombinant DNA molecule; it consisted of the entire genome of SV40 and a segment of bacterial DNA containing the three genes responsible for metabolizing galactose. It was created by synthesizing complementary cohesive ends on the respective linear DNAs, annealing the two into a circular molecule and then sealing the gaps with appropriate enzymes. That accomplishment opened a new era of fundamental and applied molecular biological research that has revolutionized the life sciences. It was for that and earlier work that I shared the 1980 Nobel Prize in Chemistry.

Baruch S. Blumberg, 2002 (photograph by I. Hargittai).

32

BARUCH S. BLUMBERG

B aruch S. Blumberg, M.D., Ph.D. (b. 1925 in New York City) was
until recently the director of the NASA Astrobiology Institute at
the NASA Ames Research Center in Moffett Field, California. Among
his many other appointments, he is Fox Chase Distinguished Scientist
at the Fox Chase Cancer Center in Philadelphia and University Professor
of Medicine and Anthropology at the University of Philadelphia. He shared
the 1976 Nobel Prize in physiology or medicine, awarded for "discoveries
concerning new mechanisms for the origin and dissemination of infectious
diseases". His fundamental contribution was the discovery of the hepatitis
B virus and the invention of the vaccine that protects against it. He
received his B.S. degree from Union College in Schenectady, New York,
in 1946; his M.D. from Columbia University in 1951; and his Ph.D.
in biochemistry from Oxford University in 1957. He is a member of
the National Academy of Sciences of the U.S.A., the American Academy
of Arts and Sciences, the American Philosophical Society, and many other
learned societies, and he has received numerous awards and distinctions.
We recorded this conversation at the Fox Chase Cancer Center in
Philadelphia on March 19, 2002.*

*Your most famous discovery is the hepatitis B virus and the creation
of the vaccine against it. I'm sure you have been asked about it on
many occasions.*

---

*In part, this has appeared in *Chemical Heritage* 2003, *21*(2), 4–7.

I have written a book about it, *Hepatitis B: The Hunt for a Killer Virus*, published by Princeton University Press in 2002.

We are recording our conversation in a building that is dedicated to the prevention of cancer. How can cancer be prevented?

The major program for the prevention of cancer has been the program for the cessation of smoking. That has had a profound effect on the incidence of prevalence of cancer, primarily in the United States. My work is connected with the prevention of cancer through the hepatitis B vaccination program. Hepatitis B accounts for about 85% of the primary cancer of the liver in the world. Primary cancer of the liver is one of the most common cancers; it's the third most common cause of death from cancer in males and the seventh most common cause of death from cancer in females. It's very difficult to treat. The life expectancy after clinical diagnosis is much less than a year, and the survival rate for 5 years — a frequent method of measuring the severity of a cancer — is 8–10%, which is extremely low. Another major cause of primary cancer of the liver is hepatitis C, either by itself or in combination with hepatitis B.

We invented a vaccine in 1969 to prevent infection with hepatitis B. That vaccine became available for general distribution in the 1980s. It is now one of the most commonly used vaccines in the world. More than a billion doses have been used; hundreds of millions of people have been vaccinated. This led to a striking decrease in the prevalence of hepatitis B infections in the world. Prevention of hepatitis B is the second most common intervention program for prevention of cancer. Cessation of cigarette smoking is the first.

At the very beginning of your career you worked on hyaluronan, as hyaluronic acid is called nowadays. This polysaccharide has since then become widely used in medicine and cosmetics. I would like to ask you about your work with this substance.

I appreciate your bringing this up, because this work was important in my early research. I remember some 40 years ago meeting with Endre Balazs, who has done a lot for expanding the field. My contribution concerned the physical biochemistry of hyaluronic acid; I extracted it from the joint fluids of cattle. At Columbia I worked with Karl Meyer, who was an excellent mentor. I worked with him from 1953 till 1955. Karl came from

Karl Meyer (1899–1990). In 1934 Karl Meyer and John Palmer described a new polysaccharide, called hyaluronic acid, isolated from bovine vitreous humor. They coined the name hyaluronic acid from hyaloid (vitreous) and uronic acid. In 1986 the name was changed to hyaluronan to be consistent with the international nomenclature for polysaccharides (courtesy of Vincent C. Hascall, Cleveland, Ohio).

that school of German biochemists of whom most left Germany and came to Britain or the United States — they were Jewish German or Jewish Austrian biochemists, who were most of my teachers in medical school. They were forced out of Germany and the rest of Europe before World War II. He was a consummate chemist, but, added to that, he had clinical training. Although he did not practice clinical medicine, he had a general feeling for human biology. He was also a very knowledgeable biblical scholar and knew a great deal about the Bible and commentaries and the literature and scholarship on the Hebrew Bible.

You were then 28 to 30 years old. Where were you in your studies at that time?

I started in physics as an undergraduate at Union College in Schenectady, New York. It was a small men's college at that time. The physics course was very good. General Electric had its research lab in Schenectady and General Electric supported the college — Irving Langmuir came to speak to us, and so forth. After I was discharged from the Navy, I continued my graduate work in physics and mathematics at Columbia University in 1947, for one year. Columbia was very strong in physics; many people who taught there had worked on the Manhattan Project.

Why did you leave it?

I didn't think I was very good at it.

Could it be that you might have been intimidated by the stellar collection of the physics faculty that gathered there after the war around I. I. Rabi?

I wasn't intimidated by the faculty, nor was I intimidated by my fellow students. But I was very impressed with how skilled they were. It was not only at Columbia — many of the undergraduates at Union College went on to a wonderful career in physics. So there was a negative aspect, but there was also a positive one that I felt that I wanted to deal with people more. I thought that medicine would be a good combination of science and people. So I switched to medical school at Columbia.

When I finished medical school, I did four years of hospital training, starting in 1951. The first two years I did in Bellevue Hospital on the Lower East Side of New York City. That was a very intense clinical experience, a rich time of my life. It was a city hospital; there was no charge — it was paid for by taxes. The hospital was very crowded at times; if somebody was sick and needed a hospital bed, we were required to take them in even if there was no space. We literally moved cots out into the hallway, and even used double-decker beds. In that hospital you developed the feeling that what you did really made a difference.

After two years at Bellevue I went to Columbia Presbyterian Hospital (where I had gone to medical school), to the College of Physicians and Surgeons. I was doing clinical work there, but I was also allowed to do research. It was different from Bellevue, where we had intense clinical care. It was recommended, because of my experience in physics and mathematics, that I work on physical biochemistry. This is how I got to work with Karl Meyer, and we worked on the physical chemical characteristics of the long-chain molecules of hyaluronic acid. We also worked on sulfonated long-chain sugars. I did a lot of light-scattering studies, centrifugation studies, using the Svedberg's centrifuge. Then, when I went to work with my mentor, Alexander Ogston, in Oxford in 1957, there was a project that he was interested in: the role of the protein in the physical characteristics of hyaluronic acid. The question is whether the protein is essential for the functional structural characteristics of the molecule. We approached this problem by using papain, an enzyme derived from papaya, to remove the protein and see whether it had an effect on the birefringence, light-scattering, and ultra-centrifuge characteristics. The conclusion of that study was that it had a profound effect on the non-Newtonian (non-shear-dependent) viscosity and the other physical characteristics of hyaluronic acid. The non-shear-dependent viscosity was a very desirable characteristic in joint fluid that allows for a certain amount of elasticity.

Did you think about its clinical significance?

Of course. Since I was trained in medicine, I always think about clinical significance. I worked in an arthritis clinic, and that was the reason I was working with hyaluronic acid. Arthritis was the subspecialty of internal medicine that I had selected at that time. Ogston was one of the major figures in hyaluronic acid, and that was the reason I went to Oxford.

Hyaluronic acid and other polysaccharides are not considered to be as important as proteins and nucleic acids.

That may be the general view, but it's not mine. I came back to work in that field again during the past 10 years. I am associated with the Oxford Glycobiology Institute. Professor Raymond A. Dwek gave me a visiting appointment — which I still have — in his department. Hepatitis B virus has the 3-glycosilation site. When I came to Oxford, a colleague of mine from Philadelphia, Tim Block, wanted to come to Oxford to work on the hepatitis virus, and I recommended that he work on the glycosilation of the hepatitis B virus. It turns out that glycosilation is extremely important for the tertiary structure of the surface protein of the virus. If the glycosilation process is interfered with — if it is altered by the use of drugs — then the virus does not assemble, and it remains in the cell and does not get exported. There are other effects as well that are currently being studied. All this forms the basis of possible therapy. I play only a secondary role in it, but it was initiated when I was in Oxford. Vaccination is very effective for *prevention*, but treatment still is not very effective. So in a strange way I came back to the very field that I worked on for my thesis.

Today it is not uncommon that a young molecular biologist goes to the frontier of research right away — without acquiring what we would call a classical training — and makes a discovery, only to disappear into a biotechnology company or into oblivion. Your path was different. You

Baruch Blumberg in Stockholm, 2001, during the Nobel Prize Centennial celebrations (photograph by I. Hargittai).

had spent many years on your training and have stayed in the forefront of science ever since. Did you chart your future consciously?

I've always believed that it's important to have a program, although I didn't always know what the outcome would be. It's like research altogether: if you know where you are going, that's the end of the discovery research phase. I had very good mentors: Karl Meyer, Sandy Ogston, Charles Ragan (whom I worked with in the clinical setting). The notion was always, "Do what you're doing, do it well, enjoy it to the extent that you can enjoy working 15 hours a day taking care of very sick people, and see what happens next." Don't forget that I grew up at a time when science flourished. There was ample funding, and there weren't enough scientists — particularly clinical scientists — to fill all the positions that were available. It was the postwar period, a period of economic growth — happy, except for the cold war. It was a period of great expansion in the American society and in American science. I came into science at a very fortunate time.

What is your day-to-day concern?

Family. But my work is very important for me. What I am doing now in astrobiology has turned out to be very stimulating.

In our prior conversation, you mentioned your emotional responses to challenges. When you accepted the directorship of the NASA Astrobiological Institute, was it another of your emotional responses to challenges?

Probably. It has opened a whole new world for me, and it is a new life, too.

When you take up a new direction, you seem to embark on the new challenges without abandoning the old ones.

In my stage you don't change careers — you add them. I am still employed by Fox Chase and come here periodically, and I am still a professor in the Department of Biochemistry at the University of Pennsylvania, although I spend a very small amount of time there nowadays. Right now my major commitment is to NASA.

Apart from your being a Nobel laureate, what did you bring to your job at NASA?

Strangely — what I had not realized — it was my skills at management, in managing a research organization. The Astrobiological Institute is a basic science institution. In this it is different from other parts of NASA, which is a mission-driven organization. NASA is primarily an engineering and technology operation. In order to do any science, in order to get into space, you have to build these incredible spaceships and stations. The International Space Station is probably the most complicated thing that has ever been built; it is more complicated than the pyramids, the high-way system, or the railroad network. It includes all kinds of pioneering engineering skills. It's a remarkable piece of work — amazing that it's there. However, the purpose of all this is to do science. Space is a great mystery. Every time we look at something, observe something, it is not only that nobody has ever been there before — nobody *could* have ever been there before. Everything is new. It's like when humans used the telescope for the first time. Every time they looked through the tube they discovered something new.

Do you expect to learn something about the evolution of life on Earth from your studies?

A lot of our work is focused on that. We're particularly interested in early evolution, the very start of life, and prebiotic chemistry. There is a tremendous amount of organic molecules of several hundred compounds that falls on Earth every year in the form of meteorites and other space dust.

May that be the origin of life on Earth?

That is a possibility.

Could we please get back to your origins and your family background? Your Nobel autobiography said that your grandparents came from Europe at the end of the 19th century but not much beyond that.

My mother was born in New York, and my father was born in Eastern Europe. He didn't have any recollection of that because he was only five years old when his parents brought him to America. He got all his education in New York. He went to a boys' high school, which was an elite high school in New York City. Then he attended City College, which was a typical route taken by immigrants. Then he went to law school at New York University. It was unusual at that time because people became lawyers after having worked as apprentices to other lawyers; it was not necessary to go to university to become a lawyer. His parents were fairly well off, and the practice was also that the older siblings helped the younger ones financially. His older brother was a very successful lawyer. My father's younger brother studied mathematics, and it was not common to get a Ph.D. in mathematics in the United States, so he went to Göttingen, Germany, for his Ph.D. His thesis in 1911 was about partial differential equations. It was a pioneering work in the field. For many years he was a professor of mathematics in Ohio. He had a big effect on me — his being an academic mathematician. There were other relatives who were scientists and lived in Switzerland, and I recall their visits with us. That was inspiring. My father thought that I would do well in science. I went to Far Rockaway High School, which was outstanding, as were other New York high schools at that time.

Then you went to Union College in Schenectady, New York.

It was during the war, and there was a program — an initiative taken by the military — to provide an education to people who later would become officers. The military policy was based on the notion that it would be a very long war. If the nuclear bombs had not been dropped, we would have fought for another two or three or four years.

Any message?

I find the work we are doing in space research very exciting. We try to encourage young people to become scientists, and we would like high-school students and even grade-school students to become interested in this area. It will take generations to get this project finished.

Arvid Carlsson, 2003 (photograph by I. Hargittai).

33

ARVID CARLSSON

Per Arvid Emil Carlsson (b. 1923 in Uppsala, Sweden) is an Emeritus Professor of Pharmacology of the University of Gothenburg, Sweden. He shared the Nobel Prize in Physiology or Medicine in 2000 with Paul Greengard and Eric Kandel "for their discoveries concerning signal transduction in the nervous system". In 1958, Dr. Carlsson and his colleagues identified dopamine in brain and proposed its agonist function in the control of psychomotor activity. He and his graduate students discovered the distribution of dopamine in the brain and proposed a role for dopamine in Parkinson's disease. Dr. Carlsson has made numerous other discoveries in pharmacology.

Arvid Carlsson[1] started his studies in 1941 at the University of Lund; he received his M.L. and M.D. degrees (corresponding to American M.D. and Ph.D. degrees) in 1951. He has been at the University of Gothenburg since 1959, as Emeritus since 1989. He spent half a year with Bernard B. Brodie in 1955–1956 at the National Heart Institute in Bethesda, Maryland.

Professor Carlsson has received numerous awards and distinctions. They include the Wolf Prize (Israel) in Medicine in 1979 (together with R. W. Sperry and O. Hornykiewicz); the Gairdner Foundation Award (Canada) in 1982; the Bristol–Myers Award in 1989 (together with J. Axelrod and P. Greengard); the Japan Prize in 1994; and the Antonio Feltrinelli International Award, Accademia dei Lincei, Rome in 1999. He is a member of the Royal Swedish Academy of Sciences (1975), the Academia Europaea (1989), the Institute of Medicine of the National Academy of Sciences of the U.S.A. (1996), many other learned societies, and The Legion of Honor, given by the President of the Republic of France (2001).

We recorded our conversation on September 17, 2003, at the Department of Physiology of Gothenburg University.

In your Nobel lecture, you gave considerable space for your teachers; more than people usually do.

Absolutely.

Whom would you single out as the determining factor in your career?

Even though he may not be my greatest hero, it was Bernard B. Brodie.

I read the book Apprentice to Genius[2] *in which he is the central character.*

Did you like it?

I found it very interesting. Was Brodie kept behind as far as the Nobel Prize is concerned?

His problem in that respect was that according to the Nobel Testament a discovery should be identified. Brodie was a pioneer, no doubt. He opened up the field of chemical pharmacology, measuring the blood and tissue levels of drugs as well as their metabolites. That was his major field. But opening up a field is not a discovery. Then there were other things, like the discovery that reserpine can cause depletion of serotonin. However, the reserpine/serotonin story never really became an issue in Nobel context.

Bernard B. Brodie
(courtesy of Arvid Carlsson).

There were some people who tried very hard to emphasize his role as a pioneer. Especially, there is Professor Folke Sjöqvist in clinical pharmacology at the Karolinska Institute, who spent one or two years with Brodie and was very much excited by all his contributions. Maybe that was a mistake that one did not try to identify one particular discovery from among his works. I think that was the reason. I myself proposed him several times, together with Folke Sjöqvist. We made a mistake in not trying to identify a discovery. It is clearly stated in the Testament.

Although it is not always followed.

That's true, but they are trying. In my case, they picked out one thing. When they called me at 11:15 a.m. on October 9 — I remember it exactly — and the Secretary of the Karolinska Nobel Committee told me that I was a winner of the Nobel Prize, I asked, "What is the reason?" I wasn't so sure how they would formulate it. I was curious to find that out.

If I may interrupt you, how would you have formulated it, the citation?

I have never thought of that. Of course, keeping in mind Nobel's Testament, they did the right thing. The way they formulated it, you could point to the discovery of dopamine, a very specific thing that cannot be argued even though this has been attempted, as you possibly know. There were a few people who said that it was not me who discovered dopamine.

When the Secretary called you and you asked about the citation, something must have gone through your mind.

I never thought of it and was just curious of what they said. There could have been other ways. The description could have been more general. To illustrate the importance of the discovery of dopamine and its role in brain function they could have emphasized that it provided the first convincing evidence of chemical neurotransmission in the brain and thus opened up for a paradigm shift in brain research as a whole. Of course, the general aspect was emphasized in the summarizing formulation for the 2000 Prize: "discoveries concerning signal transduction in the nervous system".

The link between dopamine and brain function was so striking even though, as I mentioned in my Nobel lecture, there was some initial resistance to it. There was one more thing in addition to the relationship between dopamine level and function; it was the subsequent work together with

Nils-Åke Hillarp (photo by Georg Thieme, courtesy of Arvid Carlsson).

Nils-Åke Hillarp, showing the occurrence and distribution of several monoamines in the brain and their neuronal pathways and synaptic mechanisms. That helped very much to convince people. This was in the early 1960s, and after that there were few who would doubt that here we were dealing with neurotransmitters. That dopamine was a neurotransmitter was most convincingly demonstrated by showing that it had the properties of a neurotransmitter and that it was important for both motor and mental functions. That has had a very great impact. It was dopamine all right, but in perspectives, it was the chemical transmission in the brain that was for the first time convincingly demonstrated.

That was the paradigm shift.

Yes. Actually, I have argued about it with Eric Kandel. He said that it had already been clarified that in the brain it's not a matter of electrical transmission, but chemical. The reason for his position was that he had read a couple of papers by a famous neurophysiologist named John Eccles. Originally, Eccles had been strongly in favor of electrical transmission ("the sparks") but changed his mind in the early fifties. Eric came into the field some years later than I did, and his perspective of the field depends rather much on reading the literature. I have a different perspective because I was right in the middle of it. I have also talked with others, my contemporaries, for example, with Göran Steg, a neurologist, who had spent some time in the Department of Neurophysiology at the Karolinska Institute. Steg had some discussion of research problems with a colleague at the Department of Physiology where Ulf von Euler was. When some members of the Neurophysiology Department learned about that they told him, "You shouldn't talk to them; they're all wrong." Neurophysiology at that time in the late fifties WAS electrophysiology. "Neurosecretion", a synonym for chemical transmission used at that time, was regarded as quite suspect.

I give you one example from as late as 1964 when Göran Steg wrote his thesis. He then wanted to include a study in which he had given L-DOPA to rats that had been treated with reserpine. He looked at the activity in fine tail muscle nerves under various experimental conditions. He recorded the Parkinson-like rigidity in these rats and then, when he gave them L-DOPA, the rigidity disappeared and the pathological alpha- and gamma-nerve activity was normalized. It was an elegant study. The rest of Steg's thesis work was purely in neurophysiology. Steg's mentor at that time was Anders Lundberg, an internationally well-known Professor of Neurophysiology here in Gothenburg, in this very building. When Lundberg learned that Steg wanted to include that work with reserpine and L-DOPA, he told Steg, "You shouldn't have that in. These things are not so well established". That was in 1964. To be fair, however, Anders Lundberg in collaboration with Nils-Erik Andén from our research group at about the same time made valuable contributions to this paradigm shift by electrophysiological work on monoaminergic mechanisms in the spinal cord. Anders Lundberg was thus clearly open to chemical transmission but presumably regarded it risky to include such work in a thesis for the doctoral degree. Incidentally, Steg later became a neurologist and was appointed Professor of Neurology at our University (now Emeritus). He is an outstanding specialist in Parkinson's disease.

Sir Henry Dale figures in your story. What was his role?

He was a fascinating person. He was one of the pioneers in chemical transmission, pursuing very successfully Otto Loewi's original discovery of chemical transmission in the frog's heart by extending it into the mammalian peripheral nervous system. Loewi and Dale shared the Nobel Prize in 1936. I met Sir Henry only once and that was in 1960 at a meeting in London. The meeting was on adrenergic mechanisms and all the big shots of the field were present. Most of them were former pupils and colleagues of

Arvid Carlsson at his first international meeting in Milan in 1957 (courtesy of Arvid Carlsson).

his. There was always a crowd around him and he was the big man. They looked upon him as schoolboys look upon their admired teacher. Nobody would argue against Sir Henry. But that was not his fault. He did not expect people to behave like that, but he was a very impressive person. This meeting was just a couple of years after we had reported our discovery that depletion of dopamine induces the Parkinson syndrome and that L-DOPA will alleviate this syndrome by restoring the dopamine level. As we came to London, Hillarp and I, we were anxious to hear what Sir Henry would say to our discovery. And he said, "No, this cannot be true, this story doesn't fit." All the others agreed, of course.

Hillarp was, of all the people I have ever worked with, the most wonderful scientist, and here I include Brodie as well in the comparison. It's only that Brodie was more important in directing my career. Hillarp had a lot of charisma but was a rather reticent person in the sense that he didn't like to talk much in public, and I was the one who was fighting with all these guys. This was a CIBA Symposium and the discussions were subsequently printed verbatim so you can read about all these fights. Sir Henry's conclusion was, "It's a funny thing, with DOPA," he said, "it's an amino acid, but a toxic amino acid," and he thought it was unusual that an amino acid was toxic. His conclusion was that this has nothing to do with physiology, rather, it is a strange toxic action. He was sure, as was every pharmacologist in those days, that dopamine was a physiologically inactive compound and only serves the function of being an intermediate in the synthesis of noradrenaline and adrenaline. That was number one. Number two, we had found that if we gave DOPA to reserpine-treated animals and they wake up and start to move, noradrenaline still cannot be detected in their brains. So noradrenaline could not be the cause of the effect of L-DOPA. In his opinion dopamine could not be the cause either; the only thing remaining was the amino acid itself. So the conclusion was that we were dealing with a strange toxic action and that was the conclusion of the symposium.

Sir Henry Dale was the President of the Royal Society and in 1945 he gave the Copley Medal, the highest recognition of the Society to Oswald Avery for Avery's seminal discovery that DNA was the substance of heredity. This was a unique recognition of a paradigm shift in science when other recognitions were not forthcoming. In your case, however, he appeared to be a conservative.

Exactly. He had had several debates with Sir John Eccles (to become a Nobel Laureate in 1963) in London, at the Royal Society. These debates were very popular because they were really trying to hit each other, these two big guys. Eccles was the "sparks" man and Dale, the pioneer in chemical transmission, was the "soup" man. But in the early 1950s, Eccles did some experiments which he interpreted to mean that acetylcholine is a neurotransmitter in the spinal cord. Eccles was thus finally converted from the "sparks" to the "soup". However, Eccles's conversion did not seem to have had a broad impact on the scientific community. Thus, at an international symposium in Stockholm in February 1965, entitled "Mechanisms of Release of Biogenic Amines", one prominent lecturer concluded that "there is as yet, no direct evidence that acetylcholine is a central transmitter." None of the about 100 participants of this symposium, many of whom were eminent experts in the field of neurotransmission, objected to this statement. At the same meeting there was general agreement that the monoamines (dopamine, noradrenaline and serotonin), are neurotransmitters in the central nervous system.

How did it feel for you being 37 years old in 1960 and being against all the authorities?

I was young and somewhat arrogant, so I sort of enjoyed it. [Dr. Carlsson is heartily laughing.] It was a good fight.

You were sure of your findings.

Absolutely sure.

Your discovery provided a cure or alleviation for Parkinson's syndrome.

Yes. I formulated the whole concept of dopamine's role in Parkinson's disease. That was in October 1958 in Bethesda, Maryland.

Why do we still see so many people suffering from Parkinson's disease?

This is not a matter of prevention; it is a matter of treatment; a symptomatic treatment. You replace what has been lost through cell death by dopamine, by giving its precursor DOPA. In the majority of patients, there is a dramatic improvement, which can last for even more than a decade. In some cases, and sooner or later in most cases, the therapeutic action of L-DOPA is weakened. Side effects show up, such as the so-called on-off-phenomena, so that the patient is Parkinsonian-stiff and after various intervals all of a sudden switching and then is moving freely. Then, all of a sudden, the patient switches off again and is back to this stiff, motionless state. There is another side effect, the so-called dyskinesia, involuntary movements. It can be very disturbing.

When you are saying that it is treatment and not prevention, something comes to my mind with a twist. When the Nobel Prize was awarded for the discovery of insulin, subsequently there was criticism that the Nobel award diverted attention from researching diabetes in order to find the cause and find a real cure. Do you think that the recognition of dopamine may dampen further research on Parkinson's disease?

My answer would be that it's exactly the opposite. Parkinson's disease didn't attract any interest by science at all. It was a chronic disease; it was due to some cell degeneration, but the cells that degenerated were not even identified. Obviously, there was some neurodegeneration. What happened with these poor patients was that the diagnosis was established, and the doctor said, "This is a chronic disorder, I'm sorry, we can't do anything about it." There was very little scientific interest in this. After my discovery, Parkinson's disease became a subject of guided brain research, both basic and clinical research. It opened up a new concept that even if you have a degenerative disorder, it's possible to treat it. That was something new. But there I also met some resistance. This time it was from the neurologists. They thought that this was a ridiculous idea, that is, if you have a neuro-degeneration, to introduce a chemical. They thought it was nonsense. Even here, when I came here in 1959, I immediately contacted the Professor of Neurology and suggested collaboration. He was not at all interested. I did the same thing with a famous neurosurgeon in Lund, where I was before I moved here, and he was not interested either. To them this idea was impossible because in their minds, they had a model for the brain where one cell communicated with another by an electric spark, so how could a chemical do anything of interest in this context. Therefore, this discovery was not only a paradigm shift from the point of view of basic research, it was a paradigm shift in clinical neurology and even psychiatry, for that matter. It had a tremendous impact, exactly the opposite of what you proposed.

I didn't really propose it.

I know, of course. You just wanted to tickle me a little bit. [Again, Dr. Carlsson is heartily laughing.]

There was brain research in Germany in the Nazi times. They killed patients to study their brains. Did anything ever come out of this?

Not that I know of. It would be strange. To my knowledge, there was no eminent scientist involved in this. In order to gain anything from post-mortem brain research, and it was not even post-mortem, it was experiments on human beings, apart from the ethical aspects, which are terrible, if you disregard that, you must have somebody who is a great scientist to be able to gain any knowledge from it. I suppose there was nobody of that caliber around there. This is not surprising because a prominent scientist would never become involved in this kind of thing.

That is not so clear because today we know that the infamous Auschwitz doctor, Josef Mengele sent his blood samples to his colleagues in Butenandt's institute and Butenandt was a famous scientist, a Nobel laureate. He denied knowledge of everything after the war but the facts remain facts that at least his institute was involved in experiments on human beings.

That's terrible.

What I am trying to say is that just because these were experiments on human beings does not exclude the involvement of prominent scientists.

Maybe, you're right. It's a possibility, I agree. I accept what you are saying about Butenandt, but even if Butenandt was a great scientist, there is the other part; to plan a scientific work in a rational way, I doubt that Mengele was competent in what he was doing.

I only know that Mengele was in fact a postdoctoral fellow of the Deutsche Forschungsgemeinschaft, a most respectable organization. My feeling is that "respectable" German science was much more deeply involved in Nazi crimes than we have been led to believe.

That makes me feel sad.

There is a book by Benno Müller-Hill about it, Murderous Science.[3] *He is a Professor of Genetics in Cologne. His findings are based on the archives of the Deutsche Forschungsgemeinschaft.*

That's terrible. German science lost so much from the loss of the Jews. They played such an enormous role and not only in German science but also in German culture. They were part of German culture and when they were pushed out of the system, what was left was not good at all.

I have long been associated with Norwegian scientists. I used to go to Norway often, much before I started coming to Sweden. In Norway I heard a lot of criticism from my colleagues about Sweden.

Yes, of course. That's understandable. They were occupied by the Germans, while on the other side of the border, we were there and we were free.

How do you look back? This is the first time I have the opportunity to ask this question from someone like yourself.

In 1943, I was 20 years old. I did some military service. I had my personal reaction to hearing about Norwegian students being taken to prison and actually being sent to Germany. I know such people; they are my friends now, and they have told me about it. It was terrible and we hated it. We very much disliked that there were the so-called permit trains through Sweden all the time. Ostensibly these trains carried German soldiers on leave from Norway to Germany and back. In reality, all kinds of cargo were being transported through Sweden between Germany and Norway.

We allowed it. It's then understandable that the Norwegians responded very negatively. Our government gave priority to keeping Sweden out of the war. They yielded a lot.

On the other hand, Sweden played a very positive role in saving thousands of Hungarian Jews in Budapest through the activities of Raoul Wallenberg and other associates of the Swedish Embassy.

Of course, Sweden played a game here, to prevent occupation by Germany, but the Swedish government didn't like the Nazis at all. There is a terrible story. Just after the Wannsee meeting, where the German leaders made the plans for the "final solution", the annihilation of the Jews, one person, a member of the SS, who was present, subsequently got in touch with a Swedish diplomat. He told him the whole story of what had occurred in Wannsee. The Swedish diplomat reported back to Stockholm of what he had learned. But that was the end of the story. They didn't do anything.

At one point, after already hundreds of thousands of Jews had been deported to Auschwitz, and the deportation of the Budapest Jews was to come up next, the Swedish King made a protest. Others made protests too and the front was fast approaching, so the Hungarian government stopped the deportation. The protest by the Swedish King may have contributed to this.

I'm pleased to hear that.

You were about 20 years old at that time.

We disliked it of course. On the other hand, when the Russians invaded Finland in 1939, we were very much aroused by that. We said, "Finland's cause is our cause" and lots of Swedes went to fight the Russians. That was perhaps the most striking response of Sweden to any of the war events. The ties with Germany were very strong before the war. The textbooks I used as a student were German, both in anatomy and in histology. By the time we came to physiology, they were English. German science and German culture was very much preferred over the English in Sweden. This has changed since the war.

Coming back to the Nobel Prize, there was some controversy and criticism about it. The Prize was divided among three scientists. Has there been any criticism concerning your share?

There has been and I can tell you about it. The protest in my case came from one single person and some people around him. That person was Oleh Hornykiewicz in Vienna. Their protest was directed against me; he didn't mind the others. He had expected that he would get the Nobel Prize together with me. I had an animal model and he continued the work and analyzed the brains from Parkinsonian patients. He found that they were low in dopamine. Then, together with a clinician, he started to use L-DOPA, as I had done in my animal work. They brought the animal model to humans. It would've been quite logical for the two of us to get the Prize. The Viennese people, Birkmayer and Hornykiewicz started this work, in the early 1960s, based on our findings, but they didn't succeed in making the treatment into a useful procedure. That came later, in the late 1960s. A Greek-American scientist, Cotzias in New York, modified the DOPA treatment by giving oral doses, climbing doses to make it into an efficient, useful treatment. Therefore, at one time, the three of us were proposed to become Nobel Prize-winners. That was probably a very strong case. Unfortunately, Cotzias died of cancer. So the two of us were left and we were not so strong anymore. If there are three, the case has a greater

The three co-recipients of the Nobel Prize in Physiology or Medicine 2000, Paul Greengard, Arvid Carlsson, and Eric Kandel (courtesy of Arvid Carlsson).

chance than if there are two or one. Hornykiewicz kept expecting that we would get the Prize and when I got it finally, he was greatly disappointed.

You would not have been surprised if Hornykiewicz had been included.

Not at all. That was an obvious alternative to what actually happened.

So his and his supporters' disappointment was justified.

Yes, from that point of view. However, what made me sad was what Hornykiewicz said, "Look at Carlsson, he had just done animal experiments and if we hadn't done this in humans, they might have even be forgotten by now."

From my conversations with Paul Greengard and Eric Kandel, separate conversations, of course, I perceived their perspective. Obviously, they did not find any fault with the Prize, but according to them, Hornykiewicz's contribution was more on the clinical side and his was not the basic discovery.

That's true.

Each of them explained to me as well that Hornykiewicz was not alone in this and if his contribution had been included in the Prize then similar contributions of others would have had to be considered as well.

Exactly. It's also doubtful now if Hornykiewicz will ever get the Prize.

From all my conversations, I have formed the following picture about this Prize, and, admittedly, it is pure speculation, and in an oversimplifying way. You were chosen for the discovery of dopamine and your choice was unambiguous. Kandel was very much favored by some in the Nobel Committee. But the contributions of the two of you were so far apart that somebody had to bridge the two of you, and that was Greengard.

How funny [Dr. Carlsson is heartily laughing].

Greengard might have received it with you, he might have received it with Kandel, but you and Kandel were too far apart.

That may well be true.

In fact, Greengard had received prizes on separate occasions with you and with Kandel. So the scheme I have been told seems to be plausible. Could it be?

Possibly.

On at least two occasions in your Nobel lecture, you were comparing foreign and Swedish discoverers and note that the foreign scientist received the Nobel Prize and the Swede did not.

I had written about this before that as well. But this issue was not mentioned in my actual Nobel lecture.

On the other hand, there has been criticism that Swedish scientists have been favored for the Nobel Prize.

Perhaps one can say that it is easier for a Swede to become a Nobel laureate, if you are located in the Stockholm area. There may be just two exceptions; I am one of them and the other one is Nils Gustaf Dalén, the great inventor, who received the physics prize in 1912. He was from western Sweden. There is one more Swede outside Stockholm and Uppsala, that is Torsten Wiesel, who lives and works in the United States and received the medical prize in 1981.

What I think is, and this is my personal bias, if you are a professor of the Karolinska, and you are not a Nobel Prize-winner, and you have a younger colleague in your field, e.g., in Lund or in Gothenburg, would you like him to become a Nobel laureate? Perhaps you wouldn't. That's human nature.

What turned you to science originally?

I grew up in an academic family even though the others were humanists. Both my parents studied in Uppsala. My father was a Professor of History; my mother studied literature and history of literature, and finally her topic became the legal status of women during the Middle Ages in Sweden. She wrote books about it and at the age of 75 received an honorary degree. My older brother became a Professor of History in Uppsala and my sister studied literature. I departed a bit because I started to study medicine and my younger brother followed me and he too studied medicine. So it was an academic environment where I heard about research all the time. My

Arvid Carlsson as a schoolboy in 1939
(courtesy of Arvid Carlsson).

father was also interested in the Middle Ages, and then especially in some prominent people living in Sweden during that time. I was fascinated by the thought of research. Why I departed from the humanistic line of my family, I don't know. Maybe it was a little bit of protest; maybe I thought that these humanists were not so useful and you could be more useful if you did something like medicine.

Did you ever consider going into clinical practice?

My initial ambition was to become a scientist, a pre-clinical scientist, but at one time, in 1954, I wasn't so sure whether I would be able to make a successful career in pharmacology. At that point, I went over for one year to the medical clinic of the University of Lund as a doctor. During that year, even though I enjoyed the experience, I realized that I would like to do pre-clinical science.

Many future Nobel laureates had to overcome various adversities, but your life is an example of smooth sailing.

Not quite because I applied in 1953 for a professorship, there were two competitors, and I didn't get it. The expert committee let me know that the kind of science that I did was not in the mainstream of pharmacology. That made me feel that I might not be able to make it as a pharmacologist. What I had been working on before, and my thesis dealt with that, was calcium metabolism. That was initiated by the availability of the radio-isotopes, which like the necessary equipment had become commercially available by then. That was in the late 1940s. It was a successful piece of work, but the pharmacologists thought that it was not in the center of their field.

Mineral metabolism is now an important topic; maybe your interest was a little ahead of its time?

Biochemical pharmacology was hardly thought of at that time. It was an odd kind of research. When finally I decided that I would like to remain in pre-clinical pharmacology, I went to Sune Bergström, later a Nobel laureate too, and he was in the same building at the University of Lund. I told him that I would like to remain in pharmacology and I needed a change; I would like to do something that the pharmacologists like better and that I would like to do some kind of biochemical pharmacology. I asked him for help because I knew that he had very fine contacts with America. Bergström wrote a letter to his friend Bernard Witkop, a prominent organic chemist

Arvid Carlsson with Sidney Udenfriend during Udenfriend's visit at the Department of Pharmacology, University of Gothenburg in 1965 (courtesy of Arvid Carlsson).

at the National Institutes of Health, Witkop forwarded the letter to Sidney Udenfriend who in turn forwarded it to his boss, Bernard Brodie. Brodie answered that he would be more than happy to have me working with him, but he had no money for me. The Swedish America Foundation and the Medical Society of Lund gave me support, not much, just enough for me to survive.

From the time of your seminal discovery to the Nobel Prize, more than 40 years elapsed. Were you in anticipation during this time?

Yes. I was on a fairly short list several times during this period of time. I knew about it because there is always gossip.

Did it make you nervous?

Not really because my notion of the Nobel Prize was different before as compared to after I received it. I understood that the Nobel Prize was the top prize, but I thought that it was not so much more than other prizes. For example, I had received the Japan Prize, which was more in terms of money than the Nobel Prize. I didn't understand the difference until I got the Nobel Prize. When I did get it, it was such an enormous sensation. All my friends from all over the world congratulated me; many of them said that when they heard about it, they did strange things; one of them was just driving his car and when he heard the announcement he was happy that the police did not see him. On the 9th of October 2000, I finally realized what the Nobel Prize was. But not before.

I know about other people who lived in anticipation of the Nobel Prize. Brodie, for example, a number of times when the Nobel Prize was announced, he became so depressed that he had to stay in bed for a couple of weeks. He was so focused on getting the Nobel Prize. Then, when his pupil, Julius Axelrod got it, that was a blow to him. I never experienced anything like that.

If you had received it 30 years ago, would it have changed your life more dramatically? Would it have helped you or would it have disturbed your work?

It's very hard to tell. It could have gone in either direction. If you are a young man, as a Nobel Prize-winner, you will be inevitably involved in all kinds of things that take you away from science. From that point

Receiving the Nobel Prize from the King of Sweden (courtesy of Arvid Carlsson).

of view it was good that my prize came later. I worked and I felt that I needed to prove myself. It helped to keep me active, that is, not having received the prize. Another thing is that with the time passing by, the amount of money with the prize had gone up very much. But that was not very important, of course.

What did you do with the money?

We founded a company earlier in the year of the Nobel Prize. I started out as a CEO of the company and I didn't take any income from that company. So I used private money for my subsistence. But I had enough money for my subsistence anyway. I invested private money into the company by buying shares from younger co-founders who needed cash.

What is the company involved in?

It is involved in dopamine. We want to improve the medicines that are available now. The compounds we work on are able to stabilize the function of dopamine, which is something new. They will probably be useful both in psychiatry and neurology.

What can you change? Dopamine has an exact chemical composition.

That's an interesting thing. The chemical structure of the compounds we use is much more like that of dopamine than those drugs that are acting on dopamine receptors as antagonists that are being used for schizophrenia and psychosis. This is probably an important aspect and the reason why our compounds have so much more favorable properties than the drugs that are available today. One can see and recognize the chemical structure of dopamine, with some modifications, of course, with the drugs that we have been working on for 30 years.

Are these chiral substances?

Some of them are and others are not. The one we have been working most on in our company is not chiral; so we don't have to worry about separation. The one we had before was chiral and it makes our life so much easier when we don't have an asymmetric carbon atom.

I have heard that in university discussions you often found yourself in the minority when there were some disagreements, even in a minority of one.

Most of the time I was. My brother used to tease me that I was a man of different opinion. I was such a man in our faculty and I was such a man in science as well. I have always been in fights. People used to say about me during these decades, "Arvid Carlsson did nice work until five years ago, but he has not been very good during the last five years." That they have kept saying all the time. Probably, I think differently and I don't mind that I do that.

This is not very Swedish, is it?

People say that Swedes like to keep a rather low profile; they don't want to be different. But I have been different. I have been involved in all kinds of fights. For example, in the mid-1990s the Swedish Medical Research Council did some bad things; for example, they started giving most of the money available to themselves within the Council members. I started an opposition there that became successful. There were also other problems, for example, discrimination against the other sex. Finally, we managed to throw them out of business. Another example was the water fluoridation project in the 1960s and 1970s. They wanted to add fluoride to the water. I didn't like that. I started to fight that vigorously. I was the main figure among those who opposed, and we won, finally.

What do you recommend?

Don't add fluoride to the water.

How about toothpaste?

That's fine, that's wonderful. That's how it should be. At one time it was believed that the effect of fluoride is important before the tooth comes out. But that is not true. The action of fluorine on the teeth is local. Having it coming through the blood is ridiculous. Also, why have fluoride in the water that is used for many other purposes? Have it in the toothpaste; then it's working.

Arvid Carlsson and his wife at an international meeting in 1975 (courtesy of Arvid Carlsson).

What does your wife do?

We went to the same school and she took her matriculation one year after me. She started to study medicine and we had one course together. We married in 1945. She took her examinations and she is what the Americans call an M.D. Then she found that to become a medical doctor was not her cup of tea, but we had five children, whom she took very well care of. After our children had grown up, she became my assistant. This was connected with an aspect of my professional life that we have not covered so far. Soon after I came to Gothenburg, I became consultant for a drug company. I proposed to them to start on a special kind of drugs that are called beta-blockers.

I made an interview with James Black.[4]

Black was ahead of us, but only by a couple of years. We came in as number 2. Actually, the drug beta-blocker that is still the main one on the market is the one that was developed here in Gothenburg. I had a patent.

I happen to be taking Betaloc manufactured in Budapest according to a license from Astra, Sweden.

Astra has a subsidiary in Gothenburg. The generic name is metoprolol. That is our drug. I am one of its inventors. The patent has expired a long time ago, so I am not getting any money from it. During a long time though I got some royalty, and from it I founded a little private company, a different one from the company I mentioned earlier. I founded that company partly for tax reasons in connection with the money from beta-blockers. My wife was employed by that company. The patent money came as salary for my wife for several years.

Did you have any interaction with James Black?

We didn't know anything about ICI (Imperial Chemical Industry) at that time. It took a number of years until we learned that they were doing the same thing and that they were ahead of us. Our feelings were very mixed when we found out. The good part of it is that this little company, Hässle in Gothenburg, a subsidiary of Astra, with its headquarters near Stockholm, could be so innovative. Incidentally, Astra no longer exists, it is now Astra–Zeneca, and Zeneca used to be ICI; thus the two pioneers in beta-blockers have fused. The fusion had nothing to do with beta-blockers, it just happened. Metoprolol is now the winner on the world market. By the way, when the patent expired, they did something very clever; they changed the product to a different salt and thus they could get a new patent. I don't get any royalty for the new patent although it's still the same medicine.

This is different from brain chemistry.

It is, but it's neurotransmitters nevertheless. At the same meeting where I presented the story about dopamine and Parkinson's disease in Bethesda in October 1958, I heard for the first time about the first beta-blocker. It was a compound that didn't make it at all. It was from the American Lilly and they soon stopped it; they were not interested in developing it further. But the pharmacologists reported about it. When I became a consultant for Hässle, they were then interested in the heart. I told them that if they wanted to create a medicine for the heart, it should be a beta-blocker.

How did you come to this idea?

It was a very simple reasoning. The nervous system sometimes over-reacts. There are conditions under which the heart is pushed too much by the sympathetic system. We knew that the sympathetic system acted via beta-receptors on the heart; that was discovered in the 1940s by an American who was born in Sweden by the name Raymond Ahlquist.

But his work was not well known.

No, it wasn't initially. It was a purely academic discovery, but it turned out to be very important.

How did you come across it?

That was at this meeting in Bethesda in 1958. The people from Lilly reported the pharmacological properties of their beta-blocker, which was called DCI, dichloroisoprenaline. Isoprenaline is a catecholamine, not very much different from adrenaline, but instead of the two OH groups they put in chlorines. DCI turned out to be a blocking agent, but it was not

Arvid Carlsson in his lab in 1981 (courtesy of Arvid Carlsson).

useful. They had some bad experience with it when they administered it to patients and they dropped it very quickly at Lilly. I thought that the idea was nice and that there must be conditions under which beta-blockers should be useful. In the treatment of hypertension, alpha-adrenergic receptor-blocking agents, which act on blood vessels, were used in those days. Now we thought that beta-adrenergic receptor-blocking agents, which act on the heart, could be used to dampen overactivity of the heart. It could be that under certain conditions the sympathetic system acts too strongly on the heart. I thought it might be a good idea to dampen the sympathetic input on the heart under such conditions. That was my simple concept.

Another important innovation came out of my collaboration with Astra/Hässle, namely the first selective serotonin re-uptake inhibitor (SSRI). It was called zimelidine. It was marketed in 1982 and turned out to be an excellent antidepressant agent. However, it was withdrawn from the market in 1983 because of a suspected serious, though very rare side effect. Zimelidine was soon followed by several other SSRIs, which were to revolutionize the treatment of depression and several anxiety disorders. The most well known SSRI is called Prozac.

Do you think you could have been included in the 1988 Nobel Prize in Physiology or Medicine?[5]

Yes, I think so.

Do any of your children follow your path?

Yes, Maria is an Associate Professor of Pharmacology at this University. She is doing research in the area of neurotransmitters of the brain. Some of her papers are co-authored with me. She also works in our company, in which she is one of the founders. Her elder sister Lena is a science writer. She has a Bachelor of Arts, mainly in natural sciences. Together with me she has written a book on brain neurotransmitters and neuropharmacology, entitled *Messengers of the Brain* which is intended for students and educated laymen. Bo graduated as an electrical engineer at Chalmers University of Technology in Gothenburg but later switched to medicine, became an M.D. and specialized in ophthalmology. He is now working at the University Hospital in Gothenburg. Hans is an Associate Professor at the Department of Economics, Lund University, Sweden. Magnus has a Ph.D. in computer science and has been affiliated with the Chalmers University of Technology and at the Oregon Health and Science University in the U.S.A. We have eight grandchildren.

I would like to ask you about religion.

My grandfather was a priest in a small and rural community. When I was a child, at one time I wanted to become an archbishop. That did not last long though. After that I have not been too much involved in religion. I am not a member of the State Church. At least, it used to be a State Church, which was Protestant, of course, Lutheran. I left that several years ago when they were arguing about female priests. I thought it was ridiculous that they should make discrimination against women. But I would like to stress that the message of Jesus Christ has made a big impression on me. To the extent I know it, and I am not a theologian, he was the first who said that you should love the other human beings as you love yourself. As far as I am concerned that was a paradigm shift. Nobody had said that before. I also believe that he must have existed. He must have been a historical figure even though it's possible that lots of things have been added to the story.

But that is not religion.

No, it's not religion. But I can hardly get farther than this. If you think about the world, the whole Universe, and everything that we experience, the whole thing is so enormously mysterious, isn't it. The whole thing is such a wonder. So, if in addition to that, there is a God, that wouldn't add much to the mystery. With or without God, the world is a fantastic thing to experience. The concept of God is very strange but one cannot say that it is impossible although I can hardly believe it. As a child you take everything as natural, but the older you get the more you wonder about it. As for our children, my wife and I tried not to do anything that would interfere with their integrity. This concerned their profession, religion, everything. Whatever their choices were, we supported them. None of them stayed in religion as far as I know.

Is there anything that we have not touched yet but you would like to mention?

I have been interested in the theory of science. I have been reading Thomas Kuhn, not a very easy read, but interesting and fascinating. I agree with the core of his message. There is a lot to what he says about normal science and about scientific revolution, how he emphasizes that discovery is usually not a eureka kind of event. Rather, it is a process. You have the observation but then you have to start thinking in a different way and formulate a new program in your mind. The discoverer has to go through a psychological process in order to realize what he has seen. Many people would see the same thing without realizing what is new in it.

You write somewhere that Brodie and you had the advantage of being outsiders in the field.

We were outsiders and that was very important. This fits well with Kuhn's teachings. He describes the solid establishment, the dogma. Then somebody comes in from the outside and says it can't be true. One of the things that impressed me about Brodie was that he could listen to a seminar outside of his field and at the end he would come up with some penetrating questions.

I find it interesting that after about 60 years in science you keep returning to those 5 months that you spent with Brodie.

He was fascinating in many different ways. I still think about him, about what really made him such a character. Let me give you a couple of examples that makes it even more mysterious. The discovery of serotonin depletion by reserpine was truly groundbreaking. But that was not his discovery; he just picked it up. The discovery itself was made by a young man in his lab, Parkhurst Shore, who by the way was my mentor under Brodie's supervision during my stay in Brodie's lab. I had a phone conversation with him just a couple of weeks ago because I wanted to find out a little bit more about the thing. It's mentioned in the book, *Apprentice to Genius*,[2] that it was Shore. Shore was put on a problem that had to do with a Brodie kind of concept, that is, how two drugs interact with each other; it was a basic problem. In this context, they were investigating also drugs that acted on the central nervous system (CNS). Shore started to read about this and he was fascinated by the finding that there are a number of CNS agents that were indoles, serotonin and reserpine and LSD. Then it so happened that there was a younger collaborator of Udenfriend, Herb Weissbach, who had developed a method to measure the metabolite of serotonin in the urine. Shore went to Weissbach and asked him if he would like to analyze urines from animals that had been given reserpine. So they did that and they were astounded by the results. They had never seen such an elevation in the metabolite level. At that time Brodie was on vacation in Florida, but as soon as he heard about it, he rushed back and immediately started talking about "our project". He took over. Shore is such a modest man. He says, that was the system.

Shore had another discovery; if you give certain drugs that are N-containing bases and then analyze the content of the stomach, you see an enrichment of the bases there from which they could derive the idea that it's a matter of distribution between two phases in the stomach and if it is a base, it will be enriched in the acid phase. That was also Shore's finding but Brodie gave it over to another guy who was more of a specialist of the stomach in Brodie's group. There was a publication from it with Shore's name as first author. It was the structure of Brodie's lab.

Then it was Julie Axelrod who was the one who went to a neighboring lab and learned to prepare and fractionate liver and discovered the role of the microsomes in drug metabolism. It also became Brodie's discovery, not only Axelrod's. There were even jokes about it, about Brodie discovering the effect of reserpine on serotonin on the beach in Miami. These are negative aspects of Brodie. But he was the one who created this atmosphere

Arvid Carlsson during the conversation
(photograph by I. Hargittai).

of creativity; he did not develop the spectrophotofluorimeter, but it was created in the atmosphere of his lab, and he may well have had the original idea. The spectrophotofluorimeter was a marvelous instrument in those days, which opened up new possibilities; for the first time it became possible to measure small amounts of many important substances with chemical methods.

All this happened in Brodie's lab. He attracted these people, who did all this. So, what can we say? He was a pioneer of a field. Even though he was a greedy kind of pioneer, he was a pioneer. If it weren't for him, these discoveries wouldn't have happened. He was a great pioneer of modern pharmacology. But taken everything together, there could also be found arguments against a Nobel Prize for him.

Do you have a message?

I don't know.

At being 80 years old, how do you keep so young?

There is perhaps a message here. I think that the role of the genes these days is very much exaggerated. I don't understand how it comes about, reaching almost hysteric proportions. If you look upon longevity of man and look back 500–600 years, there has been a tremendous increase in longevity in certain areas, for example in Europe. At the same time the gene pool has remained the same. Doesn't it tell you about the enormous impact of the environment? This is being disregarded these days. This is very bad because the environmental factors are so much easier to handle than the genetic factors. It hampers our development if we put too much emphasis on the genes. We have to realize the importance of the environment and this is what I ascribe my good health to as well. I am in good health apart from the chronic bronchitis that makes me sound as if I was a smoker, which I am not.

Did you ever smoke?

I used to be a heavy smoker, but I stopped in 1959. It was not easy because although I wanted to stop, my wife smoked too. However, she became pregnant with our fourth child and for the first time in her pregnancies she had a severe nausea. This made her stop smoking and I could do it with her. We never smoked again.

What else?

Then, of course, it is important what kind of food you eat although I am not a believer in any special food; everything should be reasonable. Moderation in alcohol is important. I have a small glass of red wine with every dinner. I am a strong believer in sleep and in exercise. I lead a sound life. And one more thing. Even though I have been involved in so much fighting and I still am, professional fighting, our family life has been, on the whole, very harmonious. There may be fights outside, but I have my home, I have my wife, who is very loyal to me; we are really a pair, the two of us are like one. From my childhood on, everything has come out successfully; I have been very fortunate.

Notes and References

1. Carlsson, A. "Autobiography", in *The History of Neuroscience in Autobiography*, Volume 2, edited by Larry R. Squire, Academic Press, San Diego, CA, U.S.A. 1988, 28–66.
2. Kanigel, R. *Apprentice to Genius: The Making of a Scientific Dynasty*. The Johns Hopkins University Press, Baltimore and London, 1993.

3. Müller-Hill, B. *Murderous Science: Elimination by Scientific Selection of Jews, Gypsies, and Others in Germany, 1933–1945.* Cold Spring Harbor Laboratory Press, New York, 1998.
4. See in Hargittai, I. *Candid Science II: Conversations with Famous Biomedical Scientists.* Imperial College Press, London, 2002, 524–541.
5. It was awarded to James W. Black, Gertrude B. Elion, and George H. Hitchings "for their discoveries of important principles for drug treatment".

Oleh Hornykiewicz, 2003 (photograph by I. Hargittai).

34

OLEH HORNYKIEWICZ

Oleh Hornykiewicz (b. 1926 in Sychiw near Lviv/Lemberg, then Poland, now the Ukraine) is Professor Emeritus at the Institute of Brain Research of the University of Vienna, Austria. He is most famous for showing that the lack of dopamine causes symptoms of Parkinson's disease in humans and for suggesting treatment with L-DOPA. In 2000, the Nobel Prize in Physiology or Medicine was awarded to three scientists (see interviews with them elsewhere in this volume) "for their discoveries concerning signal transduction in the nervous system". Oleh Hornykiewicz was not among the awardees and 250 neuroscientists wrote an open letter protesting his omission.

Oleh Hornykiewicz and his family moved from Lemberg to Vienna after the outbreak of the Second World War. He studied at the Medical School of the University of Vienna between 1945 and 1951, completing his studies with an M.D. degree. In 1951, he joined the Pharmacological Institute of Vienna University. In 1956–1958, Hornykiewicz did research under the supervision of Professor Hermann Blaschko in Oxford. Between 1968 and 1976 he was Professor of Pharmacology and Head of the Psychopharmacology Section of the Clarke Institute of Psychiatry at the University of Toronto. After his return to Vienna, he headed the Institute of Biochemical Pharmacology. He was prominent in establishing the Institute of Brain Research at the University of Vienna, which was opened in 1999. Oleh Hornykiewicz has received many important awards, including the Gairdner Prize (Canada) and the Wolf Prize (Israel).

We recorded our conversation in Professor Hornykiewicz's office at the Institute of Brain Research in Vienna on October 25, 2003.

First I would like to ask you about your story. You were born in the Ukraine and have lived in Vienna.

My reply sounds like a riddle. I was born in that part of Europe that was part of the Austrian Monarchy when my parents and their forebears were born there; it was Polish when I was born there; then it became part of the Soviet Union; and now it's the westernmost province of the Ukraine. For middle-Europeans this describes precisely the area where I was born. It's Eastern Galicia. I was born near Lemberg by its German name, Lwów is the Polish name, the Russian spelling is Lvov, and now it is called Lviv. My parents were Ukrainians, they regarded themselves as Ukrainians, so I am Ukrainian by birth. On my father's side we had a Hungarian connection and several Polish relatives, some of them well known in Polish pre-war politics and diplomatic service. On my mother's side, who was the descendant of landed gentry, there were Austrian ancestors as well as a Czech connection; one of my mother's uncles was a notable political figure here in Vienna during the reign of Emperor Franz Joseph. The ethnic history of my family

The Hornykiewicz family in Toronto, 1976 (Oleh, wife Christina, daughter Maria, sons (from left) Nikolai, Stephan, and Joseph, courtesy of Oleh Hornykiewicz).

has been like the history of many families in that area, and, in a way, like the history of the country itself.

We came here, to Vienna, when the Second World War broke out. My parents decided that it was better to move out of that area, which has always been an area of wars. My father's brother lived in Vienna so we had a base. We left after the Russians had occupied Lemberg according to their non-aggression treaty with the Germans. When Poland was defeated, those Polish territories became Russian, or, rather, part of the Soviet Union. Since my mother had Austrian ancestors, we were permitted to leave.

I turned 13 in November 1939, and three months later we arrived in Vienna; I spent most of my school years in Vienna; I started with elementary school at age 7, and I lost one year of my schooling due to the move. I started and finished my medical studies here.

Despite my Austrian ancestors on my mother's side, I did not speak German when we came to Vienna; I spoke Ukrainian and Polish. Back in Lemberg, I went to a school where the languages of instruction were Ukrainian and Polish, depending on the subject; mathematics in Ukrainian, but history, significantly, in Polish. So when we came to Vienna, I had first to learn German.

Have you been back to Lvov since?

I have been back after 50 years. When the Soviet Union disintegrated, I went back, and have since been back three times. I still have many relatives there. I could not remember anything from my childhood in terms of the architecture or the esthetic characteristics of the city, and I was impressed. Lemberg is a beautiful city.

Very Central European. I have also visited it.

It reminds one of the almost two hundred years of Austrian rule and of the Polish and the Ukrainian influence. It's a beautiful place.

You finished your medical studies when you were 25 years old; so you regained some of the lost years.

I finished Medical School in 1951 here in Vienna; the times were hard and uncertain; I tried to be as quick as possible.

What did you do afterwards?

At the Pharmacological Institute in Vienna, 1958. From left to right: Otto Loewi (Nobel Prize 1936), Hans Molitor, Ernst Pick, all former members of the Institute, and Franz von Brücke, head of the Institute, 1947–1970 (courtesy of Oleh Hornykiewicz).

I immediately joined the Pharmacology Department of the University of Vienna. The reason for that was my teacher in pharmacology, Professor Franz von Brücke, the Head of the Pharmacological Institute. Brücke came from a family of famous scientists-physiologists; his grandmother Brücke was a née Wittgenstein, so he was related to Ludwig Wittgenstein, the philosopher. Brücke had a brilliant gift of speech, and was an excellent lecturer. He impressed me very much by the wide range of his knowledge, his classical education, and his ability to communicate to us students the excitement of the pharmacological research and its relevance for the patient. I wanted to do something in his institute. Since my position was without a salary, I also worked half days in a hospital in order to earn some money. In the afternoons, until late at night, I worked in Pharmacology.

In 1955, I applied for a British Council scholarship at the advice of my supervisor, Docent Adolf Lindner, who later became the Professor of General and Experimental Pathology at this University. I went to work with Dr. Hermann Blaschko in the Department of Pharmacology of Oxford University. Blaschko was an authority on catecholamine metabolism. I chose

Hermann (Hugh) Blaschko, in Oxford in the 1980s (courtesy of Oleh Hornykiewicz).

this more biochemical side of pharmacology because of the influence of Professor Friedrich Wessely, my teacher on the subject "Chemistry for Medical Students" in 1945/1946. He was not a member of the medical faculty originally; he was replacing a professor there who had been dismissed for political reasons. Wessely was a high-level, hard-nosed, hard-boiled organic chemist who apparently did not know how ignorant we medical students were when it came to the exact sciences.

He must have been the same person about whom Max Perutz talked to me warmly. He said that Wessely was most encouraging and supportive to him. It was also Wessely, who after the war recommended Hans Tuppy for postdoctoral work in Cambridge and, at Perutz's recommendation, Fred Sanger took on Tuppy in his laboratory. Tuppy later had a great career in Vienna, becoming Professor of Biochemistry and Minister of Science and Research. What kind of a teacher was Wessely?

Wessely was a wonderful teacher. He was very fond of electron pushing at that time. It was quite new in 1945/1946. He tried to teach us, medical students, how to handle the electron theory of chemical compounds, and to me he made it the most exciting subject in the whole world. Of course, he had problems with our being so ignorant, but I was very much impressed

Hans Tuppy (left) and Oleh Hornykiewicz at a "Hornykiewicz Symposium" in Vienna, 1997 (courtesy of Oleh Hornykiewicz).

Friedrich Wessely (front row) at a lecture at the University of Vienna, 1962 (courtesy of Oleh Hornykiewicz).

by him, by his absolute dedication to teaching. He was the reason why I chose a more biochemical approach in my pharmacological studies.

Blaschko was an M.D., a Jewish émigré from Berlin, and he specialized in the enzymology and the metabolism of catecholamines. He asked me to look into the possibility that dopamine had its own physiological role in the body. Until 1956, dopamine was thought to be a mere intermediate in the formation of catecholamines from L-DOPA.

He asked you to look into the role of dopamine in the body, not in the brain.

In the body, in the periphery. At that time, in 1956, the brain was not yet a field of dopamine research. Blaschko was the first, together with Peter Holtz in Germany, who, in 1939, postulated the biosynthetic pathway of catecholamines in the body, with dopamine as the immediate precursor in the formation of noradrenaline. Until 1956, until Blaschko decided to do something about it, dopamine was thus regarded as just an intermediate metabolite in the formation of the two at that time known biologically active catecholamines, noradrenaline and adrenaline.

Why do you think Blaschko wanted further studies on dopamine?

That's an interesting question. In the early 1950s, dopamine was discovered to occur in mammalian tissues, in the adrenal medulla, in heart tissue and also in adrenergic nerves. There was, however, something odd about the quantitative results from the adrenergic nerves. There, the amount of dopamine was about 50% of the amount of the total catecholamines present. Now, Blaschko had a very acute, very penetrating mind. He reasoned, and he published this in writing, that as a true metabolic intermediate, dopamine should not accumulate that much in the body. Metabolic intermediates by definition are not expected to accumulate in the tissues.

Interestingly, the observation that in several tissues dopamine occurred in higher concentration than expected from a mere intermediate was known to other prominent researchers, for example to Heinz Schümann in Germany; actually, Schümann was the first to observe that discrepancy. Ulf von Euler in Sweden also knew about it. He worked in the same area. They all knew about it, but it was only Blaschko who clearly expressed the idea that because of that discrepancy, dopamine must have "some regulatory functions of its own, which are not yet known". I am now quoting more or less his words. He expressed his idea at a meeting in Switzerland in the fall of 1956. At exactly the same time, I joined Blaschko's laboratory in Oxford. He immediately started talking to me about dopamine. He wanted me to do something with dopamine. The name was new at that time. Only a few years before, in 1952, Sir Henry Dale had coined the name dopamine to replace the unwieldy chemical name, 3,4-dihydroxyphenylethylamine or — as Dale claimed — the misleading 3-hydroxytyramine.

What did Blaschko ask you to do?

Blaschko referred me to an experiment, done some 15 years earlier, by Peter Holtz in Rostock in Germany, a work published during the war, in 1942. Holtz experimented with dopamine after discovering the enzyme dopa decarboxylase, which catalyzes the formation of dopamine from L-DOPA in the body. He found, like Henry Dale before him, in 1910 in London, that dopamine was a sympathomimetic amine in the cat; that means, dopamine acted like adrenaline, it increased the blood pressure in the cat. However, Holtz also found that in the guinea pig, and also in the rabbit, dopamine had the opposite effect; instead of raising the blood pressure, as adrenaline did, dopamine lowered it. That was a discrepancy that had to be explained. Unfortunately for Holtz, he explained it in the wrong way. He thought that it was an unspecific effect of the aldehyde formed from dopamine by the enzyme monoamine oxidase. Aldehydes were known to lower the blood pressure in an unspecific way. This was the explanation Holtz gave. To Blaschko, this explanation appeared very weak at most.

When, in 1956, Blaschko conceived the idea that dopamine may have its own function in the body, he asked me to test this possibility experimentally.

Fortunately, it was the right time for these experiments. At exactly that time the first in-vivo-effective irreversible monoamine oxidase inhibitor became available. That was ipronazid. By the inhibition of the monoamine oxidase, it was possible to test the validity of Holtz's hypothesis. So I used ipronazid in those experiments and what I found was just the opposite to what Holtz had suggested. The presence of ipronazid did not abolish the blood pressure lowering effect of dopamine as would be expected if Holtz were right, but, on the contrary, ipronazid increased dopamine's blood pressure lowering effect. That was a clear indication that it was dopamine itself that produced the effect, the fall in the blood pressure. There was thus a clear biological difference between the other catecholamines and dopamine. Dopamine had its own role in the periphery that could not be explained by its being merely a sympathomimetic, noradrenaline-like amine or a catecholamine precursor.

In those experiments I also used L-DOPA, which is the precursor of dopamine and is easily converted in the body to dopamine. L-DOPA had the same blood pressure lowering effect as dopamine. Ipronazid, that is, inhibition of monoamine oxidase, also potentiated this effect of L-DOPA. These were clear-cut results and they convinced Blaschko and me that dopamine must have some own physiological functions in the body. This was in 1957. A few months later, at the beginning of 1958, I returned to Vienna and Blaschko gave me the advice to continue working with dopamine. He said he saw before dopamine a bright future.

Did you take his advice?

In Vienna, I continued working with dopamine, but I soon changed to the brain. The reason was that at about the time of my returning to Vienna, dopamine was for the first time found to occur in the brain; in animal brain and also in human brain.

But let me here digress a little and say a few words about my student courses in brain anatomy and my first postdoctoral work. My teacher in neuroanatomy and brain development, back in 1946, was Friedrich Ehmann. He was a highly competent lecturer, and a master of the topographic description of the human brain. He used examples from brain phylogeny and ontogeny, to show us the basically simple plan that underlies the seemingly bewildering structural complexity of the human brain. His lectures fascinated me. They were very clear and precise. Thanks to him, I continued being intensely interested in the human brain and its diseases.

I turned my interest in the human brain into practical work immediately after joining the Pharmacological Institute in the fall of 1951. I did my first postdoctoral work there together with Gustav Niebauer, who later became the Professor of Dermatology at the Vienna University. We both lived in Vienna's Second District, and we had the same way home, after working sometimes until after midnight. We then would walk home together, all the way talking about our experiments. We were measuring in the human blood serum the activity of a polyphenol oxidase as we termed it, following Otto Warburg in his work on copper enzymes. We found that our copper-dependent enzyme activity was very low in patients with Wilson's disease. Wilson's disease is a brain disease due to disturbed copper metabolism. In the brain, it affects especially severely the basal ganglia.

Well, that has been a digression in my account, but I think it appropriately illuminates the background of my interest in the human brain and my first steps in that research area.

Thus, when dopamine was found in the brain, I immediately turned my attention to the brain. The first report on brain dopamine came out in *Nature* in August 1957. I was still in Oxford at that time. I remember my sudden interest when I saw it in the latest issue of *Nature*, in the library of the Pharmacology Department. The author of that report was Kathleen Montagu. She worked in the research laboratory at the Wickford Hospital near London. The head of that laboratory was Weil-Malherbe, like Blaschko an émigré from Germany. He followed up Montagu's paper by publishing in November 1957, also in *Nature*, on the intracellular distribution of brain dopamine. Half a year after Montagu's report, a third paper on brain dopamine came out, in the last February 1958 issue of *Science*; the report was by Arvid Carlsson in Lund. At the same time, in the fall of 1957, two behavioral observations related to dopamine were published. Holtz in Frankfurt showed that dopamine's precursor L-DOPA had a strong central excitatory, awakening effect, and he speculated on dopamine's role in the brain; and Carlsson reported that in mice and rabbits L-DOPA reversed the reserpine tranquilization, as he called it in his paper.

So, by 1958, a whole group of publications came out more or less at the same time showing that dopamine was at least of some interest in the brain. Carlsson's February 28, 1958 *Science* article also presented the important observation that reserpine depleted the level of dopamine in the brain and L-DOPA restored its level; a finding also reported in May 1958 in *Nature* by Weil-Malherbe. Actually, Weil-Malherbe's study contained a more complete account than Carlsson's. Unfortunately, after publishing another full paper on brain dopamine in 1959, Weil-Malherbe did not continue these studies. He had problems with some critics of the chemical assay procedures he used in his work. Especially Marthe Vogt was very critical of him. She was very influential in British neuropharmacology at that time. In 1959, Weil-Malherbe left England for the United States. He went to Joel Elke's NIMH laboratory in the Saint Elizabeth Hospital in Washington, D.C., and he died in the U.S.A.

I started my first study on dopamine in the rat brain in 1958. I studied the effects of monoamine oxidase inhibitors and cataleptogenic agents, as well as cocaine, on the levels of dopamine in rat brain. I had hardly finished that study, when at the beginning of 1959, another paper came out of Carlsson's lab in Lund, this time by Åke Bertler and Evald Rosengren. For me, it was the most exciting paper of them all. It dealt with the regional localization of dopamine in the brain. You have to remember that until then, it was only known that dopamine occurred in homogenates of the whole brain, but nothing was known about in which brain center it was localized.

Bertler and Rosengren patterned themselves on the landmark study of Marthe Vogt on the localization of noradrenaline in the dog brain. She was the first to show that noradrenaline was localized in specific parts of the brain, especially the hypothalamus, and not, as previously thought, diffusely distributed all over the brain. It's a classical study, which she published in 1954.

Bertler and Rosengren found, like Marthe Vogt also in the dog, that dopamine was localized completely differently from noradrenaline. The highest dopamine concentrations were in the caudate nucleus and the putamen, in the so-called basal ganglia. In these locations, the noradrenaline concentrations were very low. On the other hand, the noradrenaline-rich hypothalamus did not contain high amounts of dopamine. I found these findings decisive for understanding the role of dopamine in brain function for a very simple reason. We all knew at that time that in animals reserpine caused a syndrome called catalepsy, which consists of immobility and rigidity of the skeletal muscles. In humans, reserpine was already known to cause a Parkinson-like condition. Parkinson's disease proper is characterized by paucity of movement, called akinesia, as well as rigidity of the skeletal muscles, and tremor. It has always been felt to be a disturbance of the basal ganglia function.

The knowledge that dopamine was concentrated in the basal ganglia and that reserpine depleted the dopamine level and L-DOPA restored it, made it easy to put the whole puzzle together. Bertler and Rosengren expressed the view in their paper that dopamine in the brain was probably involved with control of motor functions. Also to me the situation was quite clear. But what should one do next? From my postdoctoral work, I was already interested in basal ganglia and knew something about its diseases; I was interested in dopamine and had already done experiments on peripheral dopamine in Oxford and on brain dopamine in Vienna. So I decided that the best thing for me to do would be to look directly into the brain of Parkinson patients and determine whether there was a change in dopamine levels or not. To me, it was the simplest and most direct way to find things out, rather than play around with animal models. All the animal models were less than perfect. Even reserpine as a parkinsonism-inducing drug was not perfect because reserpine depleted not only dopamine, but to the same degree noradrenaline and serotonin in the brain, so the relative importance of these biochemical changes remained obscure. For nearly a decade, starting in about 1956, fierce battles were fought about that question.

I started collecting Parkinson brains in February–March 1959 together with my postdoctoral collaborator, Herbert Ehringer. Very soon, in April, we were already analyzing our first Parkinson brain. All our studies were done on fresh post-mortem brains. We first analyzed several control brains. We had to show that dopamine was still detectable in non-Parkinsonian control post-mortem brain tissue and that its amounts and regional distribution were similar to what we knew from animal experiments. All this we could show very easily.

Did you cooperate with pathologists?

The pathologists were crucial in supplying the brains, but there was no formal cooperation at that time. We received the control brains from the University's Pathology Department, which was located at that time in this very building. However, they did not have Parkinsonian material. Parkinsonians rarely died in the General Hospital. They were looked after in chronic care institutions and also died there. We received our Parkinsonian material from the largest city hospital in the periphery of Vienna to which a home for old-age people was attached. They had a neurological ward there, and there were many Parkinson patients there.

We divided the work between Ehringer and me in such a way that he would go to the hospital to get the brains and dissect them immediately under the guidance of the chief prosector of that pathology department. He would take out the areas of interest and bring them immediately to the laboratory. We would homogenize the samples and Ehringer would estimate noradrenaline in the brain and I did the dopamine estimations. The method for the detection of brain dopamine I had already developed in my rat brain studies. I could not use one of the sensitive fluorometric methods. At that time, we did not have an Aminco-Bowman spectrofluorimeter in our laboratories. Instead, I adapted the von Euler-Hamberg iodine oxidation method from Stockholm, which is a colorimetric method. In the presence of dopamine, the reaction gives a pink color. A simple method. I had learned it in Blaschko's laboratory in Oxford. Even before putting the samples of our first Parkinson brain into the colorimeter, I knew the result. The controls, which we always did in parallel, showed the beautiful pink color in the caudate nucleus and putamen, which indicated high amounts of dopamine. The Parkinson brain showed hardly any discoloration. Even before putting the samples in the colorimeter, I could see, for the first time ever, the brain dopamine deficiency in Parkinson's disease literally with my own naked eye.

However, it took us more than a year to collect six cases of Parkinson's disease. After a year we wanted to publish the whole thing with three cases. The head of the Department, Professor von Brücke felt that we should collect more cases. He found that three cases were too small a sample to draw important conclusions. He made us continue the collection and it cost us an additional 10 months or so.

Why did it go so slow?

Before we started our studies in freshly autopsied human brains, the pathologists had rarely, if at all, received such a request. Biochemists had been traditionally skeptical about using post-mortem material for neurotransmitter studies. Catecholamine transmitters in particular were regarded as too unstable in post-mortem tissue to produce meaningful results. Ours was an unusual request. Then, there was the problem of communication. We were totally dependent on receiving a call from the Pathology Department telling us that a case was available. That did not always work. In addition, patients dying on weekends or work-free holidays, such as Christmas, New Year, etc., had too long post-mortem intervals to be suitable for our study. All these factors contributed to loss of valuable patient material.

Is it still the case today?

Today there are other complications. The requirements that have to be met before an autopsy is done are now more complicated. There are stricter

Oleh Hornykiewicz in the laboratory in 1960 at the Pharmacological Institute in Vienna. He is seen with the manometric "Warburg Apparatus" (courtesy of Oleh Hornykiewicz).

regulations, especially about the consent for an autopsy. At that time this was not as strict. On paper, there were strict regulations, but the pathologists had a greater degree of freedom to decide.

We finally published our paper at the end of 1960 with six Parkinson cases; four of them were postencephalitic and two of them were idiopathic.

Our paper has become textbook material and has withstood the test of time. It is still valid. All following studies have been based on our findings of the severe loss of dopamine in the caudate nucleus and the putamen, especially the putamen. For that study we had collected not only something like 20 control brains and analyzed them as well, but we also had collected other cases with basal ganglia disorders, that is, two cases with Huntington's disease, and six cases of basal ganglia symptomatology of unknown etiology. Thus we had a total of 14 pathological basal ganglia cases in addition to the control cases, but it was only the 6 Parkinson cases that showed the characteristic loss of dopamine. The other basal ganglia conditions had normal concentrations of dopamine.

The singular thing about the study was, and still is, in my opinion, that it was so straightforward. The idea was clear, the method simple, and the results were well defined. This absolute simplicity had all the qualities of the true and durable.

You also suggested a treatment.

When I saw that result, it was easy for me to conceive the dopamine replacement idea. I remember exactly when it was. Before the paper came out, in December 1960, I was visiting Blaschko again, in Oxford, in October–November 1960, and I received the proofs of our paper sent after me from Vienna. As I was correcting the proofs, the idea occurred to me, "Why not try L-DOPA in those patients?" I had already worked with L-DOPA in Oxford when I was doing those blood pressure experiments with dopamine in guinea pigs, and L-DOPA had the same effect as dopamine. The idea of trying L-DOPA in human Parkinson patients was for me straightforward. Of course, I knew all the literature on reserpine and L-DOPA; I knew that L-DOPA also counteracted the reserpine syndrome in animals. At the beginning of 1960, there was already a paper on the anti-reserpine effect of L-DOPA in human patients. It was published by Degkwitz in Germany. He was the first to show that L-DOPA counteracted reserpine sedation, as he called it, in humans, but his patients must have had a strong akinesia which is a prominent symptom of the reserpine-induced Parkinsonism in humans. Interestingly enough, Degkwitz, who was a neuropsychiatrist, did not think of giving L-DOPA to Parkinson patients. To me, the idea to try L-DOPA occurred immediately. Of course, I was prepared for that idea. Was it then all due to chance only? Well, wasn't it Louis Pasteur who said, chance favors the prepared mind?

Why L-DOPA rather than dopamine?

At that time it was already known that dopamine does not penetrate the blood brain barrier. Also, other monoamines, such as noradrenaline, adrenaline or serotonin do not easily get into the brain. It was a well-known fact. I didn't even try to think of using dopamine itself. We used it later to show that the antiparkinson effect of L-DOPA was not due to the formation of dopamine in the periphery.

You were not a clinician, so you had to find a clinician partner.

Yes, of course. I asked Walther Birkmayer, who was the neurologist in that city hospital from which we had received the Parkinsonian brains.

As I understand, Walther Birkmayer had a complicated past. He had been a member of the SS even before Hitler annexed Austria, when

Oleh Hornykiewicz and Walther Birkmayer at the Fifth International Symposium on Parkinson's Disease, Vienna, 1975 (courtesy of Oleh Hornykiewicz).

such a membership in a Nazi organization was still illegal in Austria.
When it turned up that Birkmayer had some Jewish ancestry, he was
kicked out of the SS, but was not harmed otherwise.

That was so indeed. After the war, he had political difficulties because
of his past, but he soon recovered. But several members of the Medical
Faculty remained skeptical of him.

At the time when I suggested to Birkmayer the L-DOPA treatment, we
were not on very good terms. So, initially he didn't want to do that. It
took some nine months until he finally did the first experiment. The reason
why he was not on good terms with me was that when I returned from
Oxford, he approached me in the spring of 1958, with the idea to analyze
the hypothalamus of Parkinsonian patients for serotonin. He thought there
was a temperature regulation problem in Parkinson's hypothalamus. I sent
him away. I didn't see any rationale for doing it. Parkinson's disease was
not a hypothalamic disorder, but a basal ganglia disorder, so I did not see
any reason why I should lose time by looking into the hypothalamus and
serotonin. I did not tell him this directly; I used an excuse. I told him that
I did not have an appropriate method for estimating serotonin. Of course,
he understood. When two years later I approached him with the request
to try L-DOPA on Parkinson's patients on his ward, it was not too surprising
that he was not very enthusiastic. First, he didn't see much reason for
doing it because he had never been involved with any dopamine work
or ideas about dopamine; all that was very new at that time, so he did
not realize the physiological importance of dopamine in the brain. Secondly,
and most important, he simply wanted to pay me back my negative attitude
to him. This he actually "confessed" in a letter, which he wrote me several
years later.

So I asked my department head Professor Brücke to work on Birkmayer.
Brücke had a strong influence in the Medical Faculty, and Birkmayer wanted
an academic career. I asked Birkmayer the first time in November 1960, he
did it finally in July 1961. I gave him about 2 grams of L-DOPA which I
had obtained from the "Roche Biochemica" for some of my earlier experiments. At that time, Hoffman–La Roche laboratories used to give samples
of rare chemicals free of charge to university laboratories. I also gave Birkmayer
the instructions how to dissolve it for intravenous injections. I took these
instructions from the paper by Degkwitz, who earlier had published on the
anti-reserpine effect of L-DOPA in psychiatric patients. After the first patients
were injected with L-DOPA intravenously, Birkmayer immediately became

a zealous convert. The effect was really dramatic. Patients who had been bedridden for years could get up and walk around after L-DOPA. It was like a miracle. Birkmayer forgot all his grudges against me and became one of the most enthusiastic L-DOPA clinicians. After four weeks we made the first film with five patients showing the effects, the differences between the various subgroups of Parkinsonian patients. There was nearly no effect in arteriosclerotic Parkinsonians and a full effect in postencephalitic Parkinsonian patients. We then submitted a paper, a short communication, which came out in the Viennese clinical journal,[1] in November 1961. Also, Professor Brücke arranged a presentation of the results at the scientific session of the Medical Association in Vienna. That was the first public presentation of both the dopamine results in Parkinsonian brains done in pharmacological research and of the film we made of the patients treated with L-DOPA. This was on November 10, 1961.

Did it then spread all over the world?

It eventually did, but it was not that simple. First, L-DOPA was a rare chemical and it was not easy for clinicians to obtain it in sufficient amounts. Secondly, for the intravenous injection, as we used it, we usually pretreated the patients with a monoamine oxidase inhibitor, which potentiated the effects of dopamine, but had its own untoward effects. This was still clinically not practicable as a treatment in a chronic condition, such as Parkinson's

George Cotzias around 1970
(courtesy of Oleh Hornykiewicz).

disease. The effect of intravenous L-DOPA in our patients was very strong, after inhibition of the monoamine oxidase, like a miracle. But it still was very short lived, one or two hours. And we could not use higher or more frequent doses so as to prolong the effect because of the acute side effects, such as strong vomiting. So it was not until 6 years later that L-DOPA really became accepted and used everywhere. This happened when George Cotzias in New York had the idea and the courage to use, on a daily basis, very high oral doses of L-DOPA given in frequent intervals. That way, the high therapeutic effect could be maintained on a long-term basis. By the way, Cotzias used for his first patients DL-DOPA because it was easier to obtain in large enough quantities, and also cheaper.

Did D-DOPA have any effect different from that of L-DOPA?

I also had a sample of D-DOPA from Roche and I asked Birkmayer to test it on the patients. It was ineffective, which was understandable because Holtz, the discoverer of the enzyme dopa decarboxylase, had shown already in 1939 that dopa decarboxylase decarboxylated only the L, but not the D isomer. The enzyme reaction was very stereospecific. There is a way to convert D-DOPA to dopamine, but it is complicated.

Does it happen in the human body?

It does and that is a funny thing. Holtz did not have D-DOPA at the time, he used only L- and DL-DOPA in his experiments. When he gave them to animals, he was surprised that the urine of those animals that had received the DL racemic mixture contained more dopamine than he found in the L-DOPA animals. But the conversion is made through several rather complicated steps, something like oxidation, deamination and asymmetric transamination reactions which I cannot reproduce in detail from memory. But there is no reason to doubt that it also happens in the human body. The reaction was studied by Ted Sourkes at McGill University in Montreal. It's a slow reaction and does not produce a high enough yield to be of any consequence. By the way, Ted Sourkes was one of the most experienced dopa decarboxylase and catecholamine researchers since the early 1950s, and he then became very active in the field of brain dopamine. At the same time when we did our clinical experiments with L-DOPA, Ted Sourkes suggested to his clinician colleague André Barbeau to use L-DOPA orally in Parkinsonian patients. They published their results in 1962 and they also found that L-DOPA had a beneficial effect. The effects were not as strong as our intravenous results; also, they had only small amounts of L-DOPA available and could not use higher doses; they used a few hundred milligrams at a time. When George Cotzias started in 1967 with high oral DL-DOPA, he used grams of it, the effect was very strong, and it persisted for longer periods of time. Different from the intravenous route as we used it, oral treatment could be repeated without practical limitations. Patients could be given the drug several times a day and continually. That was the best approach to the L-DOPA treatment, the regimen which is still used today.

But it is still not a cure.

It was clear from the beginning that L-DOPA was a symptomatic treatment. It is replacing dopamine, the missing substance, like insulin is used for diabetes.

Using this parallel, I would like to ask you the following question. When insulin was discovered, subsequently some people said that the discovery of insulin, which is just a treatment, in a way hindered further research on diabetes, because now there was a treatment. I wonder if you have heard of such a consideration. What would be your comment on this? Would your discovery divert attention from finding the cure for Parkinson's disease?

On the contrary. If you look at the history of research in Parkinson's disease, until the discoveries about the importance of dopamine for the disorder and the replacement treatment with L-DOPA, the research was very modest. Before the use of L-DOPA, Parkinson's disease was regarded as an essentially untreatable disease. Parkinson's disease is a very severe progressive degenerative brain disorder. There were a few drugs, the so-called anticholinergics, which had a very modest effect, not more than 20 percent of improvement, and that was about all. Then the neurosurgeons developed surgical procedures in Parkinsonian patients, showing that they were to some degree effective, mostly on tremor only, and certainly not strong and persistent enough to be of use as a routine procedure. The research in Parkinson's disease was at a very low activity level, and there was no research worth mentioning on the possible causes of the disorder.

This changed dramatically after the loss of dopamine as a neurotransmitter-like substance in the Parkinson brain was discovered and it was demonstrated that the replacement of that substance with its precursor L-DOPA showed a full therapeutic effect; that was when the real research on Parkinson's disease and its possible causes actually started. When it started, it exploded, and those findings that I just mentioned stimulated research in other brain diseases. The hope emerged that similar changes could be found in other degenerative brain diseases and that treatments could be found for similar until then untreatable diseases. Today, it is difficult for people, even for the young neurologists, to realize what the situation of the neurologists was before the L-DOPA era. The Parkinson patients were hopeless patients; they were crowding the chronic wards in the hospitals and they ended up completely stiff and bedridden, they could not get up, they could not feed themselves, they could not move, and had to be cared for until they died. The doctors were powerless. They could not do anything for those patients.

The discoveries brought about a change in all that and showed that it was possible to treat these patients and that even a chronic degenerative progressive brain disease could be treated. It has stimulated further research on new treatments and the causes of such diseases. It brought about an explosion in human brain research. This research continues at a high rate. So what has happened in Parkinson's disease research is just the opposite of what you mentioned about the possibility that the discovery of insulin may have slowed down diabetes research.

Cotzias has died. Do you think if he had not died, there might have been a different composition of the winners of the Nobel Prize? I am referring to the Nobel Prize in the year 2000, which was given to Arvid Carlsson, Paul Greengard, and Eric Kandel "for their discoveries concerning signal transduction in the nervous system". Following the announcement of that Nobel Prize, 250 neuroscientists published an open letter protesting the decision. It was conspicuous not only that you were omitted from the prize but that your contribution was not even mentioned in the announcement.[2] In this sense the Nobel Prize has a watershed effect; it introduces a sharp division between those who receive it and those who do not.

It's difficult to speculate about the Nobel Prizes. In distinction from other prestigious awards, the Nobel Prizes seem to happen almost in an imaginary, unreal world and not in the everyday world we live in.

Carlsson, Hornykiewicz, and Cotzias could have been a reasonable composition.

As far as I know, this composition had been proposed on occasion. Cotzias had been working hard on that. He pushed very much and in my opinion, too much. I don't think one should push for a Nobel Prize. Prizes are not part of the scientific quest. Only a fool would try to work explicitly for a prize. Then Cotzias died and his efforts came to nothing. But the facts certainly did not change because of that. Therefore, to speculate today about the chances of those constellations and nominations of 30 years ago looks, to me, conspicuously like trying to use Cotzias's death as an excuse for the controversial decision on the prize in the year 2000.

You received a tremendous expression of solidarity from your peers.

I was really impressed. And grateful. The decision of the Nobel Committee provoked the neurological community, who are well aware of the contributions of the people involved in the field. They felt that they should set the historical record straight. The discoveries in Parkinson's disease have become the center and the point of departure for all research in human degenerative brain disease, regardless of the nature of the disorders. Brain research in general has received a tremendous impetus and encouragement from the unprecedented success of the dopamine and L-DOPA research in patients with Parkinson's disease. Naturally, people are very much aware of who was involved in starting the whole research and who set it in motion. That is the reason perhaps why so many people felt that they should say something about it.

You must have thought about what might have been the reason that you were left out.

My thinking was not for long revolving around that question. I soon realized that there was no rational answer. But I heard a lot of comments from others. Some people said that in the year 2000 the European Union had imposed political sanctions on Austria because of that unfortunate business about our government coalition at that time. They said that it would have looked very odd if an Austrian from Vienna had received a Nobel Prize amidst all those sanctions against Austria. Some other people told me that if I had remained in North America — as you know, I spent 10 years, between 1967 and 1977 at the University of Toronto — I would have been included. It looks as if it were easier for North Americans to be included.

People also pointed to the lack of consistency in the combination of the chosen research fields as a possible answer. Of course, I know all three laureates personally. They are excellent researchers. Each of them deserves the prize. Carlsson and Greengard come from the catecholamine and dopamine field. Kandel comes from a different research area, with no direct connection with dopamine. Because of this inconsistent combination of research fields, some colleagues expressed the view that the Committee had first the three people ready for the prize, and only afterwards came up with a common theme to fit them in. By the way, a not very convincing theme. It is easy to guess that a body such as the Nobel Committee is exposed to strong pressures, some scientific and some very personal. They have to compromise. In my opinion, the Prize Committee would have been better advised to give Kandel for his work the Nobel Prize as the sole recipient.

Do you think that the Nobel Committee was ill advised?

I don't think that one can possibly quarrel with any of their decisions as such. The Nobel Foundation has the right to give their prizes to whoever they choose. But if you want to hear what I thought was unfair about the Nobel Prize 2000, and indeed wrong, I would say it is the citation specifically referring to the discoveries of the dopamine deficiency in Parkinson's brain and the L-DOPA treatment. I heard many colleagues comment on that. To everybody reading the citation, it simply says that it was Carlsson who discovered the dopamine deficiency and that he also suggested the treatment. The text does not say that in clear terms, but it says so by implication. The citation is not accidental, it is very carefully worded. It is not outright wrong, but it has the clear potential of misleading, of being only too easily misunderstood. Only a very unusual motive indeed could have forced the Committee into choosing that kind of way of substantiating their prize decision. Whatever the intention behind all this was, it has produced a distorted view of the historical facts.

In connection with this year's Nobel Prize for magnetic resonance imaging, which is again tainted by controversy, a newspaper quoted the representative, or was it the Chairman, of the Nobel Committee as saying that they did not solely rely on their own judgment, but "had their experts". Yes, of course. But how do you choose the experts? As they used to say, in the old days, a good king chooses good advisors, but what happens when the king chooses bad advisors?

One should never write science history on the basis of the Nobel Prizes.

Yes, and that holds for all prizes, not just the Nobel Prizes. But the reputation of the Nobel Prizes is so high that they do influence science history. Although they are not the result of strict historical research, they are generally taken as being the last word on who discovered what. And the damage to science history, once done, is likely irreparable. One could easily quote several examples for that. Take the *Brockhaus* volume entitled *Nobelpreise*, which came out recently in Germany. It says explicitly that the 2000 Nobel Prize was given to Carlsson because he discovered the dopamine loss in Parkinson's brain and suggested a treatment for it. Quite naturally, many historians of science rely on traditionally serious sources of information, such as the big encyclopedias are usually thought to be.

Of course, everybody agrees that today it is not so simple anymore to decide who has contributed what, and how much, to a piece of original research; writing science history as well as deciding on research prizes has become a difficult and highly responsible undertaking.

In the course of my research career, I have received many prizes, smaller prizes and bigger ones. I felt just as grateful for the smallest as for the biggest of them. I believe that all of them were based on evaluations done by the experts among one's fellow scientists, who honestly tried to do the best job of it. In essence, the procedure in Stockholm must be the same. However, the selection of the candidates had been certainly easier a hundred years ago when research was the activity of much fewer people and when most of those involved were involved in one thing or two at most, with much less overlap among them than is the case today. Nowadays everybody tries to be involved in many things, there are considerable overlaps among the activities of people and nobody is ever alone in a discovery. Today's research has become so much more a matter of joint enterprise and joint effort.

The problem faced today by the Nobel Prizes is that their high reputation, and especially their extraordinary publicity, impose on them such a high responsibility that cannot be fulfilled realistically. As you mentioned, their decisions are watershed decisions. The difficulty with that is that today's research is so highly interconnected, and simple "Yes" or "No" answers are very difficult to find. Every prize, big or small, faces nowadays that dilemma; for none of them is there any elbowroom left for easy decisions. If, however, the visibility and publicity of a prize is very high, as is the case with the Nobel Prizes, any discrepancy between their decision and the historical facts becomes so much more obvious and so much more damaging.

If a prize committee in, let's say, Japan or Israel makes a decision to give their biggest prizes to this or that person, nobody talks about it much because those prizes don't have, and don't seek, that kind of publicity. They are wonderful signs of recognition, and no harm to historical fact is done by them. It's the inordinate publicity surrounding a prize that is potentially harmful to science history.

You have received the Wolf Prize.

Yes, together with Carlsson for our work on dopamine.

Cotzias was dead by then.

Yes, and there was no clinician with us. The Wolf Prize Committee did not seem to have had any problems with that.

You have also received the Gairdner Prize.

In Toronto, yes.

I did not find much information about you on the web.

I don't have a web site and I don't go into the Internet at all. They had put a computer in my office but I threw it out. I prefer to use my brain. I guess, people regard me as a strange old man. I don't even use a mobile telephone. My wife has got one because one of our sons gave it to her recently for Christmas, but I have asked her not to call me using that machine. At home I hardly ever switch on the television. I am addicted to reading, mostly non-fiction, and thinking.

You have emphasized the importance of good teachers in your career.

I had good teachers. Friedrich Wessely in chemistry, Friedrich Ehmann in brain anatomy, and Franz von Brücke in pharmacology — these three.

Do you have heroes?

No, I don't have heroes, but I do have people whom I admire as examples to follow, be they saints or scientists. In science, in my young research days, such an example was for me Otto Warburg. When I started with polyphenol oxidase in the human blood serum, it was necessary to read something about enzymes. We did not learn much about them in our medical courses. At that time, it was in 1951, I found in our library in the Pharmacological Institute two books by Otto Warburg, which he wrote shortly after 1945. He was the director of the Kaiser Wilhelm Insitute for Cell Physiology in Berlin and although he was half-Jewish, he retained his position under Hitler. In 1945, the Soviets packed up all his equipment and sent them to Moscow. Warburg was staying on one of the islands off the north coast of Germany, without a lab, so he wrote those two books. One, on heavy metals as catalytic groups of enzymes, and the other, on hydrogen transferring enzymes. I have them right here. They are the most admirable books.

Otto Warburg (Nobel Prize 1931) in his laboratory in 1966 (by permission from H. Krebs, *Otto Warburg*. Wissenschaftliche Verlagsgesellschaft, Stuttgart, 1979, courtesy of Oleh Hornykiewicz).

I started reading them and he instantly became an example for me in terms of research. I was impressed by the simple means he made his fundamental discoveries. The straightforwardness of his reasoning and the simple way of proving things by experiment greatly impressed me. What I also admired so much was the clear, simple and plain German of Warburg's papers, an absolute exception in the German scientific literature. I remember Szent-Györgyi mentioning that when he asked Warburg what the secret of his papers was, Warburg replied: "I rewrite them sixteen times."

Warburg gave me a stimulus for my own research and it was probably also his influence that turned me to my research problems and helped me to find the simple ways to solve them.

Did you ever meet him?

Unfortunately, or perhaps fortunately, I never met him. From what I have heard about him, he could be rather unpleasant on occasion. I might have not liked him had we met in person. Many people were doubtful of him, not the least because of his strange indifference towards the Hitler regime. Our department head, Professor Brücke, had stayed in his lab, for a short time, before the war, and he was not very fond of him.

Warburg's letters have recently been published in Germany. I have not read them, but I have read a review of the book in *Nature*. I remember one episode, as retold by the reviewer: Warburg's distant cousin of the same name was working in Jerusalem at one time, as Professor of Botany, I think, before the war, some time in the 1930s. When the cousin died, *The Times* in London thought that it was Warburg who died and ran an obituary about him taken up also in *Nature*. When Warburg read the obituary, he got very upset. He did not mind that they thought that he had died but that the obituary did not appreciate his achievements sufficiently.

I would like to ask you a little about the war years here in Vienna. How do you remember them?

I remember the generally gloomy mood and the anxiety of the people. The war was taking on threatening proportions. I myself was still a teenager, and my impressions are those of a susceptible growing-up youngster.

Although they knew that I was a Ukrainian — my lack of German was too obvious — at school I did not have any problems with my Slav background. As you know, Hitler's ideology was not friendly towards the Slavs. He called them collectively "Untermenschen". In the beginning, I felt like not belonging there, but the boys and teachers in our school accepted me the way I was, with all my outlandishness and my broken German. I am still amazed about that. I remember my father trying to excuse my ignorance of German by assuring the teacher of German that from now on he and my mother would speak only German at home. But the teacher replied: "Don't do that, he is going to forget his mother tongue; leave it to us here in school to teach him German."

I believe that my school was rather an exception at that time. It was located in the Second District of Vienna that used to be a Jewish ghetto. Before Hitler annexed Austria, the majority of the pupils in that school were Jewish, a multi-ethnic crowd from many central and eastern European countries. Although by 1940 the Jewish pupils had disappeared, the multi-ethnic atmosphere had largely remained. Many of the teachers and of course all the non-Jewish pupils from pre-Hitler time were still there.

In my class there still were a few so-called half-Jewish boys, as defined at that time. I would not have been able to tell them from the others. I did not notice any hostility against them. But they were not allowed to go higher than the fourth or fifth grade of Gymnasium. Of course, I started school in Vienna in the fall of 1940, two and a half years after Hitler annexed Austria and one year after the outbreak of the war. By that time, the Hitler-euphoria and with it the racial excesses of the annexation year 1938, about which I had heard people talk, had obviously subsided. On the whole, I was kept, like the other boys, quite busy with learning, which apart from the so-called race biology, was mainly directed towards the acquisition of solid knowledge. It was a good school. To this day, I remain grateful for the knowledge I acquired there and the fair treatment they gave me despite all my strangeness.

Do you remember Jews living in Vienna during the war years?

In the district where we lived, in 1940 there were still many Jews living. I found it very distressing to watch the situation. There was an open market nearby. The Jews were not allowed to go there while the non-Jews, the Aryans, were doing their daily shopping. Only after the ringing of a bell were they permitted to rush there and buy whatever was left, mostly very little under those war conditions. We could watch those people waiting, leaning against the walls of the surrounding buildings, waiting for the bell to ring. Sometimes, when walking past them, my mother would leave for them, as if through forgetfulness, bread, or some other provision, on the windowsill of one of the street-level apartments. Giving them any help openly would have been regarded as a provocation.

I also remember an episode from the time when the Jews already had to wear the David star. On my way home from school, I saw one of those eastern Jews in traditional hat and caftan walking slowly down the street, an old bearded man with the yellow star on his caftan. A man in a workman's overall stopped him, blocked his way and started mocking and cursing him, and spitting at him. I was just passing them at that moment and I can still see those two people as if they were standing here before my eyes. If they came in here now, I think I would recognize them.

Then there was an event, one or two years later, thinking about which still disturbs me, and has remained imprinted in my mind. The SS started deporting the Jews. There was a place not far away from where we lived where they had to report with their suitcases, bags and all. They were ordered to board open trucks, and they were taken away. Their bags and suitcases, however, were left behind. The SS opened those bags, and if they found valuables or something like that, they removed them. They left the rest there, and people would come and take what they liked.

There is another unpleasant topic about which I would like to ask for your opinion. Here we are in the Institute of Brain Research of the University of Vienna and my question is not an unrelated one. In Nazi Germany, many thousands of mental patients were murdered and their brains were sent to various institutes ostensibly for research. Possibly, brains of victims killed in concentration camps were also sent back to German institutes for research. Benno Müller-Hill, the German geneticist who has investigated the science of Nazi Germany, describes that a Professor Hallervorden received hundreds of brains. At least on one occasion, Hallervorden went to one of the extermination centers, was present when the children, whose brains he wanted to examine, were being killed by gas, and showed the personnel how to take out the brains from the victims' bodies fast, after the killing. After the war, Hallervorden lived the life of a respected professor, a member of the Max-Planck-Gesellschaft. The brains of victims were preserved for decades in Germany and also in Vienna. Why?

There had been discussions going on and on for many years about what to do with those brains. Most of the brains were kept stored in the Psychiatric City Hospital in the periphery of Vienna, where the victims had been housed and killed. But a few brains were also found in the Neurological Institute that does not exist anymore. Because of public pressure, two or three years ago the question was investigated again, and it was decided to put the brains to rest in a semi-religious ceremony in an honorary grave site at Vienna's Central Cemetery. But until the end, there were people who advocated keeping the material in a more visible form as a public memento, a reminder of the atrocities and the criminal and inhuman eugenics-madness of the Hitler regime.

Personally, I would find it unacceptable to use brains for research acquired in the way you described. When I started collecting brains, some 40 years ago, some pathologists asked me about the desired time interval after death. They were trying to be helpful to my research, offering me their special services. They thought that the interval could be shortened if I would so desire. My reply was and has remained, "Follow your regulations and normal routine. Don't make any exceptions for me. I will not accept too long intervals, but until 24 hours it is fine for me." During the four decades of my occupation with fresh post-mortem brains, I had been variously offered extra-fresh brains, removed from the skull 30 or 50 minutes after death, whatever the definition of that was; or the possibility to receive larger amounts of fresh striatal tissue, up to 1 gram (!), taken during neurosurgical interventions on other brain areas. I turned them all down. Maybe I was wrong, but I felt such offers were going against medical ethics. You will even in vain look for a research paper of mine using brains of artificially aborted human fetuses, dozens of times offered to me. I believe in the very practical value of ethical barriers: they help to keep us human.

In conclusion, is there any message that you would like to convey to the people who will read this conversation?

I am not someone who would send messages to mankind, and your question has taken me by surprise.

In today's brain research, many prominent scientists stress the vast complexity of the human brain. In nine out of ten instances, this is another way of saying, "We don't yet understand." The human brain is said to be by far the most complex structure in the Universe. Some claim, extravagantly, that once we know everything about our brain, we shall also understand everything there is to understand. My guess is that the final answers to those very complicated problems that we are trying to solve, if at all in our reach, will be simple answers. It seems difficult for us to accept the idea that things are fundamentally simple. Someone once said, centuries ago, that we are composed of two opposite natures, different in kind, mind and matter; whatever we try to understand, we color with our composite qualities, always perceiving it as a mixture and never the simple thing that it is. Perhaps that's why to us our brain, or more correctly the image we have of it, appears so unimaginably complex. Blaschko and Warburg, the examples I tried to follow in research, found simple answers to not so simple problems. Whatever I have contributed to research of the brain, it was by finding simple answers to the questions that I found before me. If what I have contributed is worth anything, then it proves this point.

Now I notice that I do have something like a message to whoever happens to stumble upon these lines. It goes:

> "Don't be overwhelmed by complexity.
> Try to be simple."

References

1. Birkmayer, W.; Hornykiewicz, O. "Der L-3,4-Dioxyphenylalanin (= L-DOPA)-Effekt bei der Parkinson-Akinese", *Wiener Klinische Wochenschrift*, **1961**, *73*, 787–788.
2. Helmuth, L. *Science* **2001**, *291*, 567–569.

Paul Greengard, 2002 (photograph by I. Hargittai).

35

PAUL GREENGARD

Paul Greengard (b. 1925 in New York City) is Vincent Astor Professor in the Laboratory of Molecular and Cellular Neuroscience, The Rockefeller University, in New York City. He shared the Nobel Prize in Physiology or Medicine in 2000 with Arvid Carlsson and Eric R. Kandel "for their discoveries concerning signal transduction in the nervous system". His mother née Pearl Meister, died giving birth to him and Dr. Greengard established the Pearl Meister Award in her honor by donating his share of the Nobel Prize money. The award is given annually to an outstanding woman scientist, working anywhere in the world, in the field of biomedical research.

Paul Greengard received his Ph.D. degree from The Johns Hopkins University in 1953, did postdoctoral studies at the University of London, Cambridge University, and the National Institute for Medical Research, England, and the National Institutes of Health, Bethesda, Maryland in 1953–1959. He worked in the pharmaceutical industry in 1959–1967 where he directed the Department of Biochemistry at the Geigy Research Laboratories in Ardsley, New York. Following appointments at the Albert Einstein College of Medicine and Yale University, he has been at The Rockefeller University since 1983. His first award was the Dickson Prize and Medal in Medicine from the University of Pittsburgh in 1977. His later awards include the Bristol-Meyers Award for Distinguished Achievement in Neuroscience Research (jointly with Julius Axelrod and Arvid Carlsson in 1989), the National Academy of Sciences Award in the Neurosciences in 1991, and the Charles A. Dana Award for Pioneering Achievements in Health (shared with Eric Kandel in 1997). We recorded our conversation in Dr. Greengard's office on November 2, 2002.

You write in your autobiography that you did not know that you were Jewish until rather late.

I knew that my father was Jewish, but I didn't know that my real mother was also Jewish. I thought my stepmother was my mother and she was Episcopalian.

Did your father ever tell you why he did not tell you about your origin?

He told me that the reason he hadn't told me was because I was such a difficult child that they had terrible difficulties controlling me, disciplining me when I thought she was my mother, so it would have been much more difficult if I knew that she was my stepmother. That was what he said.

Was it a special occasion when he told you about this?

Yes. I was going to college and in those days many of the colleges, including the one I went to, had fraternities. Some of my closest friends were in one particular fraternity and they wanted me to join. I had asked what their restrictions were and they said that you just had to say that you were a Christian. I have been raised as a Christian, we have observed Easter and Christmas, but not any of the Jewish holidays, so I thought I would say that I was a Christian. When I went home over the vacation, I told my parents that I was thinking about joining this fraternity and that's when my father had this discussion with me in which he said that I was not half Jewish and half Christian at all. He said that I was Jewish and I shouldn't join the fraternity. In those days I was naïve enough just to go along with that, with what he said. It was not a religious fraternity, you just had to say you were a Christian and that was all. So, I didn't join it.

Looking back, should you have joined it?

Yes. I am an agnostic; my family had been agnostics for five generations and those friends were a charming, intelligent, lovely group of people and I deprived myself of a certain amount of joy during my college years just because of an abstract theory. It was a strictly social group. They ate there; they lived there.

Then why did they restrict themselves to Christians?

The people who had set up the rules for the fraternity were probably as foolish as my father was. And I was an agnostic and had no reason to believe in anything even remotely resembling what people call God. I see great harm and damage being done by some religious extremists. You could argue that this would be done anyhow and that human nature is what it is, and religious organizations are used to justify the actions of people like that, like those Moslem fanatics today. Even in the case of Israel, some of those extremists are doing horrible things that every reasonable thinking Jew decries. The fraternity would have been strictly a social thing for me even though some of the students went to church. This college where I went, my father had gone there and my uncle had gone there and it was a very old college and they all attended the convocations where the whole college had to assemble and they didn't just talk about He, who is above us, but Jesus Christ was mentioned. So there was a certain amount of hypocrisy on my father's part. He had obviously participated in those ceremonies without any religious inclination. Yet he prevented me from joining the fraternity. By doing this, he deprived me of what would have been a couple of happy years in college socially, which I didn't have. You could argue that it was to keep up the traditions of the Jews, but I don't subscribe to that. Had it been his purpose, we could have gone to the Synagogue on Rosh Hashanah and Yom Kippur, but we never did that. It was Hamilton College in Clinton, New York.

Do you consider yourself Jewish?

First, you have to define that.

This is why I am asking.

The simplest answer would be, yes. But it is more complicated than that. I was brought up by this anti-Semitic stepmother.

In your article, you didn't say she was anti-Semitic. You said she was Episcopalian.

She was very anti-Semitic; she was always making anti-Jewish remarks.

She married your father.

She did. Nonetheless, these anti-Semitic remarks kept coming out of her. She was a paradoxical person. During the Second World War she joined a Jewish women's organization and they supported Israel after the war. You don't seem to be a very good Jew either. Do you consider yourself Jewish?

Yes.

In what way?

Culture and tradition.

I didn't have a Jewish culture. All the Yiddish I learned was in graduate school.

I didn't have Yiddish at all. Where I live, in Hungary, you need not much to consider yourself Jewish.

You live in Hungary?

Yes.

I have some Jewish friends in Nashville, at Vanderbilt University, who sent their children to a so-called ethical culture church because they didn't want to appear to be atheists. My friends' daughter had a friend who asked her what religion she was and what church she went to. My friends' daughter said we go to ethical culture and the other girl replied, "We are Jewish too."

I suffered from anti-Semitism as a child. They sent me to Hebrew school for a few weeks, but when I came out of Hebrew school the kids would beat me up so I asked my parents if I could stop going and they said it was okay. I went only for three or four weeks. But I got into trouble there, too, as most bright kids do. I said Adam and Eve were all right, but where did everybody else come from? I asked many questions that any reasonably thoughtful child might ask. The teachers in Hebrew school did not like that at all. So I was in trouble in the school and I was beaten up when I came out. My mother, my stepmother actually, persuaded my father to let me not go. I do identify with the Jewish race more than with any other race, but I am probably the most ignorant Jew that ever lived. I only learned during the last few years what all these holidays represent. So, you understand why I find it difficult to answer the question as to whether I consider myself Jewish. I do and I don't. I do in the sense that I have great admiration for the Jewish race, the scholarship and many of the values, those values are admirable, but it is hard for me to identify with it. When I was in the Navy during the Second World War, I found it difficult to decide whether to go to the Catholic mass on Sunday, to the Jewish service on Friday, or to the Protestant one on Sunday. I opted for the Protestant one on Sunday for two reasons. There was a lot of anti-Semitism and I didn't want to be exposed to that but it was also that I wouldn't know what to do at a Jewish service. I didn't read Hebrew, I didn't read Yiddish, and I didn't know anything else. The Protestant service was simple.

Do you have children?

Yes.

How do they consider this question?

They are from my first wife. Their mother was Hungarian, so you are extremely fortunate that I am willing to give you an audition [laugh]. I don't speak to Hungarians anymore. That's not quite true. Her professional name is Olga Greengard; she is a biochemist. When I was a postdoc, she was a graduate student. Although she was Jewish, she wanted the children baptized in the Church of England where we were living at the time and we did that. We had two sons and one of them married a Jewish girl and the other married a Moslem girl. She comes from a very educated family, her father was a Professor of Mathematics at the University of Tehran in Iran until Khomeini came in when they left. Their children — my grandchildren — have been brought up without any religion too.

But they know about their origin.

It certainly hasn't been hidden from them. They live in Chappaqua in Westchester County. My second wife was half Christian and half Jewish and my third wife is Catholic. I joke with my friends that my first marriage was bad, my second marriage was a good one, and my third marriage is a great one. If you plot these three points on a curve, you can draw your own conclusions.

Incidentally, my wonderful wife was born Polish Catholic. We were visiting in Poland and the question came up whether I would like to visit Auschwitz and I just couldn't do it. I felt that it would be emotionally too shattering for me. If it would do any good for anybody — me, my wife, mankind, the Jews, anybody — I would have done it. But it would have been an incredibly traumatic thing for me and I didn't see any benefit from it for anybody. I have very mixed feelings about not having gone and sometimes, I wish I had done it.

When I was in Israel, I visited Yad Vashem and it was very moving, but it also shook me up incredibly and I barely managed to deal with it emotionally.

When General Eisenhower (as he was then), visited a concentration camp in Germany in April 1945, he made it a point to examine every corner of it because the Nazi brutality appeared on such an unbelievable scale that he expected a notion to develop one day that the stories of Nazi brutality were just propaganda. He wanted to be able to testify at first hand about it. It appears that he had tremendous foresight.

For Eisenhower to succeed in the Army, he had to have enormously good interpersonal skills, he had to understand how the human mind works. He did not have the intellectual brilliance of General MacArthur, but he had what it took to be the Supreme Commander in the European theater, the understanding of human nature. He understood that people might soon forget this or deny it.

You participated in anti-kamikaze research. What did you do?

At that time, the Japanese Kamikaze planes were flying very low above the water, about 20 feet above the water. They were coming in and performed suicide attacks on our ships. It was very difficult to detect them because from the top of a battleship you could see only about 20 miles because of the curvature of the Earth. If the plane was coming at 200 miles per hour, it took about 6 minutes from detection to the collision. It was hard to get our planes off our aircraft carriers and shoot the Japanese planes down in that short time. The strategy was developed, and I didn't develop it because I was just a kid, to have planes in the area at the height of about 20 thousand feet with radar systems, which could look out 200 miles. That gave us an hour instead of six minutes. So the idea was to have the planes up there and they would relay the information from the air to the aircraft carrier so that the Japanese planes could be intercepted in time. Translating it into practice wasn't that simple because in those days radar was rather primitive. It was the foundation of modern day television. There were three pulses in the signal sent from the plane. The first two told us where on the 360 degree map the object was and the third pulse told us its distance. There was a shortage of people who had the background and the skills to do this job. We had to take tests and they selected those of us who would go for further education. There were three levels of school and I got through them all. By the time we graduated from the third school, the people from MIT, who were developing the system called Early Warning Radar, wanted to have some enlisted men to run these things on the aircraft carrier at sea. I was chosen and one of my classmates was chosen to go from the third school to MIT. Those physicists at MIT looked old to me although they were in their late twenties. They were some of the most brilliant physicists of the United States and several went on to win Nobel Prizes. I remember only one of them by name, Isidor Rabi, who was older than the rest, I believe in his forties. I spent a few months at MIT and then we went out with this equipment to the Pacific.

I would like to ask you about your Nobel Prize-winning research. The citation quoted discoveries concerning signal transduction in the nervous system.

Carlsson was the first who recognized what we call today a slow-acting neurotransmitter. He discovered that dopamine was a neurotransmitter. I found out

Paul Greengard, 1987 (photograph by Ingbert Grüttner, courtesy of Paul Greengard).

how dopamine and other slow-acting neurotransmitters work and that is what I got the Prize for. That's called slow synaptic transmission. Kandel obtained evidence that these pathways are involved in learning and memory. This is how they packaged the whole thing together. The citation was for the mechanism of slow synaptic transmission.

I noticed that one of your other awards you received together with Carlsson and another prize you received together with Kandel. So you seemed to be a link between the two.

That is correct and I have been told that I was a link between the other two.

Where does Oleh Hornykiewicz come into this picture? There was some criticism that he was left out of the Prize.

There was a protest after the prize, which originated from a misunderstanding. The protestors said that this prize was for Parkinson's disease. However, it was not given for Parkinson's disease at all if you read the citation. They mention the medical relevance, but they list medical relevance whenever there is such relevance in every prize. Carlsson had discovered the neurotransmitter dopamine and showed that when dopamine was destroyed, Parkinson-like symptoms developed and that they could be relieved by giving the patient Levodopa, a precursor of dopamine. That then suggested to some clinicians that Levodopa should be used for the treatment of Parkinson's disease. Three groups that I know of did some studies showing that Levodopa treated Parkinson's disease effectively: one group in this country led by Cotzias, another in Japan led by Sano, and a third in Austria led by Birkmayer. What happened in Vienna is controversial. If you speak with Hornykiewicz supporters, they say that he did everything. If you speak with the Birkmayer supporters, they say that this clinician had to persuade Hornykiewicz to do his experiments. These Austrian investigators not only showed that Levodopa treated Parkinson's disease, they also showed in post-mortem tissue that the level of dopamine was much lower in Parkinson's patients than in controls. The Birkmayer people said that in these clinical trials they wanted Hornykiewicz to assay the brains and it took them a year to persuade him before he finally did it. The Hornykiewicz people said that he had the idea, he wanted to assay the brains and that he had to persuade Birkmayer to do the clinical study. There is no question that Carlsson's publications made it very clear that Levodopa should be tried in Parkinson's patients and that the dopamine level in the brains of Parkinson's patients should be looked at. This is absolutely clear. If they had given the prize to Carlsson, Hornykiewicz and Birkmayer, that would have raised questions about these other two groups, one in America and the other in Japan. Carlsson could have received a prize for Parkinsonism with any of those four people, the three clinicians and Hornykiewicz. I don't subscribe to the argument of a lot of the neurologists that it was outrageous to leave Hornykiewicz out of the prize. To summarize my view of the Hornykiewicz story: the prize in fact was not given for Parkinson's disease. It was given for slow synaptic signaling.

Coming back to your work during World War II, you might have become a physicist, but you did not continue along those lines and the reason may have been that you were against the atomic bomb. I would like to ask you about that.

My stepmother didn't want me to go to college. I supported myself in college through the G. I. Bill of Rights. When I wanted to go to graduate school, I needed a fellowship. At that time most, if not all, of the fellowships to do graduate work in physics came through the Atomic Energy Commission. My feeling was that I did not want to be involved in research, which could be used to destroy people, rather, I wanted to help people. At that time my roommate in college was the son of two distinguished pediatricians. I discussed my problem with them and they told me about this emerging field of medical physics. I considered it and I read about the Biophysics Department of the University of Pennsylvania, which did electrophysiology of the brain. Interestingly, a lot of the people who went on to do distinguished work in electrophysiology of the nervous system had also carried out radar research in the Second World War. The people involved in this work at MIT were physicists, but their counterparts in England, Alan Hodgkin and Andrew Huxley, who were considerably older than me, won their Nobel Prize for their work on the nervous system. So there was an obvious connection between electronics and being able to study the electric properties of nerve cells. When I was a first year graduate student at Pennsylvania, the chairman of the Biophysics Department, Detlev Bronk moved to Johns Hopkins to become President and he took a few of us with him to start a new department there. At the time when I finished my Ph.D., I went on to do postdoctoral studies in England and Bronk moved to New York to be President of The Rockefeller University.

At Johns Hopkins, you came across quite a few big names.

There were Hubel and Wiesel as postdocs with Stephen Kuffler at Johns Hopkins at the time. I don't quite remember Hubel and Wiesel, but I knew Kuffler extremely well. Rumor has it that had Kuffler not died in 1980, he would have shared the Nobel Prize with Hubel and Wiesel in 1981.

What kind of person was Kuffler?

He was an absolutely delightful person. He was Hungarian, a Hungarian Jew. He married a Catholic girl. He was very intuitive, imaginative and creative. He was not mathematically inclined. He had great intuitive skills and great experimental skills. He developed many interesting preparations at a time when not many people were doing this sort of thing. When I was a graduate student, he was an Assistant Professor or a little higher than that. He was one of my closest friends there.

Did you talk about his background?

Not much. All his children were raised as Catholics.

What was the atmosphere like at Johns Hopkins?

Bronk could have been a great physicist. He had a great personality; he always had people who followed him blindly. Frank Brink, in biophysics, was one of my two advisors. He is still alive. My other advisor was Sidney Colowick, a biochemist. I decided that the next step for me should be understanding the biochemical basis of physiological processes of the nervous system, but there was also an interruption as I went to work at a pharmaceutical company for a few years.

When I came out of the pharmaceutical industry, the NIH was sponsoring a program of setting up Centers of Excellence. The concept was to take some second-rank schools and make them into first-rank schools. There was a fallacy in this thinking because in order to make a second-rank school into first-rank, you bring some top people there. But by doing so you are depleting other schools and moving them from first-rank into second-rank. In the framework of this program, Sidney Colowick, who was at Vanderbilt University, invited me to spend a semester there in that Centers of Excellence program. Colowick had been trained by the great Coris, Carl and Gerty Cori at Washington University in St. Louis. Many great biochemists came out of Cori's laboratory over the years, including several Nobel laureates. Vanderbilt attracted me because the great Earl Sutherland was there by then. He had been in Cleveland before. I read his papers while I was a graduate student at Hopkins, a postdoc in Europe, then in the pharmaceutical industry. Sutherland had elucidated the mechanism of action of the hormones glucagons and epinephrine in breaking down glycogen to glucose in liver and muscle. I wanted to see whether his approach might be applicable to the nervous system. At that time, it was a heretical idea, that is, that intracellular biochemical signaling could actually be involved in brain function. When Colowick invited me to go down there, to Vanderbilt, I was about to move to Yale, but I couldn't resist the idea of spending a few months in Sutherland's laboratory at Vanderbilt, and that is what I did. I published an important paper together with Colowick and Osamu Hayaishi about the enthalpy of hydrolysis of cyclic AMP. I spent a lot of time talking with Sutherland. By then he was a severe alcoholic and died of sclerosis, but he was still a very bright guy, a very intuitive person.

Sutherland had established a joint M.D./Ph.D. program at Cleveland. Once he left Cleveland, why didn't he go to one of the best-known schools?

Sutherland received the Nobel Prize in 1971 and he died in 1974. By the time his work was widely accepted, he was more or less out of it; he had no ambition anymore. When I was at Vanderbilt, in 1967, he used to come in for half a day, twice a week. So while he was still fully functional the importance of his work was not adequately appreciated.

What would you single out as your most important contribution? Was it for what you received the Nobel Prize? The citation for the three of you said, "for their discoveries concerning signal transduction in the nervous system," and you, in particular, for the mechanism of dopamine. I may also put the question this way, what is what you feel proudest of among your research?

Krebs and Fischer received their prize for discovering that protein phosphorylation is a biological regulatory mechanism involved in mediating the actions of glucagon and epinephrine. They showed that this reaction was involved in carbohydrate breakdown in the liver. I then showed that protein phosphorylation has much broader-ranging effects. I showed that it was important in the brain and in the function of many other cell types. So that's one of the things I feel proudest of. Another thing that I feel proud about is applying the concept of signal transduction to the nervous system. Everybody was investigating the electrical properties of the nervous system from the 1930s through the 1960s and most people in the field were reluctant to believe that there might be some biochemical molecules underlying those electrical signals.

As a chemist, I also find it hard to understand how simple reactions like phosphorylation can be connected with such a complicated phenomenon as memory.

That is why our work was very, very slowly accepted. This is why such a bright person as Sutherland told me back in the sixties that if I am only 90 per cent correct, I still made a phenomenal contribution to science. This was a few years before he won the Nobel Prize. But most people didn't accept the possibility, even after we had published many papers on the subject, that communication between nerve cells might involve intra-cellular biochemical reactions.

Was it frustrating for you?

Yes and no. It was frustrating, but on the other hand there was a certain joy I had in not having a lot of competition. For a long time, nobody was competing with me. I remember in the late seventies I gave a lecture at Harvard. It was organized in a seminar room with 20 seats in it. It was announced for 4 o'clock, we got there at 20 to 4 and the room was already full. They moved it to a larger room for a couple of hundred people and it was soon filled too. So they moved it again to an auditorium for 500 people and that was filled too — people were sitting on the floor. At that time, although our work was so controversial, and people thought I was wrong, they wanted to hear the story anyway. People didn't believe my work, but they knew it. That was gratifying, to get up there in front of such a crowd. I gave a lunch seminar earlier on the same day. We started at noon and it lasted three and a half hours. They kept asking me provocative questions until 20 to 4, when I had to leave to go to my lecture. That day illustrates the good and the bad things, the lack of acceptance and the tremendous interest which was also recognition.

But recognition was slow because your first award came when you were 52 years old. They then came in large numbers after you were in your sixties.

I did publish a very important paper as a postdoc, it was in *Nature* in 1956, and then I went into industry for nine years. I was already 42 when I really started my academic career. And you remarked that I got my first award when I was 52. Getting my first award at 52 was not so bad, because I really started my career only ten years before.

Did your appointment mean that Yale had tremendous trust in you?

First I went to the Albert Einstein College of Medicine in New York. There were two colleagues in England while I was a postdoc there, who had been the two finalists for a job of the chair of pharmacology at the Albert Einstein College. One of them won the chair, but he said he would take the chair only if the other two also got professorships. You are right, Yale had a tremendous trust because I didn't have a distinguished academic record at that point. I had the reputation of being a very talented person but I didn't publish much and hadn't opened up a new field at that time. I started at Yale at 42, in 1968. Usually people start their academic careers at the age of 29 and get their first award at 39. I lost about a dozen years in that sense.

When was the point when you first thought that your work might be considered for a Nobel Prize?

In the early seventies, even though almost no one at that point believed in our conclusions. The concept that puzzles you, how can a simple chemical reaction, protein phosphorylation, play such an important role in the nervous system, also bothered others. I remember when one of my children asked me what I did at work and I tried to explain it to him and during my explanation I suddenly realized how important it was. It was a real high for me. But then it took years and years to prove that. I have recently been told that I was a serious candidate in Stockholm since 1980. Many of the Nobel laureates have been serious candidates for many years. I remember one meeting in Stockholm around 1980, where there was a woman sitting next to me at dinner and she told me that her husband told her that I was going to win the Nobel Prize that year. I heard rumors again and again that I was a strong candidate and then I heard that I was on a backburner and so on. Three days before the prize was announced in 2000, somebody told me — it was on a Friday evening — that I was going to win the Nobel Prize the next Monday. Because I had heard this for so many years, I completely forgot it, so when one early morning the

Paul Greengard receiving the Nobel Prize (courtesy of Paul Greengard).

call did come, it was a total surprise. Only later did I remember that Friday evening prediction. But I did feel in the early seventies that I deserved it.

Phosphorylation related to chemical energy was discovered in the 1930s, but here we are talking about a different significance.

That is right. Here we are talking about its role in signaling. So it is not only the universal currency of energy but also a universal signaling molecule.

Did you have any doubts?

No, and I do not understand why I didn't have any more doubts about it.

There are cases when somebody made a great discovery, yet did not recognize its significance. Your case is different.

When Eddie Fischer was interviewed by the press, he said that he and Ed Krebs did not appreciate the significance of the protein phosphorylation reaction. My initial contribution in this area was a conceptual one. I realized that the way a hormone causes the breakdown of glycogen in the liver might occur in all kinds of other cells, including the brain and be related to all kinds of functions. High school children learn that now, but at the time, nobody believed it. I don't understand why they didn't believe it, at least they should have said that it was an interesting idea, but it was just not accepted. It happens a lot of times. Most Nobel Prize-winning work is not accepted at first. If it were, it would not be that important a change in thinking in the field.

Have you continued your research since the Nobel Prize?

Yes. You see, I am here on a Saturday afternoon and I will be here tomorrow.

How long did the award derail your life?

You mean whether the prize disrupted my work? It did and it still does, but to a lesser extent. The first few months were pure insanity. Telephone calls, emails, letters, about honorary degrees, about coming to give a lecture, to address this international congress, to give this annual talk at this medical school, and so on. Just saying no is a full-time job. In addition it might have been somewhat worse for me than for some other Nobel laureates because of the practical relevance of my work. My work is related to various neurological and psychiatric disorders. It has practical implications. This added a number of demands on my time. But I have raised the threshold to talking to people and I am reluctant to give interviews, for example.

To me you sounded very nice when I called you.

I am a very polite person and, also, I felt, this poor guy came over from Budapest; you can't be that arrogant all the time. And I must admit that I greatly enjoyed this.

My impression was that you have been a very lucky person who always found himself among very good people. Also, always the right positions came about at every move you have made.

Lucky means being in the right place at the right time, but I don't think that my environment played a major role in what I did. I went to high quality schools, good departments, and I got good training. Most Nobel laureates are derivatives, not in a bad sense of the word. They work close to the work of their mentor. I, on the other hand, came from nowhere. When I started my work, there were two kinds of people working on the nervous system, biochemists who studied chemical reactions in the brain but didn't care about function and then there were the electrophysiologists who refused to consider that there may be some underlying biochemical machinery. But there were brilliant people I admired, to be sure.

When I read your autobiography, there are so many big names in it, including several Nobel laureates that you came across during your studies.

It is true that I was in places where there were many talented people who provided inspiration.

There are others who never come across a Nobel laureate.

That is true.

Paul Greengard and friend at the time of the interview (photograph by I. Hargittai).

Do you have heroes?

I have some, like everybody else, Einstein, for example. Earl Sutherland, Ed Krebs and Eddie Fischer are heroes to me. They influenced my career. Just reading their papers influenced me a great deal.

Did you consider giving up research after the Novel Prize for some other challenge?

I didn't have such a challenge and I wouldn't consider it unless it had some truly great social significance. Running the Gates Foundation, something where one can really make a difference for society would have been something I would seriously have considered.

You live in New York City. You are the 21st Nobel laureate in the history of Rockefeller University. Do you live a sizzling intellectual life?

I would say I do. I meet a lot of amazingly interesting people, more now than before. My wife is a very distinguished artist; she is a sculptor. Her pieces are in many museums all over the world. So we meet many interesting people both in science and in the art world. But I love my work. I spend a lot of time in my lab. My work, family, friends; my life is fun.

Eric R. Kandel, 2002 (photograph by I. Hargittai).

36

ERIC R. KANDEL

E ric R. Kandel (b. 1929, in Vienna, Austria) is University Professor at the Center for Neurobiology and Behavior, Columbia University and Senior Investigator, Howard Hughes Medical Institute, Columbia University. He received the Nobel Prize in Physiology or Medicine in 2000, jointly with Arvid Carlsson and Paul Greengard "for their discoveries concerning signal transduction in the nervous system".

Eric Kandel received his M.D. degree in 1956 from New York University School of Medicine. In 1960–1965 he was at the Harvard Medical School in Boston. In 1965–1974 he was Associate Professor at the Department of Physiology and Psychiatry, New York University, and he has been at Columbia University since 1974.

His many awards include the Albert Lasker Basic Medical Research Award (1983), the Gairdner International Award for Outstanding Achievements in Medical Science (1987), the National Medal of Science (U.S.A., 1988), the Harvey Prize (1993), the Charles A. Dana Award for Pioneering Achievement in Health (1997), and the Wolf Prize in Biology and Medicine (Israel, 1999).

We recorded our conversation in Dr. Kandel's office at Columbia University in October, 2002, and the text was revised by Dr. Kandel in July 2004.

Yours is probably the longest autobiography on the Nobel website. You write in great detail about your childhood in Vienna.

Yes, I wrote at length about my youth in Vienna because Vienna has a great significance for me in both a positive and a negative sense. On the one hand I love many aspects of Viennese culture, its music and its art. On the other hand I find Viennese hypocrisy and anti-Semitism difficult to accept. To give you but one example, after I won the Nobel Prize, the Austrians, the very people who expelled me, all of the sudden said, how wonderful it was for <u>another</u> Austrian to have won the Nobel Prize. I had to remind them that this was not an Austrian Nobel Prize, this was an American Nobel Prize. A bit later the President of Austria, Thomas Klestil wrote me a nice letter and asked, "How can we recognize you?" I answered that I didn't need any recognition, but it would be nice to have a symposium in Vienna on the attitude toward national socialism of the Austrians, both before, during and after the Hitler period. He agreed and with the help of Fritz Stern, the distinguished German historian at Columbia, who is a friend of mine, we're organizing something in June 2003 designed to contrast the honest and transparent German response to the Nazi period to the persistent denial of complicity and guilt on the part of the Austrians.

As I pointed out in my essay, my year in Vienna 1938–1939 following the annexation of Austria by Germany was extremely difficult. The anti-Semitism that was always present in Vienna came to the surface in a vitriolic and vicious manner. Moreover, it was not only anti-Semitism that drove the Viennese to steal from the Jews it was also their opportunism. They were eager to advance their own academic and financial position by taking advantage of the fact that when Jewish people were being evicted from their homes, they could take their paintings, their furniture, and of course their jobs.

Coming back to your autobiography...

The reason it's so lengthy, and I'm somewhat embarrassed by it is because I wanted to give it to my children and to my students. I did that and they said, wow, we didn't know this. This made me realize that my generation is coming to an end and we have to remind people of what had happened.

Of course, your Nobel Prize is a great excuse to talk about it.

That's right.

You are studying the molecular basis of memory. What does it mean? What is the molecular mechanism of us remembering an observation?

As a result of learning we strengthen preexisting synaptic connections between nerve cells and sometimes grow new connections between nerve cells. To support that strengthening and that growth, neurons need to turn on genes, protein synthesis, and specifically the synthesis of proteins involved in forming new synaptic connections.

Does it happen very quickly? Do proteins get synthesized right away?

The immediate effect is fast. It does not involve protein synthesis, but rather covalent modification of pre-existing proteins by phosphorylation, for instance.

Isn't covalent modification a drastic change?

No; covalent changes go on all the time. They are enzymatic reactions and they happen reversibly in the nerve cells all the time.

Are these changes stable since they will have to develop into long-time memory?

No, the initial changes are labile, but with time, if the learning is repeated sufficiently, the enzyme that produces this covalent modification, such as the catalytic subunit of the cyclic AMP dependent protein kinase, moves into the nucleus where it turns on genes. The genes produce proteins that move out of the cell body to the synapse over a period of six to twelve hours. These new proteins give rise to an anatomical change that persists.

For many years?

We so far have studied memory only over a period of days and weeks. We're just now beginning to look at how to perpetuate it for longer periods of time.

What's the connection between memory and intelligence?

A prerequisite of intelligence is a certain amount of memory. If you didn't remember anything, you would not seem very intelligent. But intelligence is a multi-varied dimension; one is not necessarily equally intelligent in everything; most people are more intelligent in some things and not others. Memory and intelligence are related but they're by no means perfectly correlated. There are many highly intelligent people who don't have a particularly good memory.

What does it mean to be intelligent?

To be able to think deeply about problems, to be able to analyze new problems, to see relationships between events, to be creative. I'm not saying that intelligence does not require a memory, only that memory need not be spectacular.

There's this view that human intelligence is the same today as it was in the Stone Age.

As far as we can tell the human brain has not evolved significantly since the Stone Age. In that sense we're probably not more intelligent. But we have many more tools; we have taken advantage of the fact that we have developed a culture and the lessons that our fathers learned they can teach us and we can teach our children. We have learned strategies through cultural evolution that primitive people did not have.

But we can't inherit that knowledge.

No, we can't. We inherit lots of things, but not the details of knowledge *per se*.

What can we inherit?

Tendencies. Intelligence, capabilities, motivations — there are a number of things that can be inherited.

There was some criticism following the announcement of your Nobel Prize that Oleh Hornykiewicz was omitted. What was the relationship between his work and the laureates' work?

Hornykiewicz is a very fine scientist and he has done some important studies on Parkinson's disease but Carlsson's work preceded his and everyone else's and the Nobel Prize Committee typically goes to the person who does it first. In Parkinson's disease, Arvid Carlsson wrote a paper in 1957 that spelled everything out. He discovered that dopamine is a transmitter; he showed that when dopamine was depleted in an experimental animal by giving reserpine, the animal shows great muscular weakness and the animal recovers if it is given a dopamine precursor (L-DOPA). All that followed, including the work of Hornykiewicz, derives from Arvid's pioneering studies. In addition to his extraordinary contribution in opening up the molecular era of Parkinson's disease, Arvid did the same for schizophrenia and depression.

Moreover, a Nobel Prize typically recognizes an area. The prize in the year 2000 did not go to Parkinson's disease. Had it gone to Parkinson's disease, Hornykiewicz might well have been included with Arvid. The prize in 2000 went to signal transduction. This is an area in which Hornykiewicz is not a central contributor.

Who were your mentors?

I learned a lot from my teachers from Wade Marshall, Harry Grundfest, Dominick Purpura and Ladislav Tauc. I also learned a great deal from Steve Kuffler who was not my teacher but a scientist from whom I learned a great deal. But perhaps most important, I learned as much from people of my own age from Alden Spencer, Jimmy Schwartz, Richard Axel, Tom Jessell, Steven Siegelbaum.

Your long autobiography is a continuous success story, but did you ever have failures and frustrations?

There were lots of them too. When I started studying the snail, it was a period of anxiety because I could not know for sure whether I would not be throwing away my career.

You seem to advocate reductionism in science.

The purpose of reductionism is to use the simplest example of a general phenomenon and try to understand it on a fundamental level with the idea that the underlying mechanisms would be conserved even in their more complex manifestations. This means working at a simple level but at a level that would allow one to try to reconstitute the more complex process again. For me reductionism is the first step in a synthesis that is to follow after that. It's really based on the idea that complicated phenomena are too difficult to handle. We need to take some components of such phenomena, some representations, and try to understand them as well as possible.

In addition to your research activities, you have also authored textbooks.

I find writing textbooks very stimulating. Jimmy Schwartz and I and later Tom Jessell; Jimmy Schwartz and I have edited a textbook in our field entitled *Principles of Neural Science*. This has allowed each of us to gain a broader view of neural science by knowing of what is going on outside of my own work. When you work on the snail, it's so easy to get lost in some specialized part of the snail without realizing that the questions you are asking of the snail need ultimately to be directed to understanding how the human mind works. That's what we all would like to understand. Therefore I get much pleasure and profit from reading human cognitive psychology, how the monkey brain works, how mice work. These are various steps in understanding human behavior, with that as a background I have to put my snail work in context. This is one reason for doing a textbook. Another reason is that I simply like to teach medical students and I thought that they didn't have a good textbook.

You use this expression, "Collecting art is recapturing hopelessly lost youth."

It's funny to hear my own words being played for me. I find, to my amazement that I continue to be immersed in German and Austrian history, literature and art. I can't tell you how many books I've read about Austrian history, about Dollfuss [Engelbert Dollfuss, 1892–1934, Austrian chancellor killed by the national socialists], about Schuschnigg [Kurt von Schuschnigg, 1897–1977, Austrian chancellor who could not prevent Austrian annexation by Germany] and about Hitler. I sometimes wonder why I'm spending so much of my time reading about those long past events. I seem to need to work through those difficult periods of my life. You have asked me whether I have had frustrations. You must know what it was like living in Vienna because we have shared that experience except that you were there in a more difficult period of time. Your life was endangered; my life was never endangered to any significant degree while I was there. Nevertheless I've spent a lot of time reliving those experiences.

Why?

Because I would like to understand these experiences better. These phenomena such as the Austrians' attitude toward national socialism, for example need to be understood on a deeper level. I wrote my honors thesis about this topic at Harvard College. I wrote about three very different individuals, Zuckmayer, an anti-Nazi, Carossa, an internal emigrant, and Junger a proto-Nazi. For the last 30 years or so I have been collecting Austrian expressionist art, Kokoschka, Klint, and others. I'm giving a lecture very soon to our local club with the title, "The Viennese School of Medicine, the Emergence of Austrian Expressionism, and the Persistence of Memory

Storage". Of course, this is a somewhat artificial connection. This is an informal club of physicians called the Practitioner's Club who get together for dinner.

The Viennese School of Medicine introduced scientific medicine; it said that you can't pay attention to the symptoms alone, you have to go below the skin and find out what's going on in the body. They introduced clinical correlations, the first study of the relation of autopsy findings and clinical findings. This desire to go below the skin you see in Freud. The idea of what's going on in the unconscious came from the Vienna School of Medicine. The Austrian expressionists Kokoschka and Schiele were doing the same thing. They were doing psychological portraits. They didn't want to stop at the surface; they wanted to describe the personality within. I say that Freud was the connecting link between the two and there was also a parallel in the person of Ramón y Cajal, who was the first one to think of how memory might be stored by changes in synaptic strength. The same idea was pronounced in the same year, 1894, by Freud. He wrote about it in a report that he never published and which was published after his death in the 1950s. In that paper, he points out that the critical problem in the mind's memory is that it works by changing synaptic strength. He didn't use that terminology, but it was the same idea. This idea that these two people developed has fascinated me and I am going to bring it up to date by what I'm doing now.

What's your present research?

It's very much as before. The Nobel Prize is a nice stopping point to think about what you would like to do in the future. I decided to continue as long as I am capable of doing it. I continue working with the snail, with the mouse, I continue working with memory. I'm investigating the mouse to explore the role of attention in memory storage and for the snail, how memory is perpetuated. I have a group of 20 people, mostly postdocs. I take very few graduate students. People bring special talents here and most of them had never worked with the brain before. They come from molecular biology, from genetics, from immunology. I try to teach them about the nervous system and try to use the particular skill that they have.

Do you write grant applications?

Yes. I get most of my support from the Howard Hughes Medical Institute. It's a wonderful organization; it also supports a number of my postdoctoral fellows. But I was also supported by the NIH and by private foundations such as The Mathers Foundation and NARSAD.

Your Nobel Prize was awarded for your studies of learning and memory on the cellular-molecular level. How long did it take from your essential publications to the award?

My first really important publications on learning and memory were a series of three back-to-back papers in *Science* in 1970 that analyzed the cellular mechanisms for two forms of learning and showed for the first time that learning involves changes in synaptic strength at individual synaptic connections. But I would guess it was not on the basis of these three papers alone but for the fact that by staying with the problem I was able to advance it to the molecular and structural level. Had I stopped in 1970, I most likely would not have received the Nobel Prize. I think my good fortune was that I love the problem of memory and I've stayed with it and tried to go deeper and deeper. I began in 1965 to develop a system which made memory accessible to cellular and molecular approaches and showed that synapses are changed and then went on delineate how this came about. I found that the changes for one type of learning — sensitization — are mediated by cAMP and the cyclic AMP dependent protein kinase for functional changes important for short-term memory and by CREB-mediated gene expression and structural changes for long-term memory. I have always worked with the snail but in 1990 I began studying memory in genetically-modified mice and that proved very helpful for showing that the finding in the snail were general. One could attack more complicated examples with the same approach. The mouse work complemented my work with the snail and probably made the Aplysia work more comprehensible for other people.

Lars Ernster the late biochemist at Stockholm University distinguished between the drilling type and the digging type of researchers ...

Isaiah Berlin wrote an essay on Tolstoy in which he distinguished between the hedgehog and the fox. The fox knows many things and the hedgehog knows one thing well. I am a hedgehog. I advise my friends not to jump around all the time. I believe science is done best when you have to put your flag down on a spot and try to analyze the terrain. I admire people who stake out an area very early before everybody else sees it as an area of great interest. But this is only my preference. The wonderful thing about science is that you can make interesting science in hundreds of ways. One of the people I most admire, Steve Kuffler, was fantastic and he did it by moving around. My role model was Bernard Katz [Nobel Prize, 1970]. He did a wonderful job going deeper and deeper analyzing physiologically how synapses work. I said to myself that I wanted to understand plasticity, which is the mechanism underlying memory storage in the way Katz understands synaptic transmission. I would say to Kuffler how much I admired Katz's stick to "itness" and Kuffler who also admired Katz enormously would say, "Poor Bernard, he has to tell the same story over time, I cannot do that." I told myself at the time I wished I had such a story to tell. Kuffler said of himself later on that he felt sorry that he did not stick with any problems longer than he did, but he just couldn't. He pointed to Hubel and Wiesel as people who stuck with a problem beautifully and soon after Kuffler's death, Hubel and Wiesel won the Nobel Prize [in 1981]. In fact Kuffler would almost certainly have gotten it with them. Kuffler did a beautiful set of studies on the retina which greatly stimulated Hubel and Wiesel. In addition, they were in Kuffler's department and he shepherded their career in the most wonderful and generous way. They went far beyond him, but because he was the founding contributor in vision and did so many other things as well, I think he would've been recognized with them.

In any case, you could have received the prize for some time.

There is no way of knowing — after 1983 when I received the Lasker Award in Basic Medical Sciences and after 1988 when I received the National Medal of Science, my scientific friends started making occasional comments. But the people in Stockholm, the people who count, never talk. I never got a clue from them.

Was it frustrating, the wait?

I don't think anyone has a right to expect the Nobel Prize. The Nobel Prize is a most wonderful award, but you can't base your life on it.

Some people do.

I didn't. Had I not received the Nobel Prize, I would've still considered myself very lucky to be a Viennese émigré to have come to the United States and have had a wonderful personal life and a wonderful career.

Eric Kandel after the Nobel Prize announcement, 2000 (courtesy of Eric Kandel).

The Nobel Prize is so much above every other recognition.

If you receive it, it changes your life completely.

How did it change your life?

It made my life more wonderful. My children, my wife and I have gotten much pleasure out of my good fortune. I have opportunities come my way that would not have come my way before. I'm recognized by people and institutions who'd not heard of me before. Moreover, institutions to which I belong such as Columbia University, the Psychiatric Institute and the Howard Hughes Medical Institute have been very gracious. In turn, I think I can now help Columbia in ways I might not have been able to help it before.

Did you know Erwin Chargaff?

Slightly. I found him really sad. A wonderful but difficult guy. He made a very nice scientific contribution, but he felt bitter because he felt he was not recognized. He was very cultured, *Heraclitean Fire* is an amazing book, but very pretentious.

He could've been more magnanimous.

Of course. He was always nasty, biting, critical, foolish. One has to step back and be bigger than oneself. He wrote how for three hours a night he read Greek in the original texts. When he wasn't reading Greek he was reading Schiller and Goethe, Goethe and Schiller.

He donated his library to Vienna.

That's really crazy. But, hope springs eternal within the human breast.

May there be a genetic component to Jewish achievement?

There may be. But one should not underestimate the great tradition of scholarship. The Jews have always been the people of the book. Most fathers have gotten a lot of pleasure out of having one of their sons be a scholar. Since the destruction of the Second Temple and the loss of a single central place for worship and praise every Jew, man and woman, rich or poor has had the obligation to read the bible on his own. Among the Greeks and Romans only the wealthy members of society were literate. Among the Jews everyone no matter how poor was expected to read and write. As a result they could move into academic professions. Jews have been in medicine since the middle ages. In the middle ages half of the physicians in Europe were Jews. Christian religion discouraged people from going into medicine and the Jews took it as an opportunity because it was an open field. Even though people could be anti-Semitic and wouldn't trust Jews with many things, they'd trust them with their bodies because they were good physicians. There's always been a great scholarly tradition, the law, medicine, science, that's attracted Jews.

Many of the American Jewish Nobel laureates have been immigrants or the children of immigrant parents. A few generations down the road this affinity for scholarship seems to diminish.

You and I are émigrés ...

I'm not, I live in Budapest.

I see, I didn't know that. The generation that left Vienna has been very successful. My children have wonderfully productive lives so I don't see any diminution of scholarship in them. But it's too early to tell.

Could it be that it was not in spite of the environment but that the hostile environment forced them to outperform their environment?

Eric Kandel in his office, 2002
(photo by I. Hargittai).

That may well be because there was tension. When they came to the United States where the tension is gone, things may turn out differently. I spoke with a former dean of Harvard University who came originally to the United States from the Free City of Danzig (now Gdansk in Poland). He is very interested in Jewish experience at Harvard. He has found that the Jews no longer get the highest grades at Harvard. Now, it's the Chinese. So it may be that Jewish creativity and hunger for knowledge has reached its peak. I doubt it.

What do your children do?

My son is an economist who is in finance. He manages a set of funds for Dreyfus Carnegie-Mellon. My daughter is a lawyer. She does public interest law and specializes in family violence. She defends women who have been abused.

You write about your studies of the complexities of human behavior. How do you feel about this quotation, "Judge not so not be judged." This is about the complexities of human behavior because you write, "A society's culture is not a reliable indicator of its respect for human life." I find this a very profound statement.

It's very sad.

But you are facing a problem that we often try to avoid. How far have you gone into the study of the complexities of human behavior?

I work on snails and mice so I have not worked on these problems. But I came from a tradition of psychoanalysis and psychiatry, so I'm sensitive to those issues.

Is there any question that I should've asked and did not?

I've never been interviewed by anyone who shares my European experiences as much as you.

NAME INDEX

Page numbers in bold refer to interviews

Cumulative Index
of Interviewees
Candid Science I–V